YANGAMBI 01°N; 24°E 490 m - 30/20°C

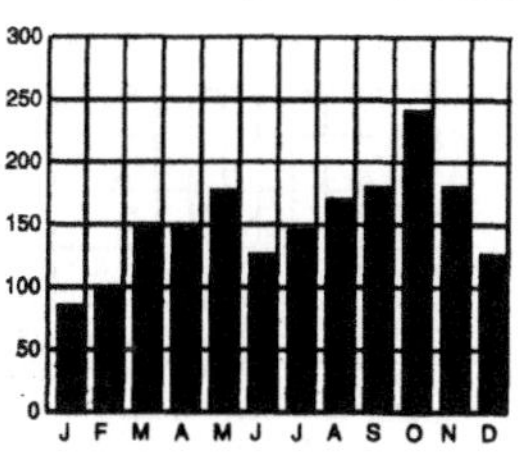

WAU 08°N; 28°E 440 m - 35/20°C

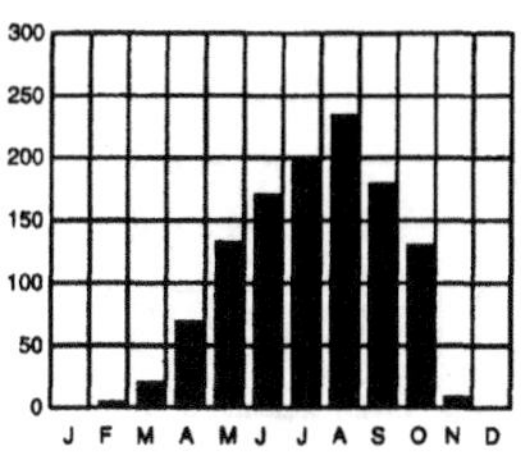

NIOKA 02°N; 31°E 1680 m - 25/13°C

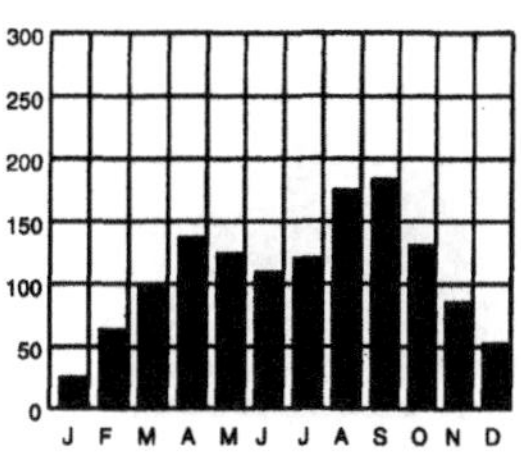

JUBA 5°N; 32°E 460 m - 31/21°C

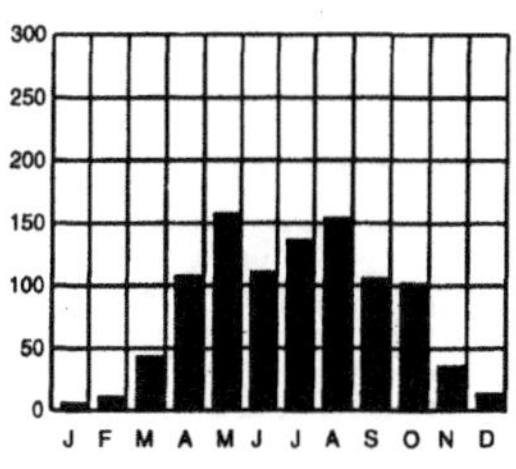

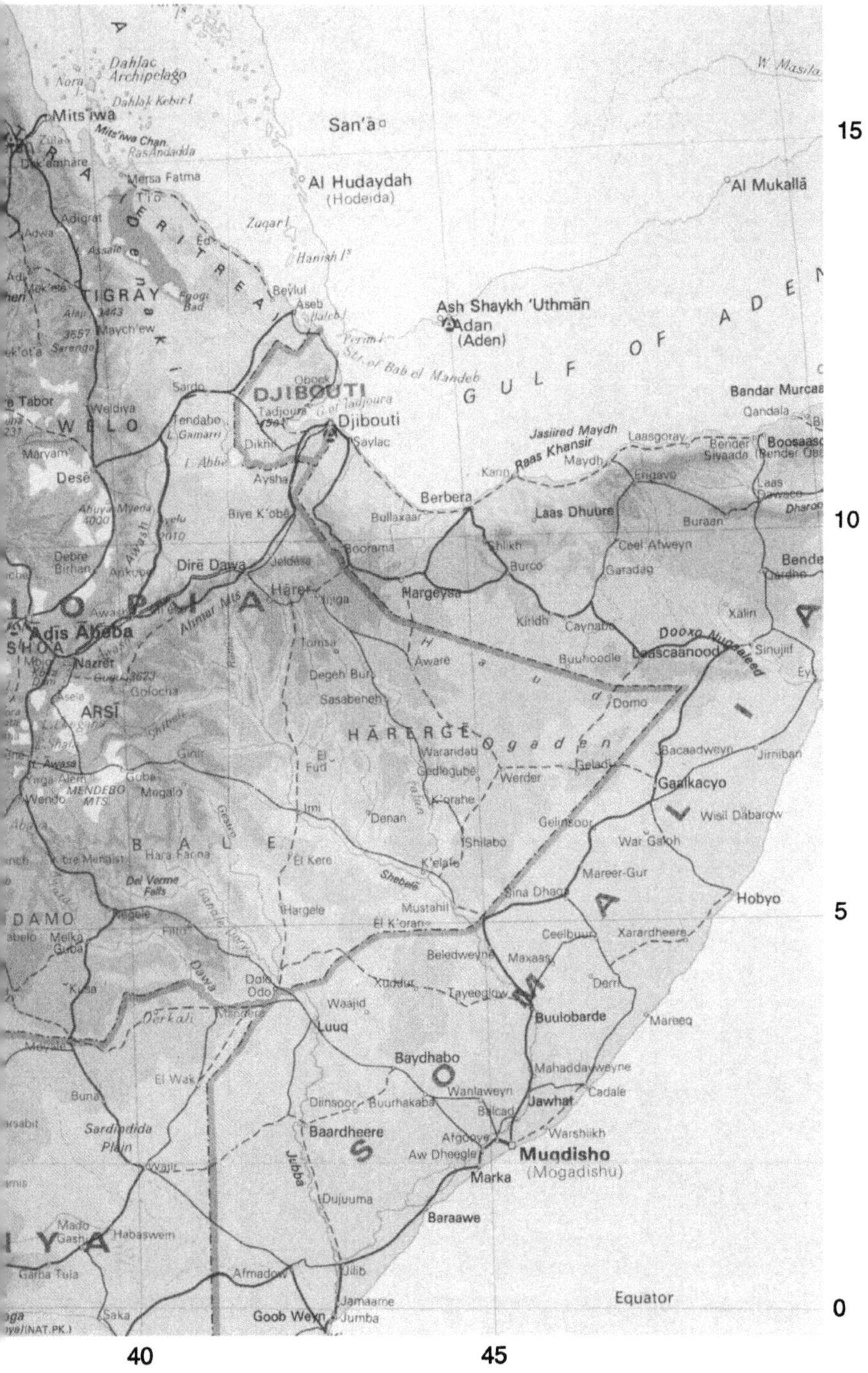

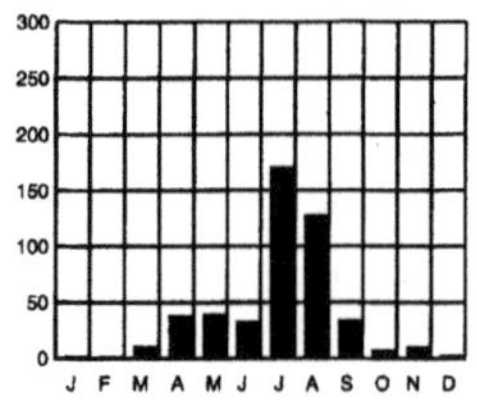
ASMARA 15°N; 39°E 2300 m - 23/10°C

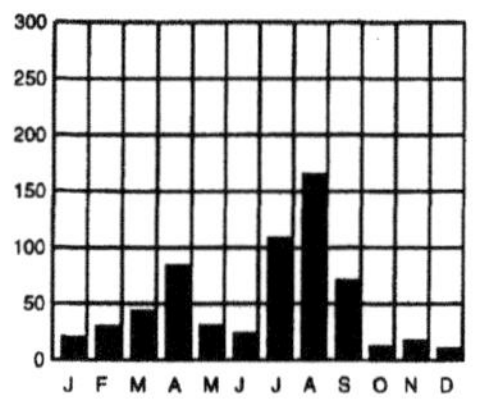
DIRE DAWA 9°N; 42°E 1200 m - 32/17°C

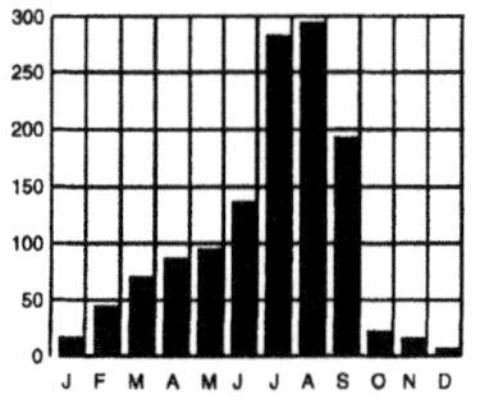
ADDIS ABEBA 9°N; 39°E 2450 m - 23/8°C

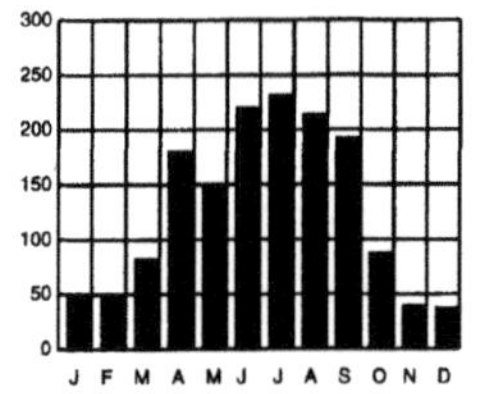
JIMMA 8°N; 37°E 1750 m - 25/13°C

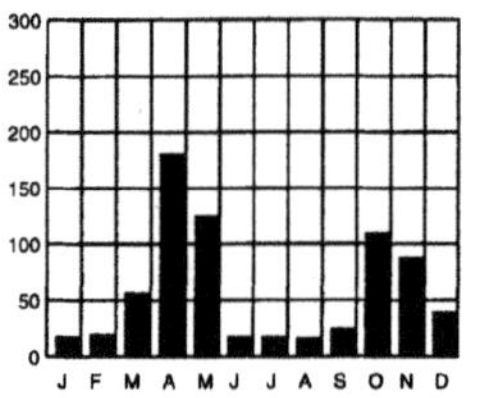
Moyale 04°N; 39°E 1110 m - 27/17°C

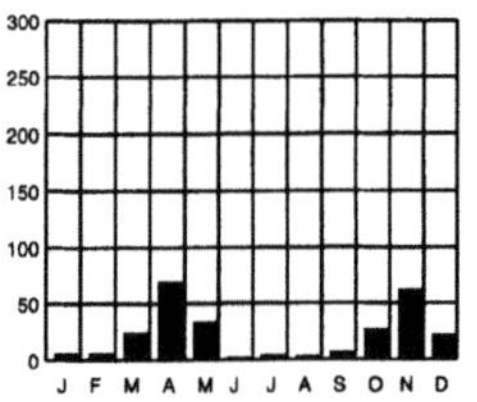
WAJIR 2°N; 40°E 240 m - 34/22°C

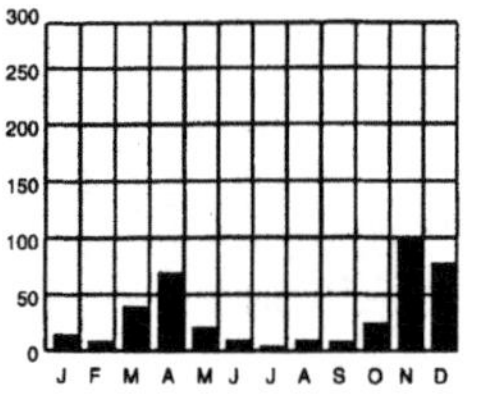
GARISSA 0°S; 40°E 140 m - 34/23°C

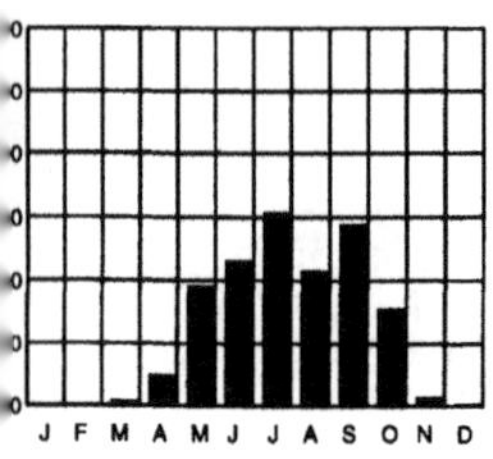
MALAKAL 10°N; 32°E 390 m - 35/21°C

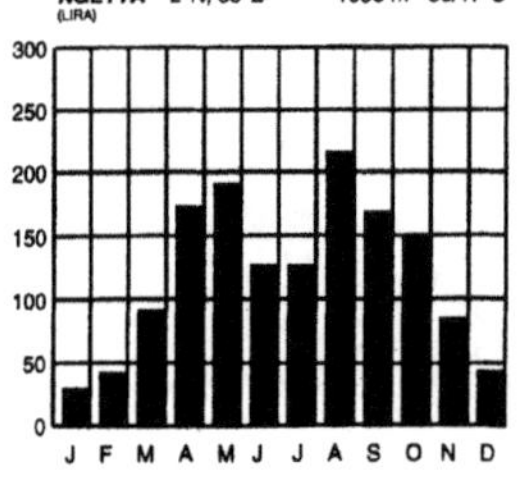
NGETTA (LIRA) 2°N; 33°E 1090 m - 30/17°C

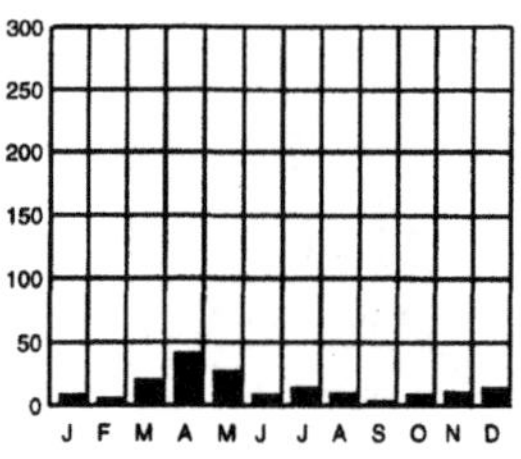
LODWAR 3°N; 36°E 510 m - 35/23°C

East Africa's grasses and fodders

Tasks for vegetation science 29

Series Editors

HELMUT LIETH

University of Osnabrück, Germany

HAROLD A. MOONEY

Stanford University, Stanford, Calif., U.S.A.

East Africa's grasses and fodders: Their ecology and husbandry

Joseph G. BOONMAN
Independent Consultant
Formerly of Kitale (Kenya)

KLUWER ACADEMIC PUBLISHERS
DORDRECHT / BOSTON / LONDON

Library of Congress Cataloging-in-Publication Data

Boonman, J. G.
East Africa's grasses and fodders : ecology and husbandry / J.G. Boonman.
p. cm. -- (Tasks for vegetarian science ; 29)
Includes bibliographical references (p.) and index.
ISBN 0-7923-1867-6 (hb : acid free paper)
1. Forage plants--Africa, East. 2. Grasses--Africa, East.
3. Pastures--Africa, East. 4. Grassland ecology--Africa, East.
I. Title. II. Series.
SB193.3.A353B66 1992
633.2'02'0967--dc20 92-23961

Published by Kluwer Academic Publishers,
P.O. Box 17, 3300 AA Dordrecht, The Netherlands.

Kluwer Academic Publishers incorporates the publishing programmes of
D. Reidel, Martinus Nijhoff, Dr W. Junk and MTP Press.

Sold and distributed in the U.S.A. and Canada
by Kluwer Academic Publishers,
101 Philip Drive, Norwell, MA 02061, U.S.A.

In all other countries, sold and distributed
by Kluwer Academic Publishers Group,
P.O. Box 322, 3300 AH Dordrecht, The Netherlands.

Printed on acid-free paper

This book was partly financed by the Netherlands Minister for Development Cooperation. Responsibility for the contents and for the opinions expressed rests solely with the author; publication does not constitute an endorsement by the Netherlands Minister for Development Cooperation.

Printed in The Netherlands

To Elly, my wife and home-editor who helped this book develop with more patience and encouragement than I had any right to expect

TABLE OF CONTENTS

Series editor's preface

The series Tasks for Vegetation Science is devoted to a variety of research aspects in vegetation science, pure as well as applied. Of the applied problems one of the most pressing is to achieve better knowledge and improvement of the pasture vegetation in tropical and subtropical regions.

As series editor I was impressed by the sheer volume of useful scientific information concerning pasture plants from East Africa collected during many years by Dr. Boonman and compiled in one manuscript.

Dr. Boonman first came to East Africa in 1963 on an overland journey along the river Nile which took him from Alexandria in Egypt to Lake Victoria in Uganda and Kenya and back again. After a brief spell as a cotton agronomist in the Sudan Gezira he joined the grassland research team at Kitale, Kenya in 1966. Improvement of seed yield of tropical grasses was his principal interest which finally led him into the field of breeding grasses. Well-known varieties from his work include Boma & Elmba Rhodesgrass as well as Clone 13 Elephantgrass. In 1979 he was recalled to The Netherlands to head a cooperative seed company involved in the breeding of grasses and cereals.

The author has focused this study on one region, Eastern Africa. Global application of theories runs into conflicts too easily with local types of farming, if not with bare economics. Very few books can be found that describe existing practices and seek local answers by digging deep in the stacks of old, local reports. There is a wealth of knowledge available that is not being used. The author intends, in this volume, to open ways to this wealth of literature.

A great deal of success in farming has been achieved with the well-known species of Elephantgrass and Rhodesgrass. It is, therefore, time that empirical evidence and research information are brought together. Much counter-productive debate is avoided if the correct information is available. It then emerges that much was done in the past on what is now known as agro-forestry or sustainable farming. Maintaining soil fertility through a temporary productive grass cover was recognized early on as a valuable legacy of conventional agriculture, especially when integrated closely with grazing livestock. This has formed the basis of the successful systems of mixed farming as we know them in East Africa.

The series editor hopes that this volume contributes to the general improvement of pastures in Africa and the other tropical regions of the world.

Osnabrück, December 1992 — Helmut Lieth

CHAPTER 1

Introduction

East Africa is the home of world's most famous cultivated tropical grasses. Buffel, Kikuyu, Guinea, Rhodes, Setaria, Star and Sorghum varieties of this region have become household names in other continents, in countries as far apart as Mexico and Brazil, Japan and Australia.

East Africa has played a leading role in tropical grassland development, world-wide, by opening up the treasure of its gene centre. What began modestly with the exchange of small samples of seed grew into a flourishing seed industry for export and local use. Especially the extent to which cultivated grasses have been adopted by smallholders is without parallel elsewhere in the tropical world.

Some 400–500 ton of certified grass seed are handled annually in Kenya alone (Fig. 8.1.), enough to sow 75–100.000 ha/year. In some years export to neighbouring countries and overseas may amount to 50% but this may be more than balanced by the amounts of home-grown and commercial seed. Many (small) farmers buy and multiply seed for their own requirements. The acreages grown to fodder crops such as Elephant-grass and Sorghum rival those of sown grasses.

How these grasses and fodders have been cultivated in East Africa itself is not widely known. Neither is there a written account of the prevailing types of mixed-farming within which these grasses function. This is perhaps surprising at a time when so much interest is being shown in sustainable farming systems and in methods of protecting the environment.

This book deals mainly with equatorial East Africa, from eastern Zaïre to Somalia, covering Uganda, Kenya, Tanzania, Rwanda, Burundi, and to some extent also southern Sudan and Ethiopia. The 15° latitudes are taken as northern and southern limits. Beyond these parallels, deserts or winters on the inland plateaux preclude the use of many perennial tropical plants. Thus, Zambia, Malawi and northern Zimbabwe are covered, though by necessity in less detail. Within East Africa all zones where cultivation of grasses or fodder plants has been practised are taken into account, the semiarid areas included.

1.1. Grasses for livestock, soils and environment

Tropical grasslands have aroused renewed interest following recent reports of the *United Nations Environment Programme.* The results show that, while so many bemoan the destruction of rain-forests, tropical grasslands turn more carbon dioxide into carbohydrates than forests do. The concern for the environment being as great as it is today, mention should be made of the fact that conservation of the soil and grazing resources were the main incentive around 1930 in East Africa to initiate grassland research within the context of the prevailing mixed-farming systems, both large- and smallscale. Crop yields had been declining under cultivation and soil erosion was rampant, also in the grazing areas.

Since it had been found that leguminous green-manure crops and animal manure were not effective enough, attention had turned to the traditional systems of shifting cultivation and the role of grasses therein. It was confirmed that fallows under grass had a great capacity of restoring the productive capacity of the soil and to raise subsequent crop yields. Uganda gave the lead to research into the role of cultivated grasses in speeding up the regeneration process and this soon became official government policy in most countries of the region. Within the rotation cultivated grasses became the pivot of sustainable farming for the benefit of livestock, soils and subsequent crops.

The two major cultivated grasses, Elephant (Napier) and Rhodes, had already been identified and put to use by farmers long before then, as the common names dating back to the turn of the century indicate. Like elsewhere in the world, today's cultivated grasses and fodders are still those chosen by farmers long ago, before the start of official research. Even the successful species of exotic origin (lucerne and other fodder crops), found their way into local agriculture mostly as the result of farmers' introductions and initiatives. A sobering thought, engaged as so many scientists still are in searching for new species.

It has been the tragedy of agricultural expansion that it meant more grassland being brought under cultivation and more livestock being kept rather than more output per unit. The point was reached that no new hectares were available for cropping. Refuge was sought in opening up marginal land. The fallow itself came equally under mounting pressure so that yields of already existing cultivated fields declined also. Farmers tend to expand livestock numbers or they cling to the few animals they have. Communal grazing resources are increasingly threatened by livestock, mostly privately owned. These natural grasslands have become the subject of great concern. In the continuing debate on how to protect the environment it may be argued that there is no excuse for the deterioration of very large areas in order to produce crops or livestock at very low efficiency per unit of land.

Cultivated grasses, well-managed in the mixed-farming context, take pressure off natural grassland. Introduced on the farm, they replace and restore arable land, not natural grassland. In fact, East Africa's natural grasslands rank among the best available and they rarely require replacement by sown species.

1.2. Schools of thought

Eastern Africa is treated here as one unit because of its:

- Special role in the world as the gene centre.
- Geographic separation from the rest of Africa by humid lowland forest in the west and by desert to the north and southwest.
- Similarities despite variation in high- and lowland areas.
- Mixed-farming as the system practised by most of its people.
- Emphasis on small mixed-farms characterized by a chronic shortage of grazing.

In southern and western Africa the focus of official research has been rather on the husbandry of natural grassland, whereas in Australia the main concept has been to replace woodland by legume-based sown pastures; farm-holdings are large and used for extensive beef and sheep rearing. Another school of thought is found in north west Europe, based on specialized small-scale dairy production, with intensive conservation of grass as silage and with high rates of N-fertilizer. Finally, in the USA a great role is played by lucerne and maize silage, but also by natural grassland (range). Every region has, however, a mixture of different schools. In some areas of Kenya, lucerne plays a major role as an irrigated crop. In Australia, dairy production is based on N-fertilized grasses.

By its geographic position East Africa became a meeting place of people and ideas from north, south, east and west. With a well-based grassland tradition of its own, it could not fail to become a testing ground of ideas from different schools. These were commonly introduced together with plants familiar to the originator, but not necessarily new to the area. Current absence of certain species or practices in East African farming is no proof of them having being ignored. To the contrary, there is hardly any herbage plant presently being tested that was not tried on more than one occasion in the past.

The 1960s deserve special mention. This was the age of buoyant expectancy about novel genotypes and technologies. The world-wide success of new wheat, maize and rice varieties has indeed been a "green revolution". It had been silently preceded, in a less publicized manner, by African grasses which had spread over Asia, Australia and Latin America. This prompted the concept that a second wave, but now of pasture legumes from Africa and Latin America, could be generated artificially. Scientists were clearly not contented with the grasses farmers had chosen. This heralded the screening of an unending list of wild species and ecotypes which took possession of the lion's share of research resources. Thus, although the choice of plants had already been fixed by farmers previously without the help of science, scientists believed they could alter this choice and introduce entirely new species from the wild or develop them by breeding. Apart from weaknesses such as ecological lability and proneness to diseases, these plants generally failed to find sympathy with farmers and are still very little used. The preoccupation with novel species has been one of the main reasons of stagnating progress in research in many tropical regions of the world.

Research circles, especially those with an international mandate, have continued to concentrate on the search for new species, mainly legumes. However, farmers made much more headway with the grasses they had grown right from the early days. It would appear as if the concepts of N-fixation and nutritive value of legumes have been overplayed in relation to the seemingly insuperable barrier of their lability in grazed swards. Farmers had an alternative already in their mixed-farming systems and crop-grass rotations, thereby tapping and restoring soil-N and plant-N (bye-products) resources.

East Africa's school has been unique in that from the early days grasses played a major role in restoring soil fertility lost under cropping. In this region, livestock and crops have been intertwined since time immemorial. Mixed-farming developed as an advanced form of shifting cultivation, with cropping and livestock increasingly confined to a specific area. Mounting pressure on available land can be countered by a concomitant strengthening of the role of grasses within the rotation. It is worth noting that in the international discussion of sustainable tropical pastures, multiple production systems involving grass, livestock and crops are gradually coming again to the fore.

Most progress is made by the grafting of improved methods onto the existing ones rather than by the introduction of completely novel ones. Novel approaches are better presented as a return to the traditional methods, *e.g.* the sowing of grass can be described to the farmer as the return to the ancestral method of shifting cultivation.

A statement often heard is that past policies neglected the smallholder and favoured cash crops for export. The early volumes of the *East African Agricultural and Forestry Journal* serve to indicate, however, that the latter comment may be correct but not the former. Uganda, where more grassland development work was started than in any other country of the region, was and is not a land of large farms.

1.3. Towards more intensive strategies

Workable solutions are needed to come to the aid of vulnerable soils, crops, grasslands and livestock, especially for the small mixed-farmer. Research organizations are increasingly being asked to provide practical answers. Problems are not solved through replacing well-established cultivated species by offering new and, if anything, marginally better grasses, legumes, herbs or shrubs. While it is desirable that some resources should always be devoted to the search for better pasture species, we should not lose sight of those we already have and we should continue to work on them for their betterment. Husbandry is paramount but this should not necessarily be taken to imply high financial inputs.

Mixed-farming or alternate husbandry, based on the pivotal role of grasses in restoring soil fertility as in shifting cultivation, is not an alien concept. It requires a moderate level of technology but is sound ecologically. However, the measures that are needed to improve the grassland and fodder resources further within this system cannot continue to be low-input.

Grass resources are still grossly underutilized. Many farms are overstocked and overexploited but even on the better farms in favourable areas yields rarely exceed 1500 kg/ha, of either milk or cereal grain. Where veterinary control is adequate, nutrition is the principal factor limiting animal production. The smaller the farm the more livestock tend to be kept per unit of land and the higher is the pressure on the available feed resources. Quantity of herbage is then the main limiting factor, making quality in grass harvested young irrelevant. On larger properties, however, seasonal fluctuations in herbage supply predominate and quality may easily be missed, unless grass is conserved when young. On smallholdings very little can be achieved unless the mostly unproductive weed-grass fallows are replaced by cultivated species and fertilized.

In response to current misconceptions about sustainable farming, it has been argued that scarce resources (land, fertilizer) are optimized when employed in conjunction and intensively rather than extensively. Low-input farming implies that much more land is required for cultivation to arrive at the same amount of produce. In many countries it is the lack of development that causes the damage to the vast tracts of land so poorly utilized. The environmental damage consists of the breakdown of organic matter and of soil structure resulting in erosion and leaching of nutrients.

Efforts towards intensification need not necessarily imply extra financial costs. Low-input strategies should primarily optimize the efforts already made. For example, correct timing of sowing or of fertilizing may be as crucial as the input itself. Conversely, inputs should not be dismissed lightly as "uneconomic" or "outside the reach of poor farmers", because it is well known that farmers do take to such inputs provided they can be shown to meet their objectives.

Dairy production is a profitable part of mixed-farming since milk commands high prices and the increase in demand is likely to continue. To have more grass and to have it distributed evenly, N-fertilizer coupled with surplus conservation is highly cost-effective and amenable to flexibility in application. Grasses have a reputation of responding well to good husbandry and inputs are rewarded in terms of livestock and crop yields.

Atmospheric nitrogen (N) is perhaps more efficiently trapped via industrial processes than via plants, but the latter have stored N-reserves in the soil over the ages; N is also available as a bye-product of arable crops, and often more cheaply than when fodder crops are grown specifically to provide protein.

Typical of the subject of this book is, therefore, the wide range of alternatives available in farming to achieve the same purpose. Sown pastures cannot be discussed without due attention being paid to the role of grassland in soil fertility, but other fodder sources available within the farm circuit must also be considered. Experience has shown that when emphasis in publicity is on one single aspect in isolation from, if not against, the rest it is bound to receive a rebuke eventually. Alternatives should always be kept in proper perspective.

1.4. This book's approach

Current practice is a living documentation of what was tried and chosen as useful by farmers. A written account of what has been tested and achieved should help to overcome the communication gaps between farmers and scientists and to avoid unnecessary repeats of past work. We should also mention the traps for the uninitiated so that they can be avoided.

This book centres on one region only, East Africa, and opts for intensive grassland production and utilization within the mixed-farm context for the reasons explained above. It provides no lists of "potential" species, but describes species and practices existing in actual farming. Although many of the "old" grasses had been the principal feed source, they became the *Cinderella* of science. Emphasis is here on their improvement, not on their replacement. To this end, the necessary attention is given to practical but crucial details. Where we can answer to his call, we should assist the farmer with firm recommendations and not make his choices more difficult than they already are. Concepts developed in other (tropical) continents may be inappropriate in East African farming while much more is known locally than is often realized.

This should not be surprising since the bibliography contains 1000 odd references to articles, from or about eastern Africa, not including annual reports. Since literature is rarely a true reflection of actual farmers' practice, personal observation is used to arrive at some equilibrium. Grasses such as Hyparrhenia and Molassesgrass are grown worldwide but receive only scant attention in the scientific literature. Conversely, to conclude that certain species are widely grown from the frequency with which they are reported on in literature would be equally wrong.

Literature when scattered and not comprehensively reviewed can prove an insurmountable barrier to the uninitiated who may be easily led off-course by some momentary impression. However, by re-examining the broad range of data available, including those which were considered anomalous at the time and not published, pieces of an apparent jigsaw puzzle fit together and a picture emerges which corroborates many of the established agricultural practices and helps to formulate recommendations for their betterment. One of the tasks of this book is to unearth such data in as much detail as possible to counter generalizations based on flimsy evidence but often avidly quoted. A special effort has been made to review the East African literature of the early part of this century. Although these sources are often ignored they are still as relevant as ever because not only many of the problems but also the answers to solve them are still very much the same. Many reports of that time deal with the interface of soils, crops, grasslands and livestock, often by authors with a broader perspective than grassland alone. The French literature of the region is also given the attention which is long overdue.

Ecological and environmental issues were hotly pursued in the 1930s and 1940s. These are described in the chapters 2–5 with a major emphasis on the conservation of natural grasslands and on the preservation of soil fertility. The main sown species and their husbandry are reported on in chapters 6–10. A separate part (II) of the book is devoted entirely to fodder crops, especially Elephantgrass (chapters 11–13), because of the different ways in which these crops are commonly exploited. Continuous *ex-situ* feeding implies intensive removal of soil nutrients which can only be alleviated by large amounts of various sources of fertilizer, not needed under direct grazing. Additional fodders and special-purpose crops, for late-season feeding, are described in the chapters 14–19.

For an introduction to the principles of grassland science and for subjects of global application, the reader is referred to other books: Mays (1974), Shaw & Bryan (1976), Bogdan (1977), Whiteman (1980), Humphreys (1981, 1991), Crowder & Chheda (1982), Heath *et al.* (1985). Livestock of East Africa has also been well covered (Musangi, 1969; Meyn, 1971; Payne, 1971; Jahnke, 1982).

In order to avoid repeating lengthy Latin names, many of which have been subject to frequent changes recently, preference is given to common names, and where necessary to avoid confusion use is made of acronyms, *e.g. Pmax* (*Panicum maximum*); see Glossary. Genus names are used if no common names are available and if only one major species is involved: Themeda, Imperata, Exotheca.

CHAPTER 2

History of grassland development (1900–1970)

Although purposeful grassland development in East Africa is of relatively recent origin (± 1900) it has been made by a wide diversity of people. This is due to tumultuous political and economic changes which took place in the region both in the colonial period and thereafter. Also, the last part of the 1800s had seen tremendous upheavels and changes. Catastrophic epidemics of rinderpest had all but decimated the cattle herds. In that same period maize had come to replace sorghum as the major staple food, in a very short time.

The types of commercial farming we have at present on the ground formed themselves gradually over decades of trial and error. Many ideas came from different parts of the world. Europe, southern Africa, USA and Australia influenced thinking in East Africa each in turn. The main events and the names of those involved have now largely been forgotten. Yet they are important to be recalled to explain why we have what we have and why it proves to be so difficult at present to introduce something that is really new and of a lasting effect.

The history of grassland development in East Africa has been one of seemingly slow, yet steady progress. Although from the turn of the century onwards, settlers, administrators and missionaries had brought many concepts and materials from their countries of origin into the East African region, the main agricultural system, mixed-farming based on crop-grass rotations, which they would finally adopt was borrowed from the local peoples. Examples are given of Uganda and Kenya.

2.1. Uganda

Uganda was the focus of early colonial interests, more so than Kenya. The growing of coffee and cotton was promoted strongly among smallholders. These cash crops introduced into subsistence agriculture coupled with a rapid population increase led to a mounting pressure on the land, with consequent reduction in the length of fallow under natural regeneration. As a result, the Uganda Department of Agriculture (UDA) laid great emphasis on restoring soil fertility lost under cropping.

2.1.1. Green-manure crops

The early work was devoted to the testing of leguminous cover and green-manure crops: (years are those of the UDA Annual Reports): 1906, lucerne; 1912, Mucuna; 1915, Leucaena; 1925, Lablab; 1930, Pueraria. Centrosema was tested in the 1920s (Nye, 1940; *10.8.5*). Many of these had been brought into East Africa because of their importance in plantation agriculture (*2.4*) and because of the role botanic gardens played in the introduction and exchange of plants and seeds (Brockway, 1979). In later years both local and exotic legumes would be tested more specifically for their suitability in grazed pastures: 1945, Stylo (*Sgui*); 1957, *Stylosanthes fruticosa* (from Uganda itself). The use of Elephantgrass for mulching was reported by W.S. Martin in UDA Ann Rep 1928.

T.D. Maitland, Government Botanist, started a grass nursery at Kampala Plantation in 1925 (Maitland & Hubbard, 1927). Other research sites in the forest zone were Bukalasa, 50 km north of Kampala, and later Kawanda and Namulonge, also near Kampala. In the savannah zone nurseries were set up in 1931 at Ngetta and later at Serere, in Teso.

2.1.2. Grass rests in the crop rotation

Continuous cropping alternated with green manures had proved to be a failure in a rotation trial in 1932 at Serere, because crop yields were lower than on neighbouring farmers' plots under shifting cultivation (*5.2.1*). By contrast, the soil regenerating effect of Elephantgrass had been appreciated by local farmers in Buganda for a long time. It was then decided to incorporate grass into the crop rotations of government farms. Rotation experiments were set up to investigate crop:rest ratios (Martin & Biggs, 1937). At Bukalasa, Elephantgrass was used in a 3:3 rotation. At Serere and Ngetta, natural regeneration was employed while observations were carried out on a range of local grasses. The need for a more rapid and effective soil cover than that provided by natural regeneration led to studies of sown grasses (Hosking & Stephens, 1941). A planted grass rest was considered to be an accelerated form of the natural regeneration developed under shifting cultivation. Martin (1944a) found that the improved physical condition of the soil after a grass rest was of overriding importance (*5.3*).

Jameson & Kerkham (1960) and Stephens (1970) summarized the results of the long-term rotation trials started in the 1930s at Serere and showed that, in the absence of animal manure or of fertilizer, land could not be safely cropped more than 3 years in 5 (*5.5*). As early as 1920 it was realized that continuous arable cropping led to rapid soil exhaustion and erosion (Brockington *et al.*, 1965). The soils are shallow and infertile with a low organic matter content in contrast to the fertile Elephantgrass areas in the south and west of Uganda. Cotton is the main cash crop and finger millet forms the main food crop over large areas. Groundnuts, Sorghum, maize and cassave are other important crops. Again this is in contrast to the Elephantgrass region where coffee and banana assume much greater significance.

Studies in the 1940s began to incorporate measurement of herbage and root production and of subsequent crop yields. This policy was effectively continued for many years as shown by the works of Stobbs (1969e) and Tiley (1970a). As could be expected, the emphasis in encouraging the sowing and planting of grasses evolved from the need to restore fertility to that of providing feed for cattle. Better grass for cattle is the motive inducing farmers to embark upon the labour intensive operation of planting grasses. In addition to mulching (*11.3.3*), Elephantgrass began to be used for grazing and stall-feeding (*11.3.4*).

Since animal manure was found not to have much effect on soil structure (*5.2.2*), it was believed at first that grazing would detract from the soil regenerative powers of grasses and that it was better to leave the grass undisturbed. These fears were dispelled when Kerkham (1947a, b) confirmed convincingly that grazing had in fact a positive influence on subsequent crop yields; he advocated grazing in order to increase the gross cash return from the land. Similar trends were demonstrated by Mills (Serere Report 1961) and Stobbs (1969e) from grazing treatments on the Serere rotation trials. After ploughing grass-legume pastures, however, yields of arable crops, notably cotton, were depressed compared with those after pure grass swards (Table 5.8). Indications that legumes like Stylo (*Sgui*) might result in decreased crop yields led to the decision not to recommend Stylo (Brockington *et al.*, 1965; Tuley, 1968).

Concern over the loss of irreplaceable soil resources prompted great emphasis on grasses. Elephantgrass was studied at Bukalasa, Kampala Plantation and later at Kawanda. Brachiaria and Setaria (both planted vegetatively) and Rhodesgrass were satisfactory for soil cover at Ngetta and Serere. Strip cropping (Kerr, 1942), bunds and contour hedges were introduced as measures to combat erosion. *Paspalum notatum* was introduced from South America and planted specifically for the purpose of soil conservation, *e.g.* on bunds (*6.16*). By the end of the 1940s (Killick, UDA Ann Rep 1948) it became the policy of the Uganda Department of Agriculture to sow Rhodesgrass as the most suitable grass to fulfil the requirements of both grazing and soil improvement. For the latter purpose, Martin (UDA Ann Rep 1946) had found Rhodesgrass to be as effective as Elephantgrass and *Paspalum notatum*. The recommendation of Rhodesgrass in place of Elephantgrass was also recorded as a change of policy by the Experimental Committee of the Department of Agriculture of Uganda (UDA, 1949). Rhodesgrass could be readily established from seed and was also easily eradicated. By contrast, Elephantgrass was difficult to remove by ploughing and required vegetative planting (*12.1; 12.3*).

Work on legumes ceased in the 1930s after further test cropping had confirmed the inadequacy of green manures. In his summary account of research done in Uganda at the Nairobi Colloquium on Grassland Research, Kerkham (1950) did not mention legumes at all. Work on grass-legume mixtures was not taken up until 1954 by Horrell (1958b; 1963) at Serere, followed later at Kawanda by J.R. Simpson, 1956–1958. Stobbs (1965) summarized the research policy: "the object of the present pasture research at Serere Research Station is to make possible a system of ley farming for the annual cropping areas of Uganda, which will prevent erosion, improve soil fertility and sustain higher profits from both animal production and also subsequent arable crops." Both legumes and fertilizers, including nitrogen, were studied.

The grassland research work in Uganda has been remarkable in the emphasis it has laid on the evaluation of grasses and legumes through both grazing trials and subsequent test cropping.

2.2. Kenya

Huxley (1935) has given an account of how agriculture developed since the end of last century. Missionaries had brought coffee to Kenya already in 1895. Pioneer farmers such as Delamere imported exotic breeds of cattle, sheep, pigs and poultry, although they lost them nearly all to diseases, the most important among cattle being pleuro-pneumonia and ECF (east coast fever). They moved to crops such as tobacco, oranges, wattle, beans, oats, barley, potatoes, but more successfully in the end to wheat. Delamere went as far as recruiting a wheat breeder, G.W. Evans, who developed rust resistant varieties such as Equator, a variety known up to the present day. This was before World War I.

However, animals were needed for draught and carting and this implied veterinary care and supplementary feeding.

2.2.1. Veterinary control

Very little was known about animal diseases and no vaccins were available. The first Annual Reports of the Departments of Agriculture of the region were filled to a third with veterinary accounts. Research was seen as the only answer. The Naivasha Stock Farm was set up. The Kabete Veterinary Laboratories were established near Nairobi and developed at great speed one vaccin after the other against pleuro-pneumonia, rinderpest and anthrax. ECF was more difficult; at one time the disease killed 80% of all young stock. It was A. Theiler who discovered in 1904 that this disease was tickborne. Measures of control were sought in preventing ticks from moving from infected to clean areas. Strict regulations were applied on fencing and on cattle movement through "corridors". Fencing alone was, however, not adequate. Together with dipping and spraying it had eradicated ECF in South Africa, but in East Africa such measures were never effectively enforced, so that this disease is the main livestock killer up to the present day.

2.2.2. Farmers' introductions

Wherever farmers decide to move, they bring seeds along or have them sent for. Seeds were often exchanged through correspondence with countries where farmers were also searching for new species. Government botanists were generally more concerned with industrial crops (Brockway, 1979). Huxley (*op. cit.*) reports that Delamere in 1904 sent for seeds of grasses, clovers, lucerne and turnips from U.K. Seeds also entered with livestock imports. In this way it is likely that teff (*6.16*) was introduced as a fodder crop from Ethiopia, with horses and mules. Kenya soon became the trial ground for seeds brought from other parts of Africa, Europe and Australia. These seeds were not imported in the first place to replace the natural grassland with sown grasses or fodders. More likely is that farmers in those days primarily wanted fodder crops to feed their working oxen and horses, and to feed livestock at night in enclosures where they were safe from predating lion and leopard. There were also new economic activities which required special fodders. Fodder rape (*18.1*) was grown especially for ostriches raised in captivity for their feathers which were exported to Europe for the ladies' hat-fashion of that time.

Australia was an important source of early imports. Bluegrass (probably *Dicanthium sericeum*),

Burnet (*Poterium sanguisorba*) (*18.6*), *Paspalum dilatatum* (*6.16*; originally ex South America) and saltbush (*Atriplex*) were reported on by J. Johnston of the Kabete Experimental Farm in 1910 (Kabete Ann Rep 1910). However, Australia has not lastingly contributed any cultivated grass species (Williams *et al.*, 1976) and has had to rely by necessity on the import of pasture plants. This helped to open the eyes in East Africa for the grasses around. An Australian article about Rhodesgrass was quoted in the "Agricultural Journal of British East Africa" of 1910:

"this heavy fodder grass succulent and highly palatable, abundant throughout the Rift Valley, drought resister of high power and south of Lake Naivasha has not yet been known to surrender entirely its green colour. When cut for hay it has a well defined aromatic odour, pleasantly suggestive of caramel."

Other sources of seed imports were also tried, but by 1930 the main grasses we have now were well established: Elephantgrass (*11.2*), Rhodesgrass (*6.6*) and Molassesgrass (*6.10*). Some of the varieties used were the result of import, but these triggered off interest in the local ecotypes.

2.3. East Africa as a "gene centre"

East Africa soon became known internationally as a centre of genetic diversity (gene centre) of the best known and most important cultivated tropical grasses not only of Rhodesgrass but of many genera such as Buffel-, Guinea-, Signal-, Stargrass, and legumes like Kenya white clover, Glycine and Lablab (*6.1.1*; *6.4*). Many introductions into Australia from East Africa can be traced back to before 1910 (Barnard, 1972). As mentioned already, farmers were often responsible for the first imports made, *e.g.* through correspondence. Kikuyugrass, for instance, was planted in South Australia with cuttings obtained from Molo (K) as early as 1908 (Quinlan *et al.*, 1975).

Not all exports and imports, however, were intentional. Many plants had travelled from East Africa to other continents by chance, first to Asia. The same happened only in the past few centuries with Guineagrass, Molassesgrass and *Hyparrhenia rufa* in Latin America. Their common and main instrument of seed transport across the ocean was probably the bedding material made of grass for ships sailing from the African coast. The introduction of Zebu cattle also helped the "Africanization" of grasses of the New World (Parsons, 1970; 1972).

It is likely that grasses growing in abundance near the ports in Africa's coastal or riverine areas were the most successful to escape, and to invade similar environments in other continents also via ports. Grasses of the East African highlands, isolated from the coast by semidesert, were less successful in escaping to other continents. Kikuyugrass is an example of an East African grass, whose escape depended on man's purposeful initiative. This grass had only a limited distribution, in the East African highlands, until about 1900. Its unique appearance would have been noticed elsewhere and reported on much earlier, had it been present (*3.4.2.1*).

While farmers were responsible for the first and, as it would turn out, the most successful seed exchanges, the supervision of seed exchange was soon regulated and taken over by Government Botanists in charge of botanic gardens and crop introduction, *e.g.* T.D. Maitland at Entebbe (*2.1.1*). In 1911–12, 1300 official consignments of seed were handled by the Kenya Department of Agriculture (KDA Ann Rep 1912). Leucaena was introduced in 1915 and met with some success in Uganda (UDA Ann Rep 1916) but somehow it failed to have an impact, already then (Tothill, 1940; *18.5.6*). In general, grasses and legumes imported into East Africa have acquired no lasting position, with the exception of lucerne (*17*), oats (*16*), Columbusgrass (*15.2.3*) and temperate grasses and clovers for the high altitudes (*6.15*). Seed imports, however, have continued up to the present day (*1.2*), often of material received via other countries, but originating from and obtainable in East Africa.

The lesson we can learn from the history of exchange of plant species is that desirable and successful species need little in the way of official research initiative because they usually have forced themselves already upon the farmer in one way or the other. This happened in Australia with exotic legumes such as Townsville Stylo, clovers and medics and with the African grasses now in cultivation. The latter did the same in Latin America.

Subsequent exploration efforts were usually less rewarding.

Interestingly, the ultimately successful traffic of herbage seed is often one-way only. Exotic legumes, though purposefully reintroduced and tested many times, do not truly volunteer in East African grassland. When sown, their main weakness is lack of persistence and of ecological stability (*4.3.5*; *10.4*).

In a review of government efforts in crop introduction in Kenya, Suttie (1970b) demonstrated that all the major species of the world had been tested in the early years of this century. He concluded that any crop which was not suited then is unlikely to be any more suitable now, particularly as it will have to compete with crops which have subsequently been subjected to intensive research and breeding.

2.4. The start of grassland research in East Africa

Consequently, the herbage species we have in cultivation to-day had entrenched themselves in farming practice already more than 60 years ago, due to the pioneering work of farmers. Official research was to follow later and it would find better ecotypes, within the same species. Some early work was done on grasses in Uganda at Entebbe and at Kampala Plantation (*2.1.1*), and in Kenya both at the Kabete Experimental Farm and at the Naivasha Stock Farm, set up before or around 1910. The Nioka Zootechnical Station in Ituri in north east Zaïre dates back to the end of last century (Risopoulos, 1966).

The main research work in those early days, however, concerned plantation crops at Kampala Plantation, Mazeras (near Mombasa), Kibos (near Kisumu), Yangambi in Zaïre, but perhaps the most famous one was Amani, in the Usambara Mountains near Tanga, founded by the Germans in 1902 (see also Eichinger, 1914). After World War I, the Amani station would reopen in 1927 as the forerunner of EAAFRO (East African Agricultural and Forestry Organization) which moved to Muguga near Nairobi in 1947.

Interestingly, many of the early research findings on grasses in Uganda were achieved by officers mainly engaged in other crops or activities. W.S. Martin was Agricultural Chemist. A.S. Thomas was the coffee breeder at Kampala Plantation and Kawanda but he also bred Uganda Hairless Elephantgrass (*11.7*; *9.7*) and did important ecological work (Thomas, 1966).

2.4.1. Early locations of grassland research

The year 1898 saw the foundation of the Botanic Gardens of Entebbe, through which plant introductions were made, especially of plantation- and associated leguminous cover crops (*2.1.1*). In 1917 the Agricultural Department took over control of the Gardens under Maitland. Martin (*2.1.2*) became Agricultural Chemist in 1924. The Kampala Plantation housed the whole Research Division and was the official trial ground for rubber, coffee, cocoa and tea, until they moved to Kawanda in 1934. Kawanda was Royal Buganda pasturage and had earlier been donated to missionaries for their services in introducing American Upland cotton. One of the first officers-in-charge of Kawanda Research Station was A.J. Kerr (1942) who would later, in the 1960s, promote Elephantgrass while farming at Kitale. Bukalasa and Serere were chosen in 1919, originally as cotton seed selection stations, but a considerable part of the work was on cultivated grasses for soil fertility maintenance (*2.1.2.*; Hosking & Stephens, 1941). Also Serere was on land that had belonged to missionaries, of the society of Mill Hill. Much of the early work on grasses in Uganda was done on a part-time basis, as already mentioned. Unlike Tanzania and Kenya (see below) full-time grassland research officers were not appointed to Uganda until 1954 (C.R. Horrell at Serere), to be followed by G.E.D. Tiley (Kawanda) and T.H. Stobbs (Kawanda and later Serere). Grassland research was also initiated at the Kabanyolo Farm of Makerere University and by the Veterinary Department at Entebbe and Mbarara.

From the early 1930s onwards, most of the grassland research work in Tanzania was concentrated in Mpwapwa (Staples, 1931; French, 1935) but would expand to Kongwa, Tengeru near Arusha, Ukiriguru near Mwanza and other stations (van Rensburg, 1969a). In Zaïre grass work was well on its way in the early 1930s on INEAC stations (with headquarters at Yangambi) by W. Robyns (1934) and later by R. Germain (1954) in the lowland and R. Compère (1960) in Kivu.

The Kivu and Ituri highlands of eastern Zaïre have a lot in common with Rwanda and Burundi and were served by the Rubona Agricultural Centre in Rwanda. This station was started in 1930 to produce seed for farmers who had depleted their stocks after the 1923 and 1929 famines.

In Kenya, full-time grassland work was based at Nairobi in 1930 (D.C. Edwards) until it moved to the newly founded Grassland Research Station of Kitale in 1951. This station continued also the work on grasses that had been started in the 1930s by the Soil Conservation Department (Maher, 1936b). It was transformed in the 1960s into the National Agricultural Research Station (now KARI, Kenya Agricultural Research Institute) to cover both pasture and hybrid-maize research at the national level. For a list of regional research stations in Kenya, see Table 6.2. On all these stations grasses were a major subject of research.

It is worth noting that in Europe, Australia and USA organized official grassland research did not start much earlier than 1930 either. Much of what had been achieved had been pioneered there by farmers also. The investment of governments in research then ran parallel to their endeavours to reorganize the economies during and after the "Great Depression". By contrast, strong research establishments had been set up much earlier in the colonies of south east Asia, to foster plantation agriculture. To these we owe the introduction of leguminous cover crops such as Puero (Kudzu), Calopo, Centro, Stylo, Desmodium and shade trees such as Leucaena and Gliricidia (Bogdan, 1977; Skerman, 1977); see also *2.1.1.*

2.4.2. Mineral deficiencies (1920s)

Veterinary priorities continued to dominate the departments' approach to general livestock and grassland problems until 1930. As disease control improved in the 1920s, attention focussed on mineral deficiencies, again following the lead given by Theiler (*2.2.1*) and colleagues in South Africa, who had identified phosphorus deficiency as the main element limiting animal performance. French (1946b) concentrated his early work on deficiences in Tanganyika. The same was done in Kenya (Gunn, 1930). Attention was drawn to the disease "Nakuruitis", cobalt deficiency (Orr & Holm, 1931; see also *12.7.5*). Most of the deficiencies known to occur in livestock are now overcome by mineral supplementation.

2.4.3. Ecological approach (1930s)

South Africa had a great impact also on the first phases of grassland research, especially in ecology. Terms such as "Kenya veld" were adopted (Hall & Allen, 1938; Nightingale & Lloyd, 1946). South African scientists spent long periods in East Africa as advisers, *e.g.* J.F.V. Phillips and I.B. Pole Evans. The former drew attention to the ecological basis of grassland development in Tanganyika (Phillips, 1929; 1931). Pole Evans (1948) made an overland expedition from Pretoria to Turkana (K) in 1938 and wrote an illustrated account of the vegetation and grasses he observed in places which can still be recognized. Great emphasis was placed at that time on plant succession and it was believed that better grasses could be induced to gain the upper hand on reverted arable land (Edwards, 1935a; Ball, 1936). It was not until later that it was realized that this process was too slow or too costly and that more efficient grasses had to be established.

D.C. Edwards was the first officer appointed specifically to be in charge of grassland improvement and plant introduction in 1930 at the Agricultural Laboratories (now KARI) in Nairobi. Edwards gave grassland research an ecological basis and regional stations were established in different ecological zones (*3.3.1*). His contemporaries were A.S. Thomas in Uganda, W. Robyns in Zaïre, C.G. Trapnell in Zambia, and R.R. Staples and M.H. French in Tanganyika.

2.4.4. Soil conservation

Much attention was given to soil conservation. The post World War I period had seen a doubling of the livestock numbers and a great increase in the area under cropping. Fertilizers were largely unknown and animal manure was not collected in any quantity and not considered effective enough (*2.1.*; *5.2.2*). When the "dust-bowl" effects swept across northern America it did not take long for the East African governments to realize that their countries were facing similar disasters unless steps were taken to stem the tide of overgrazing and overcropping. The role of grasses in soil conservation was well appreciated (*2.1.2*). However, single crop farming was general and where stock was maintained they were often grazed outside the arable land. In many parts of the region livestock were, and often still are, divorced from the arable land.

Incentives to establish cultivated pastures were limited in the early decades when farming units were relatively large. The war periods and the Great Depression interrupted whatever development had taken place and they placed by necessity great emphasis on arable cropping. The market for fresh milk was small. Mixed-farming based on crop-grass rotations began to be firmly established in the 1930s in Kenya (Maher, 1936b; Ball, 1936) with dairying as part of the entreprise. Rhodesgrass, already put to use by farmers, became the grass recommended throughout East Africa in crop rotations, especially for the maize, wheat and cotton growing areas of Uganda, Kenya, Tanzania and Zambia as well as for tobacco rotations in Malawi and Zimbabwe, where it was valued especially for nematode control (*6.6*).

Grass establishment was the central part of soil conservation measures as developed by Maher (*op. cit.*) and colleagues at Kitale. The development of fast-establishing species and varieties became the main duty of the new Grassland Research Station in 1951.

2.5. New varieties and commercial seed multiplication

Many new species and ecotypes were introduced and screened as a result of which new varieties reached fame, also outside East Africa, such as Nandi and Kazungula Setaria as well as Likoni and Makueni Guineagrass (*Pmax*). Not surprisingly, the species adopted by farmers, *e.g.* Elephant-, Rhodes- and Molassesgrass, as well as Grass Sorghum (*15*), oats (*16*) and lucerne (*17*), maintained their position at the top of the list, but at Kitale A.V. Bogdan and R. Strange managed to identify better ecotypes such as Mbarara and Masaba Rhodes from the East African region and they also secured better clones of Elephantgrass (*11.7*). Work was resumed in the 1950s, at Serere by Horrell (1958b) and at Kitale by Strange (1958a), on local and exotic legumes, although Edwards (1950a) had already sounded a note of caution due to the many earlier attempts which had been unsuccessful (*10.6*).

An important landmark in pasture development was the start of the Kenya Seed Company at Kitale in 1956 (*8.1*). It was set up by seed growers to pool efforts of producing, processing and marketing of the grass varieties released by Bogdan and Strange of the Kitale Station. The Kenya Seed Company later ventured equally successfully into hybrid maize, cereals and sunflowers, thereby providing an essential link between research centres and farmers, while developing an outlet for export of seed at the same time.

The introduction of Kitale maize hybrids in 1965 became an instant success with the smallholder. Smallscale dairying was also given a strong impetus. Grass seed production and export developed to a great height (Fig. 8.1).

In the colonial period agricultural reseach in East Africa was well co-ordinated but highly diverse, long-term but not self-perpetuating because of the relatively frequent staff transfers. The 1960s brought the end of the colonial era and the departure of expatriate officers (Table 2.1). Within less than a decade, a complete generation of research and extension workers left their place of work. A heavy burden was laid on new workers to assist in the agricultural revolution that took place, as large farms changed ownership or were transformed into settlement and village schemes, needing a more intensive land use in the mixed-farming context. The late 1960s saw the start of donor-assisted programs, focussing attention on specialized components, mostly individual fodders (see also *5.6*).

2.6. Summary

Uganda and Kenya are described as two examples of how grassland development took shape in the period 1900–1970. In Uganda the activities were concerned mainly with the search for methods and species to restore soil fertility lost under cropping. By 1930, animal manure, leguminous greenmanure and cover crops had proved unsatisfactory. The study by Martin of the traditional methods of cultivation showed that local grasses were more effective to restore fertility. Elephant- and Rhodesgrass were identified in the 1930s as

Table 2.1. A historic overview of grassland research in East Africa until 1970.

Period	Research priorities	Centres	Leading resident specialists and year of main publication
1910–1925	Veterinary	Kabete (K)	
		Entebbe (U)	
		Mpwapwa (T)	
1920–1935	Mineral deficiencies	Mpwapwa (T)	M.H. French (1957)
		Kabete/Naivasha (K)	H. Gunn (1930)
1930–1950	Ecology	Mpwapwa (T)	R.R. Staples (1931); P.J. Greenway (1933)
		Nairobi (K)	D.C. Edwards (1940); P.J. Greenway (1943)
		Kampala (U)	A.S. Thomas (1940)
		Zambia	C.G Trapnell (1953)
		Zaïre	W. Robyns (1934); R. Germain (1954)
	Soil conservation	Uganda	W.S. Martin (1944)
		Kitale (K)	C. Maher (1936)
		Muguga (K)	H.C. Pereira (1959)
1945–1970	Varieties and agronomy of sown pastures	Kitale (K)	R. Strange (1958); A.V. Bogdan (1965)
		Serere (U)	C.R. Horrell (1963); T.H. Stobbs (1969)
		Kawanda (U)	G.E.D. Tiley (1965)
		Mpwapwa (T)	H.J. van Rensburg (1969)
		Zaïre/Rwanda	R. Compère (1961); S.A. Risopoulos (1966)
	Pasture chemistry	Kitale (K)	H.W. Dougall (1960)
		Mpwapwa (T)	M.H. French (1957)
		Entebbe (U)	R.M. Bredon (1961); R.S. Musangi (1966); B. Marshall (1967); M.I.E. Long (1967)
	Range management	Nairobi	D.V. Pratt (1969)

the best solution to replace the natural fallow by a more productive grass rest. Grazing trials and testcropping figured prominently throughout, at Kawanda and Serere. In Kenya, settler farmers set the tune in the search for suitable fodders and pasture grasses to supplement the natural grazing. The emphasis was first on exotic species but local varieties of Elephant- and Rhodesgrass were well established in farming by 1930. Mixed-farming, based on livestock and cereals in rotation with sown grasses, became the main agricultural activity. Soil conservation and maintenance of soil fertility were major issues. In the 1910s, government efforts had been concerned with veterinary control and vaccin development and in the 1920s with mineral deficiencies. The interest in conservation of grassland and cropped land became paramount in the 1930s and, as a result, Kitale became the main centre of activity for the development of new varieties and seed multiplication. However, the main species, Elephant- and Rhodesgrass, had already been put to use by farmers and they have remained the most important ever since. East Africa established its name as the main gene centre of tropical grasses in cultivation and as a source of commercial seed supply. Of the exotic legumes, only lucerne found a lasting place.

CHAPTER 3

The grassland environment

The unique feature of a "gene centre" such as East Africa is that natural grasses and their cultivated variants may be found side by side. For a proper understanding of the functioning of grasses it is essential to consider the environment in terms of climate, soil conditions as well as of vegetation. In particular, those elements which are relevant for classification or zonation purposes will be discussed before specific grassland types and their husbandry (*4*) can be analyzed.

3.1. Climate of eastern Africa (15°N – 15°S)

The climates of the regions with which we are concerned have been described by Griffiths (1972) in "Climates of Africa". Climate in relation to grassland in East Africa has been discussed by Pratt & Gwynne (1977), whose description of "Eco-climatic Zones" will be largely followed. Details will be restricted here to rainfall, altitude and temperature as the most relevant parameters (Fig. 3.1). Typically for large areas of the region, highland plateaux of 2000 m are crossed by lowlands of 1000 m altitude.

3.1.1. Rainfall

Very low rainfall is experienced in the northern region towards the Sahara, and on the eastern flank. The latter arid or semiarid region, often referred to as the Horn of Africa, pushes its way south towards and beyond the equator, stretching along the seabord from Eritrea through Somalia and the Ogaden region of Ethiopia to northern and eastern Kenya, and extends well into central Tanzania. In this large belt the average rainfall only exceeds 500 mm on a few higher ranges.

The better rainfall in the rest of the region is characterized by a considerable variation from place to place, from year to year and from season to season. Kenya receives a reliable 750 mm/year in only 15% of its territory, compared with about 50% of Tanzania and nearly 80% of Uganda (Table 3.1).

Brown (1963) defined high-potential areas as those receiving more than 900 mm rain, while cropping is risky in the range of 650–900 mm if rain is scattered over many months. In areas with only 500 mm rainfall, *e.g.* Kongwa in eastern Tanzania, cropping is possible since rain is concentrated in a period of only 4 months. By contrast, rainfall efficiency is reduced in areas with 2 maxima in opposite seasons; bimodal patterns of "long and short rains" are most marked in the region north east of Nairobi. For a discussion of seasonal distribution of rainfall in Kenya and Tanzania, see Braun (1986).

It is with these considerations that the rainfall histograms of Fig. 3.1. should be interpreted. The recording locations are mostly research stations, mentioned in this book. Apart from those near the capital cities, research stations were usually located at sites, representative of major agricultural zones.

3.1.2. Temperature

The variation in monthly air temperature through the year is very small at the equator (Fig. 3.3) and increases to about 6 °C units at 15°N and °S. Some exceptions are found; the range is 8–9 °C at certain locations such as Hargeisa (northern Somalia; 9°N) and Mongu (western Zambia; 15°S), due to cold winds from desert areas in the cool season.

In East Africa only few stations have recorded screen frost and even ground frosts seldom occur

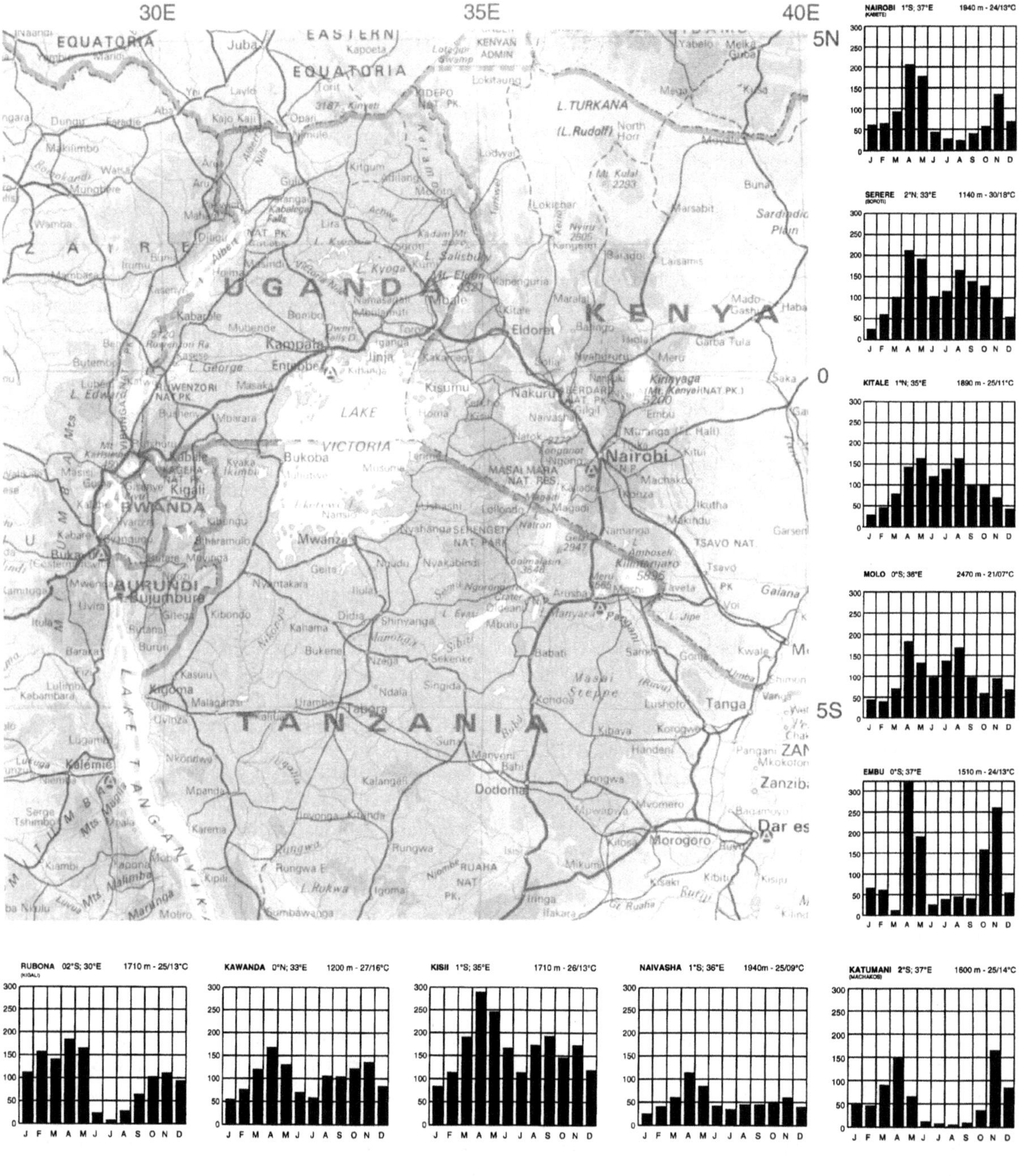

Fig. 3.1. Rainfall, altitude and temperatures at selected stations in Eastern Africa (after Griffiths, 1972; East African Meteorological Department, 1975). See also end papers and inside cover pages for further data.

Table 3.1. Proportion (%) of land area receiving selected amounts of annual rainfall in 4 years out of 5 (after Griffiths, 1972).

Rainfall (mm)	Kenya	Tanzania	Uganda	East Africa
< 500	72	16	12	35
500–750	13	33	10	20
750–1250	12	47	72	41
> 1250	3	4	6	4

below 2500 m, except during July and August in the southern highlands of Tanzania and occasionally in cold-air drainage hollows down to about 1800 m (Griffiths, *op. cit.*). In the Mount Kenya region frost occurs down to 2250 m. Further south, screen frost is observed each winter near the Zambia/Zaïre border, at Solwezi (12°S; 1330 m). Really severe frost in the region, occurring south of the Zambezi/Kafue river system, is limited to occasional incursions of very cold air, which reach Zambia and Zimbabwe from the southwest across Botswana. In Ethiopia frost is experienced above 2250 m during the cool season and is not uncommon occurrence in local areas, such as valleys with drainage of cool air. For instance, the town of Awash, 160 km east of Addis Ababa (10°N) in the Awash valley, often has frost, yet is located at less than 1000 m. Locations above 2500–2700 m can have frosty nights almost any month of the year. According to the map of climatic zones of the highlands of Ethiopia (le Houérou & Popov, 1981), about 40% of the country in the montane zone experiences light frost.

The naturally high average temperatures expected in the region are moderated by altitude. Lapse rates given by Griffiths (*op. cit.*) are: for Malawi, Zambia, Zimbabwe 8–9 °C/1000 m, for Ethiopia 7°C. The specific equations of best fit for equatorial East Africa are (h = elevation in 1000 m):

- Mean annual maximum temperature: 34.0–5.6 h
- Mean annual minimum temperature: 24.5–6.3 h

However, systematic deviations from the altitude-corrected means occur, which are relevant in the context of the grassland environment to explain the altitudinal ranges and limits of certain zones. The areas, in which temperatures deviate by more than 2 °C from these equations are shown in Fig. 3.2; the coastal belt is not included due to the sea's moderating influence.

A

B

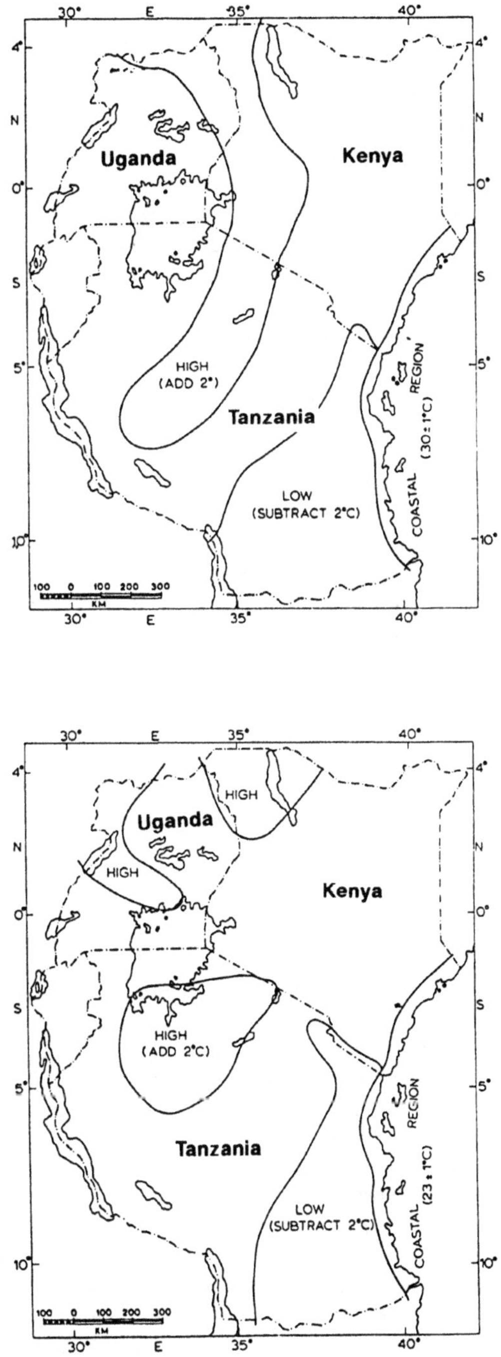

Fig. 3.2. A. Deviations of the mean annual temperature from altitude corrected mean. B. Deviations of the annual minimum temperature from altitude corrected mean (after Griffiths, 1972).

Fig. 3.2 illustrates that deviations are not correlated in a systematic way with either latitude or longitude. For instance, temperatures are 2.5–3 °C higher in western Kenya than at comparable locations east, especially in July-August (Braun, 1985). Rice cultivation is feasible at Ahero (1220 m; near Lake Victoria), but suffers from cold damage at Mwea (1160 m; near Mount Kenya) in at least 1 out of every 3 July-August seasons. Similarly, although at about the same altitude and rainfall, Nyeri (east) is situated in the Kikuyugrass zone (*3.4.2.1*) whereas Kitale (west) is not. However, not only temperatures but also solar radiation and hours of sunshine reach lower values at Nyeri than at Kitale, while the soils differ also.

How great the deviations from the altitude- or latitude-corrected means can be is evident when a comparison is made between Soroti (eastern Uganda) and Songea (southern Tanzania) as illustrated in Table 3.2. Although at similar altitude and annual rainfall regimes, the mean monthly maximum temperatures are 4.5 °C lower at Songea, 10°S of the equator, than at Soroti.

Ethiopia, the corresponding altitude is 1500 m (le Houérou & Popov, *op. cit.*) and this also applies to similar southern latitudes.

The 2250–2500 m altitude range is usually regarded as the limit above which cultivated tropical grasses are replaced by temperate species in Kenya (Morrison, 1969). This 2250–2500 m range encompasses the eastern and western deviation (Fig. 3.2) having 13–15 °C as the mean annual temperature. On Mount Elgon, on the Uganda-Kenya border, Nandi Setaria and Masaba Rhodes grow well up to 2500 m altitude, but on Mount Kenya it is common to grow temperate grasses already at 2250 m. At higher latitudes the temperature of the coolest month of the rainy season determines the altitudinal limits in the choice of species (*6.5*).

Unless specified otherwise, altitude indications refer to those typical of the Nairobi region, because it is intermediate between the western regions (add 250 m) and the southern or northern areas (subtract 250 m).

A set of climatic data, including solar radiation, is presented for Kitale (Fig. 3.3), together with data on annual fluctuation in soil temperatures (Fig. 3.4). Under bare soil, temperatures drop from nearly 26 °C at the end of the dry season to below 18 °C in the middle of the rainy season. Thus, the variation in soil temperature exceeds that of atmospheric temperatures and is, therefore, a major factor affecting plant growth (Fig. 6.1; *7.2.2*). Soil moisture and temperature relationships in a Uganda soil catena were reported by Banage & Visser (1967).

Table 3.2. The effect of temperature deviations in relation to latitude (East African Meteorological Department, 1975).

Town	Altitude (m)	Rainfall (mm)	Mean max. temperature		Mean min. temperature	
			Annual	Warmest month	Annual	Coolest month
Songea 10°41′S	1153	1127	25.7	28.8	16.0	12.4
Soroti 1°43′N	1127	1311	30.3	33.1	18.0	17.1

3.1.2.1. Temperature zones

There has been much discussion as to how to classify temperature zones occurring on and near the equator without confusing typical terminology of temperate with that of tropical (*i.e.* within the tropics) climates. How to define the limits of "highland" has also been subject of debate.

A delineation between "cool or warm" areas has been proposed by le Houérou & Popov (*op. cit.*), based on the 10 °C isotherm of the mean minimum temperature of the coldest month. This corresponds with the 1750–1800 m altitude of Nairobi-Dagoretti (1798 m; mean minimum temperature in July 9.7 °C). Taking into account the deviation from altitude-corrected means of Fig 3.2, the altitudes corresponding with the 10 °C isotherm are 2000 m in western Kenya and Uganda. The same applies to Rwanda and eastern Zaïre. For

3.2. Soils

When soils are discussed in relation to vegetation, the emphasis is usually on soil classification and on maps of a geological and topographical nature. A 1:4.000.000 soil associations map of East Africa

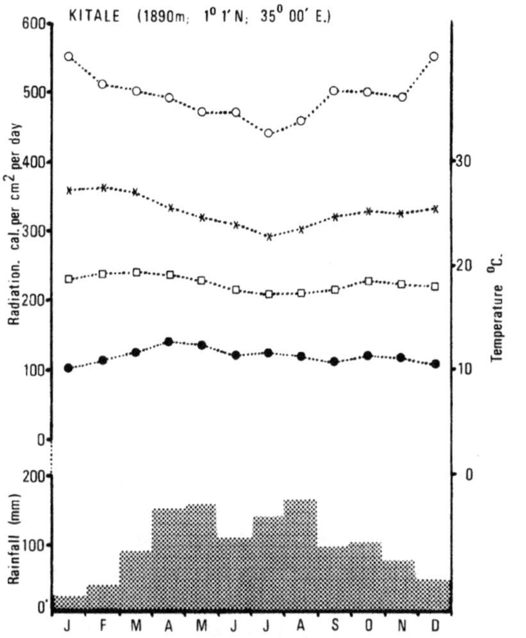

Fig. 3.3. Kitale climate. Seasonal variation in energy input, temperature and rainfall; ○···○ incoming solar radiation; ×···× mean maximum temperature; ●···● mean minimum temperature; □···□ mean temperature (Boonman, 1973b).

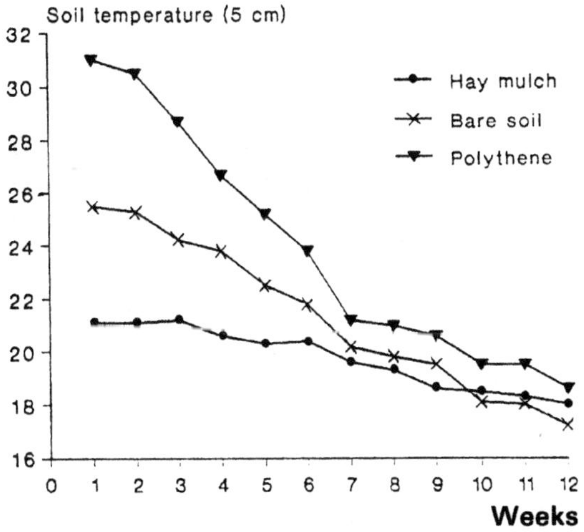

Fig. 3.4. Soil temperatures at the onset of the rains at Kitale (after Law and Cooper, 1976).

was prepared by Scott (1969), which followed the catena concept, *i.e.* a regular sequence of soil types down a slope; 36 types were arranged in 6 groups according to drainage conditions. A soil map of East Africa at scale 1:5.000.000 formed part of the "Soil Map of the World" (FAO-UNESCO, 1974). An "Exploratory Soil Map of Kenya" at scale 1:1.000.000 was prepared by Sombroek *et al.* (1982); some 350 map units were described and arranged first by landform and subsequently by parent material, in accordance with the FAO-UNESCO legend, already quoted. For a recent study on "Soils, Physiography and Agroecological Zones of Tanzania", see de Pauw (1984). For Uganda, results of a reconnaissance survey were published in six Memoirs (Harrop, 1970).

Although the main vegetation zones are generally less related to soils than to climate, within zones importance should be attached to topography and variation in soil conditions. Examples are the catena and soils of impeded drainage, respectively. How extensive the latter are is epitomized in the many common-names used: vlei, mbuga (Tanzania), dambo (Zambia, Malawi), dembo and dilungu (Shaba, formerly Katanga). Vlei is a general term denoting swampy areas inundated all or part of the time and in which water is normally in a state of slight flow (Scott, 1967). A botanical description of African wetlands is given by Denny (1985). Impeded drainage can mean continuously swampy conditions in humid regions, but lush grassland in the dry season in less humid areas. Edaphic (soil) factors often dictate why grasslands are found where the climate would otherwise indicate forest. As for the catena, contrasting soils are often associated with crest, slope and hollow according to a sequence that is repeated over the whole landscape. Well-drained soils are found on the ridges and heavier soils, often black clay, in the bottomlands. This recurring pattern of soil and topography, indicative of older land surfaces, is recognized as the catena. The catena concept has also been applied to classification of vegetation, for instance of valley grassland within catenas dominated by Miombo woodland (*3.3.1*; *3.4.3*). This sequence of distribution is then referred to as the soil-vegetation catena; see also Lang Brown & Harrop (1962).

Temperature and radiation regimes being equal, grassland productivity is determined by rainfall or moisture index (Table 3.4) rather than by soil types, except when extreme. Vertisols ("black cotton soils") are heavy textured soils with a poor physical structure, in sharp contrast with Nitisols ("Kikuyu red loam"). Soil fertility problems have generally been studied on the basis of individual crops. A review on a country-wide basis is that presented for Uganda by Stephens (1970) (*5.2*). The importance of soil fertility for the savannahs has been described by Walker (1984) and Menaut *et al.* (1984). The latter postulated that the potential amount of nutrients (N and P) in savannahs is not so low, but their release is limited by the lack of easily assimilable carbon sources (*5.3.2*).

Finally, land use itself has an effect on the relationship between soils and climate. Natural grassland can be sustained in dry areas when edaphic conditions do not deteriorate. Overgrazing can lead to bareness and loss of topsoil by erosion to such an extent that the vegetation assumes a drier appearance than the rainfall data suggest (*4.3.2.2*). Conversely, cultivated grasses can have a great ameliorating influence on the soil (*5.3*).

3.3. The natural grassland environment

Grasslands are in many respects an integration of prevailing climate, soil, animal and man-made conditions. Much can be learnt from the natural grassland, and from the vegetation in general, that is relevant to the subject of cultivated pastures also. Vegetation may often provide a better guide to describe the agricultural environment than the instruments of meteorologist or geologist (Whyte, 1974). Vegetation is an expression of the environment, including factors such as burning, grazing and human activities and their interactions. This is valid only when it is assumed that the characteristic species have been present from the onset to express themselves. The latter aspect is important when considering the indicator value of for example Kikuyugrass, which is a relatively recent arrival in some of the highland areas outside East Africa (*3.4.2.1*).

The grasses used in cultivation today are those that were found growing in the wild yesterday. The natural grasslands of eastern Africa are the world's major gene centre of cultivated tropical grasses (*2.3*; *6.4*). These species are still important components of the natural vegetation and the natural grassland in any of its forms is bound to remain a major grazing resource of great value. The threat to the natural grassland does not lie in its being replaced by cultivated pastures but rather in grazing abuse and usurpation by arable cropping. It is only logical, therefore, that we deal with East Africa's natural grasslands in this treatise. There are additional ecological reasons. Even on land where grasses have been sown, a continuous resonance of the natural grasses remains in the soil and the environment, however changing. Natural and cultivated grasses are indeed very close.

Its varied topography, interspersed with open country, is the main characteristic of East Africa, in sharp contrast with the vast plains to the north (Sudan) and the lowland forest to the west (Zaïre). Natural grassland is a major determinant of the East African landscape. Habitations and cultivations are often only few and far between, over large parts of the region. Even in densely populated areas relatively large areas are covered by natural grassland.

"Natural" should not be taken to mean primitive or unproductive. It can approach the best grassland attainable. Neither should "natural" be equated with climax or static. Grassland is subject to fluctuations in composition, although certain recurring patterns can be recognized. The major features, physiognomy and species, are fairly constant or they appear to oscillate around a certain equilibrium, both between and within woody and herbaceous plants. Major changes are usually a long-term affair. Only where grassland is artificially drained of excessive water or altered dramatically by repeated cropping can changes in the vegetation become irreversible (*3.3.3*).

Vegetation as a source of information about the environment is important to the agriculturalist interested in land use and development. Because of the diversity of climate and potential land use, a need was felt to define a number of zones which could be classified and treated accordingly. However, also within grassland science itself the need was felt from the early days to describe and to classify vegetation.

3.3.1. Grassland mapping

Edwards (1940b) proposed an ecological zonation for Kenya and it was on the basis of this zonation that research stations were allocated to different regions. Popular terms such as Bracken-, Kikuyugrass- and Stargrass zone came into use, especially in central Kenya (Bellis, 1958) but also Elephantgrass zone, in Uganda. This was done so that ecological potential could be related to a recognizable type of vegetation, which these "indicator" species presented.

When the East African Range Classification Committee (EARCC) was formed in 1964 at Kitale, there was no standard approach to classification (Pratt *et al.*, 1966). Many specialists had tended towards definitions based on grassland composition (Table 3.3).

Table 3.3. Grassland descriptions of East Africa.

Uganda:	Thomas, 1940a,b; Snowden, 1933; 1953; Langdale-Brown *et al.*, 1964; Horrell & Tiley, 1970.
Kenya:	Edwards, 1940b; 1956; Edwards & Bogdan, 1951; Heady, 1960; Trapnell *et al.*, 1966; 1970.
Tanzania:	Greenway, 1933; Staples, 1931–1939; Gillman, 1949.
Zaïre, Rwanda, Burundi:	Robyns, 1934; Germain, 1954; Léonard, 1962; Risopoulos, 1966; Compère, 1968.
Ethiopia:	Pichi-Sermolli, 1955; Fröman, 1975.
Somalia:	Edwards, 1945; Pichi-Sermolli, 1955.
Zambia:	Trapnell & Clothier, 1937; Trapnell, 1953.
East Africa:	Pichi-Sermolli, 1955; Trapnell & Langdale-Brown, 1969; Lind & Morrison, 1974; Pratt & Gwynne, 1977; Vesey-FitzGerald, 1963; 1973.
Africa:	Rattray, 1960.

An example of a map based on this approach in Uganda is that of the natural grasslands of Uganda (Fig. 3.6; Horrell & Tiley, 1970). Remarkable also are the photographic illustrations by Shantz (Shantz & Turner, 1958) who returned after a third of a century to sites he had visited in 1920 and which he photographed again in 1957 to show how small the changes had been in the vegetation of some particular sites.

The first physiognomic (the aspect in terms of height, density and cover) classification was proposed by Shantz & Marbut (1923). Terminology such as: "tall grass – low tree savannah" was used. An attempt at a classification, specific to East Africa, was made as a follow-up of the Nairobi Pasture Research Conference of 1940 (Greenway, 1943). Gillman (1949) used it to produce a vegetation map of Tanganyika. For Kenya the map produced in 1940 by Edwards had already become a standard work (Edwards, 1940b; 1956; Edwards & Bogdan, 1951; Atlas of Kenya, 1959). Edwards' categories had significance as ecological units and vegetation types (*3.3.1.3*). For Uganda an equivalent map was produced by Snowden (1953) and later by Langdale-Brown *et al.* (1964; Atlas of Uganda 1962) who followed the recommendations of the International Conference on Phytogeography held at Yangambi (Scientific Council for Africa South of the Sahara, 1956).

3.3.1.1. Ecological classification

Obviously, the general trend in grassland mapping had been towards definitions of vegetation types based on grassland composition. Vegetation had become, however, to many observers a poor indicator of ecological potential and possible land use. Indeed, there are areas which have been so degraded by human activity that it is hard to believe that they once had, and could again have, rich natural forest. The problem was how to reconcile actual vegetation with ecological potential. Classification of grassland means one thing to the specialist in phytogeography, but quite another to the planner concerned with grassland development. The Yangambi Conference represented the former whereas the EARCC used vegetation in a much broader context. The EARCC's proposal as presented by Pratt, Greenway & Gwynne (1966) and described further by Pratt & Gwynne (1977), is followed in this book, with some modification (Tables 3.4 & 3.7).

The central theme in the EARCC classification is site potential, *i.e.* the vegetation which ecological factors indicate should be found. Many derivatives occur with forest reduced by felling and burning to grassland, and the latter by overgrazing to barren land. With the exceptions of edaphic and semiarid grasslands, most of the open country in East Africa has been induced, by a past history of fire, either natural or man-made. However, a "climax" vegetation type can be reconstructed based on natural succession towards an equilibrium with the environment. This process may be interrupted and vegetation may then be described as fire subclimax or grazing subclimax.

Pratt & Gwynne (*op. cit.*) have postulated that the concept of climax should relate to site potential as determined by physical factors of the environment. These include topography and soil type, but over East Africa as a whole it is the climate, with rainfall and temperature, which presents the most important variable. The combination of climate and soil, modified by time but also by vegetation (recycling of nutrients, evaporation) determines the potential of a site in terms of vegetation.

The EARCC approach as described by Pratt & Gwynne (*op. cit.*) is a dual classification which seeks first to clarify site potential and secondly to describe the existing vegetation in physiognomic terms. The latter can be subdivided by composition, nature and contribution of woody plants and grasses. Within such a framework is it relevant to consider the details of grassland composition.

The EARCC recommended that the zones be determined by survey, using plant species and associations as indicators of certain climatic regimes and so delineate climatic boundaries more accurately than is possible by solely relying on data from the relatively sparse scatter of metereological recording stations. It was in this way that the maps, presented by Pratt & Gwynne (*op. cit.*) have been prepared. In this, great weight was placed on the work of Trapnell & Langdale-Brown (1969) and Langdale-Brown *et al.* (1964).

The natural vegetation in East Africa is attributable in large measure to the one compound factor of moisture availability. A moisture index was adopted by the EARCC to delineate 6 zones. This index was derived from data on monthly rainfall and evaporation, the latter being estimated on the basis of measures of radiation, temperature, saturation deficit and wind speed and weighed for altitude and latitude (Dagg, 1965).

Thus, 6 broad zones (Zone I–VI) were delineated (Table 3.4; Plate 3.1). Due to the overriding importance of climate, the zones were referred to as "eco-climatic" zones. Pratt & Gwynne (*op. cit.*) anticipated that, ultimately, in the process of refinement the present Zones II and III were to be subdivided further and that the coastal forms of Zones II to IV would be differentiated as separate subtypes.

The advantage in separating the description of land potential from that of present vegetation is recognized, provided this can indeed be adhered to. The Kenya Soil Survey, therefore, preferred to base agro-climatic zones only on moisture availability and temperature (Braun, 1980). In Braun's view, coastal areas which were mapped as Zone II by Pratt & Gwynne since they carry a closed-canopy tree vegetation (in fact edaphic woodlands of *Brachystegia spiciformis* on coastal sands),

Table 3.4. Eco-climatic Zones of East Africa (after Pratt & Gwynne, 1977); see also Plate 3.1.

Climatic zone		Vegetation and land use
Afro-alpine	I	Afro-alpine moorland and grassland, or barren land, at high altitude above the forest line; of limited use and potential, except as water catchment.
Equatorial to dry subhumid Moisture index not less than −10	II	Forests and derived grasslands and bushlands, with or without natural glades. The potential is for forestry, intensive agriculture including pyrethrum, coffee and tea at higher elevations.
Dry subhumid to semiarid Moisture index −10 to −30	III	Land not of forest potential, carrying a variable vegetation cover (moist woodland, bushland or savannah), the trees characteristically broad-leaved (e.g. Brachystegia or Combretum) and the larger shrubs mostly evergreen. The agricultural potential is high, soil and topography permitting, with emphasis on ley farming.
Semiarid Moisture index −30 to −42	IV	Land of marginal agricultural potential, carrying as natural vegetation dry forms of woodland and savannah (often an Acacia-Themeda association) but including dry Brachystegia woodland and equivalent deciduous or semi-evergreen bushland. This is potentially productive rangeland.
Arid Moisture index −42 to −51	V	Land only very locally suited to agriculture, the woody vegetation being dominated by Commiphora, Acacia and allied genera, mostly of shrubby habit. Perennial grasses such as *Cenchrus ciliaris* and *Chloris roxburghiana* can dominate, but succumb readily to harsh management.
Very arid Moisture index −51 to −57	VI	Rangeland of low potential, the vegetation being dwarf shrub grassland, or shrub grassland with *Acacia reficiens* subsp. *misera*, often confined to water courses and depressions with barren land between. Perennial grasses (e.g. *Chrysopogon aucheri*) are localized within a predominantly annual grassland; productivity is confined largely to unreliable seasonal flushes and grazing systems must be based on nomadism. Populations of both wild and domestic stock are restricted severely by the environment.

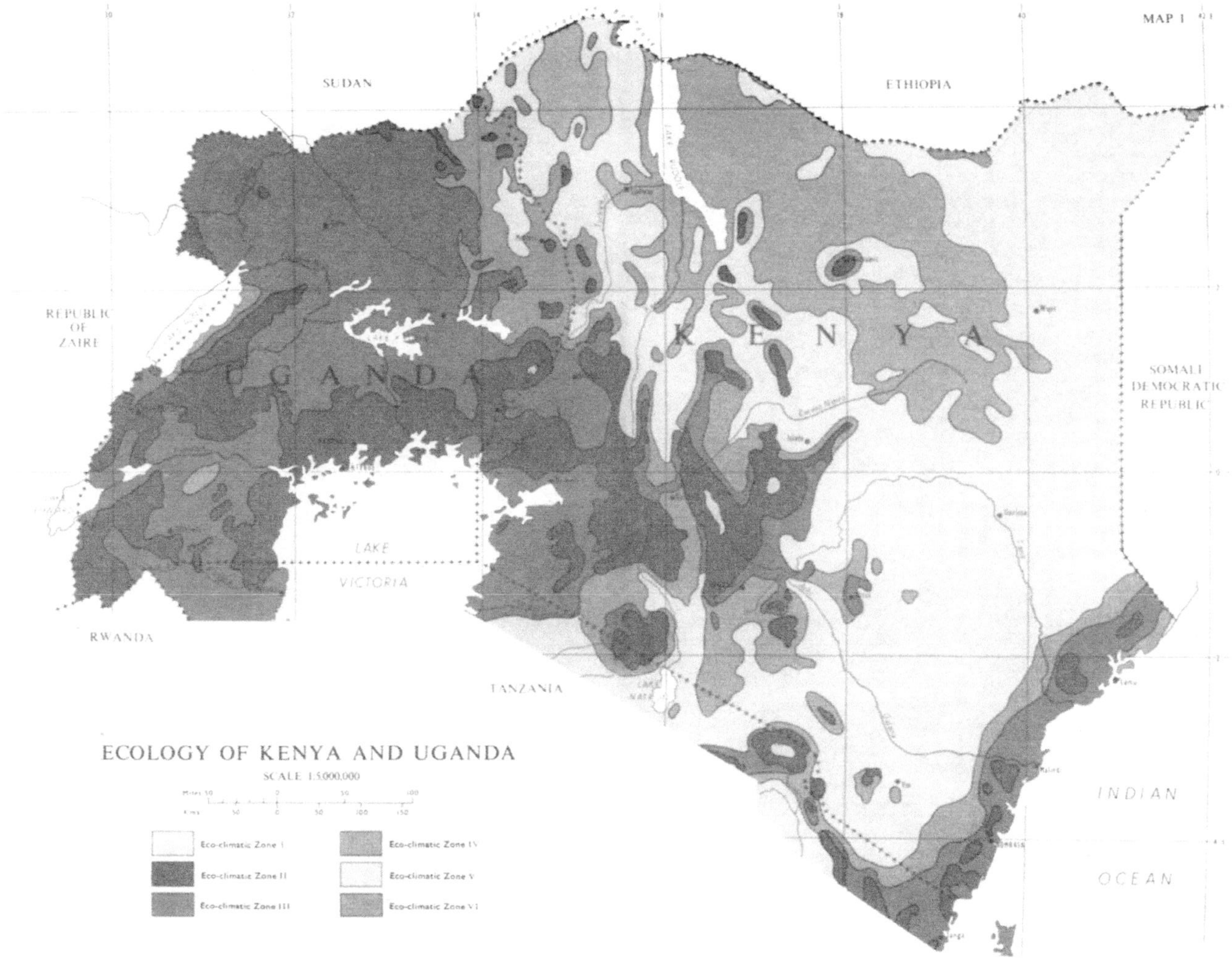
MAP 1
SUDAN
ETHIOPIA
REPUBLIC OF ZAIRE
UGANDA
KENYA
SOMALI DEMOCRATIC REPUBLIC
LAKE VICTORIA
RWANDA
TANZANIA
INDIAN OCEAN
ECOLOGY OF KENYA AND UGANDA
SCALE 1:5,000,000
Eco-climatic Zone I
Eco-climatic Zone II
Eco-climatic Zone III
Eco-climatic Zone IV
Eco-climatic Zone V
Eco-climatic Zone VI

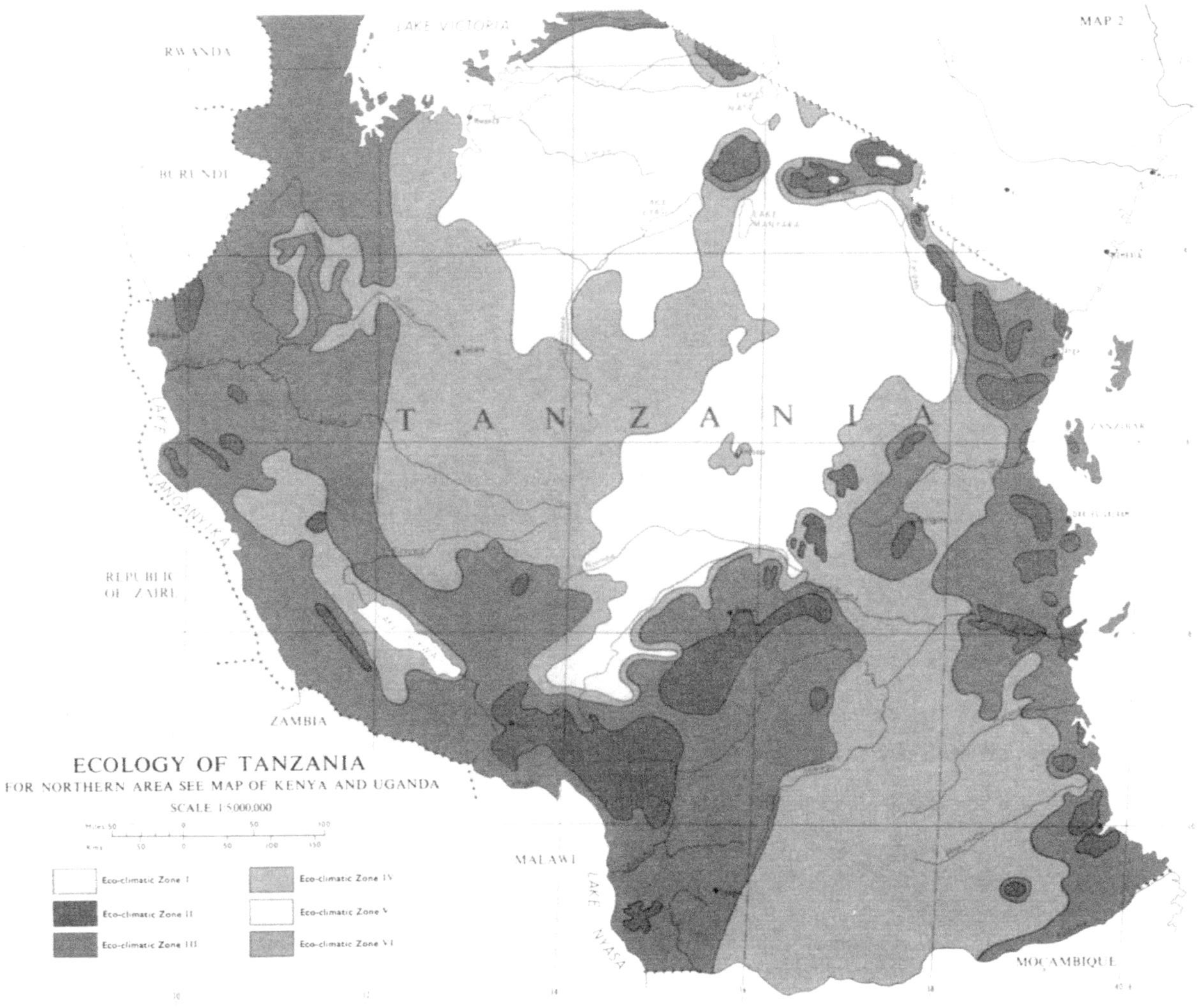

Plate 3.1. Ecology of East Africa.

should not be classified as Zone II, because this cannot be justified on the basis of meteorological data. The presence of dense trees does not necessarily imply a high site-potential. Hard as it may be to believe that certain sites were once forest, it is equally true that forest does occur in areas where climate alone would not have qualified them as "forest-potential".

Braun prepared his agro-climatic zone map for land evaluation purposes in combination with the Exploratory Soil Map of Kenya. A reduced version of Braun's map was adopted as an appendix by Jätzold & Schmidt (1982) who proposed a very detailed zonation (1:100.000) especially for crops in Kenya. Based on the rainfall/evaporation ratio (r/E_0), 7 moisture availability zones were presented (Table 3.5), which were further subdivided in 9 temperature zones on the basis of altitude.

Table 3.5. Braun's agro-climatic zones of Kenya (Braun, 1980; Sombroek *et al.*, 1982).

Zone	Climate type	Moisture availability (rainfall/E_0*)	% of country
I	Humid	> 80%	5
II	Subhumid	65–80	5
III	Semihumid	50–65	5
IV	Transitional	40–50	10
V	Semiarid	25–40	15
VI	Arid	15–25	20
VII	Very arid	< 15	40

*E_0 = mean annual potential evaporation.

A weakness of purely climatic classifications is that the criteria used are somewhat arbitrary; seasonal distribution and variation in rainfall and deviations from the expected pattern (Fig. 3.2) are not accounted for. Although Braun's climatic approach has much to commend it, Pratt & Gwynne's ecological zonation is followed here, because it has been used widely since the 1960s; it is the only one which covers Kenya as well as Uganda and Tanzania and which was written with grassland in constant perspective. However, the main reason why an ecological approach to zonation should be preferred is that climatic and soil data, even when available, cannot be condensed in a formula that accounts for the relevant necessary detail (see also Moreau, 1938; Trapnell & Griffiths, 1960).

The zonations of Pratt & Gwynne (Table 3.4) and Braun (Table 3.5) assume that site potential decreases with availability of moisture. Although their estimates varied manyfold, both proposed data to postulate that the livestock-carrying capacity of Zone II is twice that of Zone III. However, zonation based on moisture index has little to do with the potential of the plant species typical of such zones. Temperature-moisture interactions are relevant also. In fact, maize and Hyparrhenia at Kitale (Zone III) yield much more than wheat and Kikuyugrass at Molo (Zone II). This makes a subdivision of zones with reference to such grass species indispensable (*3.4*). The term "Eco-climatic Zone", coined by Pratt & Gwynne, will be referred to in the following as "Zone".

3.3.1.2. Vegetation types

In the EARCC approach separate maps of ecological potential and vegetation were presented. The vegetation types were physiognomic ones, based on the relative contribution of woody plants vis-à-vis grass and herbs. They can be subdivided according to species composition and special characteristics. The vegetation types were defined as follows (Atlas of Kenya 1970; Pratt & Gwynne, 1977):

Forest. Land covered by a closed stand of trees of one or more storeys, with an interlaced upper canopy, rising to 7–40 m or more in height. The ground cover is dominated by herbs and shrubs, and lianes and epiphytes are characteristic. Subtypes should be prefixed by the dominant genera.

Woodland (Fig. 3.5). Land supporting a stand of trees, up to 20 m in height, with an open or continuous but not thickly interlaced canopy, and a canopy cover of more than 20%. Grasses and other herbs dominate the ground cover. Shrubs, if present, contribute less than one-tenth of the canopy cover. Subtypes should be classified by reference to the genera or species of the dominant woody plants. Examples: *Brachystegia-Julbernardia* (Miombo) woodland; *Acacia seyal* woodland with Themeda ground cover; Combretum coppice woodland.

Bushland (Fig. 3.5). Land supporting an assemblage of trees and shrubs, often dominated by plants of shrubby habit but with trees always conspicuous, with a single or layered canopy, usually not exceeding 10 m in height except for occasional emergents, and a total canopy cover of more than 20%. The ground cover is poor and fires infrequent. Bushland thicket is an extreme form. Examples of subtypes, classified as for woodland: Lannea-Commiphora bushland thicket; Carissa-Acokanthera evergreen bushland; *Acacia tortilis-A. nilotica-A. mellifera* bushland.

Grassland. Land dominated by grasses and occasionally other herbs; sometimes with widely scattered or grouped trees and shrubs, the canopy cover of which does not exceed 2%. Subtypes should be classified by reference to the following criteria, in the order given: height (giant, tall, short) dominant grass genera or species; degree of swampiness; dominance by annual grasses or other herbs. Examples: Giant *Pennisetum purpureum* grassland; Eriochloa-Sorghum seasonally waterlogged annual grassland; Portulaca-Eragrostis annual herb grassland.

Wooded or tree grassland. Grassland with scattered or grouped trees, the trees always conspicuous, but having a canopy cover of less than 20%. Often subject to periodic burning. Subtype classification is by reference to grassland type and genera of the dominant woody plants, indicating, in the order given: grass height, whether the grasses are annuals, dominant genera and degree of swampiness. Examples: tall Hyperthelia-Combretum wooded grassland; *Pennisetum mezianum-Acacia drepanolobium* seasonally waterlogged wooded grassland; short annual Aristida-*Acacia tortilis* wooded grassland.

Bushed grassland (Fig. 3.5). Grassland with scattered or grouped trees and shrubs, the shrubs always conspicuous, and with a combined canopy cover of less than 20%. Subtype classification as for wooded grassland. Examples: Short Cynodon-Tarchonanthus-*Acacia seyal* bush grassland; Panicum-Commiphora bush grassland.

Shrub grassland. Grassland with scattered or grouped shrubs, the shrubs always conspicuous, but having a canopy cover of less than 20%. May be subject to periodic burning. Classification of subtypes as for wooded grassland. Examples: Lintonia-*Acacia mellifera* seasonally waterlogged shrub grassland; tall Hyparrhenia-Carissa shrub grassland.

Dwarf shrub grassland. Grassland, often sparse grassland, set with dwarf shrubs not exceeding 70 cm in height, sometimes with widely scattered larger shrubs or stunted trees. Fires are rare. Sparse dwarf shrub grassland, with a total ground cover of less than 2%, can also be recognized. Subtype classification as for wooded grassland. Examples: short annual Cenchrus-Indigofera; Chrysopogon-Duosperma. Dwarf shrub grassland has been isolated for individual mention because it is representative of a distinct plant formation of arid regions in northern Kenya. Moorland is the afro-alpine form of dwarf shrub grassland.

Permanent swamp. Land covered by permanent standing water and supporting various plant communities, including reeds, sedges, rushes, sometimes trees or shrubs, and aquatic species. Subtypes are classified by reference to the genera of dominant plants and physiognomy. Example: Typha herbaceous swamp.

Barren land. Land naturally devoid of vascular plants.

3.3.1.3. Ecological and physiognomic classification

The lasting influence of Edwards' work (1940b) shows the value that many agriculturalists attach to having maps reflecting major ecological zones. In the absence of detailed ecological surveys guidance in the planning of agricultural development had to be sought instead in available vegetation maps. The works of Edwards, Gillman, Trapnell, Langdale-Brown and Vesey-Fitzgerald for East Africa as well as of Germain, Lebrun, Léonard and Robyns for Kivu, Rwanda and Burundi confirm their importance (Table 3.3). The maps of Edwards and Gillman tended, in fact, to compromise between ecological potential and actual physiognomy which added to their value. The following types of grassland were recognized (Edwards & Bogdan, 1951):

1. Highland grassland and highland forest.
2. Coastal forest.

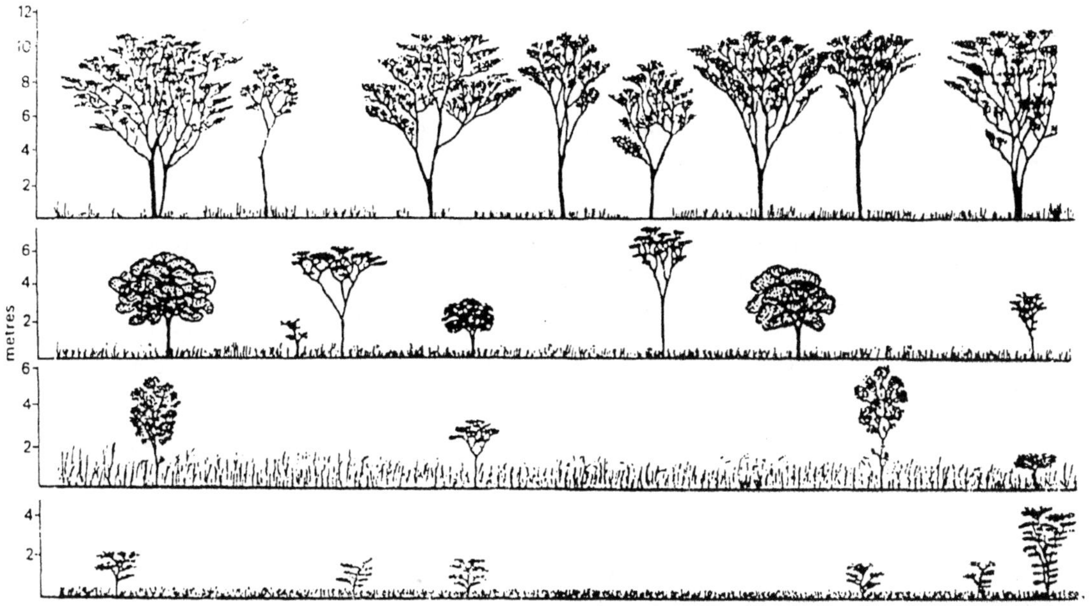

Fig. 3.5. Profiles of bushland types (Pratt *et al.*, 1966).

3. Scattered tree grassland (low tree-high grass).
4. Grouped tree grassland.
5. Coastal bush with open high grass glades.
6. Scattered tree grassland (Acacia-Themeda).
7. Scattered tree grassland (other types).
8. Desert grass-bush (dry bush with trees).
9. Desert shrub.
10. Desert grass and desert shrub.

The classification by Trapnell & Langdale-Brown (1969) was meant as an adjustment between those proposed by the Pasture Research Conference of Nairobi in 1940 (Greenway, 1943) and those accepted by the Yangambi Conference of 1956. This is illustrated by the terminology used:

Woodland	–	Miombo woodland (Brachystegia-Julbernardia; the largest single type of vegetation from south east Kenya across Tanzania, south east Zaïre and Zambia to Zimbabwe, Malawi and Angola).
Savannah (wooded grassland)	–	Combretum savannah (with Hyperthelia).
	–	Acacia savannah.
	–	Savannah–like communities, *e.g.* with Elephant- or Guineagrass.

Elaborating on the question of "climax" or natural grasslands, the authors state that "pure" natural grasslands, which would exist in the absence of fire, are probably confined to montane areas and to conditions of impeded drainage, flood plains and certain black-clay plains. Vesey-Fitz-Gerald (1970; 1973) pointed out that valley grasslands and grasslands of impeded drainage are very extensive in East Africa. The drier clay plains of impeded drainage in Kenya are typified by Themeda-*Pennisetum mezianum* and in south Masailand (Peterson & McGinnes, 1979) by *Digitaria macroblephara* and *Panicum coloratum*. Edaphic vegetation types in eastern Kenya at Kiboko (Zone V) were described by Bogdan (1958b) with Rhodes- and Buffelgrass on slightly waterlogged and alkaline soils and with *P. mezianum* on seasonally waterlogged soils.

Typical highland-grassland representatives of these kinds are *Exotheca abyssinica*, *Loudetia simplex* and *L. kagerensis*. The Russet grasses (*Loudetia* spp.) are often dominant over vast expanses of badly drained terrain in the valleys of the Zaïre river catchment and on the shores and islands of Lake Victoria and Lake Nabugabo in Uganda (Thomas, 1940b; Lind & Morrison, 1974) (Fig. 3.6). *L. simplex* and *L. kagerensis* are, however, very common in open grassland around

Eldoret and Kitale, respectively (Bogdan, 1958a) (Table 4.2).

Many experts consider that "nothing is stable; there is no climax" (Thomas, 1966). During the last few hundred years much of Uganda had been under grass, and not under the forest that we have seen in this century. Some areas were much more densely populated than at present. Many forests, formerly described as climax, are not virgin and are comparatively young. Tse-tsefly was one of the factors (Thomas, *op. cit.*) but the rinderpest epidemics must also have had an effect on grazing pressure and therefore on the spread of woodland. By contrast, the highland forests of Rwanda and Burundi were already destroyed in the 14th century (Compère, 1968).

3.3.2. Semantics

The word "grassland" is used in this book to denote land that is dominated by grasses and occasionally herbs, whatever the plant succession. Since we are concerned mainly with Zones II, III and IV (Table 3.4), less difficulty is experienced with the term grassland than when it also has to cover the arid land of Zones V and VI with so little grass in evidence. The term grassland is further justified by the fact that in many of the lands of our concern edaphic effects are as significant as those of fire. We prefer to use "savannah", in order to collectively group wood- and bushland, above "rangeland". Both terms are reminiscent of the New World, but "savannah" is more appropriate since the Latin American savannahs have been largely "Africanized" (*2.3*). The term "veld" is commonly restricted to southern Africa (*2.4.3*).

Much confusion exists as regards the many different, practical interpretations of the term "natural" grass. In current use "natural" may mean any of the following:

1. Grass, *unsown, non-cultivated* (*e.g.* Themeda), as opposed to sown (*e.g.* Rhodesgrass).
2. Grass *sown, but long ago* (and now run-down) opposite to recently sown.
3. Grass vegetatively established from *local* (*i.e.* collected in the vicinity) material (*e.g.* Stargrass) as opposed to those established from seed (*e.g.* Rhodesgrass).
4. Grass vegetatively established from *local* (*i.e.* collected in the vicinity) ecotypes (*e.g.* Stargrass) in contrast with recommended, introduced or bred varieties.
5. *Local* (*i.e. native*, cultivated or uncultivated) grass as opposed to exotic (*e.g.* Perennial Ryegrass).
6. *Secondary* or derived (colonizing, "naturalizing" grasses such as Kikuyu-, Star-, Elephant- or African Couchgrass) following forest clearing or cropping as opposed to the original, less disturbed habitat or grassland.
7. *Primitive*, opposite to improved and cultivated grasses.

It has become customary to equate natural grassland with anything that can be grazed but is not, or not known to have been, sown from seed. Uncultivatable land such as waterlogged, steep and rocky land, and homestead yards, but also abandoned plantations, tumble-down (fallow) pastures or even pasture that was sown too long ago for people to remember is called natural grassland. Pasture planted to Kikuyu- or Stargrass has been referred to as natural pasture also, perhaps because the vegetative transplanting from the local flora was seen as "natural". Or it is meant to denote the grazing land as it was presumed to have been since time immemorial.

Conversely, naturally occurring species have been described, even in scientific journals, by the name of the popular sown cultivar of that species. It is obviously an error to record naturally occurring Setaria (*Ssph*) under the name of Nandi Setaria, neither is it correct to describe naturally occurring Rhodesgrass in the Rongai/Nakuru (K) area as Rongai Rhodes. However, in the latter case the dividing lines are already vague; seed harvested from natural Rhodes in the Rongai/Nakuru area has been commercialized under the name Rongai Rhodes (*6.6.3*). A similar problem exists with the distinction between Napiergrass and Elephantgrass. Although "Napier"grass was essentially the name of a cultivated variety, it became synonymous with Elephantgrass and was subsequently almost dislodged in Kenya by the name "Bana"grass, although this is a misnomer of the Ghana (Gold Coast) variety (*11.7.3.1*).

Evidently, with all the different meanings in popular use, no single definition of "natural" seems possible. *Themeda triandra* is perhaps the best known single species that would come closest to what most will describe as a typical "natural" grass (ad 1.) and to what in Australian terminology is described as a "native" grass. "Native", *i.e.* belonging to the locality, is a stronger notion than "indigenous" (belonging to the region) or "natural". It seems to imply that the grass had been a constituent and even a dominant one of the vegetation before any major change in the environment occurred or before other grasses arrived. Conversely, colonizing grasses such as Kikuyu- and Stargrass would fall into the category of "indigenous" or "secondary" but not of "native" because of their low frequency in the original, undisturbed habitat. Some authors classify them as "secondary" grasses rather than as "natural" grasses.

3.3.3. Primary and secondary grasses

The best compromise seems to be to refer to natural grass as *non-cultivated* and not to apply the term to grasses, whatever their origin, that may be cultivated. The grasses of derived grassland are not dominant constituents of the original native vegetation, but they become dominant rapidly after a disturbance of the soil, cultivation or fire. They are, therefore, commonly called secondary and the main examples are Kikuyu-, Star- and Elephantgrass, all of Zone II, in which they fulfil the role of indicators or "zone-grasses" each in their own subzones which are, in fact, altitude belts (*3.4.2*). They also reappear rapidly in fallow (post-cultivation) communities to become dominant, even as pernicious arable weeds.

Grasses such as Themeda and Hyparrhenia/Hyperthelia may then be classified "primary" to distinguish them from secondary grasses. "Primary" is coined here to denote native grasses, whose abundance fluctuates with the intensity of tree cover, but which had been present significantly or even abundant before changes in the habitat occurred. In Zone III these otherwise so dominant natural grasses such as Themeda, Hyparrhenia/Hyperthelia but also Rhodesgrass (to extend the principle equally to stoloniferous species) rarely reappear in fallow communities after cropping, or only after a long delay. This role is taken over by annual grasses and subsequently by secondary perennial grasses like African Couchgrass (*Daby*), Povertygrass (*Rrep*) and Paspalum (*Pscr*). The latter group forms, in fact, the common constituents of fallow grasslands on farms of Zone III (*4.5*). The primary grasses are in fact increasers, not colonizers.

Consequently, the principal grasses of Zone II (Kikuyu-, Star- and Elephantgrass), are secondary and have such indicator value as to be rightly termed zone-grasses in their respective altitude belts (*3.4.2*). The main grasses of Zone III are primary (Themeda, Hyperthelia and Hyparrhenia) or secondary (African Couchgrass) but they traverse the altitude belts from coast to highland. Transitional types, between Zones II and III, like Guineagrass (*Pmax*), Brachiaria and Setaria also occur across the altitude belts and these will be discussed separately or jointly with the primary or secondary grasses where appropriate (*4.2.4*; *4.5.2*; Table 4.4). Similarly complicating species are the types of the edaphic grasslands, *e.g.* Coloured Guineagrass (*Pcol*), although most of these come close to the primary grasses, ecologically.

An additional difference of great practical consequence is that the botanical stability under grazing is better in Zones III and IV than in Zone II, which is reflected in the behaviour of the zone-grasses. Since the grazing pressure in the forest zone had been understandably low, colonizing species are not stable under grazing. Elephantgrass is the best example of a grass that is susceptible to hard grazing, but also Kikuyugrass and Stargrass and many of the grass communities in the wet coastal zone are subject to decline and/or to incursion by undesirable species, unless they are given appropriate husbandry. There is obviously a need to differentiate between the stability of a single species and that of a whole grassland community. That stability under grazing has little to do with productive capacity under cutting regimes is illustrated by the two extreme examples of Elephant- and Kikuyugrass (Tables 6.1 & 6.2). The combination of productive-plus-stable under grazing is rare among secondary grasses.

Incidentally, growth habit is not associated with colonizing behaviour and not with the distinction between primary and secondary grasses nor, for that matter, between zones. Stolons and rhizomes are well developed in Kikuyu- and Stargrass, but Rhodes- and Buffelgrass are also stoloniferous and rhizomatous, respectively. In East Africa, the latter two cannot be regarded as colonizing, secondary grasses. Little can, therefore, be explained by the pathway of vegetative perennation. A related but generative distinction, however, is that primary grasses are generally much better seeders. It is noteworthy also that no basis for the observed difference between primary and secondary grasses can be found in the taxonomic position (tribes) to which the different genera concerned belong.

While Zones II and III/IV can be distinguished on the basis of the behaviour of secondary and primary grasses, the distinction between the latter is of paramount importance to understand their ecology (*3.4.2*) and, consequently, their husbandry as natural (*4*) or cultivated (*6*; *7*) grasses, respectively.

3.4. Proposed zonation for grassland improvement

The "eco-climatic" zonation as defined by Pratt & Gwynne (1977) was primarily meant to define rangeland potential. On the other hand, Braun (1980) and Jätzold & Schmidt (1982) used their "agro-ecological" mapping for the purpose of cropping and land use.

For grassland improvement in the mixed-farming context we are in this book mainly concerned with the Zones II, III and IV. The necessary refinement will have to come from subdivision and perhaps some redivision. The reasons are the following:

1. The zonations of Pratt & Gwynne (Table 3.4) and Braun (Table 3.5) assume that site potential decreases with availability of moisture. It has, however, been pointed out that moisture-temperature interactions of plant species (*3.3.1.1*) cannot be ignored. Maize and Hyparrhenia in Zone III (Kitale) are more productive than wheat and Kikuyugrass in Zone II (Molo).
2. Grassland improvement can have an effect on site potential itself and, therefore, on the very criteria on which zonation may be based. Ploughing and sowing alter the environment itself so that accepted boundaries between zones may become irrelevant. Some grasses are sown in East Africa from Zone II down to Zone IV and they are grown in other continents also.
3. The vegetation maps of Pratt & Gwynne (1977) allocate "wooded and bush grassland" as the characteristic physiognomy to areas as different as Kitale, Mombasa (both Zone III) and Wajir (250 mm rainfall; Zone V). This may seem too radical in the eyes of the general agriculturalist.
4. The agriculturalist will also argue that it is unrealistic to place Molo (2477 m, 13.0 °C), Kampala (1150 m, 21.5 °C) and the coast (26.7 °C) all in one undivided Zone II. Obviously, site potential at Kampala is far superior. Even with similar moisture indices, highland must be differentiated from lowland (le Houérou & Popov, 1981).
5. Kikuyugrass is commonly referred to as the major species when the ecological (or site-) potential of highland-grassland communities of Zone II is described. These communities are, however, often dominated by other, much coarser grasses. Equal mention could be made of other grasses, which are applicable indicator-grasses of their respective subzones: Elephantgrass at the lower altitudes, especially in Uganda and Zaïre, and Stargrass in a zone roughly half-way between that of Kikuyu- and Elephantgrass.

It is for these reasons useful to reintroduce the old popular concept of grass zones: Kikuyu- and Stargrass zone as commonly applied in central Kenya (Bellis, 1958; Strange, 1963) and the Elephantgrass zone, typical of the fertile crescent in Uganda (Horrell & Tiley, 1970). A combination of this with the zonation of Pratt & Gwynne (Table 3.4) gives the following:

Zone	Indicator grass	Altitude (±)
Zone II_{KI}	Kikuyugrass	> 1750–3000 m
Zone II_{ST}	Stargrass	1250–1750 m
Zone II_{EL}	Elephantgrass	< 1250–1500 m

It should be noted that although these popular notions of zone- or indicator grasses seem to have arisen spontaneously, the grasses are not primary but secondary (*3.3.3*). They are colonizers with strong weed potential and they tend to achieve and maintain dominance as long as soil fertility is high.

When these grasses do also occur in Zones III and IV it is without exception in isolated, well-defined pockets, where moisture and fertility, especially organic matter, are (temporarily) high. This should not, however, detract from their significance as indicator grasses of their respective subzones in Zone II. The Atlas of Kenya (1970) recognized only 2 grasses among the indicator species delineating eco-climatic zone boundaries: Kikuyugrass (Zone II) and *Chloris roxburghiana* (Zone V). Since the number of zones we are concerned with here is smaller but more concrete, other relevant indicator grasses can be employed where relevant and useful.

Grasses play an important part as members of distinctive floras of copper-bearing soils. Duvigneaud (1958) in Katanga and Wild (1969) in Zimbabwe found that certain communities indicate the presence of copper, when they contain *Eragrostis boehmii* (syn. *E. racemosa*), *Loudetia simplex* and *Andropogon schirensis*. Cuprifuge species, those which by their absence indicate the presence of copper, include species of Hyparrhenia; see also Woolhouse & Walker (1981). The distribution of C-3 and C-4 grasses, which both occur in semiarid areas, was studied in Kenya by Tieszen *et al.* (1979) and Maranga *et al.* (1983). C-3 grasses are generally found under trees, not in the open.

Table 3.7 proposes a zonation based on the climatic boundaries delineated by Pratt & Gwynne (Table 3.4). Zone II is subdivided as indicated and Zone III is split for high and low altitudes, in agreement with the conclusion of le Houérou & Popov (1981) that it is essential to differentiate between lowland and highland, even if they have similar moisture indices. Table 3.8 lists the best available cultivated herbage species and crops for each subzone. No attempt is made in these tables to distinguish between the various farming types, intensity or scale of farming. Such differences are more important in the socio-economic than in the technical sense of agriculture, exactly because cultivated grasses travel across boundaries. Classification may inherently carry the risk of simplification, but it must at all times avoid to divide there where it has good reasons to unite.

The altitudes given in Tables 3.7 & 3.8 are only indicative and apply to the situation on the equator in the Nairobi region. The altitude limit of 1750/1800 m, at Nairobi, corresponds with that of 2000 m in western Kenya, Uganda, Rwanda and eastern Zaïre, and with 1500 m at 10°N and S (*3.1.2.1*).

In the following a description is given of the proposed grassland zonation, with reference to characteristic grasses and crops. The grassland husbandry aspects will be discussed separately (*4*).

3.4.1. Zone I (Afro-alpine)

The upper-mountain areas with an afro-alpine climate are only put up for mention here. The climate is governed by temperature, not by moisture. In Ethiopia barley is grown on a considerable scale above 3000 m. Apart from fodder oats (*16*) and *Brassica* spp. (*18*) temperate grasses are useful herbage species.

3.4.2. Zone II (Humid to subhumid)

Together with Zone III this is agriculturally the most important zone. A substantial part of the otherwise mostly semiarid Masai pastoral grazing land is in Zone II: the Loita Hills region and areas north of Lake Manyara and west of Mount Kilimanjaro. The Loita Hills are classified by Edwards & Bogdan (1951) as "Highland Grassland and Highland Forest", but it is frequently grassland as the soils in the lower parts are mostly Vertisols. Edaphic factors often dictate why grasslands are found there where site potential, based on moisture index, would indicate forest (*3.2*).

A description is given of the 3 proposed, major indicator- or zone-grasses of Zone II, which are the characteristic secondary grasses of the altitude belts as indicated. These belts are not necessarily adjacent because other, mostly primary grasses such as Themeda or Hyparrhenia/Hyperthelia occur both between and within these zones (*3.3.3*).

3.4.2.1. Zone II_{KI} (Kikuyugrass zone, 1750–3000 m)

Kikuyugrass is associated with the highlands of forest potential of Zone II in the altitude belt of 1750–3000 m, coming down progressively lower as it moves away from the equator, *e.g.* in Ethiopia, southern Tanzania and Malawi. Productive Kikuyugrass is found as low as 1750 m on the equator when associated with highly fertile soils (e.g. Kisii in Kenya and Masisi in Kivu, Zaïre).

The zone is popularly known, in central Kenya in particular, as the Kikuyugrass zone because Kikuyugrass springs up rapidly and spontaneously as a colonizer after a disturbance of soil, *e.g.* in cultivated land. It spreads quickly along new cattle tracks and new roads. Outside Zone II, it may appear in Zone III and even in Zone IV in night paddocks and other favourable sites with high fertility and moisture (*4.2.1*).

The fact that Kikuyugrass suddenly appears in isolated pockets where it had not been conspicuous previously makes it unlikely that its distribution until the beginning of this century was strictly limited to the Kikuyu land of central Kenya. Seed is easily transported by herbivores (livestock) so that plants can spring up and colonize new sites. According to Stapf & Hubbard (1934), Kikuyugrass was mainly found in the highlands of East Africa, bordered by Eritrea in the north, Kivu in the west and the Lake Malawi region in the south. Its man-made spread to other cooler regions of Africa dates back to the 1910s in southern Africa (Melle, 1918) and the 1930s in central and West Africa. On Mount Cameroon itself no Kikuyugrass was reported by Maitland (1932), but in the Cameroon highlands the spreading of Kikuyugrass dates back to the 1930s due to import (L. Onderwater, private communication). In most overseas regions Kikuyugrass became a popular import in the 1920s and spread north and south beyond 30° (Roseveare, 1948; Mears, 1970). In South America, Kikuyugrass is still extending its demarcation lines and is continuing its march up the slopes of the Andes (Parsons, 1972).

Kikuyugrass is typical of disturbed sites of the Kikuyugrass zone but is not necessarily an important constituent of open grassland in these highlands. Edwards & Bogdan (1951) listed as principal common grasses in Kenya: *Themeda triandra, Andropogon chrysostachyus, A. amethystinus, Exotheca abyssinica, Eragrostis racemosa* (syn. *E. boehmii*), *E. olivacea* (syn. *E. lasiantha*), *Setaria sphacelata, Loudetia simplex* and *Digitaria abyssinica* (African Couchgrass). Under grazing, however, coarse grasses such as Wiregrass (*Pennisetum sphacelatum*) and Manyattagrass (*Eleusine jaegeri*) can increase rapidly in the absence of burning (*4.2.1*).

For most of the forest already destroyed centuries ago in Rwanda and Burundi at 1800–2500 m, a plant succession of *Exotheca abyssinica* via *Hyparrhenia bracteata* and *Eragrostis boehmii* to *Loudetia simplex* was reported (Compère, 1968). Mathieu (1960) listed the species of the area in the following order: *Brachiaria brizantha, Themeda triandra, Exotheca abyssinica, Hyparrhenia bracteata, Eragrostis boehmii, Loudetia simplex, Andropogon schirensis, Striga asiatica* (see also *6.7.2*) and *Monocymbium ceresiiforme*.

In the southern Tanzania highlands, on the Kitulo plateau (9°S; 2750 m; basaltic and phonolitic lava) Kikuyugrass is common, yet unproductive, only on disturbed ground and Themeda is rare also (Prins, 1969). The grasses on the ridges consist for over 80% of unpalatable species: *Koeleria cristata, Festuca abyssinica, F. caprina, Exotheca abyssinica, Anthoxanthum nivale, Eragrostis pseudohispida* and *Panicum lukwangulense*. The remaining grasses comprising less than 20% are generally grazed well: *Andropogon chrysostachyus, A. dummeri, Pentaschistis minor, Eragrostis racemosa, Panicum eickii, Bromus leptoclados* and *Digitaria myurus*.

In south west Uganda close to Rwanda and Zaïre, Snowden (1933) recorded Kikuyugrass as characteristic of "subtropical moist meadowlands" on volcanic mountains. Thomas (1940b), however, reported that it had not been recorded in the wild state from the Ruwenzori mountains which he attributed to the lack of livestock rather than to the soil type (Basement Complex).

In the Kivu highlands of Zaïre, Kikuyugrass is the common and dominant grass on the fertile "sols bruns" (presumably andosols) of volcanic origin (personal observation, 1969), although Risopoulos (1966) hardly mentioned it in his description of the grasslands of the region. Compère's work (1960a) on the rehabilitation of degenerate Kikuyugrass in Kivu referred to pasture planted to this grass. Kikuyugrass is not mentioned by Mathieu (*op. cit.*) either in his description of the vegetation of Rwanda-Burundi, referring rather to fallowland consisting of *Digitaria vestita* (syn. *D. abyssinica*).

For more detailed information on Kikuyugrass or montane grasslands in various countries of Africa the reader is referred to the following:

Ethiopia	Huffnagel, 1961; Hedberg, 1971; Fröman, 1975.
Kenya	Edwards & Bogdan, 1951; Dougall, 1953b; 1954a; Morrison, 1966c.
Uganda	Snowden, 1933; Thomas, 1940b; Horrell & Tiley, 1970.
Malawi	Young & Brown, 1962.
Zaïre (Kivu; Ituri)	Compère, 1960a; Lebrun, 1960; Léonard, 1962;
Rwanda & Burundi	Lebrun, 1956; Mathieu, 1960; Scaillet, 1960; Compère, 1968.
Cameroon	Maitland, 1932; Keay, 1955.

Livestock concentration, *e.g.* along cattle tracks, in kraals and night paddocks, as well as soil disturbance appear to be the main factors causing an upsurge of Kikuyugrass in highland areas. The effect of soil disturbance is most evident in recently cleared forest and in cultivated land.

On cleared forest land, Kikuyugrass constitutes the dominant grassland and can be productive. Land developed from highland grassland has a lower production potential than that from forest (Morrison, 1966c). Kikuyugrass can be most impressive on land that has been cropped and on sites with a temporary increase in moisture and fertility, also in Zones III and IV. It would appear, however, as if Kikuyugrass swards are stable only where high moisture and fertility can be maintained. Where such conditions cannot be met, it usually degenerates into a lawn-type mat which may be persistent but which is of little productivity ("Kikuyu decline"). Dougall (1953b) pointed out that "Kikuyugrass may fail to provide more than a few days' grazing annually", for no apparent reason of abuse. Much of the grazing value of the grasslands is reduced further by the advance of coarse grasses, already mentioned: Wiregrass and Manyattagrass. Fire is essential to alter the equilibrium and make the grassland switch towards Themeda (Fig. 4.1).

The degeneration of Kikuyugrass swards is more apparent at the higher altitudes, *e.g.* on the Kinangop (2500 m; Planosol) and in the Bracken belt (poor, acid soils; *4.2.1.1*), both in central Kenya. It is evident that Kikuyugrass when it occurs at lower elevations (1750–2000 m) in high rainfall areas, as found in western Kenya (Kisii and Mount Elgon), is more productive than at higher altitudes, although the invasion of bracken-fern (*Pteridium aquilinum*), sedges, Star-, African Couchgrass and Sporobolus is commonly observed in such areas. Volcanic soils, such as the Andosols found north west of Lake Kivu (Masisi) and the Nitisols in central Kenya (Limuru), usually carry the best Kikuyu grasslands there, where a high moisture index prevails throughout the year (with high rainfall or with relatively cool and cloudy weather and with a short dry season). The lower-altitude limits above which Kikuyugrass can be expected to occur in the western regions are some 250 m higher than in central Kenya (*3.1.2.1*), except where temperatures are already moderated by nearby water surfaces or in valley bottoms.

Kikuyugrass should be valued primarily for its environmental significance as it colonizes and protects land that is laid bare. It is valued as a stabilizer on shoulders of new roads and of water-gullies and as an amenity grass for sports fields and lawns. The grass can stand a great amount of wear and tear, being very tolerant to treading. It is also very difficult to eradicate mechanically. This is the reason why Kikuyugrass is regarded as a mixed blessing in farming systems based on crop-grass rotations. It is a pernicious weed in cultivation and although it provides useful grazing in derelict arable land, it is very hard to eradicate when a new cycle of cropping is anticipated.

The Kikuyugrass zone is also known as the tea belt, because of the tolerance both Kikuyugrass and the tea bush have to soil acidity. Suitable arable crops are wheat, potatoes and pyrethrum and temperate vegetables; maize is often poor in this zone, due to temperature constraints. Ethiopia has its own traditional choice of crops with barley, oats and teff; little teff is grown at altitudes above 2500 m. Commercial mixed-farms grow wheat as the major crop. The area lends itself ideally to sheep farming and fat-lamb (Molo-lamb) production. The best cultivated grasses are temperate grasses such as Ryegrass and Cocksfoot, with Nandi Setaria and Rhodesgrass at the lower elevations (*4.2*; *6.5*). Giant Setaria, Elephantgrass and sweet potatoes vines are the main fodder crops.

Fodder oats are commonly grown either as break crops or as main-season crops.

3.4.2.2. Zone II_{ST} (Stargrass zone, 1250–1750 m) Stargrass has colonized significant areas on the eastern, lower slopes of Mount Kenya and Mount Aberdares which are commonly known as the Stargrass zone (Bellis 1958; Strange, 1963). Compared with the Kikuyugrass zone the climate is warmer. The Stargrass zone may be equated with the border areas between the coffee and sugarcane belts of Jätzold & Schmidt (1982). The soils are often Nitisols (Sombroek *et al.*, 1982).

The Kikuyu- and Stargrass zones are not necessarily adjacent or nearby. Secondary Stargrass is typical of certain belts, for instance in western Kenya (Goldson, 1967), in eastern Uganda and the Beni area of Kivu. They are at a similar elevation, level of soil fertility and climate as those of the more typical Stargrass zone (Nitisol) of central Kenya, where it is situated on mountain slopes, below the Kikuyugrass zone. Considerable overlapping between zones occurs. In secondary grasslands, Stargrass may be associated with Kikuyugrass at the higher, with Imperata at the lower and with African Couchgrass at both altitudes. In the humid and lower altitudes, Elephantgrass takes over. In the wetter parts of Uganda, Stargrass occurs little in natural grassland, but is abundant as a weed and is typical of the sides of paths and roads and of the disturbed soil of banks and ditches, yet absent from natural grassland a few yards away.

Stargrass colonizes isolated parts in the drier Zones III and IV, from sea-level to 2250 m, where organic matter has accumulated or where soil fertility is high. In semiarid areas on Vertisols in Karamoja, Stargrass is the first to colonize former kraal sites and well-trodden pathways (Thomas, 1940b). Thus, like Kikuyugrass at higher elevation, Stargrass is commonly found in kraals, grazed paddocks under shade trees, horse paddocks with their typical separation of grazed and ungrazed (dung) patches, perennial legume fields, and in general as a weed in cultivated land. Also at the coast, *e.g.* in the Tanga area, it is not unusual to find paddocks dominated by Stargrass close to dairy buildings, due to the accumulation of animal manure. A stray haybale may within a few months be overgrown with Stargrass, not seen in the vicinity. In Zone V Stargrass is mostly confined to riverbeds.

Stargrass has, therefore, less indicator value for zonation purposes than Kikuyugrass, except within Zone II towards the higher elevations. Little mention is usually made of Stargrass in the Arusha and Kilimanjaro regions (T) in spite of their high soil fertility (Naveh & Anderson, 1966). In Ethiopia it is widely distributed below 2000 m in valleys, flats and in abandoned land (USA, 1954). It is also common in the Nile- and other river valleys in Sudan (personal observation, 1990). In the Rift Valley on ancient lakebeds Stargrass is associated with densely scattered Leleshwa (*Tarchonanthus camphoratus*). It consists of short herbage strongly dominated by *Cynodon plectostachyus.* Ball (1936) recorded that in the Rongai valley in Kenya, Themeda-dominated grassland was converted in Stargrass by treading and heavy grazing, while derelict arable land changed to Rhodesgrass (specially on lava ash) and Stargrass (where soils contained soda). In the nearby but somewhat wetter Subukia area, Kikuyu-, Star and African Couchgrass were found together.

It was hoped that the taxonomic revision of what was used to be known only as *Cynodon dactylon* and *C. plectostachyus* in (East) Africa (Bogdan, 1949b) would help to define a more satisfactory ecological classification to explain the distribution of Cynodon over the different zones. For details of the revision see Clayton & Harlan (1970), Clayton *et al.* (1974) and Bogdan (1977). Only *C. plectostachyus* (Naivasha Stargrass) has retained its name but *C. dactylon* has been differentiated into *C. aethiopicus* and *C. nlemfuensis* for the stoloniferous species without rhizomes in Africa. The name *C. dactylon* is restricted to the rhizomatous type, known as a cosmopolitan weed-, grazing- and lawngrass. It is found in the subtropics and the warmer regions of temperate climates, *e.g.* southern France ("chiendent"), and known also as French grass (in Uganda), Bermudagrass or Dhub.

It would seem that, where a Stargrass of a rather robust type is found colonizing in moist and fertile pockets in Ethiopia, Sudan, Zaïre and Uganda, mostly *C. aethiopicus* (Giant Stargrass) is involved (*4.2.2*). Conversely, the finer types of Stargrass in

Kenya and Tanzania mostly refer to *C. nlemfuensis*, common on light-textured soils at the bottom of the Rift Valley (Harlan *et al.*, 1970; Bogdan, *op. cit.*). The floor of the eastern Rift, with ancient lake beds and river courses, from Ethiopia to Tanzania is the range of distribution also of *C. plectostachyus*, particularly abundant at Naivasha, to which it owes its common name; it can be easily distinguished by its arching stolons, short glumes and curling racemes. *C. nlemfuensis* and *C. aethiopicus*, however, are far less easy to distinguish (Bogdan, *op. cit.*). It should be noted that in the agro-ecological zonation of Jätzold & Schmidt (1982) the name *C. dactylon* is used throughout. Although the confusion has still not ended and old names keep being used, it can be concluded that the taxonomic revision has not contributed much to a better understanding of the distribution of Stargrass in the different areas.

Stargrass pastures have an open sward which may lead to botanical instability. Common and troublesome weeds are *Ageratum conyzoides*, *Conyza* (syn. *Erigeron*) *bonariensis* and *Dichondra repens*. Contrary to general expectation, herbage yields can be very low. Many herdsmen believe that Stargrass is disliked by livestock because of the toughness of the stems. Where animals have a choice, Stargrass is generally not well eaten. Stargrass is also commonly attacked by many pests and diseases.

Like Kikuyugrass, Stargrass is most useful for environmental protection purposes (*4.2.2*).

3.4.2.3. Zone II_{EL} (Elephantgrass zone, < 1250–1500 m)

The principal Elephantgrass zone is found along the northern and north western edges of Lake Victoria (fertile crescent) where this grass colonizes forest clearings and cultivated land (Fig. 3.6). A second major region is found on both sides of the Zaïre-Uganda border and north east of Mount Ruwenzori, up to 2600 m (Thomas, 1940b). The same author reported that west of the Ruwenzori it occurs in a belt between the lowland and highland forest. Here Elephantgrass competes successfully with forest and, with the aid of fire, has kept its hold (Lind & Morrison, 1974). It would be of interest to know why the expanses of Elephantgrass in western Uganda have remained in that state and have not gone back to forest as generally and as quickly as anticipated. Earlier authors such as Eggeling and Robyns (quoted by Lind & Morrison, 1974) believed that the spread of forest is prevented by fire, but also by elephants, which trample and eat tree seedlings and debark older trees.

The Elephantgrass zone along the Uganda-Zaïre border continues into the southern parts of Equatoria in the Sudan (Ferguson, 1948). Kenya and Tanzania have very few areas with similar forest potential at the altitude of < 1250 m. Elephantgrass is mostly restricted to riverbeds or to fertile valleys such as Kilombero (T). In Ethiopia it has been recorded in Illubador (Fröman & Persson, 1974). In cultivated form Elephantgrass is generally better known as Napiergrass (*11.2*).

In contrast with short Kikuyu- and Stargrass, Elephantgrass vegetation produces bush-like thickets which are in that state useless for grazing. When burnt, cleared and ploughed such grass produces a good soil tilth for cropping. It is the zone of the true tropical food- and fruitcrops such as sweet potatoes, cassave and banana, but the most important and widespread cash crops in this area are coffee and sugarcane. Many of the special attributes of Elephantgrass are understood better when it is realized that this grass and sugarcane share similar ecological requirements of warm and humid conditions. The zone is an area of dense human settlement so that in some places very little of the grass may be in evidence. On poorer soils and those exhausted by overcropping Elephantgrass may be replaced by Citronellagrass (*Cnar*), a grass of very low grazing value. Imperata (*Icyl*) is also common on swampy grounds, or following cultivation and burning, both in this zone and at lower altitudes (Kasaï), although its weed potential is weaker than that found in south east Asia. Both species are pernicious weeds, in grassland and in cropping, together with African Couchgrass (*4.5.1*).

Elephantgrass is susceptible to frequent grazing which leads to sward instability (*4.2.3*).

3.4.2.4. Zone II (Coastal areas)

This Zone is too small for our purpose to warrant a discussion different from that under Zone III (*3.4.3.1*).

3.4.3. *Zone III-high (> 1000 m)*

This is the subhumid savannah zone (Table 3.4), drier in appearance than Zone II, because of a marked dry season. Characteristic vegetation types are Combretum-Hyperthelia wooded grassland and Miombo woodland (*3.3.1.3*). *Hyparrhenia filipendula*, *Hyperthelia dissoluta* and *Themeda triandra* are the most common and abundant primary grasses across this zone. In contrast with the secondary zone-grasses of Zone II (*3.4.2*) these species are increasers and not colonizers, nor arable weeds and they traverse the altitude belts (*3.3.3*). Unlike its Australian counterpart (*T. australis*), *Themeda triandra* is a highly stable constituent of its characteristic communities in Zones III and IV (*4.3*) and occurs over a wide range of soils (Heady, 1966).

Hyparrhenia is a genus of a broad ecological potential, but referring simply to *Hyparrhenia* spp. is risky, as the species differ in ecological requirements and in value. Descriptions of the succession of various *Hyparrhenia* spp. are given for eastern Zaïre by Risopoulos (1966) and some general conclusions can be drawn from Edwards & Bogdan (1951), Bogdan (1958a), Lang Brown & Harrop (1962) and Langdale-Brown *et al.* (1964). The following Hyparrhenia species are associated with (seasonally) swampy conditions, in order of declining altitude: *H. collina*; *H. pilgeriana*; *H. papillipes* (*H. lintonii*) and *H. nyassae*. In the moist zones and in forest edges, more robust types such as *H. diplandra* and *H. cymbaria* are found, respectively. Hyparrhenia species of the more open grasslands are *H. filipendula* (in Miombo woodland) and *H. rufa* (Coast, Uganda). The latter is the most cosmopolitan of all and the only species used as a sown grass, in Uganda (*6.13*). *Hyperthelia* (syn. *Hyparrhenia*) *dissoluta* is common and often abundant in Combretum and Miombo woodland. The last 3 species are normally referred to when *Hyparrhenia* spp. are mentioned. They are spread right across the East African region. Although it is not uncommon to find references in Uganda to "Hyparrhenia zone", mostly to distinguish the north of the country from the central part, dominated by the Elephantgrass zone, 3 subdivisions are recognized, each dominated by different species of Hyparrhenia (Fig. 3.6; Horrell & Tiley, 1970):

1. *H. rufa* is the dominant grass in areas with higher rainfall, shorter dry season and relatively more fertile soils. Minor areas of grassland dominated by Common Guineagrass, Rhodesgrass, *Brachiaria decumbens*, *Pennisetum unisetum* and *Bothriochloa insculpta* occur within the same climatic region.
2. *H. filipendula* is more abundant in areas with lower rainfall, more severe dry season and poorer soil and is associated with *Hyperthelia dissoluta.*
3. *H. cymbaria* is prominent in limited areas of higher ground, under moister conditions in south west and north east Uganda.

Unlike those of Zone II, the grassland communities of Zone III – and their principal constituents – are botanically relatively stable under grazing and do not alter their composition in response to fertilizer either. This has been convincingly demonstrated by Smith (1964a) on Hyparrhenia (*H. dissoluta* and *H. filipendula*) veld in Zambia. Keya (1973) reported on the botanical composition, based on the frame-quadrat method, of a Hyparrhenia-dominated grassland at the Kitale Station in June 1969 and December 1970 (Table 3.6).

Little change was found between the two dates of recording in spite of the application of high levels of N and P fertilizer. It should be noted that an earlier description of the same site also referred to *Hyparrhenia* spp., *Loudetia kagerensis*, *Rhynchelytrum repens*, *Sporobolus pyramidalis*, and *Eragrostis* spp., so as to indicate that the soil fertility status was low (Keya *et al.*, 1971b). For the botanical description of a nearby site and the effects of grazing management, see Table 4.2.

Zone III forms the main maize and Sorghum belts, Sorghum being more important on the heavier soils, especially in northern Uganda and Tanzania. The zone has sunflower and also cotton, at lower altitudes.

3.4.3.1. *Zones II + III – low (Coastal areas)*

The East African coastal belt, when left undisturbed, normally produces a thick almost impenetrable vegetation of low bush. After clearing, the vegetation abounds with a wide variety of grasses and legumes, which can look attractive

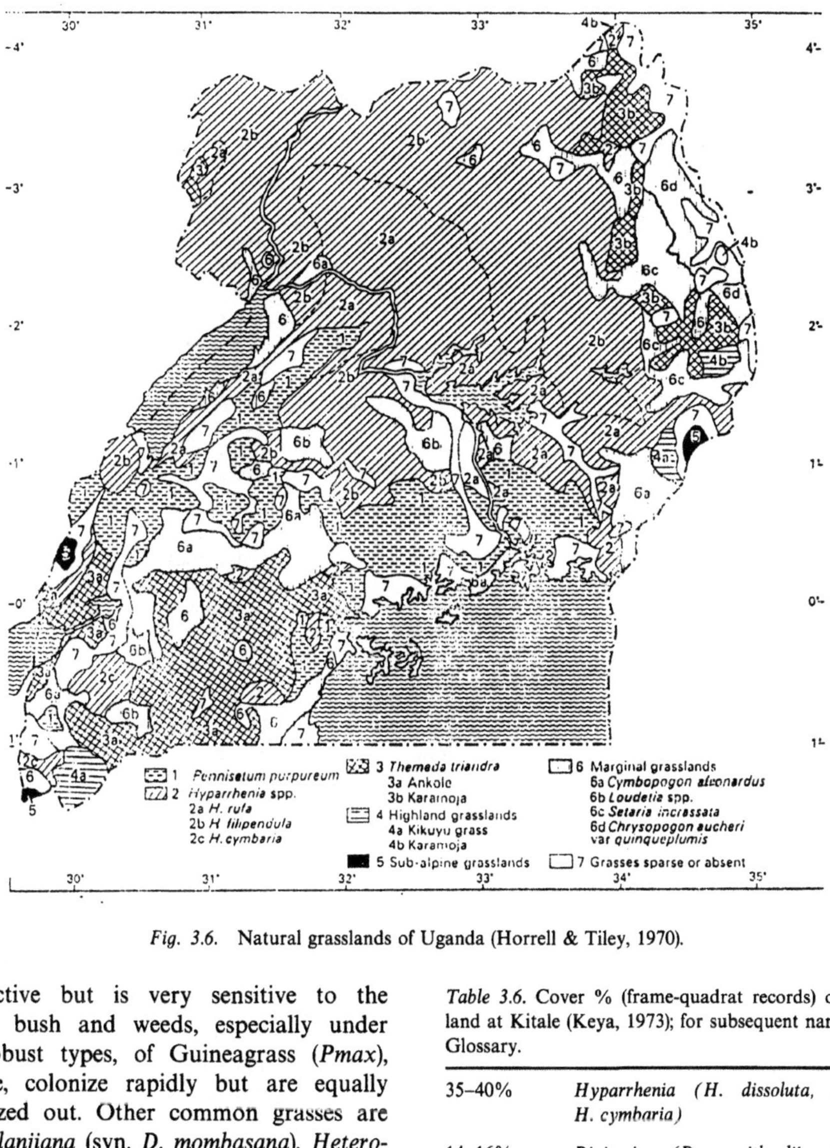

Fig. 3.6. Natural grasslands of Uganda (Horrell & Tiley, 1970).

and productive but is very sensitive to the invasion of bush and weeds, especially under grazing. Robust types, of Guineagrass (*Pmax*), for instance, colonize rapidly but are equally quickly grazed out. Other common grasses are *Digitaria milanjiana* (syn. *D. mombasana*), *Heteropogon contortus*, *Dactyloctenium geminatum*, *Hyparrhenia rufa* as well as Stargrass and Themeda. The red loamy soils near Tanga provide more scope for improvement than the coral sands which are more common in the Mombasa region (*4.4*).

3.4.4. Zone IV

This area is mostly below 1000 m, apart from the Kenya Rift Valley. Characteristic vegetation is

Table 3.6. Cover % (frame-quadrat records) of natural grass land at Kitale (Keya, 1973); for subsequent name revisions see Glossary.

35–40%	*Hyparrhenia (H. dissoluta, H. filipendula, H. cymbaria)*
14–16%	*Digitaria (D. maitlandii, D. diagonalis, D. scalarum)*
4–9%	*Eragrostis racemosa, E. tenuifolia, Brachiaria soluta, B. brizantha, Paspalum scrobiculatum, Sporobolus pyramidalis, Loudetia kagerensis*
< 3.5%	*Melinis minutiflora, Rhynchelytrum repens, Setaria trinervia, S. sphacelata, Cynodon* spp.
< 4%	Herbs: *Commelina benghalensis*
< 2%	Sedges: *Cyperus rigidifolius*
< 1.5%	Legumes: *Trifolium semipilosum; Glycine wightii.*

Table 3.7. Grassland zonation in relation to major natural grasses.

Climatic zone	Climate	Altitude range (m)	Vegetation	Natural grasses	Acronym
I	Afro-alpine	3000	Alpine Moorland and Grassland		
II_{KI} (Kikuyu)	Humid	1750–3000	Forest	Exotheca Andropogon Andropogon Wiregrass Manyattagrass (Kikuyugrass	*Eaby* *Achr* *Aame* *Psph* *Ejae* *Pcla*)
II_{ST} (Star)	Humid	1250–1750	Forest	Setaria Guineagrass Povertygrass African Couch (Stargrass	*Ssph* *Pmax* *Rrep* *Daby* *Cynodon*)
II_{EL} (Elephant)	Humid	< 1250–1500	Forest	Citronella Imperata Guineagrass Brachiaria (Elephantgrass	*Cnar* *Icyl* *Pmax* *Bdec* *Ppur*)
II_{LOW}	Humid (see III)				
III_{HIGH}	(Sub)-Humid	> 1000	Moist Woodland/Bushland	Hyparrhenia Hyperthelia Themeda Rhodesgrass Loudetia Guineagrass Povertygrass Paspalum (African Couch	*Hfil* *Hdis* *Ttri* *Cgay* *Lkag* *Pmax* *Rrep* *Pscr* *Daby*)
III_{LOW}	Sub-Humid	0–1000	Moist Woodland/Bushland	Digitaria Hyparrhenia Heteropogon Guineagrass Stargrass	*Dmai* *Hruf* *Hcon* *Pmax* *Cynodon*
IV	Semi-Arid	Mostly < 1000	Dry Woodland	Themeda Pennisetum Bothriochloa Guinea	*Ttri* *Pstr* *Bins* *Pmax*
V	Arid	Mostly < 1000	Dry Thorn Bushland	Cenchrus Chloris Digitaria Chrysopogon	*Ccil* *Crox* *Dmil* *Cauc*

Acacia (*A. gerrardii*, *A. hockii* and *A. seyal*) and Leleshwa (*Tarchonanthus camphoratus*). *Themeda triandra*, *Cymbopogon pospischilii* and *Eragrostis superba* (*6.16*) are common grasses. Where Themeda occurs on black clay, the edaphic effect is at least as important as fire. Such soils also carry *Bothriochloa insculpta* or *Pennisetum stramineum*, *P. mezianum* and *Panicum coloratum*. On volcanic plains such as in the Rongai area in the Kenya Rift Valley, Rhodesgrass can dominate, which it rarely does on other soils nor in Zone III except in Ankole (Mbarara) in Uganda, where Themeda-

Chloris communities occur (Langdale-Brown *et al.*, 1964). Langdale-Brown suggests that this association may be a natural climax. The most striking difference to the savannah areas further east is the absence of trees. Incidentally, both Rongai and Mbarara are the sites where some famous cultivated varieties of Rhodesgrass originated (*6.6.3*). Another useful grass, *Digitaria milanjiana*, occurs also in Zone IV but is more typical of Zone V, *e.g.* at Kiboko and Buchuma (Cassady, 1973) and of the coast (*3.4.3.1*).

Larger farms in Zone IV will concentrate on ranching, but cropping is becoming increasingly common. Early-maturing hybrids of maize are grown, however marginally. Sorghum, cotton and sunflower are grown at mostly lower elevations. This is also the main sisal growing region.

3.4.5. Zones V and VI

Although Zone V encircles the Ethiopian highlands, covers about half of Kenya, and extends well into central Tanzania, we now enter the zone where arable agriculture is not possible. The reader is referred to the description of Zones V–VI by Pratt & Gwynne (1977).

This is the home of Buffelgrass (*Ccil*). Both its arid zone of origin and its strictly apomictic mode of reproduction make this grass exceptional among the world's famous cultivated grasses. It differs from the above grass species of the (sub)-humid zones zones by an unsual combination of contrasting ecological characteristics and is, therefore perhaps, less sown in East Africa than its drought tolerance and persistence would suggest (*6.9*). Buffelgrass could perhaps be grouped in a separate class together with the edaphic grasses of Zone IV, just mentioned.

3.5. Summary

The diversity of climate and soils and some deviations from the normal patterns are indicated. Vegetation, including grassland, is a reflection of climate, soil, animal and man-made factors. Environment tells us about the grasses we want to use and vice versa. Previous classifications have been based on ecological although primarily climatic zones, on characteristic physiognomy of the vegetation and on indicator species, depending upon the purpose. These classifications are now joined where possible for a better understanding of the role of grasses in mixed-farming. Moisture availability is the main determinant of the 5 Zones (II–VI) of the classification by Pratt & Gwynne (1977). Sustainable agriculture, in the form of mixed-farming based on crop-grass rotations, is possible in Zones II, III and IV. Zone II is subdivided, according to altitude belts, in 3 indicator zone-grasses: Kikuyugrass, > 1750 m; Stargrass, < 1750 m and Elephantgrass, < 1250–1500 m. These natural grasses are coined "secondary"

Table 3.8. Zonation related to crops and recommended cultivated grasses and fodder species.

Zone	Crops	*Lp/D*	*Ssph*	*Ppur*	*Cgay*	*Ccil*	*Pmax*	Fodder/Legume*
I		+	–	–	–	–	–	Oats
II_{KI} (Kikuyu)	Cereals Pyrethrum Tea	+	+	–	+	–	–	Oats clover
II_{ST} (Star)	Maize Coffee	–	+	+	+	–	+	Desmodium
II_{EL} (Elephant)	Maize Coffee	–	+	+	+	–	+	Desmodium
II_{LOW}	(see III)							
III_{HIGH}	Maize	–	+	+	+	–	+	Desmodium
III_{LOW}	Cassave Cotton	–	–	–	+	+	+	Glycine
IV	Sorghum	–	–	–	+	+	–	Lucerne (irrigated)
V	–	–	–	–	–	–	–	–

*Lucerne is recommended universally.

grasses as they are rare in the undisturbed forest vegetation, but behave as colonizers of disturbed habitats, and as weeds in cultivated land; they are difficult to eradicate although their seeding properties are weak. The zone-grasses have in common that they colonize new sites in response to, however temporary, improvements in fertility or moisture, even outside their immediate zone. Their stability under grazing is low, except when soil fertility is high. Elephantgrass dies out when repeatedly grazed hard. Kikuyugrass becomes matted and is vulnerable to the invasion of weed grasses. Stargrass is also liable to weed invasion and is susceptible to pests and diseases. Conversely, primary grasses traverse these zones and, although not necessarily dominant, are principal constituents of the savannah vegetation and form relatively stable communities. Following ploughing, they reappear in cultivated land only after a long delay, following a succession of weed grasses, in spite of their good seeding properties. Examples are Themeda and Hyparrhenia/Hyperthelia, as well as Rhodesgrass. A transitional type is presented by Guineagrass (*Pmax*), *Brachiaria* spp. and Setaria (*Ssph*). The grass communities at the coast with (robust types of) Guineagrass are also unstable.

CHAPTER 4

Husbandry of natural grassland

The natural grassland has generally been the subject of neglect, both by farmers and by pasture research officers. Perhaps these grasslands seem not spectacular enough: they lack the dramatic response to management as displayed by cultivated pastures, react slowly to the onset of the rains and cease growth earlier at the end. Natural grassland is often associated in a negative sense with one or more of the following characteristics (Table 4.1):

Table 4.1. The image of natural grassland.

Characteristic		Condition applicable to
Low site potential:	–	Infertile, non-arable land.
	–	Swampy, rocky, steep or otherwise inaccessible land.
Low control over the land:	–	Communally grazed land.
	–	Lack of grazing control.
	–	Overgrazing and continuous grazing.
Low frequency of good grasses:	–	Bush and weed encroachment.
	–	Plants known as weeds of arable land.
	–	Coarse, unpalatable plants.
	–	Poisonous plants.
Low productivity of existing grasses:	–	Stemminess and lack of quality.
	–	Poor performance in the dry season.
	–	Short growing season.

Much as these shortcomings are valid, they do not apply to all types of natural grassland and some of them may equally be valid for cultivated pastures. Important is that the natural grasslands occupy 95% or more of the grazing land in the region and that they present considerable opportunity for development.

It was proposed (*3.3.3*) to define natural grassland as any grassland that is non-cultivated, including all fallow land or all secondary grassland that has regenerated following cropping. The latter are formed by "secondary" or derived grasses, whereas the term "primary" grasses was coined to denote those native grasses that do not colonize in East Africa, but fluctuate with the intensity of tree cover. Consequently, the estimate of 95% is much higher than the estimate amounting to two-thirds of the mixed-farming areas of Kenya (Poultney, 1959), which referred mainly to natural grassland in the narrow sense of primary, *e.g.* Hyparrhenia grassland.

Secondary grasses such as Kikuyu-, Star-, Elephant- and African Couchgrass have colonized large areas, as the result of increased agricultural activity. Their value for soil conservation is significant, but these grasses pose grave problems for subsequent cropping as their weed potential is so strong that they often force land to be abandoned to fallow. By contrast, due to their forest origin their stability under grazing is low, they themselves being susceptible to ingression. In crop rotations secondary grasses are a blessing and at the same time a curse. A blessing because they colonize disturbed habitats rapidly, guarding against land abuse and erosion, and a curse because they are so difficult to control when not wanted.

Primary grasses such as Themeda, Hyperthelia, Hyparrhenia and others (Table 3.7) are typical of Zone III, of the savannah country cutting right across the altitudinal belts. Savannahs have always carried the largest concentration of grazing animals and the grasses originating in them are adapted to grazing. Their stability is relatively high under grazing. These grass species, however, are not among the colonizing species of disturbed soils. They are easy to plough out and are no weeds, although they seed well and can even be stoloniferous, *e.g.* Rhodesgrass (*3.3.3*).

Apart from work by Walker & Scott (1968a) little has been done to compare the productivity of natural and sown grassland. In a way this can be justified. In East Africa sown pastures are meant to replace cropped land, not natural grassland, so that any comparison that may be made is not really between 2 alternatives. Secondly, what should be chosen to represent "typical" natural grassland? Themeda- or Hyparrhenia grassland may be as productive as 3rd-year sown pasture, whereas Kikuyu- and Stargrass can be surprisingly unproductive (Table 6.2).

Works describing the husbandry of natural grassland in various parts of Africa are: Bille *et al.* (1969), Boudet (1978), Compère (1978), Germain & Scaut (1960), Kennan *et al.* (1955). In addition to those mentioned in the previous and this chapter, specific studies on semiarid parts of East Africa include: Bogdan (1954), Jordan (1957), Heady (1966), Naveh (1966bc), Ndawula-Senyimba (1972), Pereira (1962), Pratt (1968; 1969). For poisonous plants, see Verdcourt & Trump (1969).

4.1. Legumes in natural grassland

Legumes can be present in a wide variety, especially at the lower altitudes and the coast ("coast cocktail" (*4.4*), but play little part in the natural grassland communities of East Africa (Thomas, 1940b). Their potential role was rarely highly regarded. Pratt & Gwynne (1977) devoted little attention to them and referred to "leguminous and other herbs". Legumes were hardly mentioned by Langdale-Brown *et al.* (1964), and not at all by Vesey-FitzGerald (1973), in their grassland descriptions of Uganda or Tanzania and Zambia, respectively. Bogdan (1949a) and Edwards & Bogdan (1951) recognized that legumes occur frequently in Kenya and are rich in protein, but they refer equally to the fact that their proportion in the herbage is not high, that many species are unpalatable and some even poisonous. Even Kenya white clover was not acclaimed unequivocally (Bogdan, 1965c). None of these authors recognized *Stylosanthes fruticosa* as being significant, although this is an indigenous species which is at present being retested in some countries in the region (Dzowela, 1988). The role of legumes when oversown into natural grassland is discussed later (*4.3.5*).

In the following some of the natural grassland communities of Table 3.7 are described with reference to their improvement, where relevant, and their role within mixed-farming systems. For descriptions of their zonation, where applicable, of associated grasses and of their natural distribution, see *3.4*. Two directions of improvement are considered:

1. Indirect improvement by reducing the frequency of undesirable plant species, through bush- and weed control. This is the most common approach.
2. Direct improvement by influencing the botanical composition towards desirable plant species. This approach, independent from the previous one, entails grazing management, mowing, oversowing and/or fertilizer.

4.2. Secondary grasses and their communities

4.2.1. Kikuyugrass

Kikuyugrass is one of the shortest growing grasses of economic importance in East Africa. It is a creeping perennial with strong, thick stolons and rhizomes. The creeping parts root freely at the nodes. The leaf canopy is relatively low but the grass is also unusual in appearance in that fertile tillers do not display stem elongation nor panicle emergence. The visible evidence of flowering is the early-morning emergence of anthers from the upper leaf-sheath. Fertile tillers only show up and flower when the sward is grazed or mown close to the ground. The anthers then spread a whitish tinge over the sward, as if it had been frosted. Seeds are large but are obviously difficult to harvest. Some seeding varieties have been developed in Australia which have been tested in Kenya mainly for amenity purposes.

4.2.1.1. Ecology of Kikuyugrass

This is the characteristic grass of the Kikuyugrass zone in the cool upper-part of Zone II (Zone II_{KI}; *3.4.2.1*). Kikuyugrass is, however, remarkably insignificant in occurrence until the soil is disturbed, after which it appears quickly staying on also when, mostly within a short time, soil conditions deteriorate. Not so much tree removal itself, but disturbance of the soil leads to a rapid upsurge of Kikuyugrass. This is most evident where new roads are cut through forest or old natural grassland, in which this grass had previously been insignificant. Herbivores (livestock) play an important role in the spread of Kikuyugrass to new potential sites.

Kikuyugrass can look attractive with its leafy, dense swards. Such is especially the case at the lower altitudinal range of this grass with high moisture index and better soils (*e.g.* Nitisols) in Kenya: lower Limuru near Nairobi, lower Tinderet (Songhor) and lower Molo (Elburgon). An exceptionally favourable area is situated north west of Lake Kivu (Masisi) with its humid climate and "sols bruns" (Andosols) of volcanic origin (personal observation, 1992). Kikuyugrass will also spring up spontaneously in isolated pockets of Zone III, and even IV, with favourable moisture and fertility conditions: in erosion gullies in arable land, around water taps, septic tanks, through stray hay-bales, under hedges and under shade trees in pastures but most commonly in night paddocks or kraals. It produces luxuriant growth in such favourable environments. Although its vigour and persistence in kraals are due not only to increased fertility but also to the refusal of animals to eat fouled grass, this has at times been wrongly interpreted as the ability of Kikuyugrass to do very well when intensively stocked.

At altitudes above 2500 m it is affected by frost (Etherington, 1943). In many environments, however, Kikuyu swards degenerate into a thick mat which may be persistent but of little or no productivity. This "Kikuyu decline" has been recorded in many countries, also in other continents; while pests and diseases have been implicated, the problem is nutritional in origin (Quinlan *et al.*, 1975). The symptoms are a pale colour, short internodes with small leaves, a dense mat and very poor yield. Kikuyugrass when degenerate and sod-bound, as Dougall (1953b) observed on the Kinangop in central Kenya, "may fail to provide more than a few days' grazing annually". The symptoms are well illustrated in photographs taken in Kivu by Compère (1960a).

By contrast, the upsurges of Kikuyugrass are often associated with an increase in favourable conditions of soil fertility and humidity. After forest clearing soil fertility is temporarily high in terms of exchangeable cations and lowering of aluminium saturation, but also of available N and P. Kikuyugrass, along with other plants, responds to such improved soil conditions, but has a time advantage in that it spends most of its energy to occupy space with stolons rather than to consolidate its rooting sites by leaf production. However, unless on fertile soils or continually replenished by intensive stock concentration with its concomitant fast recycling of nutrients, unfertilized Kikuyugrass will rapidly deteriorate as soil-N becomes less available and the grass becomes sod-bound.

A remarkable feature of Kikuyugrass pasture is, undoubtedly, its capacity as holding yard, *i.e.* its tolerance to heavy treading. Even degenerate Kikuyu pasture will not quickly lose its cover, however large the concentration of animals on it. Again, this holding capacity should not be understood to be equivalent to carrying capacity. It means that Kikuyugrass can hold more stock than it can actually feed. Similarly, Kikuyugrass holds well, both in Zones II and III, in sports fields and lawns, and is suitable in soil stabilization programmes.

Serious problem grasses are Wiregrass (*Psph*) and Manyattagrass (*Ejae*) (*3.4.2.1*). Wiregrass produces dense small tufts and can be dominant over large areas, with *Andropogon chrysostachyus* forming a dense cover between the tufts. It is coarse and unpalatable, and is grazed to a limited extent and then only when very young. Equally unpalatable and troublesome is Manyattagrass, the name referring to colonies of broad tufts found in old manyatta (kraal). The coarse grass stage is very persistent and can render such grasslands next to useless as no stock, except donkeys, will touch them. Both grasses have claimed large proportions of the potential Kikuyu grassland, also in Tanzania (van Rensburg, 1948b; O'Rourke *et al.*, 1976) and in Ethiopia (Fröman, 1975).

Edwards (1935a; 1936) postulated that this retrogression happens in the absence of fire. Wiregrass is often found close to farm buildings because of stock concentration and/or the exclusion of burning. If fire can be made to play its role the pasture will increase its content of highly valued Themeda. Where Themeda is present, failure to burn leads to its decline and an encroachment of the two coarse grasses. Fig. 4.1 shows the scheme as proposed by Edwards & Bogdan (1951).

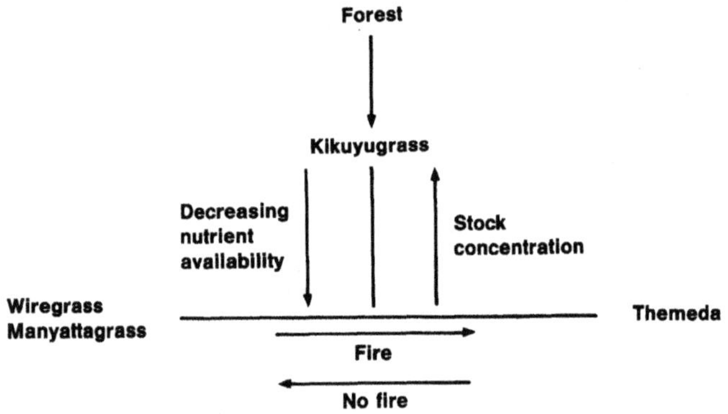

Fig. 4.1 Succession and retrogression in Kikuyugrass communities (after Edwards & Bogdon, 1951).

Edwards (1935a; 1936) regarded most of the Kenya highlands as potential Kikuyu grassland. This reflected the widespread opinion of the time that the dominance of more desirable grasses could be steered in the right direction by careful management (*2.4.3*). Strict rotation of night herding and heavy concentration of stock were recommended to encourage Kikuyugrass.

Looking at the more practical aspects, Ball (1936) confirmed that good Kikuyu grass was indeed found in Limuru and Elburgon, areas on rich forest soils. In Themeda-dominated highland areas, however, stock concentration was ineffective to establish Kikuyu. Themeda was trodden out to the coarse Wiregrass without carrying it to the Kikuyu stage. Ball (*op. cit.*) postulated that it may require more cattle to carry the Kikuyugrass than the Kikuyugrass will carry the cattle. Of the various fertilizers tested, only N-fertilizer had a positive, but limited effect on the balance between Kikuyu- and Wiregrass (Hall & Allen, 1938). Instead, reverted arable wheat land was found to go back more readily to satisfactory mixtures with Kikuyugrass. Before the use of cultivated grasses became more widespread, the reversion of derelict arable land to weed grasses was a common policy (*5.1*).

The suggestion contained in Fig. 4.1 that the decline of Kikuyugrass was due to OM-loss was challenged by Dougall (1953b). Where stock concentration and deposition of manure had taken place, the return of Kikuyugrass was found to be lacking in permanency, involving a debatable system of nutrient transfer. Complete destruction and replacement of the sward were recommended. Comparisons of Kinangop swards with more productive swards at Elburgon demonstrated that the Elburgon soil had a similar % OM and C:N but had a higher pH, more Ca, more Mg and a high base-exchange capacity (Dougall, 1954a). In a review of the work done on Kikuyugrass, Edwards (1956) concluded that in the "more restricted area of the typical community" the maintenance of a productive Kikuyugrass sward demands carefully controlled and intensive management, while in the transitional belts Kikuyugrass cannot economically be maintained.

The carrying capacity of good Kikuyu pasture should not be overrated. Even in the Masisi area of Kivu it rarely exceeds 1 L.U./ha (personal observation, 1992). How disappointing the herbage yields of Kikuyugrass can be, also when planted and N-fertilized, is shown in Table 6.2, summarizing yields over 9 locations across Kenya over 3 years. Excluding Guatemalagrass, Kikuyugrass came lowest in all locations except Naivasha (Zone IV, under irrigation). The dry season performance is poor. At Embu, in the Stargrass zone (Zone II_{ST}) at 1460 m, it produced no more than 1000 kg/ha/year. Yet, at each location the best nearby known ecotype had been planted.

Kikuyugrass has relatively high levels of crude protein but the levels of digestibility (*6.3.4*; Reid *et al.*, 1973) and intake (Table 13.4; Said, 1976) are not or only slightly above average, so that livestock performance is very similar also (Mears, 1970). Nutritive value was found to be variable in

Kivu (Compère 1961b). DM % (Table 6.3) is about average but higher than what one would expect from a grass when its appearance is lush. In grazing observations at Kabete, Edwards (1943) recorded an increase in milk yield when cows were moved from Kikuyu- to Rhodesgrass; a decrease was observed when cows were moved from Kikuyu- to Stargrass (*4.2.2.1*). At Kisii, milk yields on Kikuyu-based pastures were less than half those of Setaria-Rhodes pastures (Table 6.13).

Where lush growth occurs, digestive bloat is a major hazard on Kikuyugrass in East Africa and often makes it necessary to be extremely careful when turning cattle in the morning into Kikuyu pasture (Todd, 1956; Said, 1971; see also *17.5*). Supplementation with hay was recommended by Ball (1936). Kikuyugrass toxicity has been reported from Australia and New Zealand (Healey, 1975). Slagsvold (1969) recorded that Kikuyugrass in the fast-growing period may possess oestrogen activity; a herd on predominantly Kikuyugrass sward near Nairobi demonstrated abortions, aberrant oestrus behaviour and temporary infertility.

Reporting on a tour of central Kenya in the Mount Aberdares and Mount Kenya region, Morrison (1968) observed that cattle grazing on Kikuyugrass were the poorest and scouring and that any other cultivated species should be more productive than Kikuyu- or Stargrass. Morrison concluded that there is a need for a clearer definition of the ecological classification of the areas in which Kikuyugrass may occur but the following could be distinguished:

1. Highland Grassland (> 2250 m). Common grasses cultivated included Mbarara Rhodes, Giant Setaria, Bromus and Elephantgrass. Louisiana white clover was mentioned also.
2. Bracken belt (> 2250 m). The Bracken belt is characterized by poor, often acid, soils in relatively moist areas (up to 1750 mm rainfall) which are sparsely populated. Kikuyugrass occurs naturally but is notoriously poor. Nandi Setaria was found growing well.
3. A more restricted Kikuyugrass zone (1750–2250 m) on the better soils where Kikuyugrass did well as semi-permanent pasture when properly managed.

4.2.1.2. Kikuyu grassland improvement

As with most other natural grassland in East Africa, efforts have generally been directed at suppressing undesirable species.

With mechanical and chemical control methods, Manyattagrass can be halted effectively. Tufts can be destumped with hoes because they are larger in size than and normally not as numerous as those of Wiregrass. Spot application with chemicals such as Dalapon, Glyphosate, Aminotriazole or Prefix should also be considered. Mowing, even with tractor-mounted rotary slashers, often does a bad job as old tufts tend to be pushed over rather than cut. Mechanical and chemical means of Manyattagrass control were tried by O'Rourke *et al.* (1976) in the Ngorongoro highlands (2600 m) of Serengeti (T). Destumping was the best mechanical method. Both Dalapon and Glyphosate were effective in reducing Manyattagrass but it was not clear from the report what the effects of these blanket applications were on the rest of the sward, *e.g.* on Kikuyugrass itself. It would appear that treatment with "weed wipers", soaked with Glyphosate, would offer more selective control (*7.1.10.2*).

A dicotyle weed which invades Kikuyugrass pastures in relatively wet seasons and survives the dry season is *Ageratum conyzoides*, suspected of being poisonous. Such a plant can be cut back or be controlled with auxin-type weedkillers such as 2,4-D (Table 7.7). MCPA is effective against Stinging Nettle (*Urtica massaica*), a problem weed in fertile Kikuyugrass areas such as found in Kivu. Brackenfern is more difficult to control; crushing and rolling help to check it, but of the herbicides available only Dicamba and Picloram were found to have some effect (Ivens, 1971); Asulam should also be effective. Sedges such as *Kyllinga* spp. but especially *Cyperus rigidifolius* are familiar companions of Kikuyugrass, also in Zone III where they are commonly found around open-air watertaps ("Watergrass"). Bentazon was found to give some control of *C. rigidifolius* at Kitale (Veenstra & Boonman, 1974).

The fact that Kikuyugrass owes its expansion partly to (temporarily) better soil fertility conditions has led over the years to a series of fertilizer experiments (Orr & Holm, 1931; Hall & Allen, 1938; Dougall, 1953b; 1954a; Holme & Sherwood, 1954; Compère, 1960a; Morrison, 1966c; Fröman, 1975). The general conclusion seems to be that the response is limited and that any fertilizer considered for use be better applied to cultivated pastures, which display generally a better response. The most extensive work on the improvement of natural grassland containing Kikuyugrass in more recent years was carried out by Fröman (1975) in the Chilalo area in Ethiopia over an altitude range from 1600 to 3000 m. Fröman advocates:

1. Strategic N-application with the aim of maximizing yield when it is needed most.
2. Conservation to deal with surplus growth and to discourage the spread of coarse grasses in an early stage.
3. Burning to suppress Wiregrass.

Reference is often made to indigenous clovers, especially Kenya white clover. However, only when the Kikuyugrass itself is in a productive condition and kept short, is this clover of any significance. Where the grass is too short, Kenya white clover can become dominant but may still remain relatively unproductive (*10.9.1*). Greater success is achieved with white clover, especially Ladino white clover, in the highlands west of Lake Kivu (Compère, 1961a). It was found to persist well in specific areas with "sols bruns" (Andosols), an altitude of 2000–2250 m and with relatively short dry seasons (*e.g.* at Ngungu). A cost-effective method is the vegetative transplanting of Ladino clover into existing Kikuyugrass swards (personal observation, 1992). That this area may be exceptional in its combination of fertile soils and long rainy seasons is borne out by the widespread farming practice of sowing peas (*Pisum sativum*) into existing Kikuyugrass swards. Pea seed is broadcast into the sward without any soil preparation or fertilizer and the resulting pea crop is harvested for human consumption 4–5 months later.

Fertilizer has been studied in attempts to rehabilitate degenerate Kikuyu grasslands (Dougall, 1954a; Compère, 1960a; Morrison, 1966c). Dougall (*op. cit.*) inferred that lack of aeration was a contributory factor and mechanical treatment of the turf horizon was suggested to partly overcome this problem (see below). Morrison (*op. cit.*) studied both cultivation and fertilizers. Cultivation depressed yields. Lime had been considered earlier as an aid to assist in the decomposition of plant residue and to raise pH, but had not produced marked effects. P had a small positive effect. The response to N ranged from 15–20 kg DM/kg N applied, which was considered low in comparison with the response obtained with Perennial Ryegrass.

Although heavy mechanical treatment, *e.g.* with rotavators, has been claimed to produce positive effects on the rejuvenation of Kikuyugrass, the effects are temporary (Compère, 1960a; Morrison, 1966c). Some of the costs of rejuvenation may be recuperated by temporary cropping. All available evidence suggests, however, that where the problem of matted Kikuyugrass is of a recurrent nature, every attempt should be directed at eliminating, not at stimulating the grass. Chemical eradication with Glyphosate and Dalapon (see below) is often the only answer. Short-term arable crops such as beans, which require repeated clean weeding, are ideal. Adequate crop rotations are subsequently needed as part of the constant battle that must be waged to keep the level of reinfestation down.

The invasion by Kikuyugrass into pastures sown to Perennial Ryegrass or Cocksfoot in its zone is a serious problem. The grazing value of Kikuyugrass both in quantity and quality may seem fair at times but is far below that of sown grasses. The latter can, however, not live up to expectations if Kikuyugrass is given the chance to compete.

Eradication could be achieved in theory by heavy grazing followed by discing and mouldboard ploughing late in the dry season and by harrowing as often as needed whenever new shoots emerge. When not well-timed, however, such cultivations are needed endlessly and the strain on the machines may prove prohibitive (Kolbe, 1961). Repeated cultivation is expensive and reduces the crumb structure and OM-content of the soil, which is the very essence of benefit which grasses produce in the crop rotation (*5.3*). A more effective method of eradicating Kikuyugrass is the chemical one, with Glyphosate. This should be applied at 1–4 kg/ha *a.i.* when the grass is growing vigorously (Terry, 1984). Dalapon can also be used, at 5 kg/ha, but its effect is slower, requiring a longer waiting interval before the field can be ploughed. Glyphosate-treated fields can be safely grazed and subsequently ploughed within a week after application. World market prices have recently fallen to levels which make this herbicide an attractive proposition.

The agricultural systems in this zone are largely dictated by the presence of Kikuyugrass as a pernicious arable weed, more so than by purely economic or technical factors. Wheat is still the most widespread crop in this zone. It is, therefore, not surprising to find a strong tendency in recent decades towards the segregation of wheat-pasture rotations. Wheat tends to be grown for as many years as possible on one part of the farm whereas the rest is left semi-permanently under Kikuyugrass. In the late 1950s some success was achieved by Morrison (1966c) by proposing a wheat-pasture rotation with temperate grasses like Perennial Ryegrass, Cocksfoot and even some clovers. The system worked well with keen farmers. The essential part is to bring levels of Kikuyugrass infestation down by cultivation or by chemical means and then hold it back at that low level by wheat-pasture rotations. However, the operation collapses as soon as Kikuyugrass is given the chance to break loose. It will then take over any pasture that is sown. This explains largely why fewer temperate pastures have been sown in these areas in recent years and why crop-pasture rotations are so difficult except in the lower areas where Setaria and Rhodesgrass can be grown more successfully to compete with Kikuyugrass.

Ryegrass or Cocksfoot should be sown as soon as the rainy season starts. If attempts at eradicating Kikuyugrass have not succeeded and new shoots keep emerging it is better to sow first a dense stand of fodder oats (*16*) or fodder rape (*18.1*), before pasture is sown. All decisions on the management of the pasture and the crops that follow should be geared to eliminating Kikuyugrass as radically as possible. Without Kikuyugrass being kept under control, there can be neither productive sown pasture nor good crops.

It should be stressed that its value in soil conservation is without parallel. For soil stabilization, for lawns or sports fields, vegetative establishment is easy and can be rapid provided that high densities of at least 50 × 50 cm are applied. Kikuyugrass is a good servant but a bad master. It cannot be recommended as a cultivated grass in pasture in East Africa. As already mentioned, in the Kenya District Trials Kikuyugrass was the lowest yielding of all species, also in its own zone (Table 6.2). Upsurges of interest in planted Kikuyugrass have been common among farmers and official alike, acting on the spur of some momentary impression. From the above, however, it should be clear that the vigorous appearance of Kikuyugrass in night paddocks and on some of the most fertile soils should not be mistaken for high productivity *per se*.

By contrast, Kikuyugrass has become a well known pasture grass in highlands of Latin America, Hawaii and Sri Lanka. In Australia it has spread over large areas along the eastern seaboard where it is an important grass in N-based pastures.

4.2.2. Stargrass

4.2.2.1. Ecology of Stargrass and general aspects

The distribution of Stargrass has been described (*3.4.2.2*). Unless indicated otherwise, the name Stargrass will here refer to both *C. nlemfuensis* and *C. aethiopicus*. The latter comprises the types with robuster and taller stems (Giant Star), purple racemes and spikelets. Conversely, *C. nlemfuensis* is less robust and the racemes are usually green or pale, but sometimes reddish. The extreme forms of the two can be easily distinguished but there is a series of transitional forms (Bogdan, 1977). Confusion in early naming was increased further by mistakenly accepting larger forms of Cynodon for *C. plectostachyus*. Weed specialists often lump *C. dactylon* and *C. nlemfuensis* together as "Couch" (Ivens, 1971; Terry, 1984). Many recent reports ignore the taxonomic revision and still refer to *C. dactylon* (*3.4.2.2*).

In spite of the taxonomic revision, inter-specific hybridization of *C. nlemfuensis* with *C. dactylon* is without difficulty (Crowder & Chheda, 1982). This procedure has produced a number of important varieties in the southern USA and Central America. The Kitale accession K51415 of *C. nlemfuensis* var. *robustus* was crossed with Coastal Bermuda (*C. dactylon* var. *dactylon*) to produce Coastcross I, possessing both higher digestibility and yield in Georgia, USA (Burton *et al.*, 1967), but with limited winterhardiness. In spite of the emphasis on hybrids, adventitious introductions are still important, *e.g.* Callie Bermudagrass and Ona Stargrass (C. nlemfuensis in Florida (Larbi *et al.*, 1990), and Puerto Rico (Vicente Chandler *et al.*, 1974).

Many reports from Zimbabwe refer to N-fertilized Stargrass of 2 cultivated varieties: No. 2 Stargrass (*C. aethiopicus*) and Muguga Stargrass (*C. nlemfuensis*), the latter originating from Muguga Station near Nairobi, but hardly cultivated in Kenya itself. Superior clones were identified in Nigeria also: IB.12 (*C. aethiopicus*) and IB.8 (*C. nlemfuensis*), the latter from Lake Manyara (T) and released in 1968 (Crowder & Chheda, 1982; Chheda *et al.*, 1983). There are very few research reports on Stargrass from other African countries (*3.4.2.2*). Cynodon varieties of East African origin have apparently done well directly or in plant breeding programmes in countries outside the region. This is in support of the theory that some herbage species do relatively better under cultivation away from home (*6.4.1*).

Stargrass is slower to become dominant after colonizing cleared vegetation than Kikuyugrass, but produces a more productive pasture, also in the dry season. Over bare soil Stargrass stolons can spread at a speed of 2–3 cm/day and attain a length of 20 m. When first planted, Stargrass forms a cover rapidly but soon its vigour diminishes and it is replaced by other grasses (Thomas, 1940b).

Stargrass yields are, together with those of Kikuyugrass, among the lowest recorded in Kenya, irrespective of zone (Table 6.2). In the Stargrass zone (Zone II_{ST}) itself annual yields were recorded of 11.0 ton/ha in Mbarara Rhodes, 9.8 in Nandi Setaria, and 6.7 in Stargrass, over a 4-year period (Embu Ann Rep 1965). Pasture types of Stargrass were unproductive at Serere (Horrell, 1963). However, % DM is higher than of any other cultivated species (Table 6.3). The nutritive value of Stargrass in East Africa is relatively low in terms of digestibility (*6.3.4*; Reid *et al.*, 1973) and intake (Table 13.4; Said, 1976). It has frequently been found to be unpalatable in various parts of East Africa (Hosking & Stephens, 1941; Fiennes, 1940; Thomas, 1940b; Edwards, 1943; Bogdan, 1954). Stargrass patches are often avoided by livestock. The stems are tough but also in the younger stages animals prefer other grasses. At Kabete, Edwards (1943) recorded a considerable drop in milk yield when cows were moved from Rhodes- to Stargrass (*4.2.1.1*).

Animal disorders are not uncommon on Stargrass. Deaths due to HCN (prussic acid) poisoning were reported from cattle grazing pure Stargrass at Kongwa ranch in Tanzania (Wigg, 1973; see also Peterson & McGinnes, 1979). Cyanogenic glucosides releasing prussic acid were recorded in Stargrass already by Timson (1943). Lambs and calves borne by mothers grazed on *C. nlemfuensis* and *C. aethiopicus* developed goitre and skeletal abnormalities while the birth rate was reduced. This was ascribed to the effect of high contents of HCN, 125–250 ppm, (Rodel, 1972) and metabolic iodine-deficiency, which can be rectified by supplements or appropriate salt-licks (Rudert, 1975). Sulphur shortage may occur also due to the dietary sulphur being locked up in HCN detoxification (*15.5.1*). Moreover, *C. nlemfuensis* contains stilboestrol, *i.e.* oestrogenic activity, at levels that cause a decline of heat in heifers (Millar, 1967).

4.2.2.2. Star grassland improvement

An example of well-managed secondary Stargrass pasture can be seen on the Mukumu Mission Farm, south of Kakamega (1600 m; 2000 mm rain) in western Kenya. This 20-ha dairy farm has thrived for more than 40 years on Stargrass (*C. nlemfuensis* ?), manured with pig-slurry and more recently with N-fertilizer. The paddocks are permanent and grazing is rotational. Weed control poses few if any problems, since all paddocks are mown at least once a year for silage. The grass is mown at a sward height of ± 30 cm, left for 1–2 hours to wilt and is then hand-loaded onto trailers and dumped into 3-m deep, round stone-brick pits. No molasses nor any other additive is mixed with the grass. The grass is not chopped but is compressed heavily by rocks stacked on top. The pits are under grass roofs to keep out rain. The procedure, developed by the late Brother Michael, is impressive. Silage is made on this farm to complement the grazing in the dry season but it provides at the same time a handy tool to cope with weeds as well as with surplus grass in the rainy season (*7.2.5*). Conversely, N-fertilizer is employed to top-up grass production in the rainy season. About 10% of the acreage on the farm is sown to maize grown also for silage. The year-round carrying capacity is 2 dairy cows per ha.

The success of the system is due to the combination of husbandry and a responsive natural grass. In fact, paddocks on neighbouring farms show little Stargrass. Stargrass is not persistent and gives way to African Couchgrass under a low plane of nutrition. The Mukumu Stargrass pastures are what they are because of a years-long system of heavy organic manuring, thanks to a large piggery on the farm. The kind of management, including grazing and mowing, explains why these Stargrass pastures remain free from weeds, which are often a problem in Stargrass elsewhere (Bogdan & Storrar, 1954). Obviously, grassland manuring with animal manure presupposes a formidable feed input from outside to be effective (*5.2.2*; *12.7.1*). The Mukumu system is an example of animal manure production based on purchased pig feed.

An example of successfully cultivated Stargrass was found on small dairy farms on sandy soils at the Kenya coast, just north of Kilifi. Dairy farmers have achieved good results with a Stargrass type collected from the local vegetation (J.W.L. Harris, private communication, 1976). This Stargrass is plugged straight into the native pasture at a density of about 150 × 150 cm. Thus, no effort is made to plough the existing sward. Within a year the sward is taken over by Stargrass, which persists relatively long (personal observation, 1979). The procedure is simple and, apart from the initial labour costs, not expensive. This practice should be studied in more detail and could be applied elsewhere, except perhaps in crop-pasture rotations. In rotation experiments at the Kenya coast, land reverted to natural grasses (*3.4.3.1*; *4.4*) gave higher subsequent crop yields than planted Stargrass (Clarke, 1962). How essential manure is in maintaining Stargrass pastures at the coast is unknown, but N-fertilizer is indispensable in any dairy system to achieve quantity and quality (see also *4.4*). At Kongwa (T) a considerable increase in Stargrass occurred when N at 250 kg/ha or 50 ton of animal manure was applied and it was believed that movable night paddocks would help the spread of Stargrass (Wigg *et al.*, 1973).

Caution is needed before such experiences are extrapolated to recommend Stargrass as a cultivated grass in East Africa, even outside the context of crop rotation. The experience with sowing or planting Stargrass in other areas, where it occurs naturally in scattered spots, is less positive. Pratt (1963; 1964) tried to have Stargrass established in denuded land in Baringo (K) but these efforts failed which was "unexpected" (see below). The performance of Stargrass can be disappointing, also in better environments, both in the natural and cultivated state (Table 6.10). Even at Kakamega Station, 10 km from Mukumu Mission Farm, yields of Stargrass were lower than of any of the cultivated species (Table 6.2). At nearly every location of the Kenya District Trials there was a problem with weed invasion in the Stargrass plots, which confirms the general experience that persistence is low. From the early days of pasture research Edwards (1940c) cautioned that the reputation of the Stargrasses in Kenya was mainly due to their vigour in natural grassland on old alluvia, in contrast with their performance outside such areas.

Stargrass has been a favoured grass for special purposes, such as gully stabilization and to smother regeneration of cleared bush as part of anti-tsetse clearing (Graham 1945a; Staples, 1945; van Rensburg, 1953; 1955). Stargrass is useful for erosion control on wash-stop contours rather than for yield (Pereira, 1959; Pereira *et al.*, 1961). In the early days it was the only grass widely available that would ensure a rapid cover after vegetative propagation. In the semiarid areas, however, van Rensburg (1955) observed that planting time is limited to a small number of days when the soil is wet coinciding with the time of the season when pressure of work on crop production and general farming activities is at its height. In spite of the low viability of Stargrass seed, farmers were advised to harvest their own seed. Failure to germinate is, however, not uncommon (Keya *et al.*, 1971b).

Various attempts have been made to develop seeding varieties of Stargrass in the hope that this would promote its use, for instance, in range reseeding (Pratt & Gwynne, 1977). The seed yield of Stargrass is very low, however, also in the absence of fungi. Stargrass produces elongated stems but at Kitale these do not come into head until the beginning of the dry season, and even then only sparsely, especially in the larger and robuster types. Seed yields are only 10% of those of Rhodesgrass. A search, initiated by Bogdan (1959a), for better seeding ecotypes and, within these ecotypes, for better genotypes was finally abandoned after a reappraisal by C.N. Mwangi in the late 1970s.

The limitations to cultivated Stargrass in East Africa are lack of productivity and persistence and its weed reputation in arable land. In eastern Uganda it was regarded as one of the worst weeds in cultivation (Horrell, 1958a). Stargrass is affected by many pests and diseases. The leaves are often attacked by ladybird beetle larvae (*Epilachna similis*). Stargrass appears to be the favourite host plant but damage to maize and other grasses is observed when heavily infested Stargrass is growing in the immediate vicinity (personal observation, 1974). Damage can be severe and the leaf canopy can change its colour to white, as the larvae leave the epidermis intact. Fungal attacks are common, more so than on other grasses. The inflorescences are host to fungi such as *Ustilago* and *Sphacelia-Fusarium-Cerebellum complex*. The latter causes spikelets to produce a slimy-sugary excretion changing later to red/greyish and subsequently to black bodies (*8.7.2.5*). *Ustilago* (head smut) causes the whole flowering head to turn into a clump of black spores while the supporting grass shoot itself turns yellow/white. References to the susceptibility of Stargrass were made already by Staples (1934), who reported a streaky white colour and stunted growth.

Naivasha Stargrass (*C. plectostachyus*) dominates in grasslands around Lake Naivasha and on ancient lakebeds. Yields are not high and weeds can be a problem, but it has proved difficult and uneconomic to replace these natural grasslands by cultivated grasses or fodder crops, except perhaps lucerne. It is often managed intensively under irrigation and with N-fertilizer (17). Naivasha Stargrass is rarely used as a cultivated pasture plant, neither in nor outside East Africa, since it lacks persistence (Harlan, 1970). Much of the literature on cultivated *C. plectostachyus* should be referred to *C. aethiopicus*, with which it had been often confused (Bogdan, 1977).

Popular Stargrass lawns consist of "Cape Royal" or "Maadi River". These are in reality *Cynodon dactylon* var. *elegans* occurring commonly in southern Africa south of 13°S (Bogdan, *op. cit.*).

4.2.3. Elephantgrass

The Elephantgrass zone (Zone II_{EL}) can be equated with the zone of forest potential at <1250–1500 m. Elsewhere it is restricted to riverbeds (*3.4.2.3*).

The potential value, as a grazing resource, of the large natural stretches of Elephantgrass is immense, but at the same time limited by their very nature. Once it grows beyond 2 m and runs to stem, it becomes virtually impenetrable and the leaves are beyond the reach of the (already small Zebu) grazing animal. Only after a fire or perhaps along walking paths does valuable grazing occur. As Langdale-Brown *et al.* (1964) pointed out, Elephantgrass shades out valuable Brachiaria, Hyparrhenia and Panicum, on which much of the livestock depends. Tse-tsefly is rampant over large parts of the Elephantgrass zone.

Little has been published in any detail about the management of Elephantgrass in its natural state. Stocking rate trials at Namulonge (U) on mixed stands of Elephantgrass and Imperata showed that under continuous grazing at 0.8 ha/steer both grasses remained dominant over much of the paddocks. At 0.3 ha/steer and rotational grazing (4 week on, 8 week off) Elephantgrass was eliminated, Imperata much reduced, whilst Setaria, Guinea-, Star- and Signalgrass were greatly increased (Thornton, 1970).

Much more interest was shown in the use of Elephantgrass for other purposes, especially soil regeneration (*5.3*; *11.3.1*), mulching in coffee (*11.3.3*) and only later as a fodder grass. Much of the prosperity and continued high population of the Buganda Kingdom was attributed to the fertility maintaining properties of Elephantgrass (Thomas, 1940b).

4.2.4. *Other natural grassland communities of Zone II*

Grasses such as Guineagrass (*Pmax*), Signalgrass (*Bdec*) and Setaria (*Ssph*) are common, frequently abundant or even co-dominant constituents of Zone II of natural grassland and as colonizers of abandoned cultivated land (*3.3.3*). However, they are rarely found in direct association with the zone-grasses (*3.4.2*; *4.2*) and occur mostly at lower altitudes, *e.g.* in Uganda and eastern Zaïre, and also in Zone III. Descriptions of these communities have been presented for Uganda by Langdale-Brown et al. (1964) and for Ituri in eastern Zaïre by Froment (1960); see also Risopoulos (1966).

Guineagrass is most commonly found in Zone II (II_{low}; Table 3.7) but can also be found in Zones III and IV. From all evidence available it would seem that the more robust and giant types abound in the warm and humid zones, *i.e.* Zone II. This also applies to the more robust species within Hyparrhenia: *H. cymbaria* and *H. diplandra* (*3.4.3*).

These species are most common in Zone II, but they traverse the altitude belts and zones in the same way as the more typical grasses of Zones III and IV do (*3.3.3*). Much of what can be said about the latter in the following sections applies, therefore, also to the former, with the exception that Zone II grasses are more truly secondary. Setaria (*Ssph*), for instance, colonizes disturbed habitats in Kivu (Zaïre) when not burnt (Léonard, 1962) and is the only species capable of dominating African Couchgrass. Further west at Yangambi, Guineagrass is the dominant colonizer and further south, between Lake Kivu and Lake Tanganyika, this role is performed by *Brachiaria* spp. (Risopoulos, 1966).

Setaria, Brachiaria and ("Giant") Guineagrass are colonizers but not as dominant as the zone-grasses and they produce seed relatively well. This indicates that they are transitional types, positioned between the primary and secondary grasses, and between Zones II and III. Many have found their way into cultivation in East Africa and abroad, introduced or adventitiously (*6*).

4.3. Primary grasslands of Zones III and IV

In the previous section dealing with the secondary grasslands of Zone II we have seen that the zone-grasses (Kikuyu-, Star- and Elephantgrass) had not been principal constituents of the original forest vegetation but became dominant in their zone due to human-related activities of clearing, fire and cultivation.

The characteristic grasses (Themeda, Hyparrhenia, Hyperthelia, Rhodesgrass etc.) of Zone III are not colonizers. This is evident in 2 respects in which they differ from the zone-grasses:

1. When the woody growth of wood- or bushland, in which these grasses already occurred as original constituents of the plant community, is reduced or burned these grasses merely increase their overall presence. They are increasers, not colonizers.
2. When the vegetation is destroyed by a few cycles of ploughing and cropping, these grasses do not reappear for a long time, neither as weeds in the crops nor as volunteering grasses when cropping is abandoned, although their seeding properties are relatively better than of the zone-grasses.

A negative shade-effect by trees is generally assumed to be the reason that grass growth benefits by minimum tree-cover. Some significant exceptions exist. Concentrations of Stargrass are found under trees in Zone III, and of Kikuyugrass under *Acacia axanthofloea* in Naivasha Star grassland in Zone IV. Reduction of both temperature and solar insolation are factors involved, but there are species such as Guineagrass (*Pmax*) and *Hyparrhenia cymbaria* which are known to tolerate shade but which do not appear on these specific sites. It is postulated that the accumulation of fertility, due to leaf- and fruit drop and animal manure, coupled with the refusal of animals to graze short, fouled grass are responsible for the presence of Kikuyu- and Stargrass under trees. Higher N-concentrations have been observed in plants grown under shade (Deinum, 1984) and increased N-mineralization has been postulated. Thus, when we refer to grasses which benefit by reduced tree cover, these exceptions should be kept in mind.

The most common useful grasses which are typical of the savannah (woodland and bushland) of Zone III and IV are:

Species	Dominant in:
Themeda triandra	Laikipia (K), Ankole (U)
Hyparrhenia filipendula	Teso (U), Ituri (Z), Tabora (T)

Hyperthelia dissoluta	Kitale (K)
Panicum maximum	Coast (K, T), Teso (U), west Kivu (Z)
Pennisetum stramineum	Laikipia (K), Nyandarua (K)
Bothriochloa insculpta	Sukumaland (T)
Sporobolus ioclados	Serengeti (T)

Of these, Themeda is the single, most widespread species. It covers large areas over a range of Zones (II, III, IV) and from sea level to 3500 m, but becomes typical above 1000 m in (sub)humid climates, *i.e.* mostly in Zone III/IV. Themeda grassland is considered to be fire-induced, *i.e.* the final stage after regular grass burning. When grazed heavily and, thus, prevented from catching fire it is gradually dislodged by coarse grasses or by bush. At altitudes above 2250 m, Themeda is often ousted by Wiregrass (*Psph*) and Manyattagrass (*Ejae*) as discussed earlier (*4.2.1.2*) and below 1750 m by the equally useless Citronellagrass (*Cnar*), which also invades Hyparrhenia grassland (Froment, 1960). Thus, regular burning is a necessity to maintain Themeda as an important constituent of grassland. Themeda is usually found to be of relatively low nutritive value (*6.3.4*). The general belief in the high value of Themeda grassland probably arises from the nutritional contribution of other associated grasses such as Rhodes-, Signal and Guineagrass (Marshall & Bredon, 1967).

Hyparrhenia spp. are the most characteristic species of Zone III, except at the lower altitudes and the coast where Guineagrass is more dominant. Hyparrhenia is typical of vast areas of grasslands of northern and eastern Uganda and eastern Zaïre. *Hyperthelia* (formerly *Hyparrhenia*) *dissoluta* is a major constituent of the Hyperthelia-Combretum wooded grassland of western Kenya. Noxious grasses are *Sporobolus pyramidalis* and *Eragrostis tenuifolia* (see also Froment, 1960; Stobbs, 1969d). These grasses are too tough to be grazed and will spread rapidly in overgrazed areas when the main grasses are weak, also in run-down cultivated pastures. However, their weed potential is small in comparison with that of coarser grasses (*Ejae, Psph, Cnar, Icyl*).

The priority in natural grassland husbandry is to prevent deterioriation. Maintenance and stability of cover is all important. In grassland of mixed botanical composition, the value of the natural resource depends on the kind of species present, on the frequencies in which they occur, on their dynamics in relation to other species with which they are or may be associated, and on the management imposed.

4.3.1. Grazing and botanical composition

The swards of Zones III and IV are generally of a more varied botanical composition than those of Zone II. Consequently, the effect of grazing and of stocking policies should be considered.

The work of Bogdan & Kidner (1967) on the effect of grazing on the botanical composition of natural grassland in western Kenya is exemplary for strategies of how to obtain maximum information without too rigid experimentation. Over a 5-year period different patterns of grazing were examined and their effects studied both on the animal- and the botanical side. For the latter, see Table 4.2.

The site of the trial was under typical *Combretum gueinzii* (syn. *molle*) woodland, other important trees including *Acacia hockii, Combretum binderianum, Terminalia mollis, Erythrina abyssinica, Heeria reticulata, Syzygium guineense* and *Rhus vulgaris.* The frame-quadrat measurements (2,000 points per treatment) were made along the same transect in May 1957 and May 1962 at a similar stage of growth. The site was not grazed in the dry season.

There was practically no difference between rotational grazing and continuous grazing, practised at a uniform rate of stocking, in their effects on grass growth and liveweight gain. Stocking at 1–1.2 ha/L.U. beast was considered appropriate in the wet season. Some decrease of grass cover occurred during the 5 years of the trial. Themeda increased under rotational grazing but *Hyparrhenia* and *Hyperthelia* spp. decreased, under both rotational and continuous grazing. *Setaria* and *Brachiaria* spp., which are among the valuable species, were not affected. Consequently, on balance a loss of some of the more valuable species was recorded. Of the less desirable grasses, *Loudetia kagerensis* decreased but this was largely offset by a marked increase of *Eragrostis racemosa*, again in all treatments. In sites with denser bush *Hyparrhenia cymbaria* and *Beckeropsis uniseta*

Table 4.2. Cover % (frame-quadrat records) at the start and end (1957–1961) of natural grassland trials at Kitale (Bogdan & Kidner, 1967).

	Grazing regime								
	Rotational			Continuous					
				4-paddock			3-paddock		
	1957	1962	±	1957	1962	±	1957	1962	±
Loudetia kagerensis	26	22	− 4	33	27	−6	30	27	− 3
Hyparrhenia filipendula	13	2	−11	9	3	−6	10	2	− 8
H. (Hyperthelia) dissoluta	15	4	−11	15	7	−8	18	8	−10
Setaria sphacelata	5	3	− 2	2	2	0	3	1	− 2
Setaria trinervia	5	11	+ 6	7	10	+3	9	7	− 2
Brachiaria spp.	5	5	0	6	4	−2	4	8	+ 4
Digitaria diagonalis & D. maitlandii	4	3	− 1	4	3	−1	3	1	− 2
Themeda triandra	5	12	+ 7	7	16	+9	9	8	− 1
Eragrostis racemosa	1	9	+ 8	1	5	+4	2	10	+ 8
Microchloa kunthii	1	3	+ 2	1	1	0	1	4	+ 3
Other grasses	7	8	+ 1	4	8	+4	4	7	+ 3
Sedges	5	3	− 2	3	2	−1	2	1	− 1
Herbs	5	8	+ 3	4	7	+3	3	7	+ 4
Trees and shrubs	2	4	+ 2	3	4	+1	1	5	+ 4
Bare ground	1	3	+ 2	1	1	0	1	4	+ 3
	100	100	0	100	100	0	100	100	0

(now *Pennisetum unisetum*) had been prominent. Bogdan & Kidner (*op. cit.*) concluded that the differences between grazing regimes were small, except that continuous grazing led to a patchy grazing pattern. A comparison with the botanical analysis of Keya (Table 3.6) suggests that Keya's site was on previously cultivated land, since it contained less *Loudetia kagerensis* and Themeda but more Povertygrass (*Rrep*).

On hardpan soils at Ukiriguru in Sukumaland (T), Walker & Scott (1968a, b) found that *Bothriochloa insculpta* maintained its dominance under continuous grazing, but also under rotational grazing, provided that the stocking rates were relatively high (0.6–1.2 ha/beast). The authors recorded a total of 124 plant species, including *Panicum infestum, Heteropogon contortus, Cynodon* spp., *Hyparrhenia* spp., *Themeda triandra, Sporobolus marginatus* (now *S. ioclados*) and *Stylosanthes fruticosa.* Although the more severe grazing treatments were not detrimental to botanical composition, the differences in animal performance between continuous and rotational grazing were not consistent and generally difficult to interpret. The authors advocated high stocking rates as the most economic way of rearing cattle for beef production.

These are 2 examples of grazing trials in East Africa which show that the botanical composition of swards with primary grasses is relatively stable under grazing-to-capacity, with little difference between continuous and rotational grazing regimes. In other regions, notably Australia, the issue of botanical composition as influenced by grazing pressure (stock per unit of herbage) and stocking rate (stock per unit area) has been a subject of hot debate in connection with the problem of persistence of tropical legumes (*10.4*); for reviews the reader is referred to 't Mannetje *et al.* (1976) and Humphreys (1991). Grazing pressure is obviously an important factor to consider in its possible effect on species and on botanical composition but also on animal performance (Fig. 10.6).

When designed to assess livestock performance, grazing experiments generally suffer from their inherent rigidity. The experimenter is faced with many subjective choices and dilemmas in order to match herbage supply with animal demand even when a wide range of fixed stocking rates is employed. In farming practice a more flexible approach can be adopted on the animal-, but especially on the herbage side. Herbage supply and quality can be regulated by mowing strategies (*7.2.5*) and, where appropriate, by N-fertilizer

(*7.2.4.2*). Consequently, when evaluating grassland preference should be given to trials with variable stocking rates of the "put-and-take" system.

4.3.1.1. Rotational grazing

Rotational grazing was assumed to be less suitable than continuous set-stocking for pasture evaluation ('t Mannetje *et al.*, 1976). This conclusion was based on a number of grazing experiments, mainly from East African natural grassland with little legume; liveweight gain per animal rather than the liveweight gain per ha was taken as the main criterion. In these experiments, however, the effect of stocking rate often overrode those of either continuous or rotational grazing, while the differences between the latter were rarely significant (Walker & Scott, 1968a; McKay, 1971; Thornton & Harrington, 1971).

An additional complicating factor was that not all the grazing policies studied can be fitted into properly defined categories of either continuous or rotational grazing. In the "continuous" grazing of Walker & Scott (1968a), animals were penned elsewhere every night. It could be argued that the latter system is an extreme form of rotational grazing (half-day in, half-day out). The continuous grazing regime of Bogdan & Kidner (Table 4.2) did not include dry seasons. Rotational grazing is usually associated with fencing and resting but in continuously grazed pasture ungrazed patches are often afforded resting over much longer periods. Pratt & Gwynne (1977) pointed out that absence of fencing does not imply that grazing is continuous. On the free range, rotations are naturally imposed by the presence or absence of surface water, and by seasonal differences in vegetation so that rotation may take the form of a grazing procession rather than of a grazing rotation (Vesey-FitzGerald, 1973). Wet-season and dry-season grazing areas may develop as separate entities. When, by contrast, animals are subject to close herding, even over a restricted area, the system of grazing can hardly be described as continuous either, unless a large number of herds is ranging over the same area.

Rotational grazing shows advantage over continuous grazing when the quantity of available herbage is low (Stobbs, 1969c). Set stocking gave liveweight gains at Kitale, both per head and per ha, which were not lower than under rotational grazing, even on sown Rhodesgrass pastures (Sheldrick & Goldson, 1978). The implementation of continuous or rotational grazing regimes is, therefore, usually not dictated by the superiority of either system, but by practical considerations. In mixed-farming, with dairying based on the paddock system, rotational grazing of sown pasture is the obvious practice, while parts of the farm under natural grassland tend to be grazed on a more continuous basis. Rotational farming and rotational grazing go hand in hand (*7.2.1*).

In dry areas with permanent natural grassland as the sole feed resource, rest periods and some kind of rotational grazing are required to make provisions for the dry season, to allow grass to set seed, to permit a regular burn to suppress bush encroachment and to rehabilitate overgrazed fields. Continuous grazing is often not conducive to animal health either due to the build-up of ticks and helminth parasites. "Block rotational grazing systems" (3-herd, 4-paddock; with a rest of 1 year out of 4) have been recommended (McKay, 1970; Pratt & Gwynne, 1977); see also *4.3.2.2* & *4.3.4*. These authors concluded that while trials do not always support the superiority of rotational over continuous grazing, rotational grazing may still be preferable. The studies of Booysen *et al.* (1963) on shoot apex development have substantiated the rotational resting patterns adopted (Scott, 1967).

As we have seen, the margins of grazing pressure within which the botanical composition remains largely unchanged are fairly wide. Whether harder grazing will result in a botanical change is largely unpredictable. Although it is tempting to suggest that such changes depend on species, environment and many other factors, no general rules apply except that a good ground cover must be maintained throughout. Even then, many swards are surprisingly resilient in the way they are able to recover from abuse.

A shift in the botanical composition in the direction of a more desirable component is rarely achieved by grazing alone. No examples are known of tropical grassland improvement comparable with that of temperate climates where world's top-ranking grass, Perennial Ryegrass, becomes dominant as grassland is subjected to more intensive grazing and treading. Exceptions, perhaps, are the zone-grasses of Zone II, Kikuyu- and Stargrass, but their desirability is not unequivocally acclaimed (*3.4.2*; *4.2*; Table 6.2).

4.3.2. Weed- and bush control

Where grassland is of mixed botanical composition, unlike the "zone-grasslands" (*4.2*), improvement in composition will be primarily a matter of suppressing the less desirable grasses. Dominance of superior constituents can, so it seems, not be achieved by grazing with cattle and goats or by burning (*4.3.3*) alone. It should be complemented in many instances with intentional elimination of the less desirable species. Thus, most of the attempts at improving natural grasslands have been aimed at weed- and bush control.

Control can be carried out on smaller units by regular conservation cuts, supplemented with destumping and herbicides. At Kitale, control of *Sporobolus pyramidalis* was tried mechanically and chemically. Spot application with Dalapon made sense only where the tufts were less than one/3 m^2 and of sufficient size (personal observation). Mowing had no effect as the tough stems were pushed over rather than cut. This weed also occurs in sown pastures. Ploughing being often the only answer should be done in the best possible manner since Sporobolus can easily regenerate.

Dalapon and Glyphosate are effective and safe to control grasses, either as a spot- or a blanket application (*4.2.1.2*). Glyphosate is also effective against most herbs. To control only herbs and bush, various other herbicides can be employed (Table 7.7). Some chemicals are cheap (2,4-D), others expensive (Picloram) and again others are a health hazard (2,4,5-T). An overview of means of control can be found in Table 4.3.

Reports on comprehensive management including grazing and fire are rare. In attempts to maintain or to the restore the balance of grasses and woody species quite unexpected and often opposite effects have been obtained with bush control. Two examples follow.

4.3.2.1. Acacia control in Ankole, western Uganda

Work was done in Ankole, near Mbarara, on checking the spread of *Acacia hockii* and Citronellagrass (Harrington & Thornton, 1969; Harrington & Pratchett, 1974; Harrington, 1974). In an attempt in 1962 at eradicating tse-tsefly in the area the method of chain-clearing was introduced to control woody vegetation, dominated by tall *A. gerrardii* and *A. hockii*; the latter had previously been kept short under routine burning regimes. The rainfall is 700–1000 mm and is bimodal with peaks in April and October/November; the altitude is 1250–1500 m (Zone III/IV). *Themeda triandra* is common in the valleys, while Citronellagrass and *Loudetia kagerensis* appear on the hillsides.

An increase in height and density of the undesirable *A. hockii* and the spread of Citronellagrass tussocks were the unanticipated side-effects of chain-clearing. Secondly, the prevention of fire had allowed thickets of *A. hockii* to interlock so strongly that some areas had become inaccessible to cattle, within 5 years after the chain-clearing operation. Tse-tse had been eliminated but now there was an aggressive and unpalatable Citronellagrass which had been only rare before.

Destumping and planned burning and grazing regimes had to be designed to reduce the density of the tussocks and to have the space filled by *Brachiaria decumbens* which could be encouraged by heavy stocking but also by burning, together with *Hyparrhenia filipendula*.

Treatment with recently developed mechanical "weed wipers", soaked with Glyphosate, is expected to offer a more selective control of Citronellagrass (*7.1.10*).

4.3.2.2. Reseeding in West Pokot, Kenya

A more general problem is bush encroachment due to overgrazing and absence of fire. In Pokot and similar semiarid areas, the grass cover has been reduced to the point of denudation. For a large part of the year the soil is bare and hard, partly due to treading and hoof impact (Thomas, 1960), so that rain does not infiltrate, while run-off is not checked by grass vegetation. Where animals have to search for scattered tufts of grass the impact of the hoof on the environment is worse than that of the mouth. Not enough grass can accumulate because of the high grazing pressure, low soil-moisture and bush encroachment; conversely, bush growth increases since there is not enough grass to provide a burn in the dry season. Rainfall, although erratic, is well over 750 m, yet conditions appear semiarid, since nearly half of the rain runs off as flash floods with the consequence that only

Table 4.3. Response of bush species to mechanical and chemical methods of control (after Pratt & Gwynne, 1977).

	Burning		Mechanical		Chemical								
					2,4-D			2,4,5-T			Picloram/2,4-D		Fenuron
Species	Single	Repeat	Cutting/ Chaining	Stumping	Basal	Stump	Foliar	Basal	Stump	Foliar	Stump	Foliar	
Acacia brevispica	R	MR	R	MS	–	R	R	–	MR	MR	–	(MS)	(MS)
Acacia drepanolobium	R	MR	R	MS	(R)	(R)	(R)	(MS)	(MR)	(MR)	R-MS	S	S
Acacia gerrardii	R	(MS)	(R)	(MS)	(R)	(MR)	–	(MS)	(MS)	(MR)	(MS)	(MS)	MS
Acacia hockii	R	MS	R	(MS)	–	–	(R)	(MR)	(MR)	(R)	MR-MS	MR-MS	(MS)
Acacia mellifera	MR	MS	MS	(S)	–	–	–	–	(MS)	–	–	–	MS-S
Acacia nilotica	R	MR	(R)	(MS)	–	–	–	(MS)	–	(MR)	–	–	(MR)
Acacia reficiens	–	–	S	(S)	–	–	–	(MR)	–	–	–	–	S
Acacia senegal	R	(MR)	R	(MS)	(MR)	–	–	(MS)	–	(MR)	–	–	(R)
Acacia seyal	MR	(MR)	R	(MS)	–	–	–	(MS)	–	(MR)	–	(MR)	MR
Acacia tortilis	–	–	R	(MS)	–	–	–	(MS)	–	–	–	–	–
Acacia zanzibarica	MR	–	R	(MS)	–	–	–	(MS)	–	–	–	–	–
Aspilia mossambicensis	R	MS	R	MS	–	–	S	–	–	MS	–	–	–
Brachystegia spiciformis	R	MS	R	MR	(MS)	–	–	(MS)	–	–	–	–	–
Combretum spp.	R	(MR)	R	(MS)	(R)	–	–	R-MS	(MR)	R-MS	MR-MS	R-MS	–
Commiphora spp.	MR	(MS)	MR-MS	(S)	(MS)	(MS)	(MS)	MS-S	MS-S	MS-S	(S)	(S)	MR
Dichrostachys cinerea	R-MS	(S)	(MR)	(MS)	(MS)	–	–	(MS)	–	(MS)	–	(S)	MS
Duosperma sp.	(S)	–	(MR)	(S)	–	–	–	–	–	(MS)	–	–	–
Euclea divinorum	R	R	R	MR	R	R	R	R	(MR)	MR	(MR)	MR-MS	–
Grewia spp.	MR	(MS)	R	(MS)	–	–	(MS)	–	(MS)	(MR)	–	(MS)	–
Julbernardia globiflora	(R)	MS	R	MR	(MS)	–	–	(MS)	–	–	–	–	–
Lannea spp.	–	–	R	–	(MS)	–	(MR)	(MS)	–	(MS)	–	MR-MS	–
Lantana camara	R	(MR)	R	MS	(MS)	–	(MS)	(MS)	–	(MS)	(S)	(S)	–
Lippia javanica	R	(MR)	R	S	–	–	(MS)	–	–	(MS)	(S)	–	–
Opuntia spp.	(R)	(MS)	R	S	–	–	–	–	–	(MS)	–	–	–
Premna oligotricha	R	(R)	R	(MS)	–	–	–	–	–	–	–	–	(R)
Psidia arabica	(MS)	–	(MR)	(S)	–	–	(MS)	–	–	(S)	–	(MS)	–
Solanum incanum	R	(MR)	R	MR	–	–	S	–	–	S	–	–	–
Tarchonanthus camphoratus	R	MR	R	MS	(MR)	–	R	(MR)	(MR)	(MR)	(MR)	(MS)	(MR)

Categories of susceptibility:
S =Good control can be expected.
MR = Some effect likely but further treatment needed.
O =Preliminary indications only.
MS = Useful but not complete control.
R =No useful effect.

about 400 mm enters the soil; the rain rarely wets the soil profile deeper than 40–60 cm (Dagg, 1962). The vicious circle can be broken by withholding grazing, but the process is obviously speeded up by cutting the bush. In neighbouring Karamoja (U) it had been shown that protection from grazing was essential (Wilson, 1962).

As part of the "Special Rural Development Programme" an area of 60 ha in lower Pokot near Kongelai was cleared of bush in 1974 and fenced off with barbed wire. Trees and thorn bush (mostly *Acacia mellifera* and *A. tortilis*) were not burnt but stacked temporarily pending removal. The ultimate aim was reseeding, a term normally used when overgrazed, denuded land is rehabilitated with the aid of seed and minimum cultivation. Reseeding efforts in other parts of Kenya had previously shown, however, that the various operations should be integrated into an overall land-management policy, embracing bush and grazing control (Pratt, 1963; 1964; Pratt & Knight, 1964; Fredeslund & Cassady, 1969; see also *4.2.2.2*). In many cases, eroded and overgrazed land need not be broken to improve pasture. Fencing, bush clearing and grazing control produce equally good pasture but may take more time (Pereira *et al.*, 1961).

Enclosing the area with wire fences turned out to be ineffective. On the traditional cattle routes which passed through the block, fences were simply destroyed. By contrast, much better protection to the grass had unexpectedly been provided by the stacks of thorn bush. Within 2 months after the onset of the rainy season these stacks were becoming focuses of grass growth. The grasses consisted of annuals, but also of perennials such as *Digitaria milanjiana*, Hyparrhenia, Themeda, Buffel- and Stargrass, which had hardly been conspicuous in the area at the time of the operation but which proved to be very resilient. The reseeding operation was cancelled and the following lessons were learnt:

1. The rainfall in denuded areas, which is often much higher than the actual vegetation suggests, facilitates rehabilitation.
2. One year of effective protection following bush clearance is enough to change this type of denuded land into a reasonable grass cover. No expensive reseeding methods need be employed.
3. Cooperation of the local graziers is a prerequisite to a lasting success. If rehabilitated land is not protected against overstocking, no effort is worthwhile.

This had already become apparent in an adjacent area in the late 1950s with the "Riwa grazing scheme" (Pratt & Gwynne, 1977). In this scheme, a large area was divided in 4 rotational grazing blocks. At any one time, only a quarter of the land was grazed. Each block was to be burnt once in 4 years to prevent bush encroachment. The results were highly encouraging technically but, eventually, the scheme failed due to lack of cooperation on the part of the graziers. The explanation was that they saw this experiment as a colonial yoke which they were glad to throw off when national independence came to Kenya in 1963.

Where a grazing spell is impossible to implement, cutting and stacking thorn bush *in situ* allows the grasses to establish themselves under almost automatic protection. This may then have the educational effect of persuading graziers to exert grazing control more generally.

4.3.3. Burning

However controversial, fire is an indispensable management tool for valuable grasses which thank their evolution to fire. Fire is equally necessary to control coarse grasses and undesirable bush. Without regular burning, useful grasses are ousted by Wiregrass and Manyattagrass (Fig. 4.1) at altitudes above 2250 m and by Citronellagrass and thorn bush at lower altitudes (*4.3.2*).

The cumulative effect of periodic bush fires carried out from 1933 to 1959 was studied by Trapnell *et al.* (1976) on soils with very low OM-content and low base saturation. Burning increased pH and base status. Fire promoted the return of bases brought down in litter from the woodland canopy, while termites depleted this return by consuming unburnt litter, the derived bases gradually becoming concentrated in large termite mounds. In either case, the OM-content of the soil was kept at a very low level.

Reviews of the role of fire in Africa's grasslands were presented by Edwards (1942), Heady (1960), West (1965) and Langlands (1967). Fire may be the only feasible management tool over larger blocks of land, but on units with a high year-round grazing pressure burning is per definition wasteful. Burning destroys what was potentially valuable herbage. It also subjects the soil to high temperatures and bareness causing OM-loss and a drying-out of the top soil. Raindrop- and hoof impact by treading on the bare surface between the remaining tufts of grass may worsen the topsoil structure still further. Thus, rain will not infiltrate into the soil and there is no herbage to act as a barrier against run-off.

Burning is also unreliable in the effect it may have on the regrowth of the herbage itself. Much depends on the amount of combustible matter accumulated before burning, the root-shoot relationships, the physiological stage of grass and the weather conditions, such as direction and velocity of wind. The tiller density of the herbage has drastic effects also. High tiller density within large tufts causes more reduction in regrowth after a severe burn than when the plants are smaller in size and more spread out. Compared with cutting, burning at the end of the dry season can delay grass regrowth, result in bare patches and in a reduced tiller density despite the "green pick" aspect; the latter can be highly deceptive.

In many instances more can be learnt from simple observation of accidental fires than of planned burnings, simply because the latter are so hard to plan.

4.3.4. Mowing for conservation

Regular mowing prevents the regrowth of bush and weeds. It is better to mow and make hay than to burn overmature grass or to let it rot. The farmer, faced with a surplus of grass in the wet season and a shortage in the dry season, has to make a choice between a rescue effort or losing valuable grass.

Haymaking is easier than many may believe (*19*). In Zone III and IV, grasses make good hay while the weather conditions with 2–3 days of dry intervals can be as reliable as in countries where hay has been made since time immemorial. Also, the mature and seemingly unpalatable stands of grass can, when turned into hay, be well liked by livestock.

Apart from the value of hay for the dry season, mowing is conducive to good grassland husbandry. Weeds are suppressed and species which are more susceptible to grazing than mowing are restored. The optimum stage at which a grass should be cut for hay is normally subordinate to practical considerations. It will be carried out whenever the weather is right and when grass is surplus to direct grazing requirements.

In semiarid grasslands, the mowing of patches for hay has been applied more widely than commonly considered economical. Concepts such as "standing hay" rarely apply (*6.3.2*). One factor affecting the nutritive value of standing dried grass is the rapidity of the cut off of the rains at the end of the rainy season. If it is a sharp cut off then the aerial grass parts are rapidly killed. Should it be prolonged, degradation of the grass takes place. Van Rensburg (1948b) argued, however, that to check the rapid growth of annuals and coarse grasses heavy grazing in the early part of the season may be necessary; a rest at this time may be disastrous for the control of unwanted grasses, since this is the only period of the year when they are well eaten by stock.

Cassady (1973) studied the effect of cutting on grass growth at sites in Kenya near Nanyuki (Zone IV, mainly Themeda) and Buchuma (Zone V; mainly Guineagrass and *Digitaria milanjiana*). The dominant grasses have properties of rapid growth, early maturity and senescence. Yields of 3 ton/ha in 100 days were not uncommon. In undisturbed growth during the rainy season, near-maximum yields were already reached within 50–100 days even when soil moisture lasted 80–200 days, but losses of more than a third of DM-yields occurred subsequently in the dry period due to leaf-drop and decay, when intermittent rainfall occurred. Guineagrass is more susceptible to decay than Themeda, because stems become dormant and leaf blades are easily broken. Similar results were obtained by Potter & Said (1986) near Athi river (Zone IV/V) on a Vertisol with Acacia savannah carrying Themeda, Guineagrass and *Digitaria macroblephara*. Harvestable yields reached their maximum (2.4 ton/ha/year) already when swards were harvested at 6-week intervals and when cut low, at 5 cm (see also *6.8*). Decay and insect removal accounted for lower yields achieved at wider intervals and higher stubble heights. No data were given on changes in the botanical composition, but the authors confirmed that this particular sward/soil complex is rather robust and capable of recovery from mismanagement (Pratt & Gwynne, 1977). In ungrazed situations with relatively low growth rates in semiarid areas, the losses due to harvester ants and termites of senescent or dead plant material can be proportionally very high (Wilson, 1962). Taking into account the rather heavy losses due to insects, a 6-weeks harvest interval is perhaps not so severe.

Haymaking, therefore, makes a great difference and although the economics may not always be clear, a stack of hay is the best insurance against losses of livestock and against damage to grassland in periods of disaster. It is one of the cheapest methods of supplementary feeding, also in the early part of the wet season as a remedy against "green flush losses" of liveweight in livestock (French, 1950; Hutchison, 1965). However, possibly negative changes in the grassland composition should be avoided.

4.3.5. Oversowing

Oversowing an existing cover of grassland with grass seed has met with little enthousiasm, since large seed rates are necessary to have any impact. Although Congo Signalgrass (*Bruz*) is established vegetatively into existing swards in Kivu (personal observation, 1992; see also *6.11* & *4.2.2.2*), reports of good results by Pratt & Gwynne (1977) of seed oversown into poor grassland at Kitale have remained unconfirmed. The same applies to reports about success with legumes, although oversowing with legumes is a more logical step to consider. It circumvents, when successful, the need for N-fertilizer. Legumes improve the nutritive value of herbage when mature, which is relevant under extensive grazing conditions (*10.10*).

Following reports from Australia and New Zealand, repeated attempts have been made in central Kenya to oversow legumes into Themeda grassland on the Laikipia plateau, as well as white clover in the highlands at Molo. Failures have been commonly attributed to conditions at seeding (*4.3.5.1*); in the highlands of Kivu vegetative transplanting of Ladino white clover is a well established practice (*4.2.1.2*). Encouraging results were also reported with Hyparrhenia and/or Imperata grasslands in Zaïre (Blouard & Thuriaux, 1962; Risopoulos, 1966), Zambia (Smith, 1963; van Rensburg, 1965; 1969c), Uganda (Stobbs, 1969g) and Kenya (Keya *et al.*, 1971a,b; 1975). Stylo (*Sgui*) was the main species deployed in Zaïre, Zambia and Uganda, but earlier also at Kitale (Strange, Kitale Ann Rep 1952). Successful establishment was achieved in most cases by broadcasting, preceded by a bush clearing, light cultivation, or a burn. No seed inoculation was needed. Essential elements of success were a long rest after sowing and repeated application of superphosphate, containing sulphur and calcium. Rates of 50 kg/ha/year P_2O_5 nearly doubled liveweight gain in Uganda on Imperata grassland (Fig. 10.1). It was observed that Imperata could be suppressed by Stylo. The method appeared so successful that it was also tried in the semiarid parts of Karamoja (U). Seed was sown on termite hills encircled with dead thorn branches as protection until seed maturation. Seed was then distributed into the surrounding area by cattle eating and conditioning the seed (Stobbs, 1969g). Given the ease and low cost of the operation a break-through was anticipated. However, very little was heard subsequently; it was later reported that oversowing legumes into Imperata swards had met with only limited practical success and that the content of Stylo in the sward dropped already in the 3rd year (Otim, 1975). When applied on a large scale in *Hyparrhenia filipendula* grasslands in Acholi in northern Uganda in the early 1970s, difficulties of establishment were encountered (H. van Arkel, private communication). At Gandajika in Kasaï (Zaïre), Risopoulos (1966) recorded a loss of 85% of Stylo after an accidental fire. Most legumes are killed by fire (Wendt *et al.*, 1970).

Mechanized sod-seeding into a *Bothriochloa* and *Sporobolus ioclados* sward on a Vertisol near Arusha led to good germination of Desmodium and Glycine but 15 months after sowing the legumes were still confined to the sowing lines (Northwood & Macartney, 1969).

4.3.5.1. Seed establishment

Since the establishment phase has often been blamed for failures encountered with legumes, a detailed study was carried out by Keya (1971–1975) with *Desmodium uncinatum*, sown into Hyparrhenia grassland at Kitale (*3.4.3*). Better persistence of legumes in the grazed sward was expected in natural grassland, because of an allegedly better competitive position. It was also assumed that some N-responsive grasses such as Setaria would increase their frequency in the sward as Poultney (1959) had suggested. Grazing was completely withheld in the 1st year of sowing. Although animal data were not recorded, the trial area was under a mild regime of rotational grazing for 2 years before it was returned to the farm grazing-routine.

Heavy rates of seed (11 kg instead of the usual rates of 2–3 kg/ha in open land) and of P (100 kg/ha P_2O_5) had been used. Provided that rates of 5 kg of Desmodium seed and 100 kg P_2O_5 were applied, no problems were encountered with achieving dominance of Desmodium and no additional measures were needed (Keya *et al.*, 1971a). By the 3rd year, Desmodium had expanded so vigorously that it was feared it would swamp out Hyparrhenia and *Loudetia kagerensis* completely. However, no sooner had the field been returned to the farm grazing-routine or it switched back to a botanical composition as before. In less than 2 years Desmodium virtually disappeared whereas no shift in botanical composition towards more valuable grasses could be witnessed. In Zimbabwe Hyparrhenia veld, when grazed in the dry season only, liveweight losses were reduced by reinforcement with Desmodium from 20 in the control fields to about 5 kg. Complete resting during the growing season had, however, a negative effect on the performance of Hyparrhenia (Clatworthy, 1984b).

Reasons of failure have been commonly attributed not only to establishment methods but also to climatic factors or to companion grasses. It is only slowly becoming accepted that very few tropical legumes can withstand the stocking rates considered normal for the grass companion growing on its own. Attempts to produce good quality grazing from young grass is incompatible with the idea of maintaining legumes. Legumes seem to be inherently incapable of withstanding the grazing regimes that are needed to prevent grasses, which they are supposed to improve, from losing their productive and nutritious state through undergrazing. Benefit accrued from the legume may be partly offset by a loss in value of the grass companion. The legume-problem is largely one of persistence or of tolerance to rational grazing, not of establishment (*10.4*).

Once the reasons of failure are diagnosed and accepted, remedies can be tested and implemented (*10.7*). Oversowing of legumes is a cheap (if phosphate rates can be kept low) and potentially powerful means of improving grassland and should be defended as such, but it necessitates a complete change in the attitude towards pasture management. Fires should be avoided at all costs. Grazing should be withheld completely in the year of sowing in order to have the legume achieve dominance. In subsequent years, when the legume is under stress, rest periods are needed, possibly followed by haymaking, at least once a year. The potential advantage of maintaining legumes by strategies of infrequent harvesting must be weighed, however, against the inevitable decline in grass quality due to ageing.

One of the leading themes of this book is that in the husbandry of tropical grasslands regular conservation cuts are among the best methods available to maintain productivity, whether legumes are involved or not.

It is concluded that the oversowing of natural grassland can be a relatively easy practice, but frequent and long spells of resting are needed. The method is cheap provided there are no additional fertilizer requirements (*10.2.1*).

4.3.6. Fertilizers

Fertilizer would, at least in theory, be one of the first inputs to be considered to shift botanical composition in the direction of more desirable grass species. P should encourage legumes, and N the more responsive grasses. Significant changes in botanical composition of Zimbabwe veld due to N-fertilizer were reported by Weinmann (1964). In East Africa, however, apart from Stargrass communities (*4.2.2.2*), evidence suggests that favourable effects are small. Little in the way of botanical changes was recorded at Kitale, as the result of N- or P-fertilizer (*3.4.3*). Earlier, Poultney (1959) had observed that an annual application of sulphate of ammonia at 55 kg/ha N reduced the frequency of less desirable *Loudetia kagerensis* and Themeda, but increased that of *Setaria* spp., 3 years after the start of the trial. However, the latter consisted of *S. sphacelata* and less valuable *S. trinervia* and *S. longiseta*, so that no clear picture can be gained whether, on balance, the botanical composition had indeed been improved. In ungrazed coconut plantations at Tanga (T), Anderson (1968b) noted an increase in legume content as the result of P-fertilizer in one, but not in the subsequent cut. Guineagrass also increased its presence as the result of fertilizer. Later studies (Schmidt & Watkins, 1968; Hendy, 1975), however, disputed the practical value of these botanical changes in the event of grazing (*4.4*).

The main effect of N-fertilizer, however, is to increase yield and protein content. Studies on natural grasslands in East Africa are rare but Smith (1964a) found linear responses up to 200 kg/ha N on Hyparrhenia veld in Zambia. No changes in the botanical composition were observed but herbage yield was increased 3-fold and crude-protein yield 4-fold. The % recovery was 35 and the N-effect was 22 kg DM/kg N applied. N-application is without doubt the most effective means of increasing herbage growth. With 50–100 kg/ha N it is quite possible to double or treble the amount of grass. Even in dry areas, grass growth is limited more by nutrients, N especially, than by rainfall. N-fertilizer is, however, expensive and whether application on natural grassland is justified depends on the level of husbandry and price ratios. The Mukumu system of N-fertilized, secondary Stargrass pasture (*4.2.2.2*) has been a commercial success. On well-managed farms, N-fertilizer may be applied to natural grassland, but is usually better diverted to cultivated pastures in the crop rotation within the mixed-farming system. Topdressing with P on grazed grasses is rarely beneficial (*7.2.4.3*).

4.4. Grassland at the coast

Grassland at the East African coast can look attractive with its many species of legumes ("coast cocktail"). In the late 1960s and early 1970s, Goldson (1968) at Mtwapa (K) achieved lactations of his Jersey herd which competed with the best in the highlands. Goldson's natural pastures were interplanted with Macademia nuts and contained *Dactyloctenium geminatum* (Crowfoot) as one of the major constituent species. In the Tanga area, the grazing of cattle under coconut palms brought rapid and beneficial results in terms of milk and coconut yields, provided bush and weeds could be kept under control (Childs & Groom, 1964).

Productive natural grasses at the coast are Guinea-, Buffelgrass, Brachiaria and *Digitaria milanjiana* (syn. *D. mombasana*). The most common type of Guineagrass (*Pmax*) is robust ("Giant"). It is most in evidence along the roadsides, but much less so on the other side of the fence because grazing is not well tolerated. Guineagrass, however, colonizes rapidly in areas of recently cleared bush, together with *Digitaria* spp., and also after cropping. *Hyparrhenia rufa*, Themeda and Setaria (*Ssph*) are also common where the grazing pressure is low. Weed- and bush encroachment are, however, major problems.

The "coast cocktail" of legumes consisting of *Tephrosia* spp., *Desmodium triflorum*, *Rhynchosia* spp., *Dolichos axillaris*, *Stylosanthes fruticosa* and *Vigna triloba* was stimulated by P-application on the Tanzanian coast (Schmidt & Watkins, 1968). To this list Anderson (1968b) added *Dolichos argentis*, Stylo (*Sgui*) and Indigofera and noted a substantial increase in the basal cover, even 3 years after the application of NPK-fertilizer, on ungrazed pasture. It was hoped that Guineagrass would be stimulated, but the sward remained predominantly *Heteropogon contortus* and *Digitaria mombasana*, depending on the grazing pressure. Schmidt & Watkins (*op. cit.*) expressed doubt about the value of an increase in the legume component of the natural grassland, since most of the legume types were not relished by cattle and contributed little to the total herbage production. The same lack of palatability was observed in Kasaï by Risopoulos (1966) over a similarly wide range of indigenous legumes.

In a subsequent study at Tanga, Hendy (1975) observed that bush and weed encroachment and loss of valuable grass species occurred even at low stocking rates of 0.3 steer/ha. Resting and slashing increased the grass component but, when grazing was resumed, weeds reappeared. Yields of fertilized natural grasses were lower than those of unfertilized cultivated grasses. Hendy concluded that coastal, natural grassland was inherently incapable of much improvement and that the value of legumes was questionable. In Hendy's view, future pasture research should concentrate on cultivated grasses and legumes; the grasses suggested were Rhodesgrass and Gambagrass (*Agay*).

Boonman & Goldson (1980) carried out a study in the Tanga region. The soils are red loamy sands with generally better development opportunities than the coral sands in the Mombasa range. Bush encroachment, *e.g.* by doumpalm on abandoned sisal land, is a major problem. Bush had developed into impenetrable thickets which rendered the land next to useless. Attempts had been made to establish Guineagrass pastures vegetatively, but the costs far outweighed the benefits. It seemed that Guineagrass pastures in their natural state be better reserved for making hay, silage, or for stall-feeding. While this is the best method to maintain Guineagrass in the sward, the same applies if the indigenous legumes of the "coast cocktail" are to be encouraged.

Stargrass has been planted successfully straight into natural coastal grassland near Kilifi (*4.2.2.2*). There is no agreement as to whether sown pastures at the coast are worthwhile as seedling establishment is difficult in coastal sandy soils subject to high surface temperatures. Rhodesgrass has been the obvious choice because seed is available and results on ranches in the Tanga region have been quite positive (*6.6*; Hopkinson, 1970), although it tends to have a more upright growth habit than at higher altitudes. Guineagrass is suitable also and it should be possible to select better types within the region that are more tolerant to grazing (*6.8*).

4.5. Fallow (post-cultivation) grass communities

Cultivation, *i.e.* ploughing, has a more drastic effect on the subsequent composition of natural grassland than tree-felling, burning or grazing. Arable land reverted to grass has often a temporarily more productive and more nutritious sward than the natural grassland not disturbed by cultivation.

As we have seen (*3.3.3*; *3.4*; *4.2*; *4.3*), it is useful to distinguish primary and secondary grasses on the basis of climate, ecological behaviour and seeding properties. A special case is presented by African Couchgrass. This grass has been referred to already on numerous occasions but since it is perhaps the most ubiquitous grass of East Africa and such an important weed in mixed-farming, it is dealt with under a separate heading.

4.5.1. African Couchgrass (Digitaria abyssinica)

African Couchgrass may be found from sealevel up to 3000 m and is known in Uganda as "Lumbugu" and in Kenya as "Thangari". As the name indicates it was first described from Ethiopia (Mount Scoloda). Together with Paspalum (*Pscr*) it forms the main constituent of fallow land, also in Kivu and in Ituri (Zaïre) where it was known as *D. vestita* (Compère, 1961b; Léonard, 1962). The former name used in East Africa itself was *D. scalarum.* It is most common in both Zone II and III, as a constituent of natural grassland at the higher altitudes (> 1500 m) but especially as a weed in arable and plantation agriculture or even as a ruderal plant (Clayton & Renvoize, 1982). It is perhaps the best known grass in the mixed-farming areas between 1000 and 2250 m. It appears wherever soil is disturbed or where vegetation is burnt. The more the soil is disturbed, short of eliminating the grass itself, the more vigorous it grows. Many acres of coffee have been abandoned when badly infested with this weed (Thomas, 1940b). Russell (1976) recalls that African Couchgrass, growing in tea and coffee plantations, affects the bushes so seriously that production falls to an uneconomic level due to acute nutrient deficiency, which cannot be rectified by normal dressings of fertilizer. Cereal crops, when badly invaded, produce no ears. Root exudates have been implicated to explain these effects. A special fork with 3–4 prongs (Lumbugu fork) was designed for the purpose of pulling up the rhizomes.

Although one of the most pernicious arable weeds, it is an effective soil protector due to its rhizomatous habits which make it also resistant to fire; it is a grass which will be the first to emerge when litter or crop debris are burnt in the field. If left unchecked, the grass is hard to control by one simple mechanical operation. Fortunately, African Couchgrass is nutritious (Dougall & Bogdan, 1960), well liked by livestock and loses vigour under grazing. It then remains close to the ground. Grazing can weaken it to a point that a timely and good ploughing in the dry season can finish it off. African Couchgrass is controlled effectively with Dalapon (Veenstra & Boonman, 1974) and now that prices of Glyphosate have come down, the latter herbicide should be equally effective, if not more, since it also copes with the many dicotyle weeds commonly found in Couchgrass communities. Dalapon should be applied when the grass is growing actively; not more than 5 kg/ha should be applied since higher rates "burn" the foliage before the herbicide is transported to the roots.

African Couchgrass shares with Kikuyu-, Star- and Elephantgrass the property that it is a valuable grazing grass and an arable weed at the same time. These weed grasses often force cropping to be suspended so that, in the fallow period, land gets the chance to restore its productivity and provide grazing. However, in this system crops are rotated with their worst enemies, the weed grasses. This problem can be overcome by replacing the fallow by sown grasses ensuring a rapid cover and good grazing without the nuisance of them returning as weeds in subsequent crops (*5.1.2*; *5.6.1*).

4.5.2. Ecology and zonation of cultivated and weed grasses

In this and the preceding chapter constant reference has been made to the effect of soil disturbance on the behaviour of the important grasses of East Africa. In the following, which can be seen as summary of what has been presented in this respect, the characteristics are brought together to explain the behaviour of these grasses in cultivation. In fact, the main difference between the grasses of Zone II and Zone III is found in their behaviour in radically disturbed habitats, under cultivation. A distinction was, therefore, drawn between primary and secondary grasses (*3.3.3*).

Hyparrhenia spp., Themeda and Rhodesgrass were described as primary grasses, common in Zone III more than in Zone II. They had always been significant components of the savannah but when the tree cover decreases, these grasses expand. Under grazing their ecological stability is relatively high. These grasses are not confined to specific altitude belts and, although typical of some communities, they are rarely found in complete dominance but more commonly in association with other characteristic grasses. They rarely colonize disturbed soil, are easily destroyed by ploughing and are no weeds of cultivated land although they seed well. In the plant succession their reappearance may take many years in spite of their seed potential. Their place in fallow land is taken over by weed grasses such as Povertygrass (*Rrep*), Paspalum (*Pscr*) and by African Couchgrass (*Dsca*) which, in fact, constitute the most common fallow constituents of Zone III.

Conversely, grasses such as Kikuyu-, Star-, and Elephantgrass are secondary because of their low frequency in the original forest habitat and because of their spontaneous appearance after destruction of the site by fire, soil disturbance and cultivation in particular, more so in Zone II than III. They culminate in dominance, each in their own altitude belt ("zone"-grasses). They are arable weeds, although their seeding properties are less developed than in the grasses of Zone III. Their stability under grazing is not high, except when soil fertility is high, and they are liable to incursion by herbs or by other grasses.

A number of transitional grass species has also been defined. These can be both primary and secondary, and can occur in Zone II or III. Grasses such as *Panicum maximum*, *Brachiaria decumbens* and *Setaria sphacelata* are common, frequently abundant or even co-dominant constituents of Zone II of natural grassland and as colonizers of abandoned cultivated land. These species are rarely found in direct association with the zone-grasses and often occur at lower altitudes, *e.g.* Uganda and eastern Zaïre, and also in Zone III. It appears that the more robust or giant types of Guineagrass are found in Zone II and the smaller types in Zone III. Imperata and Citronellagrass may also be ranked among transitional types, especially at low altitudes. They differ ecologically in that they are grassland- and arable weeds but are largely ignored by livestock. The transitional Zone II grasses are colonizers, while it is tempting to suggest that they tend to be more dominant in their respective areas of distribution, though less so than the true zone-grasses. Examples from Zaïre were given of Setaria (*Ssph*) in Kivu, of Guineagrass to the west and of Congo Signalgrass (*Bruz*) to the south (*4.2.4*).

Thus, 3 major characteristics are apparent:

1. Climatic zone of origin.
2. Colonizing power, weed potential and tendency to dominate.
3. Seeding properties.

It is on the basis of these distinctions that Table 4.4 has been compiled. Zone II has no primary

Table 4.4. A dual classification of cultivated- and weed grasses in East Africa according to zonation and ecology.

Ecology	Zone II	Zone III
Primary	*Psph*	*Hyparrhenia spp.*[a]
	Ejae	*Hdis*
		Cgay
		Mmin
Secondary[b]		
Zone grasses	*Pcla, Cynodon, Ppur*	
Weed grasses	*Cnar, Icyl, Daby*[c]	*Pscr, Rrep, Daby*[c]
Transitional	*Bdec, Bruz, Pmax, Ssph*	

[a]Including *H. filipendula, H. rufa.*
[b]All the secondary grasses are potential arable weeds.
[c]*Daby* is ubiquitous.

grasses that are economically important as cultivated, but only as weed species if we exclude those of an edaphic nature.

The secondary grasses have a greater self-regeneration potential after ploughing than primary grasses. This is not necessarily linked with growth habit (*3.3.3*). Rhodesgrass for instance is both a good seeder and stoloniferous, but rarely reappears after ploughing. The merit of seeding or of having stolons does, therefore, not explain why grasses return after ploughing as volunteering plants. Consequently, the ability to regenerate vegetatively must come from dormant rootstock buds; rootstocks can stay dormant in the soil for a long time. In temperate countries an escape from decay is found in Red Fescue (*Festuca rubra*); non-decayed pieces of rootstock are turned up in large numbers by ploughing, 2 years later, after the following arable crop (Plate 4.1) (personal observation). Under the humid and hot conditions of Yangambi (Zaïre) intact structures of grasses were found 10 months after ploughing (Meyer, 1961). Resistance to decay does, of course, not necessarily imply that the material has remained alive. However, it goes without saying that this phenomenon has a major effect on maintaining soil-OM (*5.3.4*; *6.3.5*).

No hard data are available on the timescale nor on the sequence in succession needed for the primary grasses of Zone III to (re)emerge in fallow communities. General observation would suggest that this is no less than 5 years after cropping and independent from the number of previous cropping cycles after the natural sward was ploughed. *Hyparrhenia filipendula* is one of the more common species of Hyparrhenia to reappear in abandoned cultivated land after some 4–6 years

Plate 4.1. Non-decayed pieces of grass sod (*Festuca rubra*), ploughed up 2 years after sward destruction.

(Edwards & Bogdan, 1951), but it is not clear whether this timescale is reduced if the grass is abundant in the area itself or if the fallow is grazed hard. As for the plant succession, Risopoulos (1966) reports that when fallow occupied by Imperata in Kasaï (800 m) is not brought under cultivation, various species of Hyparrhenia appear: *H. rufa*, *H. filipendula*, *H. dissoluta*, followed shortly afterwards by *H. familiaris* and *H. diplandra*. No timescale was given.

4.6. Summary

Natural grassland is rarely spectacular in its response to husbandry. Primary grassland is usually productive enough not to need a replacement by cultivated grassland, unless after an intermittent cropping cycle. On the other hand, secondary grassland (Kikuyu-, Star-, Elephant- and African Couchgrass) is often not very stable under grazing, poses a weed problem for the next arable cropping cycle and is often left to itself in farming practice since it is hard to eradicate when in active growth. Secondary grasses should be valued as protection against land abuse, more so than for grazing. Kikuyugrass can deteriorate rapidly ("Kikuyu decline"). Destumping and herbicide treatment of invading coarse grasses are helpful but it is often a better solution to destroy the Kikuyugrass itself, crop the land to clean it of remnants and put it back to more productive sown pasture. Destruction of the sward with Dalapon or Glyphosate is the best practical solution. Secondary Stargrass pasture is generally not high yielding either but may be quite satisfactory on specialized dairy farms, with an ample supply of manure or fertilizer. Elephantgrass, on the other hand, is highly unstable when grazed hard. The characteristic primary grasslands (Themeda, Hyparrhenia, Hyperthelia, Rhodes) are generally more stable, even though their botanical composition is more diverse. Fertilizer does not seem to shift sward composition towards more valuable constituents. Stocking policies have also little long-term impact on botanical composition, except in the negative sense when extreme. The priority in natural grassland husbandry is, therefore, to prevent deterioriation. Bush encroachment is a major problem, which can be controlled by burning but, where possible, regular mowing is more effective, less wasteful and an aid towards both grass regrowth and fodder conservation. Cutting and stacking Acacia thorn bush in places denuded by overgrazing is effective protection to restore grass cover. Reseeding with grass seed is often not necessary, if at all possible or effective. Legume oversowing can give good establishment but legume persistence under grazing is low, even when the accompanying grasses are not aggressive. The theory is put forward that the ecological differences between primary and secondary grasses culminate in their different performances in radically disturbed habitats, *i.e.* under cultivation.

CHAPTER 5

From shifting cultivation to crop-grass rotations

Soil fertility and productivity are closely associated with the efficiency which grasses display in the grass-soil-crop complex (*2.1.2*). Livestock's grazing role herein is paramount.

In East African languages or rituals reference is constantly made to livestock. So dominant had its role been in the day-to-day lives of the peoples in the predominantly savannah environment. Livestock and cropping had been intertwined in varying degrees since time immemorial, but the dependence on crops became stronger. The pressure on land rose not only due to population growth, but also because of alienation of land and introduction of cash crops. Although less and less land remained available for grazing, livestock and cropping continued to coexist in different ways, many of which became characteristic of alternate husbandry. It was flavoured perhaps with social more than with economic elements; farmers want cattle but they need crops. Down to the smallest unit and when put to the choice, farmers rather part with cash crops than with their livestock.

The special role of grasses in the crop rotation on farms with little or no input of artificial fertilizer needs to be examined with a view to improving soil fertility. This is compared with that of (green) manures (*5.2*; *5.3*) and of pasture legumes (*10*).

5.1. Steps in crop intensification

Systems of shifting cultivation have been described by de Schlippe (1956) and Jurion & Henry (1967). For a description of fallow- and "unregulated ley" systems, in which the fallow vegetation is systematically grazed in areas not infested by tse-tsefly, see Ruthenberg (1976). Clement (1961) defined ley farming in Europe (*5.4*) as essentially grass farming with the judicious use of the plough, while alternate husbandry is arable farming with a grassland break lasting several years. Examples of the latter are the bush- and weed fallows (*5.1.1*), later replaced by sown pastures. Thus, the concept of grass under alternate husbandry is distinct from ley farming, a term which is therefore not used in this book (*5.6*). Another general term is mixed-farming, employed also here to indicate the integration of cropping and livestock, with the aid of either grass fallows or cultivated pastures. Crop residues (bye-products) play a major role in livestock feeding (*11.1*; *11.3.4*).

In the following, some of the more decisive, chronological steps in the sequence of increasing cropping intensity in conventional alternate husbandry are discussed, with particular reference to soil fertility. It is assumed that livestock are around and that crops consist of annual food crops such as maize, as an example of a crop which responds clearly to changes in soil fertility, without interference from major pests or diseases. It must be realized, however, that even where livestock and arable farming are in close proximity, their integration is frequently quite limited (Parsons, 1958; McCown *et al.*, 1979).

5.1.1. Conventional bush fallow (shifting cultivation)

With plenty of land around people grew crops on the same land only as long as they found the rewards of their efforts worthwhile. The work load of clearing new bush, by "slash-and-burn", was weighed against the decline in yield and against the increasing weed burden in the current plot. Much of the sowing was done straight into the ash. Since the natural vegetation was only partly destroyed, bush and grasses were quick to return and to replace weeds when cultivation was shifted again and when the plot was abandoned to go derelict.

Soil carrying natural vegetation, such as forest or grassland, has a higher OM-content than the same soil when cultivated. It has a higher rate of addition of OM to the soil while OM is not broken down by cultivation. The content in fallow land left to go derelict will increase for the same reason. Russell (1973) has pointed out that grassland soils may contain 50% more OM than forest, and it is also distributed through a greater depth of the profile. Grassland produces more carbon (C) in the subsoil because of the decomposition of deep grass roots (Sanchez, 1976). The amounts stored in the 0–30 cm layer of a number of soils under forest and savannah can vary as follows (kg/ha):

Total N	1800–8000
Soluble P (Truog method)	4–36
Exchangeable K	190–710
Exchangeable Ca	100–28000
Exchangeable Mg	19–5700

Data, also due to Nye & Greenland (1960), are used in Table 5.1 to illustrate the large amounts of nutrients added annually by forest and planted grassland in humid, equatorial Africa.

Table 5.1 Nutrients involved in shifting cultivation in the humid, equatorial zone in kg/ha/year (after Nye & Greenland, 1960).

	Plant content (kg/ha)				
	N	P	K	Ca	Mg
High forest (timber + litter)	248	14	294	317	71
Grassland (herbage)	270	22	300	45	85
Removal by maize (1 ton/ha)	15	3	3	0.2	1
Removal by banana (10 ton/ha)	22	4	55	0.6	–

The data of uptake by grassland were comparable with forest for all nutrients except Ca (see also Table 12.5). The fate of the nutrients depends on how the clearing is done. When burnt, almost all N is lost. P and nutrient cations are contained in the ash (*4.3.3*) but, when the debris are heaped before burning, distribution is irregular and ash will be blown by wind or washed away by rain.

Ahn (1974) pointed out that savannah soils are generally less weathered than forest soils, less leached and of higher pH. They are usually lower in OM than forest soils due to a sparser vegetation and the prevalence of annual burning. Soils at high altitudes differ fundamentally from those at low altitudes in that lower temperatures reduce the rate of humus mineralization so that, from the point of view of shifting cultivation, the overriding fact is perhaps that OM-levels decline much more slowly. Nye & Greenland (*op. cit.*) reported that crop yield declines are more marked in forest than in savannah areas.

Under shifting cultivation the nutrients involved in growing crops are only a tenth of the amounts in more modern agriculture. Yields of maize rarely exceed 1 ton/ha (Table 5.1). Periods of cropping and fallowing are irregular and depend on the factors mentioned above. It should be noted that in most systems of shifting cultivation density of livestock and its impact on soil fertility is relatively small.

5.1.2. From bush- to weed fallow

No external inputs were required in the systems of shifting cultivation but it was prodigal with nutrients, land and labour. Mounting land pressure, already quoted, led to more sedentary agriculture and stricter confinement of livestock. The pressure came to bear on the fallow duration itself, with consequent reduction in the length of rest under natural regeneration.

The nutrient balance changes. The necessity of fire, as a tool of clearing, and its concomitant losses of OM and N are greatly decreased. Livestock is more confined in its movement, so that nutrient recycling by the grazing animal becomes greater. Conversely, the diminished duration of fertility-generating vegetation leads to less OM and nutrients being added to the soil, while more of it is lost due to a higher frequency and intensity of cultivation. In brief, less fertility may now be added, but also less of it is lost, except when cultivation is intense.

However, changes in cultivation systems have consequences which reach much further than soil fertility alone. Concomitant with the decline in the physical and chemical status of the soil, which are most often quoted, is also the increase in pressure of arable weeds, of perennial grass weeds in particular (Table 4.4), in the cropping phase. When the weed grasses become too strong the land is by necessity abandoned to fallow. African Couchgrass (*Daby*) is the most widespread in East Africa. In fact, it is the main grass found on small farms in various zones (*4.5.1*).

Indirect weed effects are also at play. Grass weeds slow down the development and ripening of the crop so that harvesting is delayed. When by that time the dry season is already far advanced, the soil has become too hard. Many farmers are then forced to delay land preparation until the onset of the rains, with the result that the principal requirement of maize husbandry, to plough and to sow as early as possible, is not met (Allan, 1972). Consequently, maize sowing is delayed further every year. In the end African Couchgrass becomes so vigorous that land must be abandoned to fallow earlier than envisaged. Incidentally, this compulsory return to grass fallow is, for the soil at least, a blessing in disguise; it may be seen as nature's revenge against the abuse of overcropping.

The direct role livestock can play at this juncture in the combat against weed grasses is highly relevant and reaches again much further than with which it has been customarily credited. African Couchgrass, a secondary grass (*3.3.3*), rapidly loses vigour under grazing which facilitates its eradication (*4.5.1*). Thus, if the farmer is prepared to wait the necessary number of years to wear the African Couchgrass down, he will find the old plot in good condition for another round of cropping. Not only has the Couchgrass been weakened, also the productive capacity of the soil has been restored physically and chemically. By that time the land has again rested long enough.

A more positive example of a grass that colonizes forest fallow and abandoned cultivated land is Elephantgrass in the "fertile crescent" of Uganda (*3.4.2.3*). Although hard to eradicate by ploughing, it is killed by repeated grazing and is not as bad a crop weed as African Couchgrass. As Tiley (1965d) recalls, the farmer in these areas had long recognized that Elephantgrass was able to restore fertility, perhaps more obviously than grasses in other areas. The continued high population of much of Buganda was attributed to the fertility-maintaining properties of this grass (Thomas, 1940b). This observation of Thomas was in recognition of the fact that shifting cultivation and weed fallows are, however wasteful in terms of land and labour, ecologically sound systems. As will be seen below (*5.5*) crop yields can be more than doubled when preceded by a productive grass rest.

5.1.3. Continuous cropping

The grass weed fallow system is sound, provided it is permitted the time it needs to complete its cycle. It is low-technology and self-contained. The pivot of the cropping system is the grass fallow with which the soil productive capacity is restored.

However, as the fallow duration comes under mounting pressure, the farmer no longer affords the time to wait for nature to take its course. In spite of declining crop yields, farmers revert to the same plot earlier and earlier. Very often the weed fallow is cut back to only 1–2 years. Since African Couchgrass is then at its zenith, ploughing rejuvenates rather than kills it. Much more effort is needed to keep it down to a low level.

To control African Couchgrass, continuous cropping may, in fact, be more effective than alternating crops with weed fallows of such short duration. In densely populated areas on more fertile soils, *e.g.* central Kenya (Nitisols), the weed fallow has all but disappeared. Stock is rarely disposed of, but is mostly herded outside the farm on communal grazing lands (vlei, hillsides) or along roadsides. Some farms have shifted to stall-feeding, with grass or other roughage collected from elsewhere. This, however, usually implies a complete cut of the nutrient cycle "grass-livestock-soil" and loss of nutrients (*5.2.2*; *13.3*).

5.2. The control of soil fertility

Forest and grassland in the tropics often compare favourably with their temperate counterparts in soil-OM contents. Birch & Friend (1956) sampled 570 topsoils (0–30 cm) in East Africa and found an average content of 3.36% OM. About half of them contained more than 4% OM in the top 15 cm.

Cultivation is usually accompanied by a decline in OM and by subsequent mineralization. Under continuous cropping without inputs, little OM is added or returned whereas losses are so much greater. Soil-OM and methods of promoting it are commonly associated with high soil fertility, because a soil rich in OM is often productive and can sustain arable cropping for long periods. Such a soil will also trap rain-water rather than let it cause erosion.

In capitalized farming systems in temperate climates with high input of fertilizers and mechanical or chemical weed control, OM is commonly regarded as less critical. OM may be left to find its own level and may be maintained by crop residues. High, albeit not the highest, yields have been obtained under continuous cropping without special attention to add OM. In no-fertilizer agriculture, by contrast, OM supplies most of the N and S and half of the P of the crop. OM also provides most of the cation exchange capacity of acid or highly weathered soils (Sanchez, 1976). Soil-physical effects can be considerable also, depending on the kind of OM (*5.3.3.1*).

In most soils in the tropics the evils of continuous cropping with annual crops without the necessary remedial measures being taken are widely recognized. Under improper land use, abundant sunshine and rainfall become a liability rather than an asset, and aggravate soil erosion, leaching and mineral depletion.

Soil erosion began to draw particular attention in Uganda in the 1920s and 1930s and various methods were investigated (*2.1*). Most of these centered around the issues of halting erosion and increasing the OM-content of the soil. Since the issue of OM is being debated right up to the present day in many parts of the world, it is useful to recall the findings of that time and put them in proper perspective with the aid of insights from both tropical and temperate agriculture (Cooke, 1967; Russell, 1973). These 2 particular accounts are referred to here due to their critical attitudes towards the pros and contras often claimed when issues concerning OM are at stake.

5.2.1. Green manures (leguminous)

Legumes drew interest originally as green manure and cover crops in plantation agriculture (*2.1.1*; Robyns, 1928). Trials were conducted in the 1920s in Uganda at Bukalasa, Serere and other farms to test which leguminous green manures could maintain soil fertility also in arable rotations. It was found at Serere that crop yields after green manuring were lower than on neighbouring farmers' plots under shifting cultivation (Tothill, UDA Ann Rep 1932; Martin & Biggs, 1937). Also in the forest zone little beneficial effect was seen (Nye, 1937; 1940). The early legumes tested included sunhemp (*Crotalaria juncea*), velvet beans (*Mucuna pruriens*), Centro (*Centrosema pubescens*) and Leucaena (*L. leucocephala*). Also at high altitudes in Kenya with temperate green-manure crops results were indifferent (Ball, 1937). That green manure left no improvement of the soil was later confirmed by Martin (1944a, b), Pereira *et al.* (1954) and Stephens (1970); see also *5.3.3.* Nye & Greenland (1960) basing their results on work from various parts of East and West Africa concluded that the worth of green manures on long-term agronomic and economic grounds had yet to be proved.

Green-manure crops are normally ploughed in within the same season. Their duration may be so short that soil-OM and N-contents are hardly increased although N-losses by leaching are reduced and nutrients are concentrated in the surface soil. Russell (1973) pointed out that green manuring may confer no benefit at all. Humus content is increased if material resistant to decomposition is added to the soil, but resistant plant material is typically low in N. The opposite also applies, because N-rich plant material decomposes more rapidly. Legumes have even been suspected of hastening soil humus decomposition (Russell, *op. cit.*; Meyer, 1961). At Kawanda, Simpson (1961) observed an increase of soil acidity under legumes because of leaching, whereas N-accumulation under grass was accompanied by an increase in pH due to the upward transfer of subsoil nitrate (*5.3.2*). The roots of legumes decompose so quickly that N is rapidly mineralized and leached out to levels beyond the reach of crop roots which in the long term leads to a lowering of the pH. This early finding by Simpson of a relationship between soil acidification and legumes has recently been confirmed by Hochman & Helyar (1989); see also *10.2.1*.

Russell (*op. cit.*) recalls that in the tropics many legumes, grown for the ostensible purpose of raising the N-status of the soil, are not nodulated for most of the growing season; and when a leguminous crop fixes N, it need not enrich the soil in N. Cereal legumes (*e.g.* beans and groundnuts) even tend to reduce soil-N, since most of the N is removed with the harvested grain. The inability of legumes to contribute sufficiently to the subsequent maize crop in Kenya has recently been confirmed by McCown *et al.* (1992).

Even if green-manure crops are effective, the question remains how to fit them in the rotation. In humid areas, green-manure crops may have a role in the late season to reduce losses due to leaching. Where conditions are dry, however, soil moisture is bound to be depleted. Conversely, the N-conserving benefits of green manures are irrelevant when rainfall is so low that leaching does not occur.

In a review based on work done in the humid lowland at Yangambi (Zaïre), Meyer (1961) questioned some of the claims made in favour of leguminous green-manure crops, other than their effect on the N-balance. Physical effects were negligible since root development and build-up of OM were small; only under the best conditions were they capable of restricting losses. By contrast, perennial grasses had a considerable effect on both OM and on soil-N; their effects were physical, chemical and biological and they increased crop yields (Meyer, *op. cit.*).

The soil structure and other effects of leguminous manure crops are discussed in more detail below (*5.3.3*; *5.5*). Tiley (1970a) recalled that the poor performance of leguminous green manures had caused damage to the image of fodder- and pasture legumes also. Experimental work on the latter was not resumed until the mid 1950s, although it has remained largely unresolved (*10*).

5.2.2. *Kraal manure*

Since green-manure crops are difficult to fit into rotations, the finding of the early researchers in Uganda of their inefficiency caused not much of a stir. However, the conclusion that animal manure was not efficient either has been more controversial. This referred then mainly to the apparent inability of organic manure to improve soil structure (Agric. Res. Conf. Nairobi, 1947).

Apart from compost, animal manure is the only source of manure in the absence of artificial fertilizer. Yet, it has been little used even where plentiful. A typical statement often heard is that farmyard manure has been unduly neglected in the tropics. Farmyard manure, however, is an alien connotation reminiscent of temperate climates where livestock is stabled in winter, with manure disposed of by necessity. In the tropics, with livestock out in the open, manure is less easy to collect in quantity, even in kraals. It is proposed not to refer here to farmyard manure but to kraal manure or to animal manure.

Evidence that manure did attract considerable attention is found in the 1935–1955 issues of the *East African Agricultural Journal.* To increase it in quantity and quality new methods of composting and bedding were designed (Beckley, 1940). However, the conclusion that animal manure was not efficient in raising crop yields or soil fertility was reached quite early (Martin, 1944a; Pereira & Jones, 1954) and this has since been confirmed by Tiley (1965c) in Uganda, and in Kenya by Wapakala (1976) following long-term manurial trials at Embu.

5.2.2.1. *Composition and effects of manure*

Evidence is growing that most of the benefits that are reported from dressings with animal manure are due to the nutrients it supplies. Instances have been few where effects on physical or other soil properties caused by manure have benefited crops and we have still few modern results where it is doubtful whether the effects were solely nutritional. Wapakala (*op. cit.*) observed no increase in the % soil-C at Embu, after more than 20 years of annual dressings of 10 ton/ha. Both he and Stephens (1969) recorded an increase in the amount of exchangeable K in the soil as the principal effect of animal manure. Where the main feed of animals whose manure is collected is Elephantgrass, such manure can be very rich in K (Table 12.5). The observation, quoted earlier, that manures were not efficient referred mainly to their inability to improve soil-physical conditions. However, lack of efficiency also applies to some nutrients, especially N. Although already poor in N, only as little as one-quarter of it may be as effective as fertilizer-N (Cooke, 1967; Russell, 1973).

Stephens' (*op. cit.*) conclusion about the K-effect of animal manure (5 ton/ha of 47% DM, containing 24 N; 8.5 P; 41 K) was based on 18 long-term trials in Uganda. Adequate K is required for responsive crops such as sweet potatoes on acid (pH $<$ 5.5) ferallitic soil (see also *18.4*). Subsequently, Stephens (1970) found in the same 18 trials a decline in yields of 13–20% and an increase in response to K of 5% after 2 years of cropping. K-effects were small except on grass when cut and removed (*12.7.4*), and on continuously cropped soil, especially with sweet potatoes. In long-term manurial trials carried out at Embu from 1952–1966 significant yield increases were observed in 1st year potatoes and maize and in 3rd and 4th year Rhodesgrass but only at the highest levels of initial application, 27 ton/ha of animal manure (Gathecha, 1970). These trials formed part of those reported on by Wapakala (*op. cit.*).

The nutritional effects are obviously disappointing when the nutrients contained in the manure are not deficient in the soil for the particular crop to be grown. Unlike artificial fertilizer animal manure is not a tailor-cut product. Root crops may benefit by the K and cereal crops by the P contained in the manure, but very little response is seen in crops which respond less to P and K but the more so to N, *e.g.* grasses (*7.2.4*; *12.7*; *13.3*). Moreover, large quantities of manure rich in K may cause Mg-deficiency, which can have serious effects in coffee (Mehlich, 1965). Banana, on the other hand, has a high K-requirement (Table 5.1) which explains why this has been one of the few crops that did traditionally receive organic manure on soils poor in K (Evans & Mitchell, 1962). Positive manure effects are generally limited to very poor soils, *e.g.* some areas near Lake Victoria and at the coast (Peat & Prentice, 1949; Grimes & Clarke, 1962; Scaife, 1971). As observed in Machakos, response to animal manure is higher in the long than in the short rains (Ikombo, 1984) which serves to indicate that manure effects may be erratic in semiarid areas.

In general, however, little animal manure has been used and this situation is continuing up to the present day also where no artificial fertilizer is available. This lack of interest must be attributed mainly to the low content of desirable nutrients, especially of N, and to the concomitant low efficiency of manure as N-source. Richness in nutrients, *e.g.* K, of which soils in the region are only rarely deficient, or allegedly residual or slow-release effects are not very persuasive to the farmer. Animal manure is also feared as a source of weed seeds.

Dairy cows may excrete over half the N and three-quarters of the K of the grass they eat. Most is in the urine. On low-protein diets, however, only 15% or less of the N is excreted in the urine, the metabolism of the animals adjusting to make maximum use of what is eaten. Clearly under such conditions recycling of nutrients via excreta can be ignored (Topps & Elliot, 1965). Values of 0.13% N occurred in animal excreta in the arid Sudan Gezira; only 1 kg/ha N was deposited in faeces, which released N very slowly (Jewitt & Barlow, 1949). Slurry of 6.3% DM sampled at Kilifi in Kenya (van der Noll & Janssen, 1983) contained 1.17 kg N per m^3, which is only 0.114% N, as against 0.042% P and 0.38% K (*12.7.1*).

A comparison of manure (faeces) in East Africa and Europe is given in Table 5.2. Oenema (1980) sampled manure from kraals in a series of villages in Kisii (western Kenya). The data of Table 5.2 due to Stephens (*op. cit.*) are particularly relevant as they referred to no less than 95 samples from 18 sites, collected over a number of years. Since manure collected from kraals consists mainly of faeces, the comparative analysis for Europe (Cooke, 1967) also refers to faeces.

Table 5.2. Faeces composition (% of DM) in East Africa and Europe.

	DM	N	P	K	Ca	Mg
Kenya (Oenema, 1980)	89	1.07	0.32	1.80	0.90	0.23
Uganda (Stephens, 1969)	47	1.03	0.36	1.70	0.55	0.27
Europe (Cooke, 1967)	15	2.75	0.75	1.50	1.10	0.80

The differences in % DM should be noted. Kraal manure often refers to air-dry material. The Uganda data agree with those of Kenya. Compared with Europe, however, kraal manure is poor in N, P and Mg, but rich in K. Although the European data seem more positive, the effective contribution of 10 ton/ha of 20% DM is on average not higher than that of fertilizer containing 18 N, 10 P and 35 K (Russell, 1973). Most of the P and K contained in manure is as effective as that in fertilizer-P and K. However, as already said, the N-efficiency is only about one-quarter. Rather than as a fertilizer, stable manure is regarded in most parts of Europe as a bulk of waste to be disposed of by necessity, as already mentioned above.

Poor excreta reflect poor feeding but also poor "storage". Feed source sets the maximum nutrient level in animal manure that is attainable, but N-losses are large due to volatilization and leaching unless special precautions are taken (*12.7.1*). In an attempt at rationalizing the poor responses of coffee to kraal manure, Pereira & Jones (1954) found that such manure may contain 40 to 75% earth and that of the dry matter as little as 15% may be oxydizable OM; mulching was found to be more effective than animal manure (*11.3.3*). Some manures may contain only a little more N than the soils they are supposed to improve (Beckley, 1940). That high-quality manure can be produced is borne out in the results of Mehlich (1965) who reported a wide range of nutrient contents of manure samples in Nairobi. Where manure is considered highly, for instance in market gardening, different sets of economic rules apply, but it is unlikely that farmers will feed and house their cattle better only to make them produce better manure.

5.2.2.2. Manure logistics

Large quantities of manure are required if a response to N is to be elicited. Manure trials have shown that the fertilizer effect of 10 ton of manure is not higher than of 10 kg fertilizer-N (Stephens, 1969), not to mention low-content slurry (*12.7.1*). Such weight ratios are prohibitive. On the other hand, where crop yields are low, nutrient removal is small also and may not exceed 15 kg N/ha (Table 5.1). Applications of 8–10 ton/ha/year manure were required to maintain yields at subsistence level under continuous cropping (Stephens, 1970). Huge dressings of 20–25 ton/ha/year are no exception in those instances where positive effects of animal manure have been reported.

Such amounts are prohibitive since they are rarely available on the farm. Manure is a form of recycled fertility but the cycle gets progressively smaller because of many losses in the process. Adequate dressings cannot be sustained, unless feed sources are added from outside. The manure used on Mukumu Farm (*4.2.2.2*) was derived from a side activity of intensive pig rearing. Bye-products of arable crops but mainly outside (purchased) feed sources account for most of the manure handled in Europe.

Very few data are available in eastern Africa on nutrient cycles with feeds being brought in from outside. In Ituri (Zaïre), van Parijs (1959) observed that 3 ha of grass could feed livestock to produce enough manure for 1 ha of crops. However, it is doubtful if there are many areas left where farmers have access to so much land outside their own unit.

From the above it would appear that the alleged neglect for animal manure has not been without good reason. In low-input farming, where artificial fertilizers may not even be on the market, animal manure is useful but is not available in quantity and quality. Where feeding is poor, manure is equally poor, not to mention the difficult task of collecting faeces from often puddled and muddy ground. If it is agreed, however, that manure can be an asset, alternatives should be explored to make the best use not only of faeces but also of urine which in permanent kraals is now completely lost. By bedding down kraals with mature grass or straw, these will be crushed in the process, impregnated with urine and coated with dung.

5.2.2.3. Night paddocks

Manure losses are minimized by utilizing faeces and urine there where they are dropped, *i.e.* by day- and night grazing and by cropping such land in rotation. Under free-grazing conditions in East Africa, dung pats disintegrate rapidly due to the activity of numerous coprophagous insects, especially in the rainy season (Hanski & Cambefort, 1991). Little is seen of rejection of herbage through faecal fouling, so commonly observed in temperate climates. Grazing behaviour studies have shown that night grazing is of utmost importance for livestock production (Joblin, 1960; Goldson, 1963; Mugerwa *et al.*, 1973; Musangi, 1965).

Night paddocks are visible proof of the fertility-increasing effect of excreta. The data of Table 5.3 are due to Goldson (1976a) who estimated that the amount of faeces accumulated in night paddocks on farms near Kitale was equivalent to 170 ton/ha/year. In his survey, the size of night paddocks holding livestock for 12.5 hours/day averaged 300 m^2/animal.

Table 5.3. Soil analysis data of night-paddocks near Kitale (on average at least 5 years old; of topsoil under grass); after Goldson (1976a).

	pH	N %	C %	P ppm	K me%	Ca me%	Mg me%
Night-paddock	6.0	0.21	2.15	51	1.93	6.5	2.9
Control	5.5	0.15	1.85	12	0.63	4.4	1.8

Although the improvement is remarkable, such a concentration of fertility appears wasteful. If night paddocks are inevitable, they are better rotated around the farm at intervals of 3 months. Shifting night enclosures were recommended for sown pastures at Kongwa (T) (Wigg *et al.*, 1973). At Serere, Joblin (UDA Ann Rep 1960) reported significantly higher crop yields on land which had been night-paddocked for 9 months at a stocking rate of only 5 beasts/ha compared with yields from ungrazed plots. Stobbs (1969e) presented additional data to show that intensive grazing produced higher crop yields over 3 years of test crops; night paddocks, one-third the size of the day paddocks, gave even higher crop yields (*5.5*; Table 5.10).

A similar effect to that of night paddocks is observed when pasture due to be ploughed is set aside to function as holding ground in the dry season for supplementary feeding (*7.2.6*) or when tethering or close-folding is practised.

5.3. The role of grass in the crop rotation

Already in 1932 a decision was made to incorporate grass into the crop rotations of all Uganda government farms in an attempt to restore the humus level, which was then regarded as the most important factor in soil fertility (Tothill, UDA Ann Rep 1932; *2.1.2*). The planted grass rest was seen as an accelerated form of the natural regeneration developed under shifting cultivation. The need for a more rapid and effective soil-cover than that provided by regeneration led to further studies of sown grasses (Hoskings & Stephens, 1941). It is worth noting that the main grasses of today had by then already been identified and put to use in the region (Maher, 1936b).

The value of a 3-year resting period under Elephantgrass was confirmed in studies at Bukalasa in 1937 (Nye, 1937) as indicated by the excellence of crops produced after opening the land. In 1938 (Martin & ap. Griffith, UDA Ann Rep 1938), evidence of a soil regenerative effect was obtained in the yields of cotton after a 3-year rest under Elephantgrass. This effect was more marked than that obtained with legumes such as Centro (Nye, 1940).

The policy decision, already quoted, to use planted grass rests in Uganda can be traced back to an observation made in Toro by W.S. Martin in 1928 (UDA Ann Rep 1931). A superior structure was observed under natural Elephantgrass compared with a powdery soil under nearby cultivated coffee. Soil monoliths taken from pits in these 2 situations in 1928 have been preserved at Kawanda and clearly show the differences described by Martin (Tiley, 1970a). His first reference to crumb structure appeared in the UDA Ann Rep 1938. This was confirmed by physical measurements of water-stable crumbs. Grasses appeared to be much more effective than legumes in structure formation, which was largely attributed to the differences in root systems (Martin, 1944a, b).

In both tropical and temperate countries, there has been much debate on the extent to which the improved yield of arable crops after a grass is due to the improved structure and associated changes (*e.g.* soil moisture) or to the increased level of organic-N, and therefore of N-supply, or other nutrients. The issue is clouded since the consequences of poor structure can to some extent be ameliorated by fertilizers with which it is often possible to get high yields even of crops with restricted root systems (Russell, 1973). In a review of soil fertility data from West Africa, however, de Ridder & van Keulen (1990) concluded that, if intensive arable systems with fertilizers are to be sustainable, soils should be well supplied with organic matter.

5.3.1. Organic matter under grasses

OM had long been regarded as the principal factor determining soil fertility, which may explain why much emphasis was placed on green- and organic manures until empirical evidence came to prove their inefficiency (*5.2*). The level of soil-OM is determined by the rates of addition and oxidation of plant residues and of soil humus. Forest and grassland soil has a higher humus content than the same soil when cultivated, both because it has a higher rate of addition of OM and because it is not disturbed (*5.1*). A much smaller fraction of the total soil-OM is that contributed by recently added plant remains, but this fraction is very important for crops. It provides the main food for soil organisms; in doing so, part of the N is released as ammonium and can be used by plants directly or when converted to nitrate. The size of this active fraction determines the amount of N, but no satisfactory methods have been found up till now to give confident advice on the N-status of a soil. In forest or grassland, plant nutrients are immobilized and accumulated in the vegetation and organisms (Table 5.1); very little N in the form of nitrate is found under grass (ap. Griffith, 1949; Bartholomew *et al.*, 1953; Stephens, 1970). Losses due to leaching are, therefore, small (Ahn, 1974).

At Yamgambi (Zaïre), Meyer (1963) made an attempt at estimating the numbers of organisms involved in the turn-over of organic litter in forest fallow and under grasses. Under a fallow of 2–3 years, with 2 ton/ha litter of mostly annual species, 40,000–50,000 fungi ("moississures") were counted and 5,000,000 bacteria per gram of soil. Three grasses were studied: *Brachiaria eminii* (syn. *B. decumbens*), Setaria (*Ssph*) and Guineagrass (*Pmax*). In the 1st year, 12–15 ton DM were produced and within 3 years some 20 ton of litter had accumulated. By that time, 20–25% of the dry matter had accumulated in the root system. Some 170,000/m^2 arthropods were counted in grass litter.

Under grassland the growth of micro-organisms is stimulated by the abundance of organic substances supplied as root exudates and decomposing roots. Decomposition of soil-OM and mineralization of N are increased by subjecting the soil to drying and wetting (Birch, 1958; 1960). It was observed that for a particular soil the amount of N mineralized depended on the length of time that the soil was dry and on the soil temperature during the dry period. Birch found that for a given drying time the amounts of N mineralized were proportional to the C-content of the soil. The extra N produced varied from 70 to 830 kg/ha depending on the humic status of the soil. Flushes between 13 and 180 kg/ha of nitrate-N were observed in 13 soils from different locations in East Africa (Semb & Robinson, 1969). The nitrates in the surface of fallow soils also increase during the dry season and may reach very high figures. Up to 110 ppm of nitrate-N were observed in the top 15 cm of a fallow soil at Kawanda, Uganda, although levels of 25 ppm were more common (ap. Griffith, 1951). The release of N following drying and wetting is usually referred to as the "Birch effect", which is of great significance especially after a long dry season (Fig. 6.5). This effect is less pronounced at high altitudes (low temperature), in humid areas without a pronounced dry season or when supplementary irrigation is given (Fig. 7.2).

The rate of OM-decomposition and N-mineralization is also increased by working the soil, particularly during the growing season when the soil is warm and moist. Crops such as maize, which are more frequently row-cultivated than small cereals, had only 35% of the organic-C present in the soil 30 years previously while the continuous wheat plots had 62% (Russell, 1973).

At Serere, Brockington *et al.* (1965) have shown that there is a marked increase in organic-C and total soil-N in the grass phase and a highly significant run-down with cultivation. Fig. 5.1 illustrates this with data from a plot which had been in continuous cropping for 30 years since 1933.

Declining crop yields and increased erosion hazards due to continuous cropping are commonly associated with decrease in soil-OM. The virtues of manures and grasses in the crop rotation are often thought to lie in their organic origin. In spite of proof that plants feed on ions taken from the soil, and grow in inorganic media, this impression is retained. Fertilizers can maintain yields on a parity with yields from applications of organic manures containing much larger quantities of essential chemical elements. The larger root growth, when fertilizer is used, furnishes organic-C required for bacterial functioning. Ironically, the use of fertilizer results also in a greater production of animal manure.

Laying arable land down to grass raises soil-OM. Grasses have been regarded as a substitute for organic manures, but their effects on nutrient supplies, on soil-borne crop diseases and pests, or on obnoxious arable weeds are often more important to crop production than their effects on the stock of OM, except when associated with improved soil-structural conditions.

Organic manures and grasses affect soil properties positively but in different ways (*5.3.3.1*).

5.3.2. Soil-chemical effects of grasses

The lower fertilizer requirement of crops following grassland is perhaps the most striking effect of crop-grass rotations on crop yields in East Africa. This especially applies to N-fertilizer. In his country-wide maize agronomy work in Kenya, Allan (1972) found little response to N-dressing following grassland. The same applied to wheat, even in the 2nd year after grass (Seitzer *et al.*, 1970).

Arable soils reverting to grass with no accompanying legumes also accumulate N, and gains of the order of 50 kg/ha annually have been found

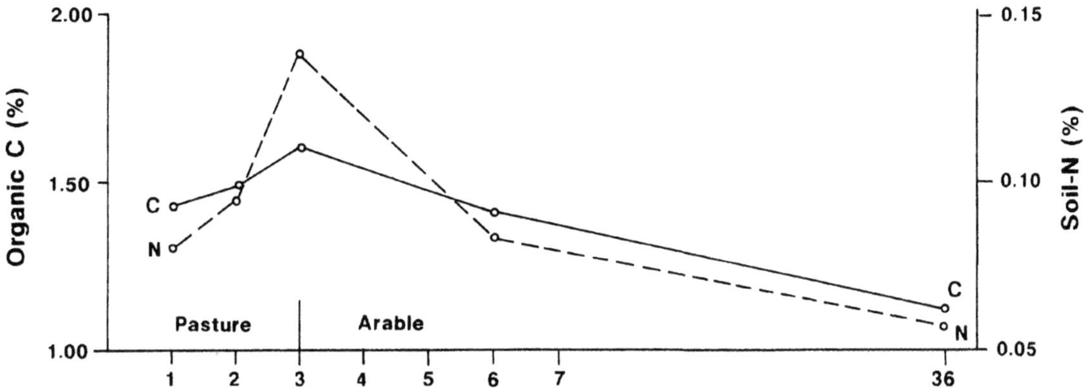

Fig. 5.1. Organic carbon and total soil nitrogen at Ngetta, Uganda (after Brockington *et al.*, 1965).

under both temperate and tropical conditions, when the grass is not burnt (Russell, 1973). The sources of N are thunderstorms and rain, and N-fixation in the grass rhizo- and phyllosphere. N-fixation by associative symbiosis between grass and bacteria (Döbereiner *et al.*, 1972) has been reported for *Azotobacter paspali*, *Azospirillum brasilense* and *Beijerinckia indica*. A review of N-fixation in tropical grasses has been given by Crowder & Chheda (1982), who state that this phenomenon holds tremendous potential and is more likely to occur with tropical grasses due to their C-4 pathway of photosynthesis. Although the reports of the amounts of N fixed by micro-organisms such as Azospirillum in association with grasses have been received with more scepticism than those by *Rhizobium* in legumes (*10.2*), there are indications that the amounts of N gained under grass can be as high as 50–100 kg/ha/year. In addition to N-mineralization and the upward N-movement from the subsoil (Simpson, 1960), N-fixation under grass helps to explain why such high amounts of N can be recovered in N-responsive crops following pasture. Grass Sorghum sown at Kitale straight into pasture after mid-season ploughing displayed a net uptake of over 275 kg/ha/year N (Table 15.1). Little evidence has been found of N-shortage in crops due to N being immobilized after ploughing well-grazed pasture. Well-structured loams hold nitrate against leaching (Russell, 1973). Percolating water moves down principally through the cracks and coarse pores between the crumbs, while most of the nitrate is formed within the crumbs.

It has been a considerable challenge to agronomists in both tropical and temperate agriculture that soils high in organic N can grow pasture that becomes so N-deficient with time. Large reserves of organic N accumulate in living remains and in soil-OM making N less available. The micro-flora competes with plants for any ammonium present and convert it into microbial protein (immobilization). In grasses immobilization approximates mineralization, because the excess of carbonaceous substrates ensures intense competition for ammonium (Simpson & Stobbs, 1981), so that there is little accumulation of available N in the soil profile under grass, except when supply outstrips demand, for instance, at the onset of the rainy season (Birch-effect).

Reduction in N-availability causes N-deficiency in grasses although soils may contain 5000 kg/ha N in the top 15 cm. The run-down of pastures occurs on soils on which crops such as maize may be grown without N-deficiency developing. Soil-N accumulated under grass is, however, released by cultivation. In addition, N is collected from the subsoil by fast-growing and deep-rooting grasses while OM-levels in the topsoil are restored; see also *7.2.4* & *12.7*. These are the basic principles underlying the benefits of crop-grass rotations in farming systems with low inputs of N-fertilizer.

A rise in P-reserves in a Busoga topsoil under a long-standing cover of Elephantgrass was observed by Chenery (UDA Ann Rep 1956). The effects were attributed to deeply penetrating roots, which drew on subsoil resources to be deposited ultimately in the upper soil. Thus, P becomes more available also. Nye & Foster (1961) showed that grasses brought up 30% of their P-uptake, about 15–30 kg/ha P_2O_5 annually, from below 30-cm depth; annual crops, such as maize and millet, only took between 7–15% from below this depth. Whyte et al. (1957) saw the value of a grass fallow largely in the ability of deep-rooted grasses to collect nutrients from the subsoil. Accumulation of not only P but especially of K, N and C in the topsoil was observed in Elephantgrass also by Stephens (1967b) at Kawanda. Where the same grass had been cut and removed, however, K-deficiency occurred and crop yields were reduced to levels below that of continuous cropping.

Increase in pH accompanied by N-accumulation was noted by Simpson (1961) in Kawanda soils under Rhodesgrass, but not under legumes (*5.2.1*); Simpson postulated that a transfer of cations and nitrate to the surface soil occurs under grass. Soil acidity is affected negatively by continuous cropping (Stephens, 1969). Cropping for 4 years depressed the topsoil values of pH by 0.28 unit (mean of 9 Uganda locations). Liming experiments were initiated by Foster (1970) on the same trial sites. The pH had dropped by 0.35 units over all unlimed plots due to continuous cropping for 7 years. The quantity of lime required to change the soil-pH by a given amount depended on % OM. Liming was considered necessary only when exchangeable Ca was very low and the soil pH below 5.25 (see also *12.7.5*).

N supplied by the grass phase dominates the effect on following crops; depletion of K-reserves when grass, high in K, is cut and removed may result in poor crops that follow unless manure or fertilizer is applied to correct these losses (*5.2.2.1*; *5.5*; *12.7.4*).

5.3.3. Soil-physical effects of grasses

The visual effects of a 3-year grass rest on soil condition can be very striking when compared with cropped land ploughed at the same time. Better rainfall acceptance and moisture status of the topsoil are the more real effects of grass rests, as observed by Stephens (1967b) and Foster (1971). Infiltration rates at Kitale were twice as high after 3 years of Rhodesgrass compared with the same period under maize (P.J. Cooper, unpublished); grass-soil had also a lower bulk density and improved aggregate stability (P.M. Ahn, unpublished). The soil tilth after grass was more robust, which visibly reduced erosion. These observations referred to conditions at the beginning of the rainy season in April following ploughing of the grass sward in December.

The obvious deterioration of soil structure under cropping and its improvement under grass had led Martin (1944a, b) to be the first to emphasize the structural condition of the soil (see above). Pereira *et al.* (1954) reexamined the techniques used by Martin. Grasses, cover crops and cultivated crops (Table 5.4) were each grown continuously for 3 years. All treatments were then given one year of intensive uniform cropping before the structure comparisons were made. Martin's soil sieving techniques for water-stable aggregation were found inadequate. Volumetric measurements of pore-space as well as percolation and rainfall-acceptance tests on soil cores showed distinct advantages of soils after resting under grass. Many more insect holes were visible in the cores from the grass rest than from the continuous-arable treatments. The results showed significantly less water-stable aggregates under cultivated crops than under grasses. These differences increased each year, the cover crops giving intermediate results. Sunhemp gave aggregation results similar to those of the cover crops rather than the cultivated crops. All of the results showed a marked downward trend in the level of aggregation. This trend was heaviest with cultivated crops and least with grasses. The physical benefits of the grass treatments were substantially better than those from the cover crops. Puero and sweet potatoes grew vigorously and gave full soil protection, but failed to leave any important improvements in physical properties. Leucaena gave far less soil cover but, although its vigorous root system improved soil aeration, it failed to resist the impact of heavy rainfall, so that there was no improvement in the infiltration rate which is critical for the prevention of erosion. The growing of 6 successive seed crops of sunhemp, although leaving a volume of larger pore-spaces greater than that of other continuous arable treatments, gave no improvement in either infiltration or percolation rates. Only with the grasses were the improvements in pore-space stable enough

Table 5.4. Infiltration and percolation tests at Kawanda; after Pereira *et al.* (1954).

Treatment	Rates in cm/hour			
	Infiltration under rainfall at 15 cm/hour		Percolation under constant head	
1 Year of uniform arable after 3 years of treatment as below:				
Elephantgrass	10.46		41.9	
Paspalum (*P. notatum*)	11.51		33.5	
Rhodesgrass	10.49		36.3	
Mean		10.82		37.2
Puero (*Pueraria phaseoloides*)	9.50		30.7	
Sweet potatoes	10.39		31.0	
Leucaena (*L. leucocephala*)	9.72		25.4	
Mean		9.87		29.0
Groundnuts	8.66		22.3	
Sunhemp (*Crotalaria juncea*)	8.71		24.4	
Cassave	9.50		25.9	
Mean		8.96		24.2
LSD at P 0.05	2.61		10.9	
LSD at P 0.05 means		1.52		6.3

to improve the acceptance and transmission of rainfall.

The results illustrated the need to measure rates of run-down as well as of build-up of soil structure. The improvement under grass-rest was considered transient, the soil returning to its original condition rather rapidly. The rate of deterioration of soil structure during the arable break is as important as its rate of restoration under grass in the planning of crop rotation for the tropics (see also Fig. 5.1). The physical condition of the soil, although of great importance under tropical conditions, was seen by the authors as only part of the effect of crop rotations. As Stephens (1967b) and Foster (1971) have shown, the improvements in topsoil structure may vanish after 3 years but yield responses remain significant for a longer period (5.5).

Obviously, crops need to be alternated with grasses in a recurrent fashion. The positive effect of grasses on soil structure is not confined to poor soils. Also on fertile soils, near Mount Kilimanjaro, the value of Rhodesgrass in rejuvenating the physical state of the soil and its effect on efficient water- and N-use by subsequent wheat crops, was demonstrated (Anderson *et al.*, 1966). Soils developed on tertiary or older basic igneous rocks (Nitisols) have a high aggregate stability (Pereira *et al.*, 1967; Sombroek *et al.*, 1982) and are less sensitive to abuse. Ahn (1974) pointed out, however, that although aluminium and iron oxides are effective cementing agents in soil aggregates, their role is secondary to that of organic compounds.

Under not only tropical but also temperate conditions, laying crop land down to grass has been claimed as an efficient method of improving soil structure (Low, 1955). Structural improvements of soil increase the amount of water useful to crops that soils can hold. They improve aeration and drainage, and encourage good root growth by providing pores of appropriate sizes in structures that do not become too rigid when dry or completely waterlogged when wet. Some soils need action more than others to maintain structure if profitable arable crops are to be grown. As the OM-content falls they become difficult to work, crumbs formed by cultivation are not water-stable, the soils run together and cap in wet weather. Apart from giving repeatedly high doses of organic manures, growing grass is the surest means of increasing OM and improving structure. Measuring the improvements in structure is, however, difficult in practical terms; structure may have only small effects on yield, and these must be measured in the presence of the often large positive effects of grasses on nutrition on the one hand and on weeds, pests and diseases on the other.

5.3.3.1. Differences between organic manures and grasses

Methods to assess differences caused by organic manure or grass have been described by Cooke (1967). Both manure and grass gave stable structure to clay soil, but permeability was increased more under grass. Heavy dressings of manure given for 18 years built up OM in sandy soil, but did not stabilize it; by contrast, 3 years of grass gave stable and highly permeable sandy soil.

Organic manures and grasses affect soil stability differently because manure sticks soil particles together, and is effective with clays, whereas sands need the fine roots of grass to bind the particles by physical ties. Overall, in the view of Cooke (*op. cit.*), grass was much more effective than animal manure.

5.3.4. Grass roots and soil structure

Soil under grass contains some of the highest concentrations of roots of any agricultural system (Rovira, 1978). Russell (1973), during his tenure as Director of EAAFRO at Muguga (K), initiated and stimulated much of the work done on grass root systems. In Russell's view, grasses differ from trees in that much of the OM they synthesize is translocated into the soil as root system. Grasses can add over 2.5 ton/ha/year DM to the soil as roots, and natural grassland soil may contain over 12 ton/ha roots below ground compared with only 2–5 ton/ha of above-ground material.

Dagg (1969) found that Buffel-, Rhodes-, Kikuyugrass and Themeda sent their roots down to 3.7 m within the first 6 months after planting, and in the following dry season the roots grew below 6 m even when grazed; these grasses can build up a deficit of 32 cm water in the top 3 m without its transpiration being appreciably affected. At the depth of 6 m this soil may hold 50 cm of available water, which happens when the seasonal rainfall is over 800-1000 mm. Conversely, Pereira *et al.* (*op. cit.*) drew attention to the fact that when Elephantgrass was dug out at the end of the dry season, the soil contained a quarter less water than a nearby bare fallow. Such a water deficit may seriously affect the next crop, except in areas of high rainfall. Elephant- and Kikuyugrass were reported to dry out the soil to a marked extent (Webster, 1954b).

Buffelgrass is one of the most drought tolerant grasses in East Africa, but there are no simple relationships with rooting patterns, as can be concluded from the work of Taerum (1970a, b) carried out at Muguga (Zone II) and Kedong (Zone IV). Rhodesgrass roots were well represented at the surface as well as at low depths compared with the other species (Table 5.5). This applied to the semiarid conditions of Kedong and the humid conditions of Muguga. Rooting was deeper at Muguga than at Kedong. That variety differences may be relevant is borne out by the fact that Mbalambala Buffelgrass did not root as deeply as Biloela (see also Table 6.9).

The data of Table 5.5 refer to the means of 6 samplings between May 1967 and July 1968. More than half of the total root volume was in the top 20 cm of soil but the proportions were found to change considerably during the season. Themeda and Eragrostis did not root deeply but these are usually regarded as drought tolerant also. The rationale of this was found by Taerum (*op. cit.*) who quoted South African work to suggest that the heavy concentration of roots of Themeda in the soil surface layer enables the species to make the fullest use of light showers of rain and that Eragrostis also has this quality. However, the data of Taerum do not point to substantial differences between the species he investigated.

Taerum also studied shoot:root ratios, the horizontal spread of roots, root morphology and gypsum resistance block readings. Apart from the observation that Buffelgrass had a low shoot:root ratio, in sharp contrast with Makueni Guinea, the data (not presented) lend little support to the hypothesis that differences in performance between species can be explained simply by rooting patterns. In this respect, rooting- and tillering patterns are analogue (*8.7.1*). Differences may be found between species, but they offer limited explanation for differences in performance in response to environmental factors. For drought studies in Sorghum, see *15.3*.

The positive role that a temporary grass cover can be made to play in dryland farming has been demonstrated by Pereira and co-workers (1953; 1958; 1959; 1961; 1967). Rather than by bare fallows with protective wash-stops, more moisture was conserved under a dense stand of fast-growing, short-season grasses. With their shallow root

Table 5.5. Dry weight distribution of roots in relation to soil depth at Kedong (Zone IV), Kenya (after Taerum, 1970a).

Soil depth	Species/variety					
	Cenchrus ciliaris		*Chloris gayana*	*Eragrostis superba*	*Panicum maximum*	*Themeda trianda*
	Biloela	Mbalambala	Mbarara	K6023	Makueni	K53302
0–20	56.5	58.1	68.5	72.2	58.3	71.1
20–40	18.8	17.9	16.2	17.1	22.6	15.1
40–60	11.7	13.5	7.3	7.6	9.9	8.6
80–100	3.2	3.1	1.5	0.4	3.8	1.5
100–120	2.6	1.3	1.1	0.2	1.1	1.0
120–140	0.8		1.5	0.1		0.6
140–160	0.5		0.5	0.1		
160–180	0.2		0.1			

system under such conditions, grasses are only able to dry the soil to a shallow depth, but maintain surface permeability, so allowing most of the water falling in heavy rainstorms to penetrate. Top-soil temperatures (2.5 cm depth) dropped from 42 °C in bare soil to 25 °C under a grass cover of 25 cm height (Pereira *et al.*, 1958); see also *5.3.5*.

Grasses may improve an aspect of structure that could be fairly long-lasting if appropriate methods of cultivation are employed to maintain them (Russell, 1973). Important in some soils is the presence of channels, down which water can move to the deeper subsoil, and in pasture soils dead roots produce such channels when they have been decomposed. Ploughing may cut and seal their tops by smearing soil over them. If only a shallow cultivation is used to produce a seedbed for the following crop these channels can be maintained, which may give a marked improvement in the ease with which the surface soil will drain. This suggestion of Russell would require a grass that is easily eradicated (*6.3.5*).

Russell (*op. cit.*) has equally drawn attention to the tensile strength of grass roots. The roots of some grasses grow into soil crumbs, holding them like beads on a string, also when the root is dead. The longer it lasts until it is decomposed, the more noticeable will be the build-up of structural units (Plate 5.1). Grass roots can be particularly strong stabilizers for reasons that have not been fully explored. The rhizosphere of some grasses can be rich in gum-producing bacteria; fine, dead roots appear to hold fine crumbs around them which could be due to the roots growing through existing crumbs or to fungal hyphae and bacterial gums holding fine granules together into fine crumbs.

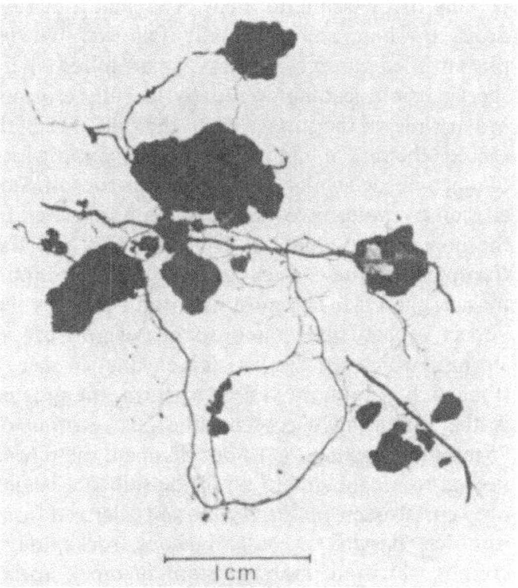

Plate 5.1. Soil crumbs held together by grass roots. (after Russell, 1973).

Roots may decompose but the outer cells may still form a sort of hollow cylinder which allows air to penetrate into the subsoil. At Yangambi intact structures of grasses were found 10 months after ploughing under hot, humid conditions (Meyer, 1961). This phenomenon has a major effect on maintaining the OM-content of the soil (*4.5.2*; *6.3.5*; Plate 4.1).

Higher grass root contents in the soil are obtained with N-fertilizer (Tiley, 1970a; Weinmann, 1964). P, K and Ca have generally much smaller effects on root content than N-fertilizer.

This effect may help to explain the widespread observation that N-fertilization increases water use efficiency if accompanied by increased yields.

5.3.5. Soil erosion

In the above some evidence has been presented to indicate that a grass break in crop rotations preserves soil structure and punctuates the nutrient drain in crop removal. Improved structure may express itself in less erosion or better plant establishment, but is difficult to evaluate in terms of crop yields. However, the direct effects on the ground itself are often most obvious.

Rainfall frequently occurs in heavy storms, especially in the semiarid areas. The action of rain drops on bare soil is usually followed by the uncontrolled run-off of excess water, followed by sheet erosion, leading eventually to gully erosion. Water runs off the surface when the rate of rainfall exceeds the rate of infiltration. This may imply loss of much-needed water and of valuable topsoil. Soil erosion by water involves 2 distinct processes: [1] the break-up of clods into fine particles; [2] the transport of these fine particles which contain most of the OM. The most important cause is the impact of fast-falling raindrops in severe tropical storms, which possess a large kinetic energy (Plate 5.2). The more violent a storm, the greater is the shattering effect of raindrops. Storms of 75 mm/hour lasting for 1 hour dissipate the power needed to lift the top 15 cm of the soil to a height of 84 cm (Russell, *op. cit.*). Some soils, derived from iron-rich basalts or basic igneous rocks, have crumbs which are more resistant to break-up by raindrops than soils derived from low-iron rocks such as granites, which are also high in fine sand. Volcanic loams are remarkably resistant to erosion (Maher, 1950).

Soils with good physical structure, as elaborated above, will permit more water to penetrate and will store or drain any excess. Clods high in OM may be softer and may break-up more easily, but will have less slaking and more stability of the pores and cracks in the subsurface soil. On the other hand, non-humified OM in the soil surface

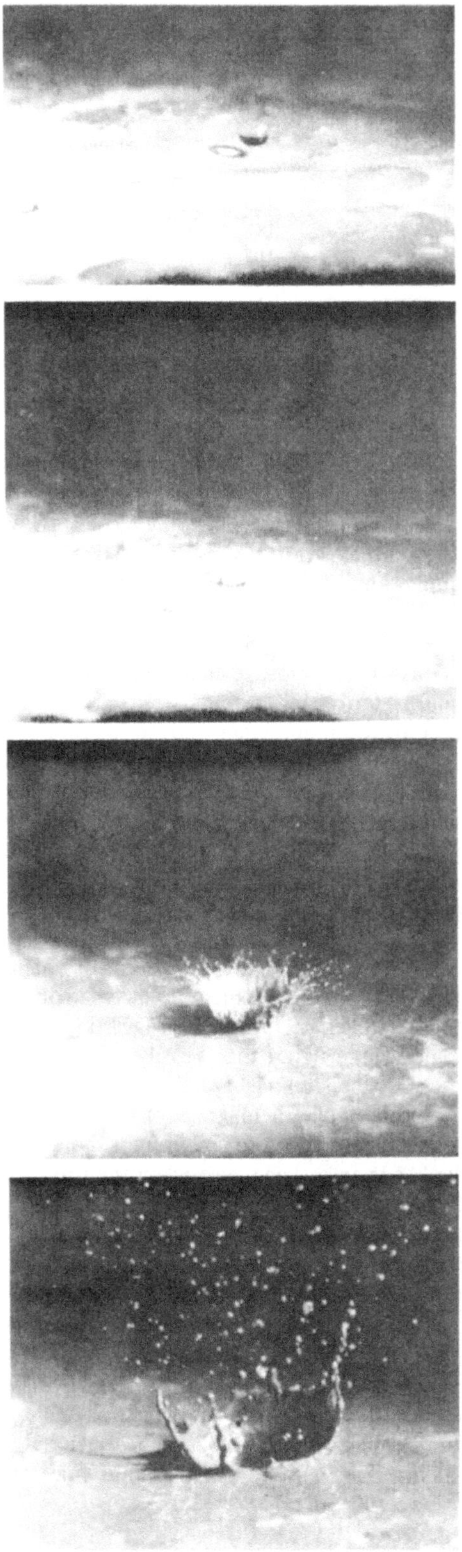

Plate 5.2. Top, raindrop about to strike the surface of wet soil. Bottom, miniature explosion caused by raindrop (Naval Research Laboratory).

will prolong the life of the surface- and subsurface channels. At Mpwapwa (T) measurements of water and soil losses over 8 years demonstrated the value of a grass cover (Table 5.6).

Methods which slow the rate of run-off and which maintain the permeability of the soil reduce the erosive power of rain water. Hutchinson *et al.* (1958) observed at Namulonge (U) that nearly one-half (600 mm) of the annual rainfall is lost through runoff when falling on bare soil, whereas from a soil under grass or a grass mulch it was only 25–125 mm.

Additional measures are often needed to hold rain water and the soil it may otherwise carry sideways, eventually to be lost from the land.

Table 5.6 The effect of grass cover on annual losses of rainwater and soil at Mpwapwa (Zone V); averages over 8 years (after van Rensburg, 1955a).

	% Run-off	Erosion (loss of soil in ton/ha)
Plot entirely cultivated	19.3	87.1
Cultivated plot with narrow grass belt across the slope	15.5	51.0
Plot half of which cultivated and half of which grass-covered	11.5	9.4
Plots entirely grass-covered	4.9	0.7

Grass strips, bunds and contour terraces have been introduced in East Africa as protective measures to combat erosion since the early 1930s (Kerr, 1942; Maher, 1950; *2.1.2*; *2.4.4*). Obviously, the greater the risk of erosion the more priority should be given to mechanical and more permanent measures to control erosion. Erosion control also applies to conditions of grazing (*4.3.2.2*). A tall grass vegetation traps more rainfall (Glover *et al.*, 1962). Overgrazed land is not only more liable to erosion, it suffers more from drought, due to loss of rain water, and from the consequent shallow soil-depth that is wetted by the infrequent rainstorms. Mention should be made of the special issue on erosion control of the *East Afr. agric. For. J*: 27 (1962).

5.4. Ley farming in Europe

The official policy endorsing the practice of alternating crops with grasses in East Africa was not made independently of trends in Europe, especially in Great Britain, at that time. The ley-farming concept in Europe was emphasized in the early 1930s with the intention of ploughing and ultimately, after a few years of cropping, reseeding inferior grassland. The image of this ley concept diminished in the 1950s in capitalized farming but is now rising again in the wake of environmental concern. It is worth recalling some of the findings.

The effect of leys varies with the type of soil. A clear example of the benefit of leys was found at Jealott's Hill in U.K. This soil is inherently unstable and Lewis *et al.* (1960) ascribed the benefits from leys to: [1] improved physical and nutrient status of the soil; [2] less disease in cereals; [3] reduced weed competition. Yields of arable crops were higher after leys. Continuous cropping led to poor crumb structure, difficult ploughing plus seedbed preparation while the soil panned after rain. Soil where 3-year leys had been grown was easier to make into a good seedbed, so that crops germinated rapidly; 2-year and 1-year leys were less effective. Low *et al.* (1963) found that the land should be under grass for at least as long as it is under arable. Grazed leys were more effective in raising yields of following crops than when the grass was cut and removed. At Rothamsted, differences in wheat yield were not related to differences in soil-OM caused by the histories of the fields or by the cropping systems tested. The effects of leys on yields of crops were mostly due to the N-residues left in the soil and to the K lost in crops removed (hay and potatoes). The structure of these soils is robust and the physical effects brought about by grasses may not influence yields (Cooke, 1967). At Hurley, an intermediate situation was observed in that leys increased the proportion of water-stable aggregates, and also increased total soil-OM by about 10 ton (from 65 to 75 ton/ha). Half of this increase was fibrous OM, mainly grass roots. Water-holding capacity of the soil at field capacity was increased by the 3.5-year leys, but the effects were not large. Improvement in structure was mostly confined to the 2.5-cm soil top, although small effects persisted to 12.5 cm deep. After ploughing improved soil may be buried while unaltered soil forms the seedbed. Perennial Ryegrass was more effective than Cocksfoot in aggregating soil. Wheat yield was highly correlated with the variation in N, not with improved soil structure.

Early exponents of ley-farming claimed that yields of leys would be much larger than the yields from permanent grass they replaced, and that yields of arable crops grown after leys would be larger than yields from corresponding crops grown in arable rotation. However, in present-day capitalized farming, in which the arable rotations are well fertilized and grow healthy crops, arable farming can be sustained except on soils of bad structure. Conversely, productive permanent grassland has since become quite possible with N-fertilizer and more persistent grass varieties. When grassland is ploughed, it is because of the ingression of inferior species, but the land is sown again to grasses rather than to arable crops. Arable and livestock farming have become separated, highly specialized economic entreprises. Instead of being integrated on the same farm as of old, arable and livestock farming have segregated, even to geographically different areas.

Environmental concern is, interestingly, leading back again to ways of integrating livestock and arable farming. More and more restrictions are being imposed on farmers in the use of pesticides, fertilizers and animal manure. This "waste" of specialized livestock enterprises, largely based on imported feedstuffs, is faced with diminishing near-by-outlets for its disposal whereas arable areas where cheap, organic manure is welcome are too far away. It is imagineable that leys will function again as an economical and effective break in arable rotations, for example to check soil-borne pests and diseases. When for this purpose livestock is redistributed from the areas where it is concentrated now, its manure will no longer be a waste nor a liability. The more recent insights that organic manure still exerts a positive effect in high-technology arable farming, for instance in potato yields, may help this reappraisal to come to fruition.

5.5. Crop-grass rotation experiments in East Africa

A review of 3 long-term experiments carried out in Uganda since the 1930s was presented by Stephens (1970), following an interim report by Jameson & Kerkham (1960). It suffices here to relate the major trends. Where no fertilizer or animal manure is used, land should be put to grass for at least half the time. Where 25–30 ton/ha animal manure is applied every 3 years, cropping may be possible continuously without loss of fertility, provided there is no accelerated soil erosion. Crop yields were better after grazed grass than when the grass was left ungrazed and there was even a tendency of crop yields being better the more grazing had been applied.

Clear effects on crop yields after grass were observed by Foster (1971) in test cropping at Kawanda following a combined fertilizer/husbandry trial of Elephantgrass (Table 5.7). A range of crops grown after 3 years of unfertilized grass gave much better yields than after 3 years of continuous cropping.

Table 5.7. Crop yields (kg/ha) following unfertilized Elephantgrass or continuous cropping at Kawanda (after Foster, 1971).

	Grain yield (kg/ha)			
	Maize		Beans	
	I(2)*	III(1)*	II(2)*	III(2)*
Continuous cropping	1580	1550	920	560
Grass grown undisturbed	3140	2290	1130	770
Grass cut and removed	3450	2060	1030	860
Grass grazed	3830	2330	1460	920
LSD	670	490	–	250

*1st or 2nd season within year I, II, or III.

It is evident from Table 5.7 that maize following grazed grass gave more than twice the yield achieved at continuous cropping. In the 3rd year the yield advantage was still 50%. By then crop yields in general had declined markedly. In the opening crop of cotton, yields after grass had been 3 times higher than at continuous cropping. Leaf analysis showed that throughout the duration of the trial, N was the main chemical difference accounting for better yields after grass. Where the soil under Elephantgrass had been depleted of K by cutting, aggravated by N-fertilizer, a significant decrease was observed in the yield of cotton, sweet potatoes and beans. Grazing gave consistently higher yields than cutting in all crops that followed, also in the fertilized grass treatments (data not presented). Higher yields of maize but not of other crops were obtained where the grass had received N-fertilizer. Similar yield differentials due to the presence of grass in the rotation had been observed under comparable conditions in Uganda by Simpson (1961) with Rhodesgrass.

Additional evidence of the benefit of grass rests also in the 4th year of cropping was produced by Stephens (1967b) at Kawanda (Table 5.8). Stephens concluded that the important effects of the grass rest were to supply N and K and to improve rainfall acceptance and moisture status of the topsoil. Improvements in topsoil structure may have vanished after 3 years but yield responses remain significant for a longer period.

The strongly negative effect of cutting and removing grass on subsequent crops should be noted because this applies directly to the practices involving zero-grazing (*12.7*; *13.3*).

legumes on yields of subsequent maize crops (P.K. Kusewa, personal communication 1989; McCown *et al.*, 1992).

The beneficial effect of paddock grazing on following crop yields (Kerkham, 1947a, b) has been enumerated by Stobbs (1969e). Sown grassland, consisting mainly of Rhodesgrass, was either grazed during the day or, on plots only one-third the size, at night over a period of 3 years. The arable rotation typical of local practice was applied: cotton and finger millet in the 1st year; beans in the 2nd year; groundnuts or Sorghum in the 3rd year (Table 5.10).

Table 5.8. Crop yields in the 4th year following grass (after Stephens, 1967b); relative to 100.

Previous treatment	Crop yields in the 4th arable year			
	Seed cotton	Bean grain	Maize grain	Sweet potatoes
Elephantgrass burnt on plots	148	118	126	121
Elephantgrass cut and removed	52	74	94	68
Continuous cropping	81	82	68	90

The (negative) effects of cutting and removing grass may seem obvious, but some legumes exert a negative influence also especially on cotton as illustrated in Table 5.9 (Tiley, 1965d).

Table 5.9. Cotton yields (kg/ha) following legumes (after Tiley, 1965d).

	Cotton yields (kg/ha)	
	Seed cotton	Stems
No legume	250	6350
Centro	260	8720
Desmodium (*D. intortum*)	280	6440
Puero	190	6290
Stylo	210	4740

Similar negative results of Stylo on cotton in other parts of Uganda were quoted by Brockington *et al.* (1965) under both cutting and grazing regimes. These negative effects of leguminous covers were later confirmed by McWalter & Wimble (1976). Although Stylo appeared to be one of the better legumes for pasture purposes, it had to be withdrawn from recommendation (Brockington, *op. cit.*; Tuley, 1968). Nutrient imbalances were believed to be involved, but root exudates are a more likely cause (*10.8.7*). Similar effects on subsequent crops were observed in Tanzania (Rijkenbusch, 1967).

More recently, experiments on dryland farming in Machakos failed to detect a beneficial effect of

The data of Table 5.10 illustrate that when grass was grazed, the effect on yield was consistently positive, amounting to about 10%. When stock was concentrated during the night on one-third of the grazing area, yields were 19% higher (*5.2.2.3*). Stobbs (*op. cit.*) further reported that N- and P-fertilizer applied to the grass had a significantly positive effect on subsequent cotton yields. Residual effects of N-fertilized grass on maize yields can be significant, but Tiley (1970a) observed that the effects of fertilizer applied to grass, when cut and removed, were generally small in terms of responses of subsequent test crops.

The available evidence suggests that cereal crops after grazed grass give twice the yield as compared with continuous cropping in the 1st year and that

Table 5.10. The effect of concentration of grazing stock on yields of subsequent arable crops (after Stobbs, 1969e).

	Day grazing	Night paddock	No grazing
Cotton (seed cotton)	1310	1380	1110
Millet (dried heads)	2080	2310	1990
Beans (dry beans)	570	570	480
Groundnuts (dried in shell)	1190	1300	1120
Sorghum (dried heads)	1320	1440	1240
Relative yield totals	110	119	100

the yield gain may still be more than 50% in the 3rd year.

5.6 Mixed-farming today

Summing-up the evidence of East Africa and elsewhere it appears that mixed-farming systems, characterized by alternate husbandry of crops and pastures, are commonly more productive than systems where part of a farm is permanently under grass, and part under continuous arable cropping. In most regions of the world the principles are similar (Cooke, 1967). N and P, supplied or drawn up by grasses, dominate the nutrient effect on following crops on soils well-supplied with K. Benefits in crop yields have been recorded from extra OM, when associated with improvement in soil physical condition and better soil-moisture status. Gains from grasses associated with erosion control, or easier working of the soil, are reported. The results emphasize that suppression of arable weeds, pests (*e.g.* nematode control by Rhodesgrass; *6.6.4*) and diseases may be equally important (Chheda *et al.*, 1983). The beneficial effects of grass on crops are at their best when the grass is grazed; depletion of nutrient reserves occurs when grass is cut and removed.

There is an implicit suggestion in the results that on more stable soils the importance of grasses as a means of increasing crop yields through the physical amelioration of the soil may have been overstressed. Where alternatives of maintaining soil fertility are available, an intermezzo with grass has even been discredited as superfluous and cumbersome. Mixed-farming based on crop-grass rotations has become a rarity in capitalized farming because of the input of fertilizer and due to rationalization which has led to the segregation of livestock and arable farming entreprises. Caution is, therefore, needed in that principles like ley farming, soil conservation and the importance of OM are not stretched too far, lest they become counterproductive.

In East Africa and the tropics in general, alternatives to deal effectively with soil erosion and declining soil fertility are not within easy reach and are often not economically justified. Besides, most farmers, small- or largescale, continue to prefer to have crops and livestock together. There have been notable exceptions. Decades-long continuous wheat farming, even with 2 crops/year, has been practised at Oldeani near Mount Kilimanjaro and in the Timau area of Mount Kenya. However, these exceptions should not detract from the general observation that continuous cropping cannot be sustained on the majority of soils at the present low levels of application of fertilizers, biocides and mechanization.

It is proposed not to use the term "ley" in East Africa. The ley-farming concept renders no justice to the local history under which alternate husbandry developed and has even had a negative effect due to its fall in popularity in capitalized farming. Also, since leys have been associated with grass-legume mixtures, the low rating given to the latter has tarnished the image of the ley even further (*10.10*). By contrast, the local grass-fallow system of alternate husbandry as described (*5.1*) is unique and it developed independently from concepts in temperate countries. Having a grassrest between cropping cycles is not a concept alien to the East African farmer (Masefield, 1946). This should be no surprise when crops after grass carry twice the yield (Table 5.7).

In present-day farming practice, however, improved production of grass for feeding cattle is the sole motive inducing farmers to plant pastures. If stock that help to improve soil fertility are kept, they should themselves be profitable. In the same way it is not profitable to grow legumes for the main purpose of fixing N; even less so if the latter is uncertain in its effect and is only brought about by adding extra P-fertilizer (*10.2.1*). However, sight must not be lost of the mutual benefit which grasses and crops may render, with livestock and the soil as assisted beneficiaries.

Soil fertility is too often taken for granted, not as something that has built up slowly over time and that may quickly be destroyed. Many small-farm settlement schemes were set up in Kenya in the 1960s by splitting-up large mixed-farms. Most of these farms had previously been for more than half of their acreage under natural or sown pasture. The newly settled farmers were greatly encouraged by the maize yields they harvested on the land they had acquired. Crops were good. Most of the remaining grassland was soon broken also and converted into arable land but little new grass was sown. It was not appreciated that the fertility encountered had been built up through grassland, a lesson that had been learnt in the 1920s and 1930s when farmers realized they could not plough natural grassland and crop it endlessly (Maher, 1936b). Molassesgrass then began to play its great role as a pioneering grass on overcropped land (*6.10*).

Similarly, in the new set-up of research and extension in the same 1960s the role of mixed-farming was no longer as clearly emphasized as it had been before. Conversely, pasture improvement was pictured primarily as a means of improving animal production. When donor-aided projects became fashionable in the 1970s, their efforts were directed at separate components, *e.g.* dairying based on maize silage or on zero-grazing, with little regard for the soil-degrading effect this practice has (Table 5.8; *12.7*; *13.3.1*). Even worse, since alternate husbandry of crops and pastures had been common practice on larger farms, it became associated with capitalized largescale farming and was thought to be beyond the reach of the small-scale farmer. That the concept of fertility restoring grass rests had in fact been developed in local, conventional agriculture was no longer realized.

5.6.1. How to project cultivated grasses?

Grasses must restore fertility of the soil for arable crops and also be a profitable source of income through the produce that is fed from it. The evidence available suggests that crop yields are doubled after grazed grass, while grass yields directly after crops are also twice those of old pasture (Fig. 6.2 & 11.2). Thus, the same yield objectives can be achieved in half the time or on half the land. Such "two-farms-in-one" systems show their greatest advantage in low-input farming and poorer soils.

No doubt, mixed-farming continues to be the best background against which to project agricultural development in East Africa (Stobbs, 1969e), since crops and grasses can be grown to mutual benefit. Soils are improved and livestock can be integrated closer within the system. In this day and age of increasing environmental concern, pasture development policies should increasingly take into account this interface of environment, cropping and livestock. The lessons of the 1930s are as relevant now as they were then. Doubling of yield (Table 5.7), while saving on fertilizer, on weed control and soil moisture, apart from better livestock through better grazing, should persuade the farmer to alternate his crops with grasses. Such yield benefits are more convincing than principles related to OM, pH or soil structure, however basic these factors may be.

Better grasses (*6*) but also better husbandry methods (*7*) are needed to make the grass phase in the rotation as effective as possible. Herbage yield is important but grasses must be easy to establish, to manage and eradicate. The outburst of production in the first 1–2 years in the life of a grass could be put to maximum advantage by having shorter rotations. The shorter the rotation, the more effectively is soil-N recycled, obviating the need for N-fixing legumes which often require extra P and other fertilizer but which disappear due to lack of persistence (*10.4*).

Land must be left bare as little as possible. Undersowing grass to maize or wheat achieves this purpose. And when the pasture is ploughed eventually, short-term annual fodder crops such as Sorghum and oats can take over the role of grass until the land is sown again to arable crops (*15.3*; *16*). These practices of sequential cropping reduce the risk of erosion and the need of cultivation. In semiarid areas more rainwater is trapped in and after the grass phase but the sward should be ploughed well in time before cropping so as to prevent moisture depletion (*5.3.4*).

The weed suppressing effect of sown grasses is a major short-term benefit when cropping itself is made impossible by grass weeds such as African Couchgrass and Kikuyugrass (*5.1.2*). Most of these weeds are rhizomatous or stoloniferous, but low-growing; they prefer open crops, *e.g.* plantation or fodder crops. Tall and dense swards of cultivated tropical grasses, however, can suppress them effectively. Land, broken from sown tropical grasses, is also remarkably free from annual broad-leaved weeds and annual grasses compared with land ploughed from cereal or legume crops (*7.1.10*). In the soil-physical studies carried out by Cooper and Ahn (*5.3.3*) it was observed that it took broad-leaved weeds 8 weeks to form a closed canopy after 3 years of Rhodesgrass, compared with only 3 weeks after the equivalent period under maize. This is a spectacular perception with which to demonstrate the value of sown grasses.

When cereal land is sown to grass, this is a small but logical step-up from the conventional weed fallow and, therefore, not an alien but a familiar concept, as already mentioned. The inclusion of cultivated pastures in the cereal rotation can maintain soil productivity without unduly high levels of mechanization and fertilizer. Mixed-farming based on the alternate husbandry of crops and grasses is a low-technology or, in more recent jargon, sustainable farming system.

The integration of cropping and livestock, however loosely linked at times, can be developed further to mutual interest. It will add to the conservation of the principal resource, the soil, on which future generations depend. The current onslaught on virgin but marginal grassland for the purpose of cropping will perhaps be curtailed as a result.

5.7. Summary

Farmers in East Africa want cattle, but they need crops. In conventional subsistence agriculture a fallow period of long duration used to be employed to restore the productive capacity of the soil. As pressure on land increased, systems of husbandry progressed from shifting cultivation to grass weed fallows which were grazed but on better soils to continuous cropping. Except on the most stable and fertile soils, continuous cropping rapidly leads to diminishing yields, nutrient depletion and soil erosion. However, fertility is more easily lost than restored so that the need to maintain fertility applies to all soils. Various methods of restoring soil fertility and especially the OM-content of the soil have been tried over the decades. Leguminous green-manure crops proved inadequate and animal manure, even when livestock is around, is generally not easily available in quantity and quality. In the 1930s attention turned to the positive role grass fallows were observed to play in conventional shifting cultivation. It was subsequently realized that cultivated grasses are able to restore fertility in a shorter time and more effectively than when land is abandoned to volunteer grasses which, by their very nature, are crop weeds (*e.g.* African Couchgrass). In fact, cultivated grasses help to suppress these weeds. Grasses add effective fresh OM and N, and can draw up minerals from lower soil depth through a deep and extensive root system. A major effect is that of improving soil structure and consequent water economy through aggregate stability, permeability and root channels. The latter aspects distinguish grasses from organic manures. More rain water is trapped on the soil surface and subsequently held in the soil profile; the risk of erosion is reduced. Grasses can double the yield of following crops, while grazing is more beneficial to such crops than when grass is mown and removed. Measurable improvements in topsoil structure may vanish early but yield responses remain significant for a long period. Crops and livestock, via better grasses, benefit directly by crop-grass rotations, but the long-term beneficiary is the soil. Alternate husbandry is typical of farming in East Africa, whether subsistence or commercial. When grasses are sown in rotation with crops, this is a step-up from the conventional weed fallow and, therefore, not an alien concept. Mixed-farming, based on alternating grazed pastures with crops, is still a solid platform from which efforts for sustained development can be launched, in the nearest possible harmony with the environment.

CHAPTER 6

Species of cultivated grasses

In the previous chapter the role of grasses was described, especially in rotation with cropping. Rarely in East Africa are grasses established for the sole purpose of replacing natural grasslands. The inclusion of grasses in the crop rotation can double crop yields and sustain soil productivity without unduly high levels of fertilizer, mechanization or weed control. The cultivation of grass in rotation with crops must be the means by which soil structure and fertility are restored while livestock, which have often been divorced from the existing forms of agriculture, are brought to be an essential part of farming practices. It goes without saying that any potential species to be considered should fit in well in the rotation and be the best available for livestock.

6.1. Potential species

Tropical, unlike temperate, grassland literature is characterized by a multitude of descriptions, botanical rather than agronomic, of species with potential use in some region. For an illustration of this contrast see Donald (1975) who demonstrated in Australia that although many species have been tested, whether introduced purposefully or accidentally, only very few are cultivated (*6.4*). As Donald concluded, the days of dramatic new species may be over, as they have long been in northern Europe.

The most authoritative species account of tropical grasses has been the world-wide review by Bogdan (1977). Working from Kitale, Bogdan himself was engaged in the exploration and identification, not primarily of "new" species but of better ecotypes of species already in use. In spite of Bogdan's work reviews keep appearing, referring to "promising germplasm" of material that is far from newly introduced (*6.2*).

The inevitable restrictions to the number of species suitable for cultivation are dictated by geographical and environmental conditions, but are frequently of a more practical nature.

6.1.1. Geographical and environmental restrictions

More than 600 grass species have been recorded for Kenya alone. For East Africa the total number approximates 1000. New ones are still being discovered and additional names appear as the result of taxonomic revision and subdivision. Eastern Africa as a whole is recognized as the centre of greatest diversity ("gene centre") of cultivated tropical grasses. The same applies to some legumes such as *Trifolium* spp., *Glycine*, *Lablab* and *Vigna* spp. The world's cultivated tropical grasses of East African origin are: *Pennisetum*, *Chloris*, *Panicum*, *Setaria*, *Hyparrhenia*, *Cynodon*, *Cenchrus*, *Brachiaria*, *Melinis*, *Andropogon*, *Eragrostis* and *Digitaria* (*2.3*; *6.4*). The main centre is East Africa proper (Bogdan, 1966a), but it extends southwards to Transvaal and neighbouring regions where the robust types of *Setaria* as well as *Panicum coloratum* and *Digitaria* spp. (*e.g.* Pangolagrass) have their greatest diversity.

Like their temperate counterparts, the number of tropical grass species of economic importance as cultivated plants on a world-wide basis is only about a dozen and their adaptation extends to most continents. There is an even further reduction in numbers when we look at species in actual commercial seed production and trade. These are Rhodes-, Guinea-, Brachiaria, Buffelgrass, Setaria, Hyparrhenia and Molassesgrass. Further still, the list of varieties controlling the lion's share in the world seed trade of tropical grasses numbers not more than 10. Most of these are no recent acquisition but date back to the first half of this century. This is proof of their wide adaptation in area and stability over time. Although exploratory work was intensified in the 1970s and although new expeditions have been mounted by international research institutes and agencies (Hanson & Lazier, 1989), very few new cosmopolitan varieties have entered the world seed trade in the past 2–3 decades. The latter is perhaps the best criterion by which to judge the impact of a new genotype. Doubts about the prospects of plant exploration have nowhere been stronger than in southern Africa, perhaps because this region used to be one of the most active in this field in the early days (Pole Evans, 1933; 1950). The recent priority given to plant collecting has, therefore, been seriously questioned (Clatworthy, 1988). Random species collecting may indeed serve little purpose, but a case can still be made for ecotype collecting for characters missing in cultivated varieties, especially when the species are strictly apomictic (*9*).

In the following both the terms ecotype and variety are employed. In practical terminology we refer to ecotype when it is the result of plant exploration and to variety when such an ecotype is brought under cultivation, with or without reselection.

6.1.1.1. Environmental interactions

The fact that East African grass varieties have become household words in countries so far apart suggests that the barriers between continents are small. By contrast, cereals such as maize and wheat are subject to genotype × environment interactions so that recommendations may differ even for adjacent areas, within one climatic zone. This concept of interaction has become so deep-rooted that it is frequently but unnecessarily extrapolated as a matter of principle to other cultivated species, also to grasses. Hence, the repeat of species- and variety trials, still so common in contemporary grassland research. Commercial grass varieties, however, are not subject to significant interactions within the major zones (*6.1.1.2*; Table 6.2); if they were, they would not be in commercial production and there is no reason to assume that this condition will change as breeding progresses. Whether in Europe or at 2500 m altitude on the equator in Kenya or in New Zealand, the best choice of the farmer may well be the same variety of Ryegrass. This also applies to tropical grasses. It also goes without saying that differences in environment and in farming systems between continents are greater than within East Africa. It should be recalled, therefore, that Strange (1958a, b) and Bogdan (1965a; 1977) who did so much to make East African varieties known all over the world, conducted their ecotype testing at one location only, Kitale, and in a simple way. In successful grasses, *i.e.* those with a wide adaptation, interactions due to environment or even husbandry are relatively small.

6.1.1.2. Zonation of cultivated grasses

An ecological zonation of East Africa in relation to the natural grassland has been proposed in Table 3.7. For cultivated grasses 2 criteria seem relevant: temperature and moisture. Although the deviations from the altitude-corrected mean temperatures can be considerable (*3.1.2*) it is advisable to allocate temperate grasses to altitudes above 2250–2500 m. Below 2250 m it is not normally possible for temperate grasses to resist the combined effects of high temperature and drought over a duration of more than 1–2 months. Also, when temperate species come near the lower limits of their adaptation they overlap with tropical grasses such as Setaria and Rhodes which are then more suitable. Fair results were achieved with Perennial Ryegrass grown on the shores of Lake Naivasha (1950 m) under irrigation, but in the end farmers switched back to Rhodesgrass. Even under irrigation, temperate grasses cannot outdo the tropicals below 2250 m. By contrast, Nandi Setaria has performed well at altitudes above 2500 m (*e.g.* upper Molo). The same applies to the Masaba and Boma varieties of Rhodesgrass. An important factor here is the ability of the sown grass to fend off Kikuyugrass, which is the natural grass that colonizes disturbed habitats, including cultivated pastures, at the higher altitudes (*4.2.1*).

The broad range of altitudes covered by species like *Setaria sphacelata* is worth noting. It has been the standard cultivated grass in Zaïre in the region of Yangambi (400 m) in lowland equatorial forest (Behaeghe & Blouard, 1962).

Drought tolerance is the main criterion determining the choice of species below 2250 m. Nandi Setaria, which is really a grass most at home in Zone II, the humid zone, from low to high altitudes, goes dormant early in the dry season and will not survive in Zone IV. The latter also applies to other typical Zone II grasses such as Elephantgrass and Brachiaria (Table 3.8). On the other hand, Rhodesgrass lends itself equally well to either Zone II, III or IV. A genuinely adapted grass for the semiarid to arid areas (Zone V) is Buffelgrass, but for various reasons this grass has only received a scanty reception by East African farmers (*6.9*).

6.1.2. Practical restrictions to numbers of species

Another reason why the number in cultivation is so limited is that the original choice of the best possible species was made by farmers, not by research institutions nor by the seed industry, and long before these took an active interest (*2.2.2*). Up to the present and especially in countries with fruitful links between farmers' organizations, seed industry and research- or regulatory agencies, it has been an established policy that species to be recommended should be small in number but properly characterized, if recommendation is to have any meaning. Recommended lists in EEC-Europe count no more than 10 grass species, to cover a continent from Norway to Crete, with more than 90% of the market controlled by Ryegrass alone. The same applies to the number of varieties within species. In fact, very few new acquisitions of lasting success have emerged in the past 2 or 3 decades. The market has been dominated for more than 50 years by ecotypes which entered the seed trade already in the 1930s, *e.g.* Melle (van Wijk *et al.*, 1993). The versatility of Ryegrass is exemplified by the existence also of a broad range of amenity types, for sports fields and lawns.

The situation in tropical grasses is no different. All the Kitale varieties of practical importance described below were registered with K-numbers in the early 1950s but most of them had been known for much longer. Although Bogdan and Strange carried on with their exploration work for another 10–15 years, no new varieties of comparable lasting value were identified. By then, the old varieties had perhaps increased their lead over new candidates by domestication under continuous seed multiplication (*9.2*).

Thus, the species were chosen by farmers and the top varieties of these farmers' introductions were recognized by the research authorities already around 1950. Similarly, also the systems of mixed-farming were those adopted by farmers (*2.4*; *5.6*). The historic choice of species by farmers was because of obvious superiority for their purposes. Borders posed by climate, farming systems and economic incentives were traversed. In the East African gene centre this is the more remarkable, since the potential choice of species was so much wider, yet only a few were lastingly taken into cultivation.

Before we define the performance criteria of species of economic importance, it is relevant to recall some of the farm practices that have been adopted across the region in cultivated pastures and fodders in the context of mixed-farming rotations:

1. Sown grasses, mainly Rhodesgrass, in rotation with maize, wheat, cotton or tobacco.
2. Elephantgrass grown for the dry season or for year-round cutting (stall-feeding).
3. Fodder crops such as lucerne, Sorghum and oats.
4. Fodder conservation, mostly hay.
5. Topdressing with N-fertilizer.

The species mentioned have been present almost from the beginning (*2.2.2*; *2.3*); they simply could not be ignored. Official research did much to confirm their value and to bring out new varieties. An overview of pasture-related government research activities since 1900 was given in Table 2.1. Although the species in use are still the same, they are much more widely used thanks to new and better varieties, efficient systems of transfer of knowledge and of propagation material (*2.5*).

6.2. Inputs alternative to genotype

No species in current cultivation is perfect genetically, but this should be no reason to shift attention to species of which much less is known.

Genetics can be complemented with husbandry to tailor crops to man's requirements. When grasses grow too fast we can mow and conserve them to be fed back in the dry season. Conversely, periods of slow growth can be anticipated by a timely boosting with N-fertilizer.

As already mentioned, it is unfortunate that in spite of Bogdan's (1977) comprehensive review, works listing dozens of genera and hundreds of species, allegedly promising, keep being published. Such extended flora's may have significance to students of alternative species but they confuse the practical worker. Listing large numbers for a broad, undefined environment is no doubt easier than detailing the limitations of a small number for a specific region. However, species which have been tested off and on for half a century but which are still not accepted by farmers cannot be listed as promising, let alone as new. Seed is often not available. Market confidence is damaged when products are publicized which are not obtainable. Progress in tropical pasture development has been hindered by wanton drives to ignore or to distract attention from well established and successful species while, by contrast, no new ones of proven ability are made available.

When long lists of alternative species appear, the impression is gained that the choice of species is paramount and that there is a "strain for every pain". A second reason why the species approach should be defused is that many still believe that the discovery of "miracle" species with some outstanding property is only a matter of extra research and time. The wealth of species available is tempting many newcomers to have a first-hand try, the more so when the species in question is not known to have failed before. From the experiments and demonstrations going on it would appear that many research programmes are still (or again) at the stage of species (re)evaluation. Most of this work is mere repetition. Absence from cultivation, however, should be seen as proof of earlier failure in actual farming conditions, not of lack of testing or research. To the contrary, those in cultivation today are those which have survived repeated testing for decades and which proved to be the best the farmer found to suit his needs (*2.2.2*).

Efforts of renewed introduction are mostly without lasting result, because to the farmer the value of a species does not depend on one single outstanding property. A species must cover in itself a whole range of qualities if it is to be acceptable in farming practice. Such universal species, when within reach, are such an obvious choice that farmers already discovered them long ago and, as indicated earlier, before the advent of government research. This was the case in Africa with Rhodesgrass, Elephantgrass, Grass Sorghum, oats and lucerne. The same applies to arable crops. Up to the present date there are no examples of lasting success with research-introduced species in agriculture. In the domain of tropical pasture legumes Siratro (*10.8.4*) could have been cited as a possible exception, had it not succumbed to diseases within a decade after its introduction.

Already in 1948, van Rensburg working from Mpwapwa (T) observed that no new grass species had been brought into cultivation over the preceding 12 years. He advocated that we "should think in terms of improved farming rather than of a wonder species that will be the automatic salvation of all our problems".

Recommended lists should not merely be a listing of possibilities and should not pass the problem of what to choose to the farmer. A provisional list of recommended grasses is given in Table 6.1. The points of strength and weakness of each grass are indicated and these are qualified in the descriptions of the species and varieties below. Yields of the Kenya District Trials are tabulated in Table 6.2.

6.3. Grass species performance criteria

A grass may be excellent in desirable properties, yet fail to reach the level of farm acceptance. If, for instance, a species is a notorious weed in subsequent cropping (*4.5.2*) or difficult to establish, chances of farm acceptance are slim. Fortunately, there are no examples of superior genotypes of which the farmer is deprived solely because of lack of ability of propagation, be it through seed or vegetatively.

To be desirable in the first place, a grass must satisfy the feed requirements and be capable

Table 6.1. Tentative rating (1–10) of grass species when cultivated.

	Zone	Herbage yield	Dry season performance	Persistence	Seed yield	Ease of husbandry	
						Establishment	Eradication
MAJOR GRASSES							
Below 2250–2500 m							
Elephant	II_{EL}	9	9	6	2	4	4
Rhodes	III–IV	8	8	6	8	9	9
Setaria	II–III	8	6	8	7	7	7
Molasses	II–III	7	6	5	8	8	8
Below 1000–1500 m							
Common Guinea	II	7	7	6	6	6	6
Buffel	IV–V	7	9	8	6	5	5
MINOR GRASSES							
Congo Signal	II	7	6	9	6	8	4
Coloured Guinea	III	7	8	8	9	9	6
Hyparrhenia rufa	III	6	6	7	7	6	8
Vegetative Propagation							
Star	II_{ST}	6	7	6	4	4	4
Kikuyu	II_{KI}	5	5	6	2	4	2
Giant Setaria	II	7	8	7	2	4	4
Guatemala	II	5	7	5	1	4	6
Above 2250–2500 m							
Perennial Rye	II_{KI}	7	6	6	4	8	8
Cocksfoot	II_{KI}	7	7	6	6	7	8
Tall Fescue	II_{KI}	7	7	7	6	7	7

of producing herbage in quantity and quality. Its ultimate value is expressed in terms of yield of milk, meat or fibre, apart from its effect on subsequent crops.

Whether a grass will grow at all in a new environment is easily determined by planting a few rows adjacent to a grass of known performance ("standard" variety). From such observations it is easy to conclude within 2 years whether it establishes well, remains free of pests and diseases, survives the dry season, produces seed and has enough vigour of regrowth under cutting.

All these characters are necessary to make a good grass. However, whether it is also a good grass under grazing can only be found out under practical field conditions. It is on field scale that the 2 most important practical characters of a herbage plant can be assessed: its capacity to establish rapidly under field conditions and to regrow well after grazing. An example of contrasting performance under nursery or pasture conditions is found in Common Guineagrass (*Pmax*) in East Africa. Guineagrass always amazes the observer, be it under nursery conditions, in seed production fields or along road sides: a mass of leafy herbage, which remains green in the dry season. However, when Guineagrass is sown to pasture and grazed, its capacity to regrow, to persist and to remain green in the dry season is severely reduced. Allrounders are rare among herbage plants. Before views are aired about the potential of a new species it is necessary to verify under grazing, however simple the technique used, whether the findings from nursery experiments still hold.

As already indicated (*6.1.1.1*), to assess the potential of new genotypes it is not obligatory to screen them in more than one location. The extra information acquired rarely compensates for the extra time and resources involved. Bogdan's testing at Kitale (Zone III) was followed

Table 6.2. Kenya District Trials. A summary of annual dry matter yields at 9 Kenya locations, 1975–1977 (ton/ha); after Boonman (1979b).

	Location								
	Kakamega	Kisii	Lanet	Embu	Kitale	Naivasha	Ol Joro Orok	Eldoret	Baraton
Pennisetum purpureum cv French Cameroons	25.3	13.4	11.2	16.9	12.9	11.6	7.5	13.9	7.4
Pennisetum purpureum cv Bana	21.4	20.0	11.2	13.3	16.0	10.4	10.0	12.8	6.6
Setaria splendida (Giant Setaria)	17.6	15.5	8.3	11.8	9.1	4.9	5.3	6.4	8.3
Tripsacum laxum (Guatemala)	10.3	10.2	5.6	4.2	7.5	2.4	7.4	6.4	8.0
Cynodon spp. local ecotypes of Star	10.2	8.4	9.3	8.5	6.6	7.5	6.5	4.7	4.8
Pennisetum clandestinum local ecotypes of Kikuyu	8.1	9.0	7.7	1.1	3.8	6.7	3.8	2.5	4.7
Panicum maximum cv Makueni	14.1	13.1	10.5	9.8	6.6	4.3	4.5	6.6	5.8
Panicum coloratum cv Solai	12.8	10.2	7.4	9.2	8.9	6.4	9.0	7.4	4.8
Setaria sphacelata cv Narset	17.4	15.3	9.5	9.5	9.6	2.4	8.5	7.4	6.5
Setaria sphacelata cv Nandi	16.4	15.0	8.5	6.9	9.2	0.5	7.2	6.7	6.6
Chloris gayana cv Boma	18.9	15.6	13.1	13.0	10.0	16.7	11.3	9.2	7.0
Chloris gayana cv Elmba	18.8	11.6	14.4	10.9	10.1	15.1	10.0	7.3	7.9
Chloris gayana cv Masaba	14.6	13.3	11.4	12.6	9.7	13.9	11.3	7.8	7.2
Chloris gayana cv Mbarara	14.1	13.5	11.2	9.4	10.1	13.7	9.6	7.7	6.6
	15.7	13.2	9.9	10.0	9.3	8.3	8.0	7.6	6.6
Number of cuts, over three years	14	14	10	10	10	10	8	10	12
Altitude (m)	1590	1716	1900	1460	1890	1950	2380	2200	2015
Annual rainfall (1975–1977)	2000	2180	850	1200*	1190	660	1020	1190	1500
Average daily temperature (°C)	20.1	19.3	16.8	20.7	18.2	17.3	13.7	15.5	17.4
Eco-climatic Zone (Table 3.8)	II_{ST}	II_{ST}	III/IV	II_{ST}	III	IV	III	III	II_{KI}

* Records not complete.

up at various locations, *e.g.* at Katumani (Zone IV, Sands *et al.*, 1970), which confirmed the correctness of the earlier choices, in spite of differences in climate. Conversely, varieties such as Makueni Guinea which originated from semiarid zones were identified at Kitale (*6.8*).

Recently, Humphreys (1991) has postulated that in most tropical areas, characterized by seasonal asynchrony of pasture supply and animal requirements, maximization of utilization rather than maximization of pasture growth should be the aim of the pasture manager. In this context he has argued that concepts of pasture management developed in Europe, the USA and South Africa have been unsuccessful in the tropics. One may wonder, however, whether the relevant barriers of application shouldn't be sought between the different grazing systems on the ground rather than between continents.

Maximization of utilization should be an obvious target at all times and all effective means available should be employed to this end. As a target it is perhaps more relevant under extensive than under intensive exploitation regimes. The latter may be smaller in surface area of grassland covered but affects a much larger number of farming families. On small farms it is rare to see any aged grass, at any time of the year, because of the high pressure of demand throughout. The emphasis will then be on increasing grass growth as much as possible, especially when needed most (see also *12.4*; *12.5*; *13.2*).

In the following a number of essential properties is listed. It is assumed that rational husbandry is applied, *i.e.* a strategy of pasture utilization with avoidance of continuous over- and under-exploitation.

6.3.1. Herbage yield

Herbage yield is the most common character of measurement, usually carried out by cutting and weighing and by determining the % dry matter (DM). Yield can be a criterion in itself, for instance the yield per cut, per year or over a period of years. Often also, one wants to know of a new genotype:

1. Within what period it is able to reach the grazing stage.
2. How the response is to adverse conditions such as drought, weeds, pests and diseases.
3. Whether it is persistent.
4. How it performs at different times of the year.
5. What the year- and location interactions are, if any.

To each of these questions it is the herbage yield, assessed in comparison with standard varieties, which can give the most relevant and accurate answer. Consequently, yield is not only a goal in itself, but also a simple yardstick to assess vitality. A healthy grass expresses its condition by yielding well. A grass with a wide geographical adaptation and great stability over time expresses itself by yielding consistingly well over a range of locations and years, respectively (Table 6.2). In contrast with other crops, *e.g.* cereals, herbage plants offer the opportunity of interim-yield measurements which are, in a sense, non-destructive.

Grasses are highly seasonal, producing most of their potential yield in a short period after the start of the growing season (Sheldrick & Thairu, 1975). Although there are considerable differences between species (Fig. 6.1), grasses which produce equal amounts of herbage each month of the growing season, even on the equator, do not seem to exist.

Fig. 6.1 shows how the growth rate of herbage species at Kitale is affected by seasonality, even within a rainy period with what may seem a constant pattern (Fig. 3.3). Also the nutritive value, in terms of % N, follows strictly seasonal patterns (Fig. 6.5). Considerable N-flushes are released due to the "Birch-effect" when drought is followed by rain (*5.3.1*). Appropriate husbandry is then required (*7.2.2*). Timing of grazing, surplus conservation and topping up by N-fertilizer are the tools needed to make herbage supply match demand throughout the year.

Herbage yield is not equally distributed over the life period of a grass either. Yields in Rhodesgrass are at their best in the first 12–18 months and then drop rapidly, followed by a rather steady state. Nandi Setaria is slow to establish and produces little herbage in the 1st year when weed competition is severe, yet together with Congo Signal (*Bruz*) it is the most persistent among the sown grasses. However, yields drop markedly

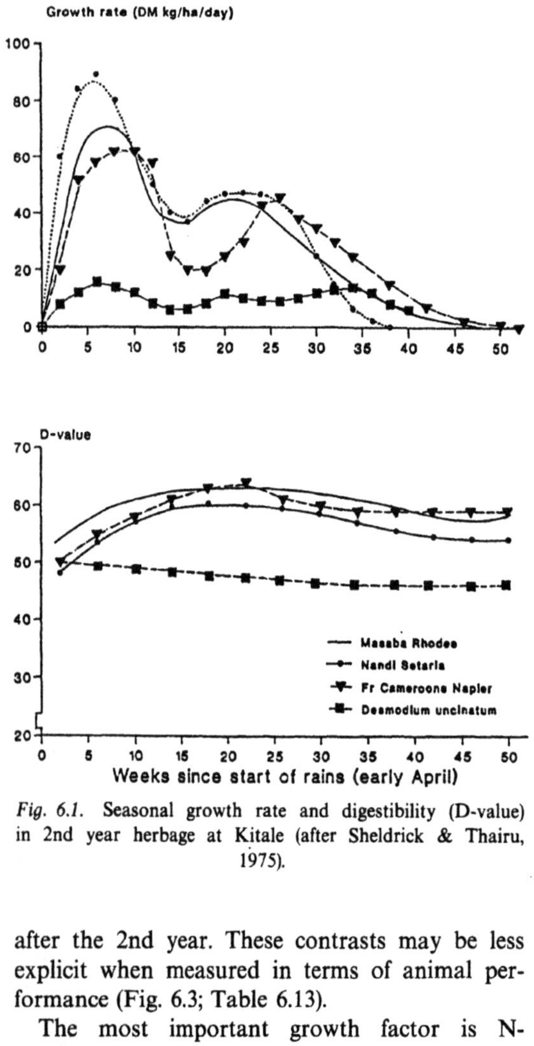

Fig. 6.1. Seasonal growth rate and digestibility (D-value) in 2nd year herbage at Kitale (after Sheldrick & Thairu, 1975).

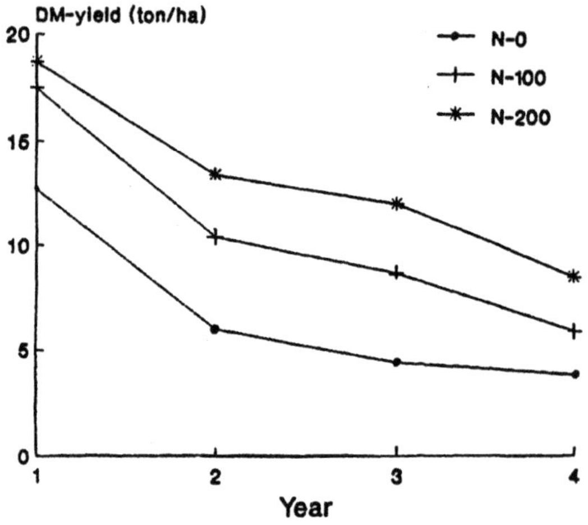

Fig. 6.2. The decline in dry matter yield of Nandi Setaria in relation to age and level of N-fertilizer (after Thairu, 1972).

after the 2nd year. These contrasts may be less explicit when measured in terms of animal performance (Fig. 6.3; Table 6.13).

The most important growth factor is N-availability. In swards of the same age, differences between years due to rainfall fade in comparison with those due to N-status. The decline of yield with age is equally alleviated by N-fertilizer (Fig. 6.2). How yields of unfertilized Elephantgrass decline in the 2nd and 3rd year is illustrated (Fig. 11.2) by data due to Tiley (1970a). Frequency and method of utilization have a bearing also. In cutting trials with 5–6 cuts/year more N seems needed in the 3rd year than in the 2nd year to maintain the same yield level. By contrast, in seed production fields fertilized and harvested twice in the rainy season, a fairly constant level of both seed and DM-yield is found (Tables 8.9 & 8.10; *8.5*).

Standing yield can be estimated visually, if one is aware of the shortcomings of the method. This procedure is more reliable when distinguishing within than between species. Cutting and weighing methods are accurate but too cumbersome when large numbers of individual plants are involved. When species are to be compared, visual appearance and fresh weights are misleading, because of differences in sward density and in % DM. Table 6.3, derived from hundreds of analyses of comparative species and variety trials across Kenya over 3 years, indicates that the % DM varies from 15.3% in Elephantgrass to 22.6% in Stargrass. Thus, species with equal fresh weight may, in fact, have almost 50% more DM-weight. Elephantgrass has the lowest % DM of all cultivated grasses in East Africa (see also Table 13.1).

Elephantgrass is often thought of as the highest yielding of all tropical grasses because of its bulky appearance. The eye is easily deceived not only by the low % DM as indicated, but also by the low sward density which is characteristic of fodder-type grasses such as Elephantgrass (*13.1.2*). It was observed at Kitale that to achieve the same DM-yield, swards of Elephantgrass had to be twice as tall as those of Rhodesgrass and to yield one-third more fresh herbage (*11.9*). "Giant" fodder-type grasses such as Elephantgrass, but also Giant Setaria (*Sspl*)

Table 6.3. The % DM of grass species; in declining order (after Boonman, 1979b).

Species	Acronym	% DM
Stargrass	*Cnle*	22.6
Rhodesgrass	*Cgay*	21.0
Makueni Common Guinea	*Pmax*	20.6
Solai Coloured Guinea	*Pcol*	20.2
Guatemalagrass	*Tand*	19.5
Kikuyugrass	*Pcla*	19.4
Nandi Setaria	*Ssph*	17.2
Gold Coast (*alias* Bana) Elephantgrass	*Ppur*	16.3
Giant Setaria	*Ssph* (*Sspl*)	15.5
French Cameroons Elephantgrass	*Ppur*	15.3

are, therefore, not as high-yielding as their appearance suggests. Conversely, Star- and Kikuyugrass are often underestimated, but their % DM and sward density, respectively, are high.

Another aspect of relevance as far as yield is concerned is the yield of utilizable herbage. In very young swards animals eat leaf and young stems, which are similar in nutritive value at that stage. In more mature grass the stems are rejected and the animal will reach for the leafy plant parts only or graze the younger tillers at the bottom of the sward. The more open the sward, the more stem will be rejected. Total yield measurements approach those of utilizable herbage only in the early phases.

Yield should preferably be assessed over 3 consecutive years including the 1st year, the year of establishment. In the 1st year, yields depend much on the speed of establishment and this is why 1st-year yields are often ignored by researchers but this can rarely be justified because:

1. Differences in vigour of establishment are largely genetic and, therefore, relevant in genotypic evaluations.
2. In mixed-farming rotations, pastures have only a life-time of 3–4 years, so that the 1st-year yields can be decisive.
3. Yields are often the highest (and cheapest) in the 1st year.

Sown grasses can be ready for the 1st cut or grazing within 2 months after sowing. However, some grasses (and legumes in particular) are so slow to establish that they hardly appear from under the weed cover in the 1st year. In Common Guineagrass and Buffelgrass seed establishment is very slow, which is one of the main impediments to the wider usage of these species. Establishment from seed is in the majority of tropical grasses faster than the establishment from vegetative material, however contrary to common belief, even in creeping grasses.

The following procedure has generally been found adequate at Kitale to assess the value of an introduction in comparisons with known standards:

1. Nursery observations to assess general characteristics.
2. Field-scale observations to assess speed of establishment and regrowth under farm-routine grazing.
3. Cutting trials to assess the potential herbage production under optimum conditions.

Procedure [3] can be combined with [2] and/or [1]. It should be noted that Bogdan and Strange confined their testing at Kitale to the procedures [1] and [2] only (Strange, 1958b). Once a species is well known for its performance it is not necessary to repeat nursery- or grazing observations. New varieties can immediately be included in routine cutting trials to determine their herbage and seed yield potential.

Much has been written about adaptation and stability of varieties (Eberhart & Russell, 1966). With yield seen as the expression of vitality, varieties with a high mean yield and a broad general adaptation (in area covered) and stability (over the years) are to be preferred. Some varieties may seem to have a specific adaptation to a relatively small area, yet such varieties usually fail to continue to yield consistently well over a long range of years. It is also a general breeders' experience that performance in higher-yielding test sites provides a better prediction of wide adaptation than lower-yielding sites, quite apart from the greater heterogeneity of the latter. Kitale is intermediate in both the altitude and rainfall range of target environments of East African tropical grasses but it would appear that its test environment seems to maximize the genotypic variation; the responses are also highly repeatable.

Table 6.2 presents data of multi-site trials. The Kenya District Trials were established in April 1975 at 9 locations across Zones II–IV. Plots of 3.5 × 2.5 m were separated by 50-cm wide paths of Congo Signal (*Bruz*) to reduce edge effects. P-fertilizer was applied only at establishment, either broadcast with the sown grasses or in the planting hole (50 × 50 cm) for the fodder-types (*Ppur, Sspl, Tand*) and for Kikuyu- and Stargrass. After each cut 40 kg/ha N was applied in the form of ammonium sulphate nitrate. Locations with higher rainfall and higher temperatures (Kakamega and Kisii) were cut more frequently and received, therefore, more N-fertilizer. This explains part of the rather large differences between sites, which are, however, in agreement with what is commonly known about the agricultural potential of the sites. Little of the variation between locations could be attributed to soil analytical data, except that the Kitale trial was on an unusually infertile site, which had been overcropped previously. The low yields recorded at the locations above 1900 m should be noted. Overall, the highest yields were achieved by Elephantgrass, closely followed by Rhodesgrass (see also Table 11.5). Setaria produced well in the Zone II locations only. Below average were the Guineagrasses and Guatemalagrass. The lowest yielding species were Kikuyu- and Stargrass.

The response to grazing should be assessed at an early date but, to some observers, measuring the economically relevant produce in terms of milk or liveweight gain is the ultimate approach to establish what a grass is worth. Despite similar DM-yields much higher liveweight gains were recorded at Kitale with Rhodes- than with Elephantgrass in the rainy season (Fig. 10.6). Generally speaking, however, it is no simple matter to conduct grazing experiments in a meaningful way. The problem is how to fit not only a variable grazing object (pasture) but also a variable grazing subject (animal) in the rigidity of grazing experiments, while still having relevance to practical farming. Results depend to a large extent on the stocking policy adopted. Policies lend themselves to much subjectivity in choice, for instance, in the date at which to commence grazing after the onset of the rains. Some species may be penalized and others favoured by the methods chosen (Stobbs & Joblin, 1966). Secondly, to be meaningful in the mixed-farming context, the economic value of a pasture through the improvement of subsequent cropping should also be assessed (5.5). Both types of trials lay a heavy claim on time and experimental resources and have been implemented simultaneously only by Stobbs (1969e) at Serere (U).

Apart from the grazing experiment already mentioned (Fig. 10.6) only few quantitative data are available on species comparisons under grazing (Fig. 6.3). The drop in the 3rd year should be noted (see also Fig. 6.2; Table 6.13; Fig. 11.2).

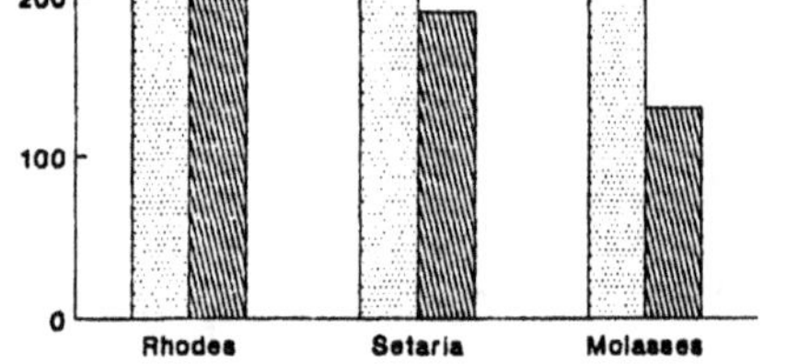

Fig. 6.3. Liveweight gains (kg/ha) of 3 Kitale grasses (Kitale Ann Rep 1960).

6.3.2. *Drought tolerance*

A species should be capable of surviving the dry season(s) and other long periods of drought without undue damage. Most grasses will do this by ceasing active growth and by going dormant. Considerable differences exist among species in the way they react to drought (Table 6.1). Setaria, Rhodes- and Elephantgrass are each in their own way typical examples. Setaria goes dormant within 2 weeks of drought. The leaves wilt and brown off rapidly. Dormancy is strict. However, with the first rains Setaria gives a sharp response and regrows very fast (Fig. 6.1). Rhodesgrass, on the other hand, does not wilt and brown off so rapidly. Herbage can stay green for more than a month. Buffelgrass is even better in that respect. Dormancy is limited so that these 2 grasses do respond to small amounts of rainfall in the dry season. Finally, Elephantgrass remains green for 3 months of drought, although active growth will have ceased quite early. When Elephantgrass goes completely dry it is usually not dormant but dead and no regrowth will occur. This is commonly observed at the coast and in Zone IV (*3.4*). Elephantgrass is a living reminder that greenness during the dry season in a zone of higher potential should not be mistaken for possible drought tolerance in a less favourable one.

Species may respond differently to cutting or grazing in the dry season. Attention has already been drawn to the example of Common Guineagrass which displays a remarkable dry-season greenness under infrequent cutting, for instance, in nurseries and seed fields, but which browns off rapidly in grazed pasture.

Not immediately related to drought tolerance is the capacity of a grass to maintain a high nutritive value of standing hay in the dry season, *i.e.* of herbage accumulated over part of the previous rainy season. This can usually be related to the degree of greenness of the leaf. Once the grass leaves begin to brown off they become vulnerable to mould and rotting, following heavy dew or the occasional shower of rain. Standing hay is "cured-on-the-stalk" only if the rains end abruptly (*4.3.4*; *19.1*).

Some grasses sown 2–3 months before the dry season display a remarkable ability of growing and remaining green deep into the dry season. This is a special characteristic of Rhodesgrass (*7.2.3.1*; *11.10*). However, in subsequent dry seasons it would be unwise to expect a similar performance. After the 1st year grasses do not produce more than 10% of the annual yield within a dry season of 3 months. A better policy then is to conserve the surplus of the rainy season to be fed back during the dry season.

Rooting patterns are often believed to be associated with species differences in drought tolerance. However, hard evidence is difficult to establish (*5.3.4*; *15.3*).

Tolerance to salinity is regarded highly in many dry areas. Rhodesgrass is usually credited with the highest salt-tolerance (Gusman *et al.*, 1954; Bogdan, 1958b; Farnworth, 1976):

Rhodes > Star > Coloured Guinea > Buffel.

6.3.3. Persistence

Persistence is the capacity of a species to maintain a dynamic presence in the sward, without giving way to other species. Persistence is not absolute. The more lenient the treatment it has received or the better the growing conditions have been, the longer a grass will persist. Differences between species are evident (Table 6.1). Of the sown grasses the following sequence prevails in Zone II and III:

Setaria > Rhodes > Common Guinea.

Some species react differently to cutting or grazing. Solai Coloured Guineagrass (*Pcol*) does not persist more than 3 years under cutting, mostly due to weed invasion. By contrast, under regimes of alternate mowing and grazing it persists quite well in comparison with other grasses as it is of a low acceptability to the grazing animal due to the sharp, spiky stubble left after mowing.

Rhodesgrass is generally regarded as less persistent and is, therefore, better replaced by Nandi Setaria for pastures designed to last 4 years or longer. One then accepts lower 1st-year yield for better yields after 3–4 years. The choice is often a matter of compromise. In short-rotation cycles, grasses such as Rhodesgrass are needed which produce a productive sward quickly after sowing and are easily and effectively eradicated by ploughing after 3 years (*5.6.1*).

Persistence should not be mistaken for productivity (Fig. 6.3). Rhodesgrass is generally not regarded as persistent, but its yields in later years may still be high. On the other hand, Congo Signalgrass (*Bruz*) maintains a dense, weed-free sward for many years, which means it is a persistent grass, yet its sward is then no more productive than that of a lawn. In fact, Congo Signal is at times used for lawns and sports fields.

A negative relationship prevails between persistence and ease of eradication of grasses and, therefore, their suitability in crop-pasture rotations. Congo Signal is persistent and hard to plough out (*6.11*). Conversely, Rhodesgrass lacks persistence but is effectively knocked out by one simple ploughing and, although a prolific seeder, it rarely produces volunteering plants in subsequent years. Most other cultivated grass species occupy some intermediate position between these extremes and there appears to be no relationship with either a tufted or a creeping growth habit. In vegetatively propagated grasses, such as Elephant- and Kikuyugrass, different relationships emerge. Although both are extremely difficult to eradicate by ploughing, their stability under grazing is such that weed ingression is quite common. These differences can be traced back to the eco-climatic zone of origin and are displayed most clearly in their behaviour following ploughing (Table 4.4; *4.5.2*).

6.3.4. Nutritive value

A grass sward is of high nutritive value when it has both the chemical and physical characters to ensure high animal intake and production. Nutritive value is subject to fluctuation, due to plant- (growth stage; herbage yield) and husbandry factors. Thus, it is not an absolute, but a relative character and should be viewed in permanent perspective of these factors. A sward is high and uniform in quality but low in yield when constantly harvested young. The measures to be considered are species and husbandry.

Although there is no published review of the extensive work done in East Africa, it is beyond the scope of this book to treat the subject in any detail. Traditionally, nutritive value has been interpreted in terms of chemical composition, of minerals but in particular of crude protein and crude fibre. Most of the analytical data of the work done in East Africa refer to chemical composition of this kind. The relevance of crude fibre, but also of crude protein, has been de-emphasized in recent decades as they account only partly for the variation in animal intake and production. Emphasis shifted to *in-vitro* digestibility (Long, 1967) and, more recently, to the efficiency of N-compounds when converted in the rumen.

The relevant works on nutritive value of grasses in East Africa include: Abate (1981), Abate *et al.* (1984), Bredon & Horrell (1961; 1962), Bredon & Wilson (1963), Bredon *et al.* (1963), Brockington (1961b), Dougall (1962c; 1963a), Dougall & Bogdan (1958a, b; 1965), Dougall & Glover (1964), Dougall *et al.* (1964), French (1940–1957), Gihad (1976), Glover & French (1957), Glover *et al.* (1957; 1960a, b), Juko & Bredon (1961), Karue (1974; 1975a, b), Long *et al.* (1969; 1970; 1972), Marshall & Bredon (1963; 1967), Marshall *et al.* (1961), McKay (1971), McKay *et al.* (1969), Musangi (1965), Odhiambo (1974), Mugerwa & Ogwang (1976), Okwaro *et al.* (1978), Reid *et al.* (1973), Said (1971; 1976), Soneji *et al.* (1971), Thomas (1977a, b), Thomas *et al.* (1980), Todd (1956), Wilson & Bredon (1963); see also *13.1*.

More specific studies on mineral composition are found in: Dougall (1962b), French (1946b), Howard (1963), Howard *et al.* (1962), Jones (1964), Long *et al.* (1969; 1970; 1972), Marshall *et al.* (1976), Orr & Holm (1931), Rudert (1975); see also *12.7*.

Intake and digestibility are related but this is not always apparent. Although the *in-vitro* technique of predicting organic matter digestibility (OMD or D-*vitro*) or dry matter digestibility (DMD) seem to offer good perspectives, considerable differences in intake may be found between genotypes or even plant parts, of similar digestibility. Differences observed between grasses and between grasses and legumes can be considerable. At Makerere University (U) Reid *et al.* (1973) studied the relationship of DMD, age and species. Three groups were defined among the grasses:

A. High initial DMD (75–85%), followed by a high rate of decline (± 0.35% per day) with advancing maturity: *Bruz, Ccil, Cgay, Mmin, Pcla, Pcol, Pmax, Ppur, Ssph.*

B. Intermediate initial DMD (70–75%) followed by a variable decline: Cynodon (*Cnle*?), *Agay*.

C. Low initial DMD (65–70%) and a low rate of decline (± 0.20% per day): *Ttri, Hruf, Pscr.*

Regression equations were calculated (Table 6.4; Fig. 6.4).

Most if not all the recommended cultivated grass species are found back in the top group A. The data confirm general opinion that differences in average DMD within this group are

Table 6.4. Regression equations of *in-vitro* dry matter digestibility of grasses and legumes at Makerere University, Uganda (after Reid *et al.*, 1973; see also Whiteman, 1980); Y: DMD; X: days of regrowth.

Tropical grasses (A + B)	Y = 80.70 − 0.31X
Tropical grasses (C)	Y = 66.02 − 0.16X
Temperate grasses	Y = 75.92 − 0.22X
Tropical legumes	Y = 68.90 − 0.11X

small, although initially high DMD was accentuated in Brachiaria, Rhodes-, Buffel- and Nandi Setaria, whereas lower values were achieved in Kikuyu- and Elephantgrass; for the latter in zero-grazing systems, see *13.1*. The low values achieved by Cynodon were later confirmed by Said (1976); see also *4.2.2.1*. Themeda is generally regarded as being low in nutritive value (Dougall, 1960; Marshall & Bredon, 1967; Karue 1974); the same applies to *Hyparrhenia rufa* (Dougall & Bogdan, 1958b). For comparisons of individual species, additional data are required. Reid's work was mainly designed to defuse claims that tropical grasses are per se inferior. High silica contents were, however, found in a number of non-cultivated grasses (Dougall & Bogdan, 1965) and in Elephantgrass (*12.7.5*).

Seasonality of nutritive value should also be considered. In the early part of the rainy season, not only growth rates (Fig. 6.1) but also nutritive values (% N) are higher than in the middle of the season (Fig. 6.5; Thairu, 1972). A similar trend is found in digestibility (Fig. 6.1), but caution is needed in attaching precise calendar dates to the data of Fig. 6.1, which refer to moving averages of 4-week periods. A 2nd note of caution is that the low values at the start of the

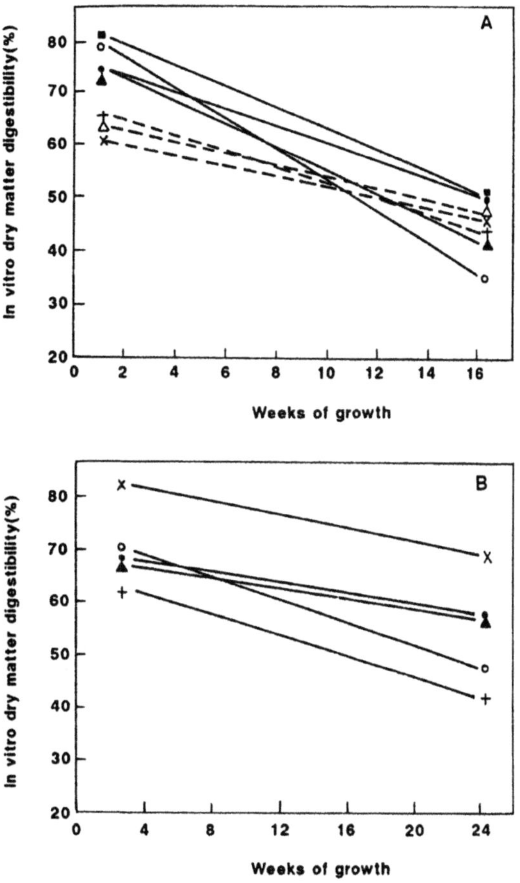

Fig. 6.4. In-vitro digestibility in grasses and legumes (after Reid *et al.* (1973).

A. "Improved" (——) and "native" (----) grasses. ■ Brachiaria; ○ Chloris; ● Setaria; ▲ Panicum; + Hyparrhenia; △ Themeda; × Cymbopogon

B. × Trifolium; ○ Medicago; ● Glycine; ▲ Stylosanthes; + Desmodium.

rainy season are due to dead material accumulated over the dry season.

In addition to chemical factors, physical sward characteristics are important. Laboratory samples of grass, when cut, are not representative of the rather selective intake by the grazing animal, except when the grass is young and consists of non-heading tillers only. The volume of intake is also determined by the ease with which the grazing animal can collect its diet. Ideally, the sward should not only be leafy but also compact, so that the animal is not forced to pick leaves one by one. It was the late T.H. Stobbs, formerly

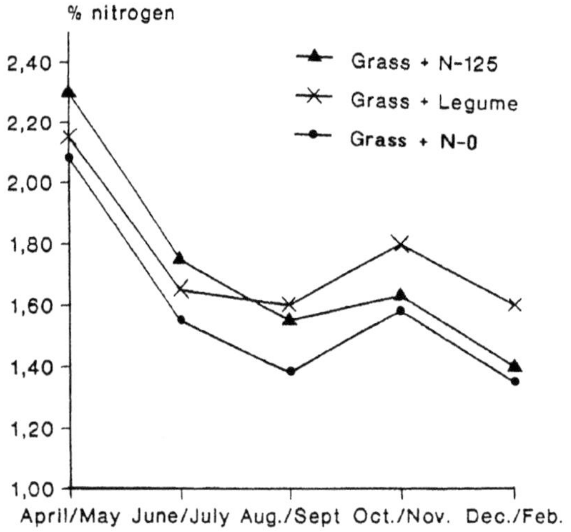

Fig. 6.5. The % nitrogen in Nandi Setaria in relation to season and treatment (after Thairu, 1972).

of Serere in Uganda, who developed in Australia the concept of bite size and number of bites (Minson *et al.*, 1976). Animals can afford only a limited number of bites per day while feeding before getting tired. The amount of herbage taken in per bite, *i.e.* bite size, then becomes important. Sward density may, therefore, be as relevant as chemical composition (Table 6.5; *13.1.2*).

Table 6.5. Sward heterogeneity (T.H. Stobbs, private communication, 1977).

	Range within swards	
	Temperate	Tropical
Nitrogen (%)	1	3
Digestibility (%)	2–4	6–9
Sward density (kg/ha/cm)	200–400	20–200
Leaf sward density (kg/ha/cm)	140–300	16–160

Obviously, tropical grasses are again the champions of extremes. Tropical grasses embrace highly contrasting representatives in terms of sward composition and density, *e.g.* Kikuyu- and Elephantgrass (*13.1.2*). When sward density is low, insufficient intake of digestible energy is the feed factor primarily responsible for limiting production although lack of protein can also be important (Stobbs, 1971; Fig. 10.6).

Van Wijk (1980) has shown at Kitale that *within* varieties, differences found in digestibility can be explained largely by botanical and agronomic characters, *e.g.* heading date and yield level. Whereas little net (true) variation remained for breeding purposes when populations were split according to heading date and yield level, digestibility was higher in early-heading than in late-heading genotypes within the same variety (*9.5.3*).

Pokot Rhodes, although late-heading and seemingly leafy, was found to be low in digestibility and in leaf:stem ratio (*6.6.3*; *9.5.2*). By contrast, Masaba Rhodes was medium-late heading but high in leaf:stem ratios and in digestibility, not only at the same physiological stage but also when sampled on the same day (Table 9.6).

These examples are important to sound a note of caution against the alleged superiority of "late-heading" and about the prospects of breeding for nutritive value. Wrong impressions are gained when the variation observed between species or varieties is interpreted as promising for breeding purposes. Differences observed between species should not be extrapolated to varieties within a species. The same applies to differences between and within varieties. Also in temperate grasses, little effective genetic variation is found and very few varieties have entered the market successfully having been bred with the specific aim of improving digestibility (van Wijk *et al.*, 1993).

In-vitro techniques are useful, but no single laboratory- or field measurement of nutritive value can adequately account for animal response to feeding. New techniques and concepts are constantly being developed, whereas old approaches are being revisited, e.g. of plant anatomy (Otieno, 1967). Secondly, there are no simple inverse relationships between herbage yield and quality aspects. They can both be at a high level in the early phases of regrowth, before heading. If nutritive value is the main objective of husbandry, then grass should be utilized rationally, *i.e.* by grazing or conserving prior to heading.

Apart from the genetic differences (Table 6.4), tropical grasses are said to be lower than temperates in chemical composition because of the tropical climate (Deinum, 1966). High temperature, in combination with adequate moisture and high solar intensity, implies rapid growth and development towards maturity. The latter is not confined to the tropics. As Bogdan (1966a) has pointed out, the situation in dry temperate areas is often no better because the bunch grasses of steppes and prairies lose their quality very rapidly in early summer also.

It is often said that tropical grasses run to stem quickly. They have a general reputation of being stemmy. In equatorial regions stems and flowering heads are in evidence at any time of the year in undisturbed grass. However, the eye is again easily deceived. Stems of tropical grasses are large and inflorescences are conspicuous in both shape and colour, which accentuates the contrast with the vegetative parts. However, heading is much more gradual than in temperate climates. Temperate grasses have a tendency to run to head all at once when their daylength requirement is met. In tropical grasses at the equator stem formation only occurs after a certain amount of herbage has developed, is gradual, and can take place at any time of the year provided moisture is adequate (*8.7*). However, in pastures which are continuously under hard grazing pressure no flowering heads may be seen at all. This is common on smaller farms. Stemminess is then not a liability, rather a luxury.

Tropical grasses may, in tropical climates, not be so nutritious but more relevant is that they are often not in a nutritious state because of their rapid rate of growth. Unless precautions are taken by adequate husbandry, they cannot be expected to be in a nutritious state when allowed to become overmature. On small farms with high stocking rates, such conditions are rare; although the grass is young and nutritious, lack of grass-on-offer is then the main factor limiting animal production.

Glover & Dougall (1961) have shown that good Kenya pastures are very similar to good temperate-zone pastures in their ability to supply similar amounts of nutriments for milk production at similar levels of appetite. As Bogdan (1966a) has pointed out, we now have grasses which can maintain their quality as long as in the warmer and drier countries of the northern hemisphere.

Rational husbandry efforts should aim at leafy, dense swards without sacrificing productivity. Adequate N-nutrition is the most important factor influencing both yield and quality. When grazing pressures are low, legumes may be a valuable addition to maintain some quality when both grass and legume are mature. However, the nutritive value of tropical legumes themselves is not high and inferior to that of clovers. Fig. 6.4 shows the low initial DMD of such legumes and the subsequent low rate of decline. Initial DMD was particularly low (64%) in stylo and Desmodium. Only when the herbage is 60 days of age or older, does the superiority of legumes over grasses emerge. By contrast, N-fertilizer has a highly favourable, consistent effect on % crude protein, % crude fibre and digestibility (*7.2.4.1*; *12.7.2*; *13.1*). Equally important is the beneficial effect N has on sward density.

6.3.5. *Ease of management*

This character is probably the most important in the eyes of the farmer. In research, however, it seems to receive less priority than, for instance, aspects of nutritive value. Whether a farmer turns a valuable grass into a success depends largely on whether he is able to put the grass to work effectively. Grasses should be reliable and easy (Table 6.1):

1. To establish.
2. To cover the ground rapidly.
3. To be maintained weed-free.
4. To manage for grazing and conservation.
5. To be eradicated when old or when to be replaced by crops.

The establishment aspects will be discussed in more detail (7). It is no surprise to find that the most commonly sown grasses such as Rhodesgrass and Nandi Setaria fulfil most if not all of these requirements, while potentially valuable species such as Common Guinea and Buffelgrass meet with reluctance on the part of the farmer because they are not easy to establish from seed. Ease of seed production is a must in sward-type grasses. In giant fodder-type grasses, *e.g.* Elephantgrass, vegetative establishment is the only alternative (*12.1*).

Cultivated grasses should be easy to eradicate, in the sense that they should not return as volunteering plants in the following crop. The best known grasses available cover the whole spectrum of "weed potential". Some grasses like Congo Signal (*Bruz*), not to mention Kikuyu- and Stargrass, are rejuvenated by cultivation, whereas others can be eliminated by one single ploughing. Most of the grasses have some ability to revive from seed, but also from parts of the rootstock. This they can do over long periods. Ploughed-in root residues can remain dormant and escape decay (*4.5.2*; *5.3.4*). No doubt, this adds considerably to the prestige grasses have earned in increasing the amount of effective organic matter in the soil.

With the exception of Elephantgrass (*11.8*) cultivated grasses are relatively free from pests. For lists of pests and diseases which have been associated with grasses, see Bohlen (1973), Bullock (1965), Ebbels & Allen (1979), Le Pelley (1959), Wheatley (1965).

6.4. Ecological significance of gene centre

It was indicated (*6.1.1*) that the East African region counts about 1000 recognized indigenous grass species. As the major gene centre of cultivated tropical grasses it became the main source from which these grasses evolved over the last few thousand years and spread to other regions and continents. As we have seen also (*2.3*), this spreading was recent in history, partly manmade but often adventitious. Many grasses escaped from Africa and intruded into Latin America or Australia via ocean traffic, some of them only this century. Farmers themselves were quite active in seed exchange and plant introduction, and nearly all cultivated species of lasting importance in East Africa were in farmers' use before official research began to take an active interest.

It is generally held that East Africa was blessed with such a wide diversity of grasses because it was populated with a high density and wide range of herbivores, wild and domesticated. Buffalo, elephant, zebra and antilopes of many kinds abounded on the savannah grassland, on plants which were capable of surviving under grazing and under treading and which evolved by adaptation. This theory of co-evolution of fauna and flora helps to explain the absence of useful cultivated grasses from the Australian continent which harboured no grazing animals before the advent of colonization (Donald, 1975). The same applies to Latin America, but to a lesser extent because colonization and import of Zebu cattle, which was an obvious means of adventitious grass seed transport from Africa, took place earlier.

The theory of a joint evolution of grazing animals and grasses needs further qualification. Pratt & Gwynne (1977) have argued that if the native grasses of Australia did not evolve the ability to adapt to grazing neither could they evolve the ability to repel the grazing animal. Australian Acacias, for example, are generally thornless. Conversely, East Africa counts not only many thorn bush species of Acacia but also many grasses which on account of their tough leaves or anti-palatability compounds (Citronellagrass, *Cnar*) are avoided by the grazing animal. The latter also applies to many indigenous species of herbaceous legumes, a point which was stressed by Edwards & Bogdan (1951).

Secondly, if the co-evolution theory is to hold, also other world regions with a similarly long history of grazing pressures should have yielded important cultivated grasses to world agriculture. For instance, the treeless prairies of North America are said to have counted more bisons prior to colonization than the total of grazing livestock now present. Yet, no major cultivated grasses have evolved from the prairies. Therefore, no direct evolutionary relationship can be drawn solely between the grazing history of a land and its content of currently cultivated grasses. Gene centres are, however, often found in mountainous areas with sharp contrasts in climates and soils at short distances, rather than in plains. East Africa meets also this requirement.

An additional factor may be suggested. The ecological origin of a grass, and its behaviour in disturbed habitats, may indicate whether it has potential as a grass to be cultivated for grazed pasture, at least in the East African home region. A distinction was observed between Zone II or

III and between primary or secondary grasses, respectively (*3.3.3*; *3.4*; Table 4.4). It is postulated that the successful cultivated grasses of East Africa are those that originated under the grazing pressures prevailing in the subhumid savannahs (Zone III) of varied topography and that they increased their frequency in the vegetation due to fire or felling, short of achieving full dominance. Their seed yield, seedling vigour and competitive ability pre-conditioned them to be eminently suitable to grow well when sown into the completely disturbed habitat which ploughed land is. These (primary) grasses form fairly stable communities under grazing (*4.3.1*) but are not, or rarely, colonizers in their home environment of East Africa. They can, however, behave as colonizers in other continents in natural or cleared vegetation when competition from the native species is relatively weak.

On the other hand, (secondary) grasses with a forest (Zone II) origin tend to achieve dominance only when their habitat is severely disturbed. They are rarely productive-plus-stable under grazing, and produce little seed (Table 6.6). Much can be learnt, therefore, from the history of origin of a particular grass.

Table 6.6. Tentative characterization of East African cultivated grass species depending on their ecological origin.

Plant property	Eco-climatic zone of origin	
	II[a] (forest)	III[b] (savannah)
Colonizing habit	strong	weak
Seed yield ability	weak	strong
Vegetative propagation	strong	weak
Arable weed potential	strong	weak
Tendency to dominance	strong	weak
Stability under grazing	weak	medium to strong

[a] Zone II: Kikuyu-, Star-, Elephantgrass.
[b] Zone III: Rhodes-, Molasses-, Hyparrhenia/Hyperthelia.
Intermediate: Brachiaria, Giant Guineagrass, Setaria, African Couchgrass.

6.4.1. Gene centre of southern Africa

Cultivated grasses of equatorial East African origin have done very well in many tropical countries over a range of altitudes and latitudes, even beyond the 30th parallels. This is the more striking because most of them originated from the 1500-2000 m altitude range on or near the equator. By contrast, equally world-famous varieties originating from higher (southern) latitudes such as Pioneer and Katambora Rhodes, Kazungula Setaria, Pangolagrass and various Panicum types have repeatedly performed poorly in Kenya or ceased to be used because of disease (see below). Some of these, *e.g.* Pioneer Rhodes in Australia, have valuable frost tolerance. Why its performance was repeatedly found to be poor in East Africa is not entirely clear. The ploidy levels differ between southern ($2n = 20$) and equatorial ($2n = 40$) Rhodesgrasses, but this cannot be extrapolated as a general explanation. In Setaria the situation is reversed or neutral, since Nandi (ex Kenya) is diploid while Kazungula (ex Zambia) but also Narok (ex Kenya) are tetraploid. Ploidy occurs at several levels in Setaria (Hacker, 1966), but also in Guinea- and Buffelgrass. A tentative explanation may be found in the inverse relationship prevailing between heading date and vigour (Fig. 9.2). If varieties from high latitudes head later at the equator they will, understandably, be relatively poor in vigour and vice versa. In fact, *Paspalum notatum*, *Brachiaria humidicola*, *Cynodon dactylon* var. *elegans* (*4.2.2*) and various *Digitaria* spp. originate from higher latitudes, produce very few heads and are used, at the equator, mostly as lawn-types.

Very few critical studies are available to explain the unexpected behaviour of some species in new environments. Donald (1975) has shown how Australia, apart from the ingress of tropical species, has been invaded adventitiously by scores of temperate and mediterranean herbage plants which naturalized. In spite of subsequent, more purposeful introductions "the list of new cultivars is disappointingly small" (Donald, *op. cit.*).

Some species are more popular as cultivated grasses abroad than "at home". East African ecotypes of Stargrass, for instance, have figured prominently among those in use in the southern USA and the Caribbean (*4.2.2*). The same may be said of Likoni and Makueni Guineagrass (*6.8*), and Buffelgrass (*6.9*); see also *6.14*. First of all, it is difficult to delineate the relationship between site of origin and the climatic or edaphic requirements of a new ecotype in a new environment. More relevant, however, is that if a grass is superior abroad, this should be seen relative to the perhaps limited choice of alternative species. This applies especially to subtropical regions with cold winters. In East Africa itself a much wider choice of adapted species is available.

6.5. Major cultivated pasture and fodder species

The species recommended in East Africa are as follows:

1. High altitudes, > 2250–2500 m:
   - Perennial Ryegrass, Cocksfoot, Tall Fescue as pastures.
   - Oats and fodder rape as fodder- or break crop.
2. Medium altitudes, < 2250–2500 m:
   - Rhodesgrass in the humid, subhumid and semiarid areas.
   - Setaria in the humid and subhumid areas when pastures of longer duration are required.
   - Elephantgrass, Grass Sorghum (Columbusgrass) and lucerne as special-purpose fodders (Part II); Columbusgrass also as break crop (*15.2.3*).
3. Low altitudes, < 1000 m:
   - Elephantgrass, Setaria, Brachiaria in the humid areas.
   - Common Guineagrass and Rhodesgrass in the subhumid areas.
   - Buffelgrass in the semiarid areas.
   - Grass Sorghum (Sudangrass) as annual fodder (*15.2.2*).

The climatic factors determining the 2250–2500 m limit range have been discussed (*3.1.2.1*). The upper altitude limit is 2500 m in western Kenya (Mount Elgon, Cherangani) and the mountain ranges of Uganda, Rwanda and Zaïre. In central Kenya, still on the equator, the limit is 2250 m. Towards the 15th parallels (Ethiopia and Malawi) the limit is 2000 m.

Table 6.1 contains species which are recommended and which are, apart from Common Guinea and Buffelgrass, widely cultivated in East Africa. This number has been strictly limited to those with an established reputation from research *and* farming practice in East Africa. Molassesgrass is added because it used to be one of the first in cultivation and the main pioneering grass. Table 6.1 lists also a number of minor species (*Bruz, Hruf, Pcol*) which are not widely grown. The reader is referred to Bogdan (1977) for botanical details.

An overview of variety characteristics, described in more detail below, is found in Tables 6.7 and 6.8.

6.6. Rhodesgrass (*Chloris gayana*)

This grass is on a global scale the most important cultivated tropical grass of the present day, mainly on account of its wide adaptation, ease of establishment, stoloniferous habit and relatively reliable seed production. Rhodesgrass was hardly known in agriculture, like so many other cultivated tropical grasses, until about 1890 when colonial settlement began in East and Central Africa. It was named after Cecil Rhodes who popularized its use. The grass quickly took to the wings and seed was introduced into USA and Australia around the turn of the century. Seed is thought to have been taken to Australia by soldiers returning from the Boer War. This is believed to be the origin of Pioneer Rhodes in Australia.

Nzoia Rhodes is another variety believed to have originated in southern Africa (*6.6.3*). Pioneer, Nzoia and Katambora are diploid, whereas varieties of strictly East African origin are tetraploid (Bogdan, 1966a). Bogdan postulated that diploid Rhodesgrass is found near the perimeters, whereas tetraploids prevail near the centre of diversification. Therefore, when the different responses of, for instance, Katambora and Masaba Rhodes are discussed it is more accurate to distinguish between southern (or subtropical) and equatorial types rather than between diploid or tetraploid, respectively.

Rhodesgrass has the typical features of a stoloniferous sward-forming grass. At the East African coast, however, an upright habit prevails. Commercial varieties have a digitate panicle with 8–12 racemes of 6–9 cm length. The seed is

Table 6.7. The major characteristics of seed-multiplied Kitale varieties (Boonman & van Wijk, 1973).

	Sequence in week of Initial Head Emergence (IHE) 1968–1971	Number of tillers at IHE (m^2) 1969	Weight per tiller at IHE (mg) 1969	Maximum head number (m^2) 1969	Maximum % heading tillers 1969	Culm length at optimum seed harvest date (OHD) 1969	1000-grain weight (mg) at OHD 1968–1971	PGS-yield per crop (kg/ha) at OHD 1968–1971
Setaria sphacelata cv. Nandi I	3	950	260	350	25	150	420	32
Setaria sphacelata cv. Nandi II	4–5 22							
Setaria sphacelata cv. Nandi III	5–6	980	220	170	9	135	420	15
Chloris gayana cv. Mbarara	2	1550	210	320	20	132	280	44
Chloris gayana cv. Masaba	4–5	1290	190	230	13	130	260	40
Chloris gayana cv. Pokot	6	1260	260	160	12	130	310	24
Panicum coloratum cv. Solai	1	1530	200	520	34	115	460	52
Panicum maximum cv. Makueni	1	1150	280	130	10	144	540	25
Brachiaria ruziziensis (Congo Signal)	9	1240	280	470	37	126	1930	23

Table 6.8. A historic overview of commercial varieties of East Africa (K: year and number of registration at Kitale).

Name of species		Farmers' introductions (< 1940)		Research introductions (> 1950)		Bred varieties	Other names
		Name	Origin	Name	Origin		
Rhodes	*Cgay*	Nzoia (2n)	South Africa (?)				
		Rongai (4n)	Rongai (nr Nakuru)				
		Endebess (4n)	Mount Elgon, K50193				
		Mpwapwa (4n)	Iringa (Tanzania), K5611				Callide, Kongwa, Giant
				Masaba	Endebess (Kenya), K50193		Endebess
				Mbarara	Mbarara (Uganda), K53166		
				Pokot	Kapenguria (Kenya), K632		
						Elmba, K7501	
						Boma, K7502	
Setaria	*Ssph*			Nandi	Nandi Hills (Kenya), K53175		Nandi (I) Original
						Nandi II, K6154	Nandi Improved
						Nandi III, K6477	
						Narset, K7503	
						Nasiwa, K7601	
						Mapema, K7602	
Molasses	*Mmin*	Kitale Commerical					
				Mbooni	Mbooni Hills (Kenya), K54524		
Coloured Guinea	*Pcol*			Solai	Solai (Kenya), K52490		
Common Guinea	*Pmax*			Likoni	Likoni (Kenya), K52128		
				Makueni	Makueni (Kenya), K6221		
Congo Signal	*Bruz*			Congo	*via* Rubona (Rwanda), K5832		

For further information, see Bogdan (1965a; 1977).

in the form of fluffy spikelets with a 1000-seed (spikelet) weight of 200–325 mg, *i.e.* 3–5 million spikelets per kg. The true seeds, caryopses, separate easily from the spikelets and have only 25% of the weight of the spikelets. Flowering takes place (at Kitale) in the afternoon (Bogdan, 1959b). Rhodesgrass is cross-pollinating (Bogdan, 1961c).

6.6.1. Adaptation

Rhodesgrass can be found naturally all over Africa in grassland, in woodland and on lake shores. Further north, it can be found up to 40°N in Spain, north of Valencia, along roadsides (personal observation) and the grass has also been used in the southern USSR (Bogdan, 1969). Roadsides are the favourite spots where grasses tend to settle first outside their normal environment.

The main zone of natural distribution in East Africa is Zone III (*3.4.3*), mostly from 1000 to 2000 m altitude, but above that limit also along roadsides. At Kitale a collection was maintained of a wide range of ecotypes from all over Africa, but the East African ecotypes ultimately proved superior (*6.4.1*).

The versatility of Rhodesgrass towards temperature ranges is remarkable. The optimum temperature lies around 35°C (Whiteman, 1980) but Rhodesgrass thrives at 40°C in the Arabian desert under irrigation, yet withstands occasional frost (Farnworth, 1976). Survival under frost depends on the duration and kind of exposure, on other climatic factors and on the condition and age of the sward itself. Although no truly suitable sown grasses exist for the East African Coast (*4.4*), Rhodes- and Guineagrass were the principal grasses sown at Tanga (Hopkinson, 1970).

Conflicting daylength responses have been reported (see also *6.7*; *8.7*). Some varieties have been found to be day-neutral (Nada, 1985), others to respond to short-days. However, most varieties produce an early- and late-season crop within the same year, also in the subtropics. Before a delay in onset of heading is attributed to photoperiodism, more information on the relation to interacting factors such as low temperatures and length of growing season is essential (Loch, 1980). In Kenya, on the equator, all equatorial varieties produce heads in abundance throughout the rainy season but the sequence in heading date is in a strict chronological order: Mpwapwa > Mbarara > Masaba > Pokot. The drier the site of origin the earlier the onset of heading when grown together at Kitale (*6.6.3*; Fig. 9.1). At higher latitudes with more variation in daylength, southern varieties such as Pioneer and Katambora head early and produce 2 crops per year. In Queensland, Callide (*6.6.3*) is late heading, but the first seed crops can be taken in early summer after an early cleaning cut and N-topdressing, as far south as Gympie at 26°S (Loch, *op cit.*). In Kyushu in Japan (33°N), where Rhodesgrass is sown as an annual crop, Katambora heads in August/September, more than 2 months earlier than Masaba, which can, therefore, not produce mature seed in time before winter.

On the equator, however, Pioneer and Katambora do not head early. As already indicated (*6.4.1*) general performance of southern material is poor on and near the equator, which makes it difficult to judge heading behaviour. Nzoia Rhodes, however, which was at one time believed by Australian authors to be "of the same stock, ex southern Africa, as Pioneer" is late-heading and was fairly vigorous in Kenya before it succumbed to disease (*6.6.3*).

Support for the hypothesis that the sequence of heading is reversed when varieties are moved from the equator to higher latitudes is provided by York *et al.* (1989) in Zimbabwe. The following sequence of heading was observed (in March):

Katambora > Callide > Giant > Samford > Masaba > Boma > Mbarara > Elmba

This is of interest because it would show that the heading order in Zimbabwe of the last 4 varieties, *i.e.* Masaba, Boma, Mbarara and Elmba (all ex Kitale) is exactly the opposite of what is commonly found in Kenya. Elmba and Mbarara may, therefore, be too late at high latitudes. It could also clarify why varieties of higher latitudes do not perform well at the equator (*6.4.1*).

A similar phenomenon of contrasting heading sequence was observed in Buffelgrass in

Queensland by Humphreys (1967) who found that Gayndah is later than Biloela if spring sown but earlier under shortening day conditions. Gayndah originates from a higher latitude in East Africa (Table 6.9).

Rhodesgrass is versatile as regards moisture requirements also. It has a reputation as a drought resister of great strength, but 500–600 mm annual rainfall is a minimum. On the other extreme, it thrives under ample rainfall and responds very well to irrigation as experience from Lake Naivasha (K) and the Arabian Peninsula (Farnworth, 1976) has shown. DM-yields of 50 ton/ha/year have been recorded but such figures should be treated with caution. The same applies to accounts that Rhodesgrass has tolerance to waterlogging. Connected with its good response to irrigation is its salt tolerance (*6.3.2*) which is of great importance in the lake areas of the Rift Valley and in the Sudan (Bunting & Lea, 1956). Rhodesgrass is rich in minerals (Henrici, 1931), in Ca (Marshall & Long, 1971) and especially in Na, of which it can tolerate high levels in the soil.

The wide adaptation of Rhodesgrass goes a long way to explain the reliable performance of this grass over a wide range of locations with different climates and soils (Tables 6.2 & 11.5). On Vertisol but also on Brigalow soils in Australia Rhodesgrass has been observed to have persisted for over 25 years and to have increased in density (Silvey & Jones, 1990), although Rhodesgrass persistence is not generally regarded as high (*6.3.3*).

6.6.2. Husbandry

Rhodesgrass is easy and quick to establish from seed. Success of a sown pasture depends largely on the success of sowing. This is an issue which can, therefore, not be over-emphasized (*7.1*).

Under Kitale conditions at the onset of the rains, seedlings emerge visibly in 8–10 days, which is fast compared with other tropical grasses. Its subsequent establishment gathers speed gradually, but a 50-cm high sward can be achieved within 50 days after sowing, yielding 2000 kg/ha DM. The recommended seed rate of Rhodesgrass is 1 kg/ha PGS (pure germinating seed) (Table 7.3). Thus, 4 kg of 25% PGS is needed. With spikelets numbering 3.5 million/kg (1000-seed weight of 275 mg), this represents $360/m^2$ potential seedlings. Mortality is, however, high so that rarely more than 100 seedlings establish, often even less. Much higher seed rates are sometimes used by farmers, especially when the seed is home-grown and the germination capacity unknown. High seed rates are not only wasteful but they also prevent seedlings from developing stolons. Hence, if Rhodesgrass is sown too thickly, also within wide rows, stolons cannot develop properly and the field has the appearance of a row crop rather than of a sward. Such rows may remain visible for many years. This is especially evident after mowing and after a burn, since regrowth begins first in the original sowing lines. By contrast, when seedlings are sparse, with $50/m^2$ or less, stolons develop rapidly and knit together tightly into a sward. However, very low seed rates aimed at $50/m^2$ seedlings are a great risk and although experiments have shown that in hand-sown rows establishment was equally good over a range of seed rates from 0.2–1.8 kg PGS (Boonman, 1972b), farming practice should stick to a rate of 1–1.5 kg. Rhodesgrass seedlings are often only lightly anchored in the soil, with no more than a thread-like elongated internode, probably of the coleoptyle, which makes them vulnerable. Evidence of pests and diseases on Rhodesgrass is small, but some damage occurs at times in late sown Rhodesgrass from maggots of a shootfly (*Atherigona* spp.) which kill the main shoot of young seedlings. The same maggots have fortunately an even more negative effect on the seedlings of Wimbigrass (*Eind*), a common weed of arable land (*7.1.10.1*); see also Bullock (1965). The Rhodesgrass scale (*Antonina graminis*) (Chada & Wood, 1960), known to damage a wide range of species in North and South America, is of no importance in East Africa.

Rhodesgrass was planted vegetatively in central Kenya (Bogdan, 1959a; 1965b) but vegetative propagation is generally an unreliable method of establishing Rhodesgrass, also in (sub)humid climates. This applies to many of the grasses of Zone III (Table 6.6), also when stoloniferous. Considerable effort has been spent at Kitale on finding optimum conditions of vegetative transplanting. Weather conditions are important but of overriding influence is the right physiological stage of the material to be transplanted. Stolon cuttings are not suitable, but instead sections (root splits) should be taken from plants in full tillering after the seedling stage. Splits dug from 2nd or 3rd year swards are difficult to be kept alive. Mowing in order to encourage regrowth, before splits are dug, does not help their survival (van der Torren, unpublished).

Rhodesgrass used to be mainly sown directly into open land but more recently undersowing to maize and wheat has again become popular. Undersowing to wheat carries the risk of the herbage overtopping the wheat, but this is likely to occur only when Rhodesgrass is sown at high seed rates of varieties with a high seedling vigour such as Boma or Elmba Rhodes.

Whereas Rhodesgrass is mostly used for direct grazing, it has been the most popular grass for haymaking (Edwards, 1934), also on smaller farms in semiarid areas such as Baringo and Eastern Province where Rhodesgrass is doing well (Pereira, 1959) and widely sown (Rossiter, 1974). In comparison with for instance Setaria, Rhodesgrass has by itself a high DM % (Table 6.3) and the stems are thinner and easier to cure. Cutting and carting of hay on the same day is not unusual.

It is generally assumed that, in Zone III, Rhodesgrass pastures can carry 2 lactating exotic dairy cows/ha throughout the rainy season producing 4000 kg/ha/year milk for the first 2 years of the pasture, without N-fertilizer. In the 3rd and 4th year high milk yields can be maintained, with N-fertilizer at 100 kg/ha N. In beef production this should mean an annual liveweight gain of 400–500 kg/ha or more (see also Fig. 10.6).

The decline of productivity from the 2nd year onwards is common to all cultivated tropical grasses (Figs. 6.2, 6.3 & 11.2), but N-fertilizer prolongs the lifetime. Rhodesgrass pastures are essentially short-term and fit ideally in crop-grass rotations promoting a rapid recycling of soil-N (*5.3.2*; *5.6.1*). Eradication by ploughing is less troublesome than in more persistent grasses. Rhodesgrass is better ploughed before weed grasses such as African Couchgrass (*Daby*), *Sporobolus pyramidalis* and *Eragrostis tenuifolia* (*4.3*) get the chance to invade. Reduced competition on the part of the grass does, however, not seem to encourage legumes, if any, in the grazed sward. Rhodesgrass is generally regarded as too aggressive for legumes to establish properly. However, it is becoming widely appreciated that the problem of legume lability in grazed swards is inherent to herbaceous legumes in general (*4.3.5*; *10.4*).

6.6.3. Ecotypes

On volcanic plains such as in the Rongai (Nakuru) area in the Kenya Rift Valley, Rhodesgrass can dominate (*3.4*), but dominance is the exception rather than the rule in East Africa. In Ankole (Mbarara) in Uganda, Themeda-Chloris communities occur (Langdale-Brown *et al.*, 1964). Rhodesgrass is more commonly associated with *Hyparrhenia* spp., *Hyperthelia dissoluta*, Guinea- and Signalgrass of the Acacia savannah of Langdale-Brown's classification.

It should be noted that Rongai and Mbarara are the sites where two of the most popular varieties of Rhodesgrass originated. The name Rongai has been associated with Rhodesgrass in the Rongai-Nakuru area (*3.3.2*; Ball, 1937). Seed of the "Rongai type" was easily obtainable (Bogdan, 1959a) and cheap. It was coded K5653 ex Dalgety (a trading company) and was one of the already existing varieties which the Kenya Seed Company undertook to process and market when it was founded in 1956. Although cheap seed has often been equated with inferior varieties (*9.5*) Rongai used to be a most popular grass also in the Mount Kenya area (Jones & Evans, 1961).

As Bogdan (1959a) pointed out, attempts to introduce varieties better than Rongai have been made since the early 1930s when seed of an "Australian strain" (probably the variety now known as Pioneer Rhodes) was imported by Edwards (1934). It was tested throughout East Africa, but without satisfactory results (Maher, 1936b; Martin & Badcock, 1937; Staples, 1938). More recently, Pioneer was confirmed to do poorly in Ethiopia (Lazier, 1989).

At Kitale in the Trans Nzoia district, another variety was grown (Maher, 1936b; Ball, 1937; Halcrow, 1938) which was later named Nzoia. It was tested by Edwards (1939) along with 2 other local varieties, Kabete and Alego, as well as Teso (ex Uganda) and Kafue (ex Rhodesia). Finally, Nzoia Rhodes was recommended (Edwards, 1943) but still no mention was made of Rongai. However, until the 1960s Rongai and, in Uganda, Teso Rhodes remained the main variety of which seed was commercially available. Rongai is still popular.

Although all the grasses sown at present are derived from ecotypes of East African origin, it took considerable time before their superiority over imported material was recognized (Maher, 1936b). Imports had dominated the scene from the very early days (*2.2.2*) and this trend continued in the early work of Edwards (1936) on sown grasses.

Nzoia Rhodes emerged from material grown in the Kitale area, but is probably of South African origin (Bogdan, 1965a). It may have been introduced by settlers in the Nzoia area, acquiring a local name in the 1930s, to distinguish it from the rest. That Nzoia was probably imported is based on the finding that all varieties found in the Trans Nzoia district are tetraploid and not diploid like Nzoia (and Pioneer) and that Nzoia was less stoloniferous and became susceptible to Helminthosporium (Bogdan, 1959a; 1965a; 1966a). This disease susceptibility is shared with varieties of other grasses which are not local, like "Kitale" Molasses (*6.10*) and Pangolagrass (*6.4.1*). Had Nzoia Rhodes been of the "same stock as Pioneer Rhodes" (see above) then the similarity with the "Australian strain" imported by Edwards (1934) would not have gone unnoticed. Although genetic shift or varietal admixture cannot be ruled out (*9.1*), Nzoia Rhodes was vigorous and late-heading (Bogdan, 1959a). The South African connection may explain why Nzoia has been the only Rhodes variety with an East African name on the Cedara (Natal) Variety List.

Although Nzoia Rhodes was the officially recommended variety, its marketing developed slowly. Government multiplication schemes were set up after World War II at Thika and subsequently at Eldoret in the early 1950s (*9.1*) but also after the founding of the Kenya Seed Company in 1956 its share never exceeded more than a few % of the Rhodes seed market and ended in 1964. It was registered by Strange and Bogdan at Kitale as K50198 at the same time as other varieties then in multiplication which included Endebess Rhodes (K50193), Alego Rhodes (K50199) and Slender Guinea (K50201).

Endebess was already a well-known farmers' introduction around the time of World War II, originating from and multiplied in the Endebess area, on the eastern slopes of Mount Elgon. Material was taken in 1950 by R. Strange to the Kitale Station. It was renamed Masaba Rhodes (Bogdan, 1961c; *6.6.4*), to distinguish certified seed from that produced and sold directly by farmers. Bogdan (1959a) noted that it had shown good results in dry areas. See also Table 6.8.

Other varieties which attracted interest at some time, but which failed to reach long-term recognition are Pokot, Mpwapwa, Kapedo and Teso Rhodes.

Pokot Rhodes (K632) was selected by Bogdan (1965a) from material (K57168) collected by R. Strange in Pokot district in western Kenya. The variety is extremely late-heading and, prior to heading, reaches a sward height higher than other commercial varieties. Leaves are large and broad, and this leafy appearance gave Pokot a popular start. Since the Pokot area is largely semiarid, Pokot Rhodes was soon credited with drought tolerance also. Its popularity was, however, short-lived. Seed yields were disappointing, which led to other characters being held suspect also. The drought tolerance of Pokot could not be confirmed and it was only then realized that the site of original collection was not in the dry and hot lowland, but in the moist and cool upper parts of the Kapenguria escarpment of Pokot district. Late-heading ecotypes are usually indicative of long growing seasons. Secondly, herbage yields were found to be lower than in Masaba and Mbarara. Pokot produced a seed yield only 60% of that of Mbarara and Masaba (Boonman & van Wijk, 1973). Also its leafy appearance proved to be misleading. The leaves are large, yet the leaf:stem ratio and digestibility are low (Table 9.6). Pokot was the first variety subjected to breeding work, in 1968, when a large intra-variety variation in heading date was detected (Fig. 9.1; Boonman, 1971a), but work was later discontinued when seed multiplication was suspended by the Kenya Seed Company (Combes, 1976). Pokot is an example of how misleading appearance and naming can be. Its place was largely retaken by Masaba Rhodes, which is also late-heading, but more productive (*6.6.4*). Another late-heading

Kitale variety (Chepararia; K51426) from Pokot district was reported on by Naveh & Anderson (1966) and Sands *et al.* (1970), but Bogdan (1959a) and Bogdan & Mwakha (1970) found it to be less persistent and low in seed yield.

Mpwapwa Rhodes, (K5611). This variety was brought to the Mpwapwa Station from Iringa (T; 8°S) in 1937 (van Rensburg, 1955; Bogdan, 1959a). Being the only Tanzania variety it continued to draw attention (Anderson, 1968a). It was sent to Australia in 1953 where it was released in 1961, and later renamed "Callide" (Barnard, 1972). Not unlike other varieties heading early on the equator (*6.6.1*), Callide is late-heading in southern Queensland (Loch, 1980). Mpwapwa has longer awns (Bogdan, 1961c) and longer hairs on the spikelets than other varieties, except Kapedo. Seed was marketed by the Kenya Seed Company in the early 1960s but Mpwapwa has not become popular. It performed relatively well in a variety trial at Katumani (Table 6.10), but was outyielded by Boma Rhodes. At the equally semiarid site of Kongwa near Mpwapwa, it gave way to Buffel- and Stargrass (Brzostowski & Owen, 1964; Wigg *et al.*, 1973). Whereas Buffelgrass performed better at Kongwa, Rhodes was more satisfactory at Mpwapwa (Wigg, 1973). Sands *et al.* (1970) ranked Mpwapwa Rhodes among the better types at Katumani and Chemogoch, but work was later continued with Chepararia Rhodes (Table 14.2). On account of their dry-area names and associated drought tolerance, both Mpwapwa and Kapedo Rhodes were re-evaluated at Kitale in the 1970s; they were early-heading but no advantages in drought tolerance or herbage potential over the existing Kitale varieties could be observed.

Kapedo Rhodes, from Kapedo, north Baringo (K), similar to Mpwapwa. Originally evaluated at Katumani. Was found to be not worth further trying. A major shortcoming of Kapedo and Mpwapwa is the difficulty in seed handling caused by the hairy spikelets. Kapedo was included in the Kenya District Trials (Table 6.2), but its performance was below average, also in the semiarid sites.

Teso Rhodes, from Teso (U). Seed of natural stands of local (Teso) Rhodesgrass was collected and processed at Serere Station and sold to farmers (Martin & Badcock, 1937). It was the main variety available in Uganda (Ledger, 1950) before the import of Kitale varieties.

The re-evaluation of Rhodes varieties confirmed that not drought tolerance but heading time can be related to the climate of the area of origin. Varieties that originate from dry (*e.g.* Kapedo) areas do not display a better drought tolerance than those from humid areas (*e.g.* Masaba). In fact, Masaba and Boma display a wider adaptation to dry and warm locations than ecotypes from drier habitats. The mechanism, found in Kapedo but also in Rongai Rhodes, is to escape drought by heading very early and going dormant early, *i.e.* by the time the main drought period sets in. The shorter the rainy season in the area of origin, the earlier an ecotype heads, also when transferred to other equatorial areas for cultivation. This explains why ecotypes from Zones III and IV such as Mpwapwa, Kapedo, Rongai and Mbarara head early in contrast with Masaba and Pokot, which originate from Zone II/III (*3.4*). Results of a range of Rhodesgrass varieties, introduced to Kitale in the early 1950s and tested in (semi)arid environments such as Katumani have been given by Sands *et al.* (1970).

It is perhaps worth noting that varieties such as Nzoia, Pokot and Mpwapwa received a lot of research publicity but their share in the seed trade was short-lived and minimal in comparison with Rongai and Endebess/Masaba. In the 1970s emphasis shifted to Mbarara and Masaba Rhodes and to new varieties bred from these.

6.6.4. Commercial varieties in East Africa

Mbarara Rhodes, (4n = 40), (K53166). Collected near Mbarara (U) by Miss Trelawny of the Kitale Station (Bogdan, 1965a). It was the most persistent variety under trial (Bogdan, 1959a). An early, high-yielding variety, with a high DM %, good seeding, reliable and very popular. Mbarara contains plants with non-anthocyanin yellow heads which is of interest in inheritance and pollen contamination studies (Bogdan, 1963c).

Elmba Rhodes, (4n = 40), (K7501). A more recent *EarLy* variety bred from selected plants of

*Mb*arara (Boonman, 1978c); hence the suffix *mba*. High in seed yield. Similar to Mbarara, but more than one week earlier in heading. Especially suitable for frequent, intensive grazing regimes. Chosen by farmers for irrigated, heavily N-fertilized dairy farms on the shores of Lake Naivasha.

Masaba Rhodes, (4n=40), (K50193). A farmers' introduction of the 1930s, identical to Endebess (see above), but the name Masaba (local name for Mount Elgon) was adopted since about 1960 to distinguish certified seed from "commercial" Endebess Rhodes. The oldest of the famous Kitale range of Rhodesgrass varieties. Medium-late heading, yet with fair yields of seed and high yields of herbage, and high digestibility. Masaba has a lower seed yield, but better persistence and digestibility (independent of growth stage) than Mbarara (Table 9.6). It is also well adapted to a wide range of temperatures, performing well both in coastal areas and at high altitudes.

Boma Rhodes (4n=40), (K7502) A more recent variety bred from selected plants of Masaba Rhodes (Boonman, 1978c); hence the suffix *ma*. Boma Rhodes was released not only because of its higher seed yield and better digestibility compared with Masaba, but also because of its wide adaptation. In the Kenya District Trials (Table 6.2) Boma Rhodes turned out to be the most productive of all sown grasses, at humid sites such as Ol Joro Orok (2380 m), Kakamega (1590 m), Embu (1460 m), but also at semiarid Katumani (Table 6.10) and the coast. Seed has been exported to South America, the Arabian peninsula and Japan and is also multiplied in Malawi and Zambia (Kulich & Kaluba, 1984). In Zambia, Boma yielded more seed than Mbarara and Katambora (Kulich, 1985).

Another variety of world repute, but from southern latitudes, is Katambora (2n=20), from near the Zambezi river in Zambia, where it was known prior to 1950. Katambora is an important variety on the world seed market and has been used in many countries, also in Australia where it is said to be finer-leaved, later-heading and more vigorous than its diploid counterpart Pioneer Rhodes (Barnard, 1972). Katambora came into use in the 1950s in Zimbabwe and neighbouring countries and is mainly grown in rotation with tobacco to control root-knot nematode (*Meloidogyne javanica*). It is less productive there than equatorial (tetraploid) varieties, *e.g.* those from Kitale (York *et al.*, 1989), which do, however, not suppress this nematode. It goes without saying that this unique feature of biological nematode control has greatly enhanced the image of Rhodesgrass.

Mbarara and Masaba are doing well in various parts of West Africa and in Indian Ocean islands (Borget, 1969). No information on the performance of the Kitale varieties is available from Australia because of a ban on the commercial imports of herbage seeds imposed in the 1960s. Rhodesgrass is the most widely sown grass in Australia (Walker & Weston, 1990). Since the release of Callide in the early 1960s no new varieties have been introduced nor developed by breeding. At Kitale, the range of varieties used in Australia, including Katambora, Callide and Samford Rhodes was reintroduced and retested (Boonman, 1978c), but none of them had a performance comparable to that of the commercial Kitale varieties. Lack of vigour was observed and in subsequent regrowth most varieties also lacked the capacity to form a dense sward. Genetic shift and admixture may have occurred, but the types were still adequately recognizable. A more likely reason is that varieties from or multiplied at higher latitudes are not adapted to equatorial East Africa or succumb to diseases (*6.4.1*).

6.7. Setaria (*Setaria sphacelata*)

The Setaria genus includes more than 100 species. Of the annuals, foxtail millet (*S. italica*) is an important grain crop, but again other annual species, *S. pumila* (syn. *S. pallide-fusca*) and *S. verticillata*, are notorious weeds in arable land. From the herbage point of view, attention will focus on *Setaria sphacelata* var. *sericea*, the grazing type. Kazungula was described as an intermediate between *Ssph* var. *sericea*, to which Nandi belongs, and var. *splendida* (Clayton & Renvoize, 1982). Var. *splendida* will be discussed separately as a fodder type (*14.1*).

Setaria sphacelata is a polymorphous species and different names were used at times at the (sub)species level. However, inter-specific hybridization with other Setaria species was easy (Bogdan, 1961d). Within *Setaria sphacelata* a whole range of ploidy levels can be found: diploid (Nandi) and tetraploid (Kazungula; Narok), but Hacker (1966) reported accessions also of 2n = 45, 54, 72, and 90 with chromosome pairing in most of the combinations that had even chromosome numbers. Between ploidy levels cross-fertilization commonly occurs.

The use of Setaria was pioneered in South Africa. The varieties Kazungula (from Zambia) and du Toits Kraal (from north Transvaal) were collected by Pole Evans (1950) and were grown together with Bua River (from Malawi) in Transvaal before 1940 (Botha, 1945). Their role remained small and they have done less well at the equator (*6.4.1*). Kazungula was stemmy at Kitale (Bogdan, 1955). The more giant types of Setaria, including var. *splendida*, have their maximum diversity in southern Africa (Bogdan, 1966a).

Setaria did well at Ngetta in north Uganda (Martin & Badcock, 1937) and also drew attention in Zaïre (INEAC Ann Rep 1948) and Tanzania (van Rensburg, 1948a). Although Nandi Setaria had been transferred by D.C. Edwards in 1935 from the Baraton Veterinary Centre in Nandi district to his Kabete nurseries, it was only after its transfer to the newly founded Grasslands Research Station at Kitale in 1953 (K53175) that its potential came to be realized. Nandi Setaria was found to be the best out of a total of 106 ecotypes (Bogdan, 1960c). It became popular with farmers, when the Kenya Seed Company started operations (*2.5*). Seed became available in 1957. Nandi Setaria has since been tested and grown in many countries (Hacker & Jones, 1969). In Uganda Nandi Setaria does particularly well in the "fertile crescent" but may not be satisfactory in northern regions with long dry seasons (Horrell, 1958a).

Various attempts have been made to exploit the variation found within Setaria varieties (van Rensburg, 1956; Bogdan, 1959a; Behaeghe & Blouard, 1962). Bogdan's aim was to select for leafy plants (Bogdan, 1959a) which, in actual practice, meant the singling out of late-heading plants. Nandi II (K6154), also named Nandi Improved, was selected from Nandi (renamed Nandi I); in a further selection cycle, Nandi III (K6477) was developed from Nandi II (see also van Wijk, 1980). However, seed yields of Nandi III were found to be less than half those of Nandi I, on field scale (Kitale Ann Rep 1968), so that Nandi II became suspect also. It was subsequently found that not only seed but also herbage yields are negatively affected by the indiscriminate selecting for late-heading, *alias* leafy plants, because of the inverse relationship between heading date and vigour of regrowth (Boonman & van Wijk, 1973). The Setaria programme was revised accordingly which led to the release of Narset and Nasiwa Setaria, also selected from Nandi (*9.5.3*; Tables 6.2, 6.8 & 11.5). Simultaneously, a renewed attempt was made at studying other ecotypes: Lugari, Subukia, Gilgil and Meru, introduced by Strange and Bogdan. A 3-year comparative variety trial carried out by van Wijk (unpublished) confirmed, however, that Nandi was superior over all 3 ecotypes in all important aspects.

Australia had become interested in Setaria around 1960, although Kazungula had been introduced from South Africa already earlier, in 1949 (Barnard, 1972). Seed of Nandi (Nandi I; CPI 28709) was imported commercially in 1961 (and in 1964 of Nandi II) and met with considerable success in the warmer and wetter parts of (sub)tropical Australia. Soon after, all commercial imports of grass seed from East Africa were banned. Subsequently, most Australian research interest went to Narok Setaria, a direct introduction made by R.J. Jones in the mountainous Aberdares region of Kenya. Narok Setaria has a better frost tolerance. When subsequently tried in Kenya itself, Narok failed to produce seed in commercial quantities, which limits its use also in Australia (Hacker, 1985).

Characteristic of Setaria are the flat stem basis, the keeled leaf sheaths and the long and dense ear-like panicles which are normally 13–16 cm long in seed fields, but often twice as long in the year of sowing, especially in the Narok variety. For more detailed botanical descriptions, see Bogdan (1977) and Barnard (1972).

6.7.1. Adaptation

Setaria is widespread in East Africa, mainly in the humid Zone II (*3.4.2*), independent of altitude (from sealevel to 3000 m), and the wetter parts of Zone III. Setaria occurs in the Kikuyu- and Stargrass zone but also in the humid lowlands in central Zaïre (Blouard & Behaeghe, 1961). In eastern Zaïre it was found to colonize abandoned cultivations (Risopoulos, 1966) and to be the only plant capable of dominating African Couchgrass (*Daby*). Clear cases of dominance of Setaria in natural vegetations are rare in East Africa, although Rattray (1960) quotes reports from Ethiopia about a transitional type of Setaria grassland. Setaria is normally found in association with other grass species having a more reliable indicator value which can be *Exotheca abyssinica* and *Andropogon chrysostachyus* at the higher and *Themeda triandra, Hyparrhenia* spp. or *Brachiaria* at the lower elevations.

Consequently, Setaria thrives in humid conditions, temperature being much less restrictive. Sealevel conditions are not humid enough in East Africa, in contrast with Malaysia where Nandi does well (D. Nauta, private communication). Setaria can withstand seasonal waterlogging and is recommended for vleis (*3.2*; Gosnell, 1963). It has little reputation as a grass for irrigated areas and has no tolerance to alkalinity or salinity.

Nandi and its derivatives will exceed the altitude limits of Boma Rhodes. It has been grown successfully as high as Molo (2477 m), where temperate grass species do well also (Table 6.1). Its natural occurrence at altitudes above 2500 m, occasionally to about 3000 m, makes it not surprising that Setaria tolerates light frosts well.

On or near the equator, Nandi Setaria produces flowering heads during active growth, throughout the year. Further north and south in the subtropics it may respond as a quantitative short-day plant and head more profusely in early autumn, but 2 seed crops of Setaria per year, in January and May, are apparently quite normal in Queensland (Hacker & Jones, 1971; Humphreys & Riveros, 1986). The issue is confused, since long-day responses have been found both in southern Australia, in Japan (Nada, 1985) and at Wageningen (Dirven, private communication, 1984).

Nandi Setaria is relatively susceptible to damage by fire. For instance, when grown in seed fields in wide rows, regrowth is greatly impaired when the grass catches fire in or at the end of the dry season.

6.7.2. Husbandry

Seed yields of Nandi Setaria, the best known cultivated variety, are relatively good, but can vary tremendously from one year to the next, in comparison with Rhodesgrass, because seed set is particularly sensitive to drought periods (*8.5.2*).

The recommended seed rate of Nandi Setaria is 1.5–2.0 kg/ha PGS, when broadcast (Table 7.3). The 1000-seed weight is about 425 mg, so that spikelets number 2.5 million/kg. The % PGS of cleaned seed may vary from only a few % to over 50% so that indications of seed yield or seed rate are irrelevant without reference to % PGS.

Setaria germinates much slower than Rhodesgrass. Under Kitale conditions seedlings may not emerge in less than 15 days. Weed competition can be so serious that little Setaria is available for grazing in the year of sowing. In the Kitale region it has been common practice to defer judgment until after the 1st dry season. Farmers are not worried when establishment appears to be poor. The resilience of Nandi Setaria after the 1st dry season is, by contrast, quite remarkable. Real failures are, in fact, uncommon and mostly arise when seed is buried too deep.

Vegetative propagation is easy (Table 6.6). Mortality of root splits is much less than in Rhodesgrass. However, since the grass is tufted many splits need to be planted to achieve a closed sward. In humid areas such as in Zaïre, vegetative propagation is a possibility when seed storage facilities are poor (Behaeghe, 1960).

Mixtures of Nandi with other grasses are not advocated (*7.1*). Results with legumes have at times been better experimentally than with other grasses, probably because Nandi Setaria is less competitive in the year of sowing. At altitudes above 1750 m Nandi Setaria mixes quite well with Louisiana white clover (*10.9.2*). In East Africa Setaria is remarkably free from leaf diseases and pests, a virtue which is shared with Rhodesgrass. White mould (*Beniowskia sphaeroidea*) occurs sporadically on the foliage, especially in Giant Setaria, but in a milder form than in Elephantgrass (*11.8.1*). A notable fungus is bunt (*Tilletia echinosperma*) which produces inflated spikelets (*8.7.2.5*).

Unfavourable growth conditions favour the development of leafy shoots in the ripening panicles. This is called vivipary or vegetative proliferation (Langer & Ryle, 1958) and occurs usually over the whole length of the panicle. When such panicles are removed and planted, root formation occurs and many new plants grow up. The phenomenon of vegetative proliferation in Nandi Setaria is common, when plants are grown in small containers, but is of little practical importance.

Rare, but rather serious where it occurs is the parasitic *Striga forbesii* which grows on the roots of Setaria and produces small flowers, usually of a pink or nearby colour. Only 2 sites are known to the author where isolated Setaria plants were affected by this parasite, growing hardly more than 10 cm tall. Affected Setaria plants were badly weakened. This Striga is different from the tall *S. hermonthica* (purple witchweed) and from *S. asiatica* (red witchweed) affecting Sorghum in particular (*15.6.3*) but also other arable graminaceous crops and grasslands. *S. asiatica* was recorded in edaphic grasslands with *Loudetia kagerensis* between Lake Victoria and Masaka (Lind & Morrison, 1974). Mathieu (1960) listed it as a component of high altitude grasslands of Rwanda and Burundi (*3.4.2.1*). The weed aspects of witchweed have been described by Tiley (1970b), Ivens (1971) and Terry (1984).

A drawback of Setaria when fields are ploughed is that seedlings volunteer in subsequent crops. Seedlings can achieve a complete cover and dominate other grasses. This characteristic is shared with Brachiaria and Panicum, but not with Rhodesgrass (Table 6.6).

The local use in Kenya of Nandi Setaria has diminished in recent years in favour of Rhodesgrass, for various reasons. Nandi Setaria is preferred by dairy farmers, especially in the wetter areas towards the higher altitudes. Such farmers admit that Setaria is slower to establish, harder to eradicate and less useful in the dry season compared with Rhodesgrass, but these weaknesses can be alleviated by appropriate husbandry. Great value is attached to its capacity to remain productive and to respond to N-fertilizer, thereby increasing the stocking rate both per year and over a greater number of years (Fig. 6.2). Responses of 65 kg DM/kg N applied were observed over a 3-year period with 2 cuts taken per year (Boonman, 1972a).

In the Kenya District Trials (Table 6.2) Setaria's performance was only about average, also at Baraton, the site of its original collection. It did relatively better in the high rainfall sites below 1750 m. Nandi was outyielded by Narset, a new selection (van Wijk, 1980).

6.7.3. Animal production and nutritive value

Animal production data off Nandi Setaria are scarce but more numerous than of Rhodesgrass. At Kitale, a close correlation was observed between milk yield and the % crude protein in the growing season (Kidner, 1963). Impressive records were achieved at the Elgon Downs Farm at Endebess (K) over a 10-year period with fertilized Nandi pastures. The impression was gained that N-fertilizer applied at the rate of 150 kg/ha/year N from the 3rd year onwards can support 2 lactating Friesians over a period of 10 months, yielding a total of 5000 kg/ha/year milk. At the Kitale Station, unfertilized Nandi pastures gave liveweight gains of 330 kg/ha in the 2nd year and only 190 kg in the 3rd year (Fig. 6.3). At Kisii, milk yields on Kikuyu-based pastures were less than half those of Setaria-Rhodes pastures (Table 6.13).

The interpretation of the nutritive value of tropical herbage is complicated because assessments frequently refer to different stages of growth (*6.3.4*; *9.5*; van Wijk, 1980). In Uganda, Bredon & Horrell (1961; 1962) rated Setaria relatively low, while Reid *et al.* (1973) and, in Kenya, Dougall (1960) observed that nutritive value of Setaria was high initially, but declined rapidly with maturity. Statistical analysis of the numerous digestibility (*in-vitro*) data available at Kitale from species- and variety comparisons over a number of years revealed that Nandi Setaria was 2–3 units lower than Mbarara or Masaba Rhodes, but its % crude protein was as high if not higher (van Wijk & Boonman, unpublished).

More clarity exists about the mineral status. Setaria is low in Ca and Na, but very high in K (Long *et al.*, 1969); see also Elephantgrass (*12.7*;

13.1). Of the different varieties, Nandi is relatively low in Na (Hacker & Jones, 1969). An additional peculiarity of Setaria and related genera is the ability to build up ammonia (Dougall & Birch, 1966; 1967). The presence of ammonia is connected with a high acidity (4.8 pH), which in turn is linked with high contents of organic acids, including oxalic acid. Oxalic acid is a common constituent of plants such as Oxalis, Portulaca and Amaranthus. It is known that extra Ca should be added to the diet to compensate for the Ca-oxalate formed in the body. Oxalate-rich fodders generally do not affect ruminants, once adapted, unless present in exceptionally high concentrations but horses can be affected by prolonged exposure to relatively low levels. Hypocalcaemia may occur in Kazungula, Bua River and Giant Setaria as these have higher levels of oxalic acid than Nandi (Jones *et al.*, 1970; Middleton & Barry, 1978). In Uganda oxalic acid was relatively high in Elephantgrass also (*12.7.5*; Ndyanabo, 1974). Buffel- and Kikuyu-grass are equally suspect (Humphreys, 1991) which indicates that the genera affected are closely related.

6.8. Common Guineagrass (*Panicum maximum*)

The role of Common Guinea in East Africa as a cultivated grass has been relatively small. As a single species, however, it is probably the most widely distributed naturally, from sea level to over 2000 m, in grassland of the dry or humid zones, in abandoned cultivations and along roadsides (*3.4*). Guineagrass is common at lower altitudes and in coastal areas where the more robust types prevail. It is also much in evidence along tree-lined avenues in towns, probably because it is favoured by shade (Staples, 1937) and can grow undisturbed on such sites. Guineagrass is well liked by livestock, but many ecotypes are stalky and coarse (van Rensburg, 1955b). Its drought tolerance in undisturbed growth can be remarkable, although misleading (*6.3.2*), but the grass shows its best performance under humid conditions.

Guineagrass is less suitable in semiarid areas (Sands *et al.*, 1970). It lacks persistence under grazing. Seeding properties are generally poor also (Staples, 1937). Yields and % PGS are among the lowest on record (Table 6.7) and establishment from seed is slow.

In South and Central America a rapid expansion has taken place of Guineagrass since it was introduced adventitiously by the way of the slave traffic from the Guinea Coast of West Africa in the 17th century (Parsons, 1972). Ocean vessels calling on African ports picked up Guineagrass as an assisted migrant in bedding and packing material to be disposed of at ports of destination of America, Asia and later of Australia. Being a grass of tropical coasts, it invaded rapidly into its new coastal environments (*2.3*). The grass left Africa in so many ways and along so many routes, that no records exist of the exact origin of the (robust) types of Guineagrass which have naturalized over large areas of these continents.

Is it possible that this rapid colonization over such a wide front for 3 centuries has helped Guineagrass to evolve new ecotypes more adapted to their new environments and, perhaps, more amenable to domestication. This may explain its acceptance as a cultivated grass in the countries of adoption rather than in its home continent and also why re-introduction of cultivated types from South America or Australia to East Africa has not resulted in more acceptable types for cultivation, a fate shared with other species and varieties of, however, mostly southern African origin (*6.4.1*); latitude of origin in Africa may indeed be the main reason. Large collections of Guineagrass were screened at Kitale, but also at warmer sites such as Serere, Katumani, Embu and at the coast. At the Serere Station seed was collected in the natural vegetation. These were mostly the more robust types, which do not form continuous swards and establish slowly giving companion legumes more time and room to become dominant.

Sward stability under grazing, especially of the robust types, is a major problem in natural grasslands at the coast (*4.4*) and further inland (Staples, 1937). Guineagrass is, however, capable of colonizing cultivated land at the coast and elsewhere (Horrell, 1963), behaving as a secondary grass (*4.4*; *4.5.2*; Table 6.6). In mixed-farming rotations it may, therefore, not be an ideal companion. A common complaint is also that soil wash occurs between the tussocks.

Strange and Bogdan left no stone unturned to locate the elusive ecotype of Common Guinea which could one day be developed for cultivated pastures in East Africa. Apart from the fodders (*14.2*) and the *trichoglume* type, represented by Slender Guinea (K50201), they identified many ecotypes of which Embu Creeping Guinea (K6237) and Makueni Guinea (K6221) drew most attention.

Slender Guinea (K50201). *P. maximum* var. *trichoglume*. The variety Green Panic is grown in Australia but originated probably from Kenya (Bogdan, 1965a). A similar type (K53838) was collected from Rumuruti in central Kenya (Bogdan, 1955). Slender Guinea was multiplied until 1962; it went out of favour because of its disappointing seeding properties (Bogdan, 1965a).

Embu Creeping Guinea (K6237). Found by R. Strange along a forest road near Embu, Kenya (Bogdan, 1965a). A remarkable type with long creeping stem bases, many-noded stems and sharp-pointed leaves, with a dark green colour. Seed yields were low. Although the grass produced a fairly dense sward, unusual for Guineagrass, the performance under grazing was disappointing and productivity was low. When the sward is mown a stubble with sharp points remains. Work was discontinued when it was observed to have been wiped out by a streak virus disease at Embu (Embu Ann Rep 1968).

Makueni Guinea (K6221). A small type, collected and planted by J. Knight at Makueni (60 km south east of Machakos) in eastern Kenya (Bogdan, 1965a). An attractive looking grass in nurseries and seed fields, especially in the dry season when it stands out green, while other grasses have already dried off. Under grazing, however, little of this is found back: swards are open, which allows weed invasion, and dry off rapidly at the end of the rains. It was tested with Slender Guinea at Katumani near Machakos and was found to cease growth fairly early in the dry season (Sands *et al.*, 1970), but Rossiter obtained better results in 1977/1978 (Table 6.10). In the Kenya District Trials (*6.3.1*) very few stations reported more than average yields (Table 6.2). By contrast, Makueni Guinea was found to perform very well in the cool season in north east Australia (Grof & Harding, 1970). Seed yields are low (Table 6.7), mainly due to a very prolonged emergence of the flowering heads which emerge very early but remain few in number (Fig. 8.5). Spikelets number about 2 million/kg. Seedling vigour is weak, so that a seed rate of 2.5 kg/ha PGS is needed (Table 7.3). Seed diseases are common in Guineagrass. Spikelets are affected by bunt, by ergot and/or honeydew disease (*8.7.2.5*). Vegetative propagation of Makueni Guinea gives higher mortality than of more robust or giant types (*14.2*). Apomixis is believed to be the mode of reproduction, but in the Kitale studies on intra-variety variation (*9.1*) Makueni Guinea proved to be very heterogeneous not only in quantitative characters, but also in its patterns of hairiness. This confirms the general observation that apomixis is facultative (*9.8*; Bogdan, 1963e; 1966a; Combes & Pernès, 1970; Savidan, 1983).

Likoni Guinea (K52128). Likoni is a town at the Kenya coast. Some seed was produced in 1964 by the Kenya Seed Company, but interest in this grass was mainly found in Uganda (Horrell, 1958a; 1963). Horrell & Tiley (1970) recommended Likoni rather than locally collected seed, while Reid *et al.* (1973) recorded a high digestibility as compared with Makueni Guinea. Likoni Guinea is often mentioned in reports from the Caribbean, especially Cuba.

Ntchisi/Tengeru/Sigor Guinea. These are fodder types of Guinea grass, which are discussed separately (Part II; *14.2*).

In the 1970s an intensive search for new ecotypes of Common Guinea was mounted by van Soest (1976) in the lower and drier parts of Kenya. Many introductions were screened at Kikambala Station at the Kenya coast. The material was later transferred to various locations and finally to Muguga, awaiting seed renewal. As yet, no new commercial varieties have emerged. Although Guineagrass is a recurrent entry in lists of recommended grasses in East

Africa, encouraging data from cutting or grazing trials are very scarce. Better varieties will have to come forward if Guineagrass is to play a major role in cultivated pastures in East Africa. The disappointing performance under grazing of material that looks so promising under nursery or cutting observations is frustrating. The optimum cutting interval is often found to be 6–8 weeks (Bredon & Horrell, 1963; Horrell & Bredon, 1963; van Voorthuizen, 1972; see also *4.3.4*). Opportunities would perhaps be better at lower altitudes and at the coast. At Serere (1100 m) better results (Horrell, 1963) were obtained than at Kitale (1890 m). Near Tanga at the Tanzania coast Guineagrass has been cultivated on government farms. These were robust types and vegetative propagation was practised. Persistence under grazing, however, left a lot to be desired (personal observation, 1980).

In West Africa greater optimism prevails as regards the future chances of Common Guinea in cultivation. Operating from the Ivory Coast, D. Combes and J. Pernès conducted an exploration in the late 1960s in East Africa which yielded some ecotypes that should be of interest in East Africa also, *e.g.* T58 (Noirot, 1989; private communication, 1989). Although various ploidy levels and apomixis are common, sexual diploids occur in the Tanga region, which should serve as a centre of variability. Sexual types were subsequently found also in the Iringa-Dodoma region of southern Tanzania (Nakajima, 1985).

6.9. Buffelgrass (*Cenchrus ciliaris*)

Buffelgrass is another example of a grass of which East African ecotypes have reached great fame in other continents, especially in Australia (Table 6.9), while its use as a cultivated plant in East Africa itself is very small. Buffelgrass is exceptional in that it is the only cultivated grass with both its area of origin and of potential application mainly in the arid Zone V (*3.4.5*). Its rooting patterns have already been described (*5.3.4*).

Intra-variety variation is negligible in Buffelgrass (9.1) which agrees with the high degree of apomixis normally reported (Bogdan, 1961b). Like the highly apomictic *Poa pratensis* in temperate climates, Buffelgrass is characterized by a wealth of natural ecotypes which can be visually distinguished and rapidly screened. Crossing programmes based on sexual plants have thus far not produced varieties of equally high performance, while it has proved difficult to restore the degree of apomixis needed in the progeny to fix the heterosis found (*9.8*).

Considerable exploration work was conducted to identify suitable ecotypes (Edwards, 1940c; Bogdan, 1961b), followed by van Soest (1976). At Katumani (1620 m) in Kenya and at Kongwa (1021 m) in Tanzania a wide range of Buffelgrass ecotypes was tested by Sands *et al.* (1970) and Wigg (1973), respectively. Both mentioned Edwards Tall (ex Kitale, K5148), and Biloela but agreed that the Kongwa type (KN531) was the best. This was also observed by Anderson & Naveh (1968) for northern Tanzania. Wigg (*op. cit.*) and Owen & Brzotowski (1966) made great efforts to find Buffelgrass a place. Buffelgrass increased the carrying capacity by about 50% over 8 years when compared with natural grassland, consisting mainly of annual grasses, but the costs of establishment were prohibitive. Comparisons of Buffel- with Rhodesgrass gave variable results (Pereira & Beckley, 1952; Pereira *et al.*, 1961; 1967).

For reseeding denuded land, Buffelgrass holds promise but reseeding is a last resort (Pratt & Gwynne, 1977). Oversowing of existing grassland as well as sowing into properly prepared strips have repeatedly proved ineffective (Pratt, 1964; Owen & Brzostowski, 1967; *4.3.5*). In Australia, Buffelgrass used to be sown into the ash of burnt woodland. This appears to be the ideal seedbed, but this practice is hard to reproduce under East African conditions and is also considered too destructive to the environment.

Buffelgrass does not perform well under high rainfall (Strange, 1958b) and leaf diseases are common (Horrell, 1963). Seed production is fairly good in dry environments, but at Kitale Buffelgrass grows poorly and only very small yields of seed are obtained. Like in Rhodesgrass (*6.6*), its growth habit at Kitale is different from that at the coast. Seed units, in the form of fascicles of clusters of spikelets are extremely fluffy and

Table 6.9. Important Buffelgrass ecotypes of eastern and southern African origin; after Bogdan (1977) and Barnard (1972).

Australian name	Origin in Africa	Year of introduction	Type
Biloela	Dodoma (T)	<1937	rhizomatous, tall
Nunbank	Uganda	<1949	rhizomatous, tall
Tarewinnabar	Kenya	<1950	rhizomatous, tall
Molopo	Transvaal	<1940	rhizomatous, tall
Gayndah	Nairobi	<1930	non-rhizomatous, medium
Mbalambala	Tana River (K)	–	non-rhizomatous, medium

voluminous, with only 200.000–400.000/kg, which makes the handling of the seed difficult. Mbalambala has spikelets which are short-awned and easier to handle. Bristles can be removed by hammer milling but care is needed not to separate the caryopses. Caryopses germinate very rapidly under laboratory conditions, often within 24 hours, as in Elephantgrass (*9.7*), but subsequent growth is slow. Seed dormancy is said to be common (*8.4.1*). Seed heads are often infected with ergot and/or honeydew disease (*8.7.2.5*). Establishment in the field is a major problem of both spikelets and caryopses (Brzostowski, 1961; Wigg *et al.*, 1973). Pelleting in lime or in other moisture absorbant material may be expected to increase the germination of surface-sown seed. Pelleting aids seed handling as well as reducing theft by harvester ants and termites but negative effects have been reported also (Bogdan & Pratt, 1967); see also *7.1.10.2*.

In spite of its drought tolerance Buffelgrass has been little used, perhaps because of its unusual combination of contrasting properties (*3.4.5*). Buffelgrass may seed relatively well, but is difficult to establish. Once well established and under fertile conditions, it is very persistent and its tufts are often too tough to be easily ploughed. There has also been controversy in the region about the relative unpalatability of Buffelgrass when maturity advances (Peat & Prentice, 1949; Barker & Kidar, 1989).

Several attempts, in successive seasons, were made to sow Kongwa Buffel at Katumani, Kiboko and at Kikambala in a series sequel to the Kenya District Trials (*6.3.1*). Seed had been of proven germination capacity but seedlings failed to establish properly before the first drought arrived. In late 1976 a similar experiment was set up by Rossiter at Katumani but the results obtained were not favourable for Kongwa Buffel (Table 6.10). At Ukiriguru (T), Walker (1969b) established the variety Molopo vegetatively from cuttings but replanting was necessary. The response to N-fertilizer was remarkable; DM yields were trebled with a dressing of only 120 kg N/ha.

Another principal reason why Buffelgrass is little used is that it is a lowland grass, requiring high temperatures for optimum growth (Humphreys, 1967). It grows poorly at medium (1500 m) and higher altitudes. Most of the lowland country in East Africa where its drought tolerance could be put to use is lagging behind in economic development. Good results have been achieved in water spreading schemes in the 1960s at Lorengippe in Turkana of north west Kenya (J. Newbould, private communication; Pratt & Gwynne, 1977). Soil texture and pH have been suggested as factors of adaptation (Bogdan, 1958b; Brzostowski, 1962). Even in seemingly suitable areas with keen farmers, however, this grass has not had much appeal. Better results are generally achieved at higher latitudes.

The drought tolerance and persistence of this grass are obviously not convincing enough. In addition to establishment and ploughing difficulties, Buffelgrass has a reputation of not being suitable in crop rotations. In Queensland, yields of a first crop of Sorghum after Buffelgrass were about one-half of the yield following Guinea- or Rhodesgrass (Humphreys, 1967).

6.10. Molassesgrass (*Melinis minutiflora*)

Molassesgrass owes its name to the distinctive sweet smell of the sticky excretion of glandular

Table 6.10. Dry matter yields (ton/ha) at Katumani (Zone V); after Rossiter (Katumani report, 1978).

Entry	Dry matter yields (ton/ha)				
	Oct	Feb	April	Total	
Boma Rhodes	2.5	3.2	5.6	11.3	
Mpwapwa Rhodes	4.1	2.5	3.9	10.5	
Makueni Guinea	2.7	2.0	5.2	9.9	
Elmba Rhodes	1.6	2.0	4.4	8.0	
Solai Coloured Guinea	0.8	2.3	2.4	5.5	
Kongwa Buffel	0.3	1.0	2.0	3.3	
Local Stargrass	–	–	2.9	–	(replanted)

hairs. Livestock needs time to get used to this odour, although it does not cause tainting of milk or meat. The sticky substance can affect the fleece of sheep but has been believed to deter ticks and vermin (Hubbard, 1927; Renson, 1928; Fiennes, 1940; Whyte *et al.*, 1959).

Molassesgrass has a disjunctive distribution, in East and West Africa (*e.g.* Angola and Cameroon). In East Africa it is confined to the 1250–2500 m altitude range. In the Kitale region, it is still commonly found on roadbanks and along fences, perhaps as a relict of past sowings. Although now almost forgotten, it used to be the common grass in the Kitale area, mostly undersown to maize, in the 1930s (Maher, 1936b) but was gradually replaced by Rhodesgrass; seed multiplication was suspended in 1968.

Molassesgrass had great value in East Africa for revegetating overcropped land, which was a serious concern in the 1930s. It is still a serious concern in the 1990s, so that Molassesgrass may well play a major role once again (*5.6*). It does better on such soils than Rhodesgrass and Setaria. Molassesgrass is easy to establish and to eradicate by ploughing. First-year stands can be very vigorous. Yields decline, however, rapidly (Fig. 6.3). It lost ground as the practice of mixed-farming, based on sown pastures, gathered strength in the 1950s. Once the productive capacity of a soil is restored Molassesgrass is better replaced by Setaria or Rhodesgrass. Molassesgrass has been widely sown in low fertility situations in Latin America, south east Asia and Australia, where it does well on steep hillsides, which underlines its value as a protector of soil. Molassesgrass does not respond to high rates of N-fertilizer (Vicente-Chandler, 1974), but it should be noted that it is susceptible to heavy cuts usually found when high N-rates are employed (see below).

"Kitale Commercial" (K52294) used to be the common variety in the seed trade. It was one of the first varieties sown around Kitale (Maher, 1936b) but, in spite of its name, probably not of East African origin. Its tufted habit and susceptibility to the "small-leaf or stunting" virus disease led Bogdan (1977) to suggest that Kitale Commercial originated from southern Africa (see also *6.4.1*; *6.6.3*). Bogdan (1959a; 1960b) referred to cultivated types from other parts of the Kitale district, one of which flowered 10 days earlier, and to commercial forms in other continents, which he considered similar to Kitale Commercial. In Brazil, for example, where Molassesgrass was recorded already back in 1823 (Parsons, 1972) and where it is a major cultivated and naturalized species, similar types occur. Renson (1928) mentioned a connection of Latin America and "Congo Portugais" so that it is possible that Kitale Commercial reached Kenya via that route. Molassesgrass is a common constituent of the flora of eastern Zaïre (personal observation, 1992).

Kitale Commercial is late-heading but a good seed producer. Seeds are small (6 million/kg) but produce vigorous seedlings with a typical ovate-orbicular first leaf. A dense sward is rapidly formed. The grass is tetraploid ($2n=36$) and largely apomictic (Bogdan, 1960a, b; 1966c). Vegetative propagation usually results in failure (Table 6.6). It will not stand a heavy hay cut, nor a fire and it tends to bunch and eventually disappear under heavy grazing (Horrell, 1958a). These inabilities may be linked with the absence of basal leaves on the developing stem. The truly local varieties Chania River and Mbooni Hills (both locations in eastern Kenya) are more prostrate and even stoloniferous, while being virus resistant. However, the disease was not widespread at the time and it is unlikely that it caused by itself the decline in the popularity of Molassesgrass. Like Pokot Rhodes and possibly Makueni Guinea, the Chania River and Mbooni Hills ecotypes are names reminiscent of lowland semiarid areas, but their actual sites of collection were mostly in more favourable habitats.

Strange (1961a) recorded high yields of Molassesgrass at Kitale (Table 6.11). Its % DM is relatively high also.

Table 6.11. Molassesgrass yields at Kitale (after Strange, 1961a).

Year	Dry matter yield (kg/ha)			
	Kitale Commercial Molasses	Nandi Setaria	Nzoia Rhodes	Star
1955	7300	4450	4250	4340
1956	5020	3220	3220	2660

If the Kitale types of Molassesgrass are to be revived again, it should be an easy task to harvest seed from the relicts that can still be found on road- and terrace banks in the region.

6.11. Signalgrass (*Brachiaria* spp.)

Brachiaria spp. are common, valuable constituents of the natural vegetation in East Africa in Zone II/III, especially in the warm, humid parts of western Uganda and adjoining countries (*3.4*). It forms a dense, rather aggressive sward, highly responsive to N-fertilizer (Compère, 1961d; Fredericksen & Kategile, 1980). Establishment from seed is easy, but eradication is a problem (Table 6.6). Signalgrass is sown on only a small scale in East Africa. For the purpose of long-term pastures, however, a reappraisal is worthwhile.

The taxonomic position of the various species is not clear and old names have been perpetuated which has caused differences to be regarded as important, whereas the similarities were probably more relevant. In Zaïre, the names *B. eminii*, *B. decumbens* and *B. ruziziensis* have been used, perhaps for the same material. Bogdan (1977) stated that *B. ruziziensis* and *B. decumbens* were closely related, while *B. ruziziensis* and *B. brizantha* were not distinguished for a long time at Kitale (Bogdan, 1958; 1959a). *B. brizantha* and *B. decumbens* intergrade on all their morphological features and the present description of the 2 species represents nothing more than the 2 extremities of the range of variation (Clayton & Renvoize, 1982). These authors regard *B. ruziziensis* as a local segregate of *B. decumbens* (see below).

Name uncertainties and confusions have practical implications. Under the name of *B. ruziziensis*, Congo Signalgrass has been in commercial seed multiplication in Kenya since 1960. While *B. brizantha* was poor in seed-set (Bogdan, 1955), *B. decumbens* is a seed producer of great strength. The latter was reported on by Hosking & Stephens (1941), Horrell (1958a) and Strange (1958b). It is used on a large scale in Brazil and is highly regarded as a cultivated grass in Colombia (Crowder *et al.*, 1970) as well as, more recently, in Australia. In coastal Queensland the grass had naturalized over wide areas, notably in and around Brisbane, in an aggressive way and was regarded as a grass which would not allow associated legume growth. It was believed to be incapable of producing viable seed in Australia until Grof (1968) showed that germination of freshly harvested seed was impeded by an impermeable seed coat. CPI 1694, introduced from Uganda in 1930, was not released until 1966, and was registered as Basilisk in 1973 (Loch, 1977). *B. decumbens* is a tetraploid ($2n=36$) and was believed to be an obligate aposporous apomict, but diploid sexual types are commonly found among tetraploid apomictic accessions. This is probably how Congo Signal (K5832) emerged.

Congo Signalgrass (B. ruziziensis) (K5832). This name is quoted here since it was under this name that Signalgrass has met with some success as a cultivated grass in Zaïre (Dewez, 1959), Uganda (Wendt *et al.*, 1970) and Kenya. The Ruzizi river, to which the species name refers, runs between Lake Kivu and Lake Tanganyika. The original seed sample multiplied at Kitale was obtained from

INEAC at Rubona (Rwanda) and the variety is known as Congo Signalgrass. This is the only variety being multiplied. Considerable intra-variety variation was observed at Yangambi (Behaeghe & Blouard, 1962). At Kitale intra-variety selection produced distinct test-varieties (*9.8*). That this grass is sexual and diploid (2n = 18) has been confirmed by Ferguson & Crowder (1974).

The spikelets are large, with a 1000-seed weight of 2000 mg, numbering some 500.000 seeds/kg. Germination is rapid, similar to if not better than Rhodesgrass, so that establishment is easy. The recommended seed rate is 1.5–2.0 kg PGS (Table 7.3). Although the spikelets are large, levels of % purity are not higher than in small-seeded tropical grasses. If larger seed size were, at least in theory, to lead to higher vigour and lower mortality of seedlings, it should make up partly for the lower number of seeds sown. No proof was found at Kitale that Signalgrass is more suitable for oversowing purposes than any other grass (*4.3.5*). Vegetative establishment into existing swards is common practice in various parts of Zaïre (personal observation, 1992).

The seed yield at Kitale is not high (Table 6.7) and quite erratic from year to year. Head emergence begins in May/June but is then delayed (Fig. 8.5) when growth stagnates in the often less clement weather of June/July (Fig. 3.3). When head emergence is delayed only one seed crop is possible per year under Kitale conditions, towards the end of the rainy season.

A dense and attractive sward is formed rapidly, ready for grazing within 8–10 weeks after sowing. Signalgrass is highly nutritious (Fig. 6.4) (Urio *et al.*, 1988). Soneji *et al.* (1971) found significantly higher digestibilities as well as animal intake in Congo Signal than in (unspecified) varieties of Setaria. Hay can be made without much difficulty as the stems are thin and easy to cure. Congo Signal is impressive in the 1st year, but in the 2nd year production can drop dramatically on less fertile soils. The sward turns in colour from dark green to pale-brown and becomes sod-bound. N-fertilizer at some 50 kg/ha N applied early in the rainy season is needed to restore production. Even then, yields are often low (Table 6.1).

Excellent ground cover is maintained also in the absence of N-fertilizer. Little herbage may be produced and the number of grazing days recorded may be small, but the sward is very persistent. The grass is, therefore, suitable for low-maintenance sports fields. It is ideal also where a grass border is required, since it is the only grass available that can be grown from seed while requiring little maintenance. After the 2nd year Congo Signal's behaviour is quite similar to that of *B. humidicola* (Lawn Signalgrass, ex Zimbabwe) mostly used for soil conservation and sports fields, but multiplied vegetatively since heads are hardly formed (*6.4.1*).

A remarkable capacity of natural regeneration of Congo Signal has been observed at Kitale:

1. A pasture was ploughed in 1967 and cropped for 6 years. In 1974 the field was allowed to revert to fallow and a complete sward of Congo Signal re-appeared.
2. In 1974, an old sod-bound pasture was ploughed and sown to maize. The maize was interrow-cultivated. In December after the maize harvest, a fire swept across the field. A month later, at the peak of the dry season, vigorous rows of Congo Signal appeared from under the ash as if they had been sown in the same rows as the maize crop.

This offers possibilities of having self-regenerating pastures in maize rotations. However, unless this volunteering capacity is kept under control, the grass can become a major nuisance in crop rotations. Congo Signal presents a good choice for N-fertilized semi-permanent pastures in the Elephant- and Stargrass subzones of Zone II (*3.4.2*), where it can be alternated with maize before being allowed to rejuvenate itself; Signalgrass is not so susceptible to maize herbicides such as Atrazine (*7.1.10.2*). This practice may give it an advantage over sown Setaria, which would otherwise be the best alternative in these zones.

In South America where *Brachiaria* spp. have become widespread, photosensitization in livestock (facial eczema) and severe attack by spittlebug (*Cercopidae*: *Homoptera*) and Rhodesgrass scale (*6.6.2*) have been reported, but in East Africa no such cases have been recorded. Instead, infestations by mites are common.

Considerable effort has been spent on exploration work in East Africa by CIAT (Colombia) and ILCA (Ethiopia) and it is hoped that new ecotypes will emerge that will earn a well deserved place for this grass in pasture improvement in East Africa.

6.12. Coloured Guinea (*Panicum coloratum*)

In Kenya only one variety (Solai, K52490) has been sown on field scale with just over 100 ha certified seed in the early 1970s (Boonman, 1980). Bambatsi Makarikarigrass (*Pcol* var. *makarikariensis*) from Zimbabwe (K5387) was planted vegetatively on terrace banks in dry areas and on waterlogged soils or black clays. For pasture purposes it was less suitable (Pereira, 1959). A better choice is Solai Coloured Guinea, partly because of its ease of establishment. The ecotype was collected by D.C. Edwards near Solai (Bogdan, 1965a), 50 km north of Nakuru, probably on a waterlogged soil in a semiarid part of the area (*3.3*). Coloured Guinea was one of the first grasses tested by Edwards (1932; 1936).

Solai is a very good seeder with a vigorous seedling, strong enough to cope with difficult soils. The seeds are relatively heavy (1000-seed weight of 450 mg) with 2–2.5 million/kg. The recommended seed rate is 1 kg/ha PGS (Table 7.3). Hardseededness was reported originally (Edwards, 1932) but the levels of purity and germination are high and fairly consistent in commercial seed production compared with other tropical grasses (Table 6.7). Although apomixis was suspected due to apparent uniformity (Bogdan, 1977) considerable intra-variety variation was observed within Solai, and several test-varieties were developed which were distinct botanically (*9.8*; van Wijk, unpublished). A considerable export of seed to Japan developed in the 1970s.

Solai can be ready for grazing or hay in 8–10 weeks after sowing, respectively. Solai has a high % DM (Table 6.3) and nutritive value. Careful husbandry including conservation cuts is necessary to prevent Solai from becoming too stemmy. Solai is drought tolerant and one of the earliest heading among the cultivated grasses (Table 6.7). Heading is relatively synchrone. A larger proportion of tillers elongates, producing more heads than in any of the other Kitale grasses (Fig. 8.5).

Solai is not a very high yielder of herbage and has no persistence under cutting regimes. An open sward is produced which is liable to weed invasion. After mowing a stubble is left which feels sharp to the touch. Under grazing regimes, however, Solai is more persistent. In cafetaria tests cattle tend to ignore Solai and choose slower developing species (*6.3.3*). Consequently, Solai has some reputation of being persistent and compatible with legumes such as Desmodium and Glycine. Goitrogenous effects were observed in Zimbabwe (Rodel, 1972) but considerably less so than in *Cynodon* spp. (*4.2.2.2*). The grass should be valued for haymaking on land that may be too wet for regular grazing.

6.13. *Hyparrhenia rufa*

Hyparrhenia rufa is the dominant grass over a large area north of Lake Kyoga, including Serere, in Uganda (*3.4.3*; Fig. 3.6). Having been introduced adventitiously 2–3 centuries ago in Latin America and having naturalized over large areas, it is a major cultivated grass (Jaraguagrass), particularly in Brazil (Ferguson, 1990).

Edwards (1940c) reported that this grass seeded very well and was issued for contour planting. At Serere considerable effort was spent to use this grass for legume-based sown pastures (Horrell, 1963; Stobbs & Joblin, 1966; Wendt *et al.*, 1970). Grass seed was collected from the local grass vegetation (Gola ecotype). *H. rufa* does not seem to require a well prepared seedbed but is very slow to establish. The heavy awning of the seed makes it difficult to sow. To maintain leafy swards, it should be grazed frequently which is well tolerated. It was believed to be the most valuable grass at Serere for peasant farming on poor soils (Horrell, 1963). Dougall & Bogdan (1958) cautioned, however, that % crude protein was very low and % phosphorus exceedingly low. Reid *et al.* (1973) listed it among the lowest in digestibility of grasses used in cultivation (Fig. 6.4). Stobbs & Joblin (*op. cit.*) recorded a 15% increase in total liveweight gain over a 30 month period when *H. rufa* was enriched with Centro (Table 6.12).

Table 6.12. Total liveweight gain (kg/ha) over 30 months at Serere (Stobbs & Joblin, 1966).

Treatment	Total liveweight gain (30 months) (kg/ha)
Hyparrhenia	940
Hyparrhenia + Centro	1080
Natural regeneration + Stylo	950

6.14. Stargrass (*Cynodon* spp.); Kikuyugrass (*Pennisetum clandestinum*)

These grasses are potentially valuable as components of natural grassland, as colonizers and for stabilizing road banks, gullies, grassing waterways, general soil conservation work (*3.4.2*; *4.2*) and for amenity purposes. Popular Cynodon lawntypes are Cape Royal and Maadi River (Bogdan, 1955). This is in fact a *C. dactylon* var. *elegans*, common in southern Africa, beyond 13°S. Kikuyugrass can also produce attractive lawn, but is harder to trim than *Brachiaria humidicola* or *Paspalum notatum*; for amenity grasses see also Graham (1945b) and Jex-Blake (1957).

With the exception of the coastal areas (*4.2.2.2*), where planted Stargrass was found to do well, neither Stargrass nor Kikuyugrass should be recommended for general use in cultivated pastures. As already pointed out (*4.2*), the negative properties of these grasses are well known: low herbage yield (Table 6.2); lack of stability under grazing; ingression of weeds; low nutritive value (Stargrass); difficulty of eradication and weed potential is arable crops (Tables 6.1 & 6.6). The fact that Star- and Kikuyugrass, of East African origin, may be important at higher latitudes and in other continents is shared as we have seen above by other indigenous species, such as Guinea-, Buffelgrass, and Brachiaria. Relatively better performance abroad is no proof that such species should also be the best choice "at home" (*6.4.1*).

Ironically, Kikuyugrass was the lowest yielding, relative to all other locations and entries of the Kenya District trials (Table 6.2), at Ol Joro Orok. Similarly, Stargrass was the poorest entry in the Stargras zone of Kakamega, apart from Kikuyugrass. At each location a local ecotype was planted (*6.3.1*). No data are available on long-term comparisons of animal performance of Kikuyugrass compared with sown grasses, but the data obtained in 1971/1972 by Munyi & Smit (1973) at Kisii are of interest. Although milk yields were very high, even for the highly favourable Kisii environment, the relatively low yields of Kikuyugrass should be noted (Table 6.13).

It should be clear that seeding varieties of Star- or Kikuyugrass, can do little to improve the image of these grasses for grazing purposes. Establishment is slower than by vegetative propagation and seed will only aggravate their weed potential. This should be remembered when seed is imported to meet the demand of curious customers. In many countries, both species are prohibited imports.

Table 6.13. Milk yields at Kisii on Kikuyugrass and sown pastures of different ages (after Smit & Munyi, 1973).

Pasture	Milk yields (kg/ha/year)
Setaria-Rhodesgrass	
2nd year	7370
3rd year	5720
4th year	4800
Kikuyugrass	2840

6.15. Temperate grasses

Temperate grasses were among the first grasses tested (Kabete Ann Rep 1910; Ball, 1936; Maher, 1936b). Bromus (*6.16*) and the annual Wimmera Ryegrass (*Lolium rigidum*) were the only grasses for the high altitudes of which seed could be grown locally. The import of Wimmera Ryegrass, which had become a weed in cereal rotations, was later prohibited (Birch, 1959b).

The obvious choice for high altitudes, 2250–2500 m (*6.5*), is Perennial Ryegrass (*Lolium perenne*) or Cocksfoot (*Dactylis glomerata*). Some 10.000 ha were established by 1960 (Birch, *op. cit.*). Perennial Ryegrass is the best choice where the dry season is not too long. Ryegrass is easy to establish, very productive in the 1st year and of high nutritive value. DM-yields were more than trebled by applying 130 kg/ha N and % crude protein was markedly increased (Birch, 1959c; Suttie, 1967). Cocksfoot, on the other hand, is more drought tolerant and persistent. The seed rates employed in Kenya are 8–15 kg/ha but when grass-clover mixtures are sown seed rates should be reduced to 2 kg, to give clover a chance to establish. Since heading in these temperate grasses is so shy near the equator (Birch, 1958b), the classical differences which distinguish early- and late-heading varieties at high latitudes become immaterial. Early-heading varieties of European or New Zealand origin are appropriate and it seems justified to import varieties that are widely available and less expensive on the world market. Italian Ryegrass (*L. multiflorum*) and corresponding hybrids are of even shorter duration than Perennial Ryegrass; when short-term fodder crops are needed preference is better given to oats (*16*). Kikuyugrass invasion is a major problem in sown pastures at high altitudes (*4.2.1.2*). Tall Fescue (*Festuca arundinacea*) is more competitive towards Kikuyugrass and varieties such as Alta were found useful (Suttie, 1967). Hybridization between Lolium and Festuca has been found to give high although erratic yields in tropical highlands, but any possible advantage should be weighed against the high cost of seed. A notable common failure among temperate grasses is Timothy (*Phleum pratense*). Severe insect damage to Cocksfoot and Fescue pasture on the Kitulo Plateau in southern Tanzania was reported by McCrae (1975).

Most of the research work done on temperate grasses was conducted at Marindas (2800 m) near Molo (Birch, 1959; Morrison, 1969) and in Kivu (Compère, 1961a) and was especially designed to develop clover-based pastures, although Birch (1960) postulated that temperatures at Molo were too low for optimum clover growth.

6.16. Miscellaneous cultivated grasses

Elephantgrass, Giant Guinea, Giant Setaria and Guatemalagrass are described in detail in Part II.

Bogdan (1977) has described over 500 species, most of which he tested himself in Kenya. The reader is referred to this work and to that of Horrell (1958; 1963), but a list follows of some 20 grass species which had drawn particular interest.

Andropogon gayanus (Gambagrass)
Common in East Africa and an important natural grass in eastern Kenya, *e.g.* around Kitui. Important as a cultivated grass in Brazil (Whyte *et al.*, 1959) and widely tested in West and East Africa (Oyenuga, 1957; Bowden, 1963). Some success has been achieved near Tanga (Hendy, 1975) on abandoned sisal plantations and at Kongwa (Wigg, 1973). It is a grass of low digestibility (*6.3.4*).

Bothriochloa insculpta (syn. *Amphilophis insculpta*)
Common in Zone IV/V, especially in plains with heavy soil (*3.4*; *4.3*), where it provides useful grazing. Seeds well but seed is difficult to handle. It was widely tested (Edwards, 1936; 1940c; Bogdan, 1955; 1958) but proved to be of insufficient productivity (Bogdan, 1977), although useful for heavy soils and for reseeding denuded pastoral land (Bogdan, 1959a; Bogdan & Pratt, 1967).

Bromus catharticus (syn. *B. uniloïdes*) (Rescuegrass)
From the pampas of South America (Roseveare, 1948). Introduced and tested early this century (Kabete Ann Rep. 1910). Was grown at the higher altitudes east and west of Nakuru as *Bromus marginatus* (Edwards, 1940c; Etherington, 1943; Bouwer & Strange, 1948; Maher, 1950; Birch, 1959b), and introduced into other countries as Nakurugrass (Natal Variety List 1978). Abandoned in the 1960s due to lack of persistence (Bogdan, 1977). The only cultivated high-altitude grass producing worthwhile seed yields on the equator.

Digitaria decumbens (syn. *D. eriantha*) (Pangolagrass)
This grass from eastern Transvaal was among the Digitaria species collected by Pole Evans (1933) and subsequently tested in East Africa, under the name of *D. eriantha* (Edwards, 1936; van Rensburg, 1948a). Because of successes overseas in the late 1950s and 1960s, it has been retested repeatedly in East Africa. The regions of possible application should be found in the Star- and Elephantgrass zones. Tiley (1970a)

only mentions it on his list of introduced species but Kusewa (1974) reported yields, in the period of 1962–1965, of less than 4 ton/ha, only a third of those of Mbarara Rhodes. This unlikely low yield may have been due to the "Pangola stunt virus" (Dirven & van Hoof, 1960), or to rust or other diseases, which so often devastate vegetatively propagated plants when grown on increased scale (*11.8*; Baker, 1978).

Echinochloa haploclada
A type from a stream margin near Songhor (K) was reported on by Edwards (1940c). Moderate performance was achieved with a good seeding dryland type (K53542) from near Mount Marsabit (Bogdan, 1963b) under grazing observations at Kitale. Further work was abandoned in the late 1960s.

Entolasia imbricata (Bungomagrass)
This grass from swampy grounds drew considerable attention at one time (Mwakha, 1971) when it was found to have a very high N- content (Bogdan, 1963b) which it shares with other grasses from swampy grounds (Stobbs, 1971). In contrast with seemingly less attractive grasses in the same nursery collection, Entolasia was ignored by army worm (*Spodoptera exempta*) during a severe outbreak in May 1961 (Bogdan, *op. cit.*). Bungomagrass lacked persistence and it disappeared completely within 2 years of grazing.

Eragrostis tef (teff)
Introduced in Kenya at the turn of the century (Kabete Ann Rep 1910), probably in the company of imported horses and mules from Ethiopia (*2.2.2*). This is an annual cereal, valued at one time as a nurse crop for slower growing grasses or as a quick growing fodder catch crop in dryland farming (Maher, 1936b; Pereira *et al.*, 1958; see also *5.3.4*). Seeds are very small (2.5 million/kg) and seed rates of 2–3 kg/ha are adequate (Bogdan, 1965c). Now preference is mostly given to Grass Sorghum or oats for fodder purposes.

Eragrostis superba (Masai Lovegrass)
Common in Zone IV (*3.4.4*) on over-cultivated soils at lower altitudes, withstanding heavy grazing and drought resistant. The Kaloleni ecotype from the Kenya coast is leafy and bulky and a heavy seeder (Horrell, 1963). It was the best grass for reseeding denuded land in Kitui and other parts of Kenya (Jordan, 1957; Bogdan & Pratt, 1967).

For the same purpose *Eragrostis curvula* was tested, introduced from southern Africa, where it is valued in tobacco rotations for the control of root-knot nematode. As a pasture grass in semiarid areas of Kenya it was less successful (Pereira, 1959; Bogdan, 1977).

Paspalum dilatatum (Dallisgrass)
Originally from the pampas of South America (Roseveare, 1948), but imported via Australia or Natal and tested at Kabete (Kabete Ann Rep 1910). Because of its lasting importance in other countries it continued to draw interest (Edwards, 1936; 1940c) together with other *Paspalum* spp. (Bogdan, 1955; Horrell, 1958a). Was found to be of no value in Uganda and was used at Kitale mainly as a border grass. A very persistent grass, producing little herbage and few flowering heads, thereby requiring little maintenance. Spikelets are often infected with ergot and/or honeydew disease (*8.7.2.5*).

Paspalum notatum (Bahiagrass)
From Central America, introduced into Uganda in 1923 (Tiley, 1970a). Widely used since then at low and medium altitudes for erosion control work, as a bund grass and for sports grounds and lawns. At Kitale seed heads are very rarely produced; at lower altitude, head production is still scarce but can be induced to some extent by burning (Horrell, 1958a); it is regarded as unpalatable to livestock and unsuitable in crop rotations. A grass of high persistence; Harker (1962b) referred to fields planted at Namulonge (U) 30 years earlier, which showed little response to NPS-fertilizer. These fields have persisted up to the present day (personal observation).

Planted vegetatively, it produces an attractive uniform lawn which requires little maintenance but rapidly browns off in the dry season. It is too coarse for fine lawns as compared with Cape Royal Cynodon (*6.14*).

Paspalum plicatulum
From Central America. Attractive but slow (Bogdan, 1955). Seeds sparsely and is insufficient-

ly adapted to tropical conditions (Horrell, 1963). Later abandoned (Bogdan, 1977).

Paspalum scrobiculatum (syn. *P. commersonii*)
Local grass, common in cultivation and in overgrazed pasture (*4.5*). It seeds well (Horrell, 1958a) and was among the first grasses tested in the Kitale area (Halcrow, 1938). A few varieties were cultivated, which persisted under grazing but formed thin swards (Bogdan, 1955; Strange, 1958b). This is a grass of low digestibility (*6.3.4*).

Pennisetum squamulatum
A local grass, tried at Kitale with a view of developing a seeding fodder-type grass (Bogdan, 1963b). Later abandoned due to low yields of herbage.

Pennisetum unisetum (syn. *Beckeropsis uniseta*)
A common grass (*3.4.3*). A very good seed producer, but too short-lived (Bogdan, 1955; 1977).

Phalaris (Ronphagrass)
An introduced grass with hybrid status (*P. arundinacea* × *P. aquatica*) and concomitant publicity in Kenya in the 1950s but abandoned when found to be almost totally unpalatable (Suttie, 1967). Previously, Australian Phalaris (*P. aquatica*) had been tested widely under the name of *P. tuberosa* (Edwards, 1936; 1940c; Maher, 1950).

Snowdenia polystachya (Abyssinian grass)
From Ethiopia. Cultivated in Kenya in the 1940s at high altitudes (Bogdan, 1955), but was abandoned because of lack of persistence. It remained as a serious weed of cultivation (Strange, 1958b).

Vetiveria zizanoides (syn. *Andropogon muricatus*)
Vetivergrass has been grown on Indian Ocean islands and in Kivu. It has been used for ornamental purposes and for the extraction of cosmetic oils from the roots (van den Abeele & Vandenput, 1951).

Potential species for reseeding denuded pastoral land.
The following is a list of species, apart from Rhodes- and Buffelgrass and the species just mentioned, holding some promise for the reseeding of denuded pastoral land (Bogdan 1977; Bogdan & Pratt, 1967; Sands *et al.*, 1970): *Chloris roxburghiana*; *Enteropogon macrostachyus* (see also Bogdan, 1963a); *Leptochloa obtusiflora*; *Leptothrium senegalense*; *Paspalidium desertorum*. As Pratt & Gwynne (1977) have pointed out, seed is usually not available if and when required for reseeding.

6.17. Summary

Since grasses are primarily sown or planted in rotation with crops, they should fit well in such systems. Only a small number out of the 1000 odd species of the East African gene centre is useful for that purpose and the number of species of which seed is available commercially is even smaller. Their superiority extends to most (sub)tropical continents. Unlike cereals, successful grass varieties are relatively insensitive to genotype × environment interactions and the same variety can be topping the list under similar moisture and temperature conditions in quite different regions. The choice of species was entirely the farmer's, already in the first part of this century; recent acquisitions of lasting success are rare. Most of the top varieties can be dated back to before or about 1950. Very few ecotypes introduced after 1960 have had the same impact as the older ones. Consequently, recommendation can validly be limited to a small number of well characterized species and varieties, of which seed or propagation material is available. New genotypes should be allrounders in herbage yield; drought tolerance; persistence; nutritive value and ease of management. East Africa's sub-humid savannah yielded most of the world's cultivated tropical grasses due to a co-evolution of grasses and grazing animals, a varied topography and climate and because their seeding ability is better than of those of forest origin.

Rhodesgrass is the most popular grass in East Africa because of its reliable performance over a wide range of climates and soils, ease of management and drought tolerance. It is both easy to establish and to eradicate. The search for drought tolerant varieties has been elusive but intravariety selection of early-heading plants out of Mbarara and Masaba produced Elmba and Boma Rhodes, respectively, which are now the most widely grown. Nandi Setaria is more persistent and adapted to humid conditions, from low to high altitude. Its herbage yield and nutritive value are lower than of Rhodesgrass. Common Guinea and Buffelgrass have been cultivated on an experimental scale only but have, in spite of a continued search for ecotypes, not had a great impact partly because the lowland areas, where their potential role is more evident, have lagged behind in development. Molassesgrass used to be grown widely and might again play a major role on soils impoverished by cropping. Other species of some importance are Coloured Guinea, Signalgrass and *Hyparrhenia rufa.*

CHAPTER 7

The establishment and husbandry of sown grasses

A distinction is drawn between the establishment year and subsequent years. "First-year" is taken to mean the year of sowing, not necessarily the year of first grazing or harvesting.

7.1. The first year: grass establishment

Rapid establishment is essential to achieve the best and cheapest pasture available: that of the 1st year. In the 1st year, soil-N is readily available due to mineralization after ploughing. N-availability declines rapidly in the following years (*5.3.2*). N is the key to grass yield and quality and should, therefore, be made use of efficiently when so readily available.

In contrast with other husbandry aspects, *e.g.* grazing policies, very few scientific reports deal with the subject of how best to establish and treat tropical grasses in the year of sowing. This is partly because this aspect is easily delegated back to the general adviser as a routine matter, subject to conditions of soil and weather, partly because it is not appreciated that establishment processes can be influenced to a great deal. A further reason is that many researchers prefer to ignore the data of primary growth, if collected at all. In their opinion, results in the 1st season are too strongly influenced by weather, by the degree of control over weeds and by the difference in speed with which seeds germinate and establish, and less so by the specific factors under investigation. In species and variety comparisons reasons of poor establishment are indeed environmental, but also genetic. The genetic capacity of rapid establishment is, however, one of the prerequisites if a species is to have a place in agriculture, especially in short-term pastures in mixed-farming rotations. The 1st-year's data of pasture experiments may be presented separately, but they should always be reported on.

Quite often an attitude of apathy is observed also among farmers. It has been argued that a 1st-year's pasture is often not worth bothering about and that management is better deferred to the following year. The whole of the 1st-year including the dry season is supposedly needed for the annual weeds to die out and to be smothered by the sown grass as the latter gathers strength. There is at least one positive point in this perhaps negative approach and that is the general experience that total failures of establishment are rare. Very little of a sown grass may be visible in the year of sowing but in the following year, after the dry season, the pasture can be as good as ever. However, this defaitist approach is neither cheap nor inevitable. It is quite well possible to establish pasture properly, with little extra cost but with a little more care.

Rapid establishment is a precondition if grasses are to serve their purpose in crop-pasture rotations, not only to provide an early grazing but also to suppress arable weeds and restore the productive capacity of the soil. However, a more direct argument already mentioned is that 1st-year pasture is potentially the best of its lifetime. A good 1st-year's pasture can represent more than 50% of the total pasture production over 3–4 years:

1. The young grass roots of a primary sward are effective in exploiting mineralized soil-N reserves. First-year pastures are potentially very productive, without N-fertilizer.
2. First-year pastures can be productive deep into the dry season when older pastures have already ceased to grow. This is partly because more water is still stored in the soil.

First-year pasture differs from older pasture in

density and size (weight) of tillers. In the 2nd and following years tillers may number 2000–2750/m^2 in well-fertilized broadcast stands (Fig. 8.6), but in the 1st year this is less than 1000/m^2 (Boonman, 1971b). In compensation, however, tillers are much heavier in the 1st year. When compared at the same physiological stage, at IHE (Initial Head Emergence), 1st-year tillers are many times heavier than in subsequent years (Table 7.1).

Table 7.1. Weight per tiller (mg), at IHE (Initial Head Emergence), and % tillers heading at seed maturity, in 1st or 2nd year sowings (after Boonman, 1971b).

Variety	Tillering			
	Year 1		Year 2	
	Weight	%	Weight	%
Nandi I Setaria	760	27	260	57
Nandi III Setaria	390	18	220	19
Masaba Rhodes	1160	53	190	39
Pokot Rhodes	1640	38	260	29
Solai Coloured Guinea	520	64	200	76
Makueni Guinea	400	12	280	45
Congo Signal	910	32	280	78

Differences in tiller size may not only explain why 1st-year tillers are more effective in resisting stress and drought (*11.10*), but also why they produce more and better seed irrespective of the weather conditions prevailing, especially in varieties with a high % heading tillers (*8.7.1*).

Many farmers prefer to use the primary growth of the 1st-year pasture either for a late hay-crop, at the beginning of the dry season, or for a seed crop or both. It is evident that utmost care should be spent on pasture establishment, so that the 1st-year's productivity, also in the dry season, can be maximized.

In the following we refer to pastures consisting of 1 grass species only. Legumes are little used (*10*) and mixtures of 2 or more grasses are not advocated. In the past, mixtures of grass species were proposed for various reasons, mostly to achieve a less tussocky sward which may be a problem with Guinea- and Buffelgrass (van Rensburg, 1955; Blouard & Behaeghe, 1961; Horrell, 1963; Anderson, 1968a). In temperate climates commercial mixtures made up of several grass species have been popular with the seed industry, but the present trend is towards single species since less competitive components tend to disappear from the sward eventually. Hard evidence that mixtures composed of more grasses have any advantage over and above the best component is difficult to establish (H. Vos, private communication).

Pests interfere with seedlings but serious attacks are rare. Armyworm (*Spodoptera exempta*), Dusty Brown Beetle (*Gonocephalum simplex*) and the Shiny Cereal Weevil (*Nematocerus* spp.) attack grasses and cereals (see also Bohlen, 1973). Maggots of a shootfly (*Atherigona* spp.) can kill the main shoot of young seedlings of Rhodesgrass (*6.6.2*) and especially Wimbigrass (*Eind*), a common weed of arable land. Seed-harvesting ants can be a problem in semiarid areas; see also *6.3.5*.

The approach to pasture establishment differs depending on direct sowing or undersowing to a cereal crop.

7.1.1. Direct sowing of grass seed

The seed is sown directly, without a nurse crop, in land prepared for the purpose of pasture establishment. The farmer normally wants to sow pasture into land that has had 2 or more years of arable crops which can be cereals, cotton, tobacco or root crops. Replacement of pasture directly by sown pasture is uncommon and not recommended, except on specialized dairy farms.

Ploughing is the beginning of all good farming and this applies the more so to pasture establishment. Early sowing in the rainy season is essential so that the land must be prepared well in time. Land should be ploughed late in the preceding rainy season to achieve maximum control of perennial grass weeds. Mechanized ploughing can be conducted under much drier conditions than ox-ploughing or hoe-cultivation. African Couchgrass can be eliminated by one timely ploughing (*4.5.1*). For Stargrass and, especially, Kikuyugrass more elaborate methods are needed. These aggressive grasses are, incidentally, often the weeds which make the farmer decide to abandon cropping (*5.1.2*). Unless they are controlled, it will be impossible to grow crops or pastures which are not overrun by these weeds.

Where the infestation is too severe it may be necessary to delay sowing and to use cultural, mechanical and/or chemical means to destroy the grass weeds (*4*). A second ploughing may be necessary. Quick growing fodder break crops such as Sorghum or oats, sown between two ploughings, can help to suppress weeds (*15*; *16*). Frequent harrowing as a means of weed control is not adequate and not desirable. Too many mechanical operations have a deteriorating effect on soil structure by exposing it to the detrimental effect of rainfall impact and sunshine (*5.3.5*). Crops such as Sorghum and oats help to protect the soil and prevent grass weeds from sprouting, provided that the fodder is harvested and ploughed in before the weeds reappear again in force. When sown in dense stands so that rooting is shallow such crops may, in fact, help to conserve more moisture of heavy rainstorms than is possible in bare fallows (*5.3.4*).

7.1.2. Date of sowing

If the ploughed land is free from grass weeds, it is best to sow as early in the rainy season as possible. Speed of germination declines as the rainy season advances. At the onset of the rainy season the soil is warmer and also of a better structure. Soil temperatures decline rapidly as the rainy season proceeds (Fig. 3.4).

An experiment was conducted in 1967 at Kitale on dates of sowing. Strips of land were sown at 2-weekly intervals from the onset of the rains in early April until August. The strips set aside for later sowing were by necessity harrowed every 2 weeks until they themselves were sown. The following was observed:

1. Early sowing gave faster establishment.
2. Late sowing gave slower establishment but fewer weeds due to 2-weekly harrowing.

If weeds are to be avoided at all costs, which will rarely be the case, this can indeed be achieved by a number of harrowings prior to sowing. In farming practice much better success is achieved by early sowing because crop seedlings are then more vigorous, while weeds may be controlled, if necessary, by chemical weed control (*7.1.10*). Tillage can be minimized also by spraying an existing weed cover with Paraquat at only 0.5 kg/ha product together with a non-ionic wetting agent; rates can be reduced when spraying is carried out in the late afternoon or in overcast weather. A light harrowing may suffice before sowing.

After an early-season arable crop late sowing may be practised to maximize grass production deep into the dry season. If sown 3 to 4 months before the dry season, grasses remain more vigorous in the 1st dry season than in subsequent dry seasons (*7.2.3.1*).

In bi-modal rainfall areas sowing is preferred in the short rains so that annual weeds are soon eliminated. In semiarid areas the requirement is a coarse tilth that traps all incident rainfall and that shields the seedlings from sun and wind (Sands *et al.*, 1970).

7.1.3. Harrowing, rolling and sowing

Seed is often sown straight into freshly ploughed land, but harrowing can be an important part of seedbed preparation after ploughing. Disc-harrows are mostly used, but too fine a tilth and heavy compaction of the subsoil are often the result. Peg-tooth harrows should be preferred. The aim is to level the land; to break the crust, if any; to aerate the topsoil and to loosen the contact between germinating weed seedlings and the soil.

Seeds, when sown onto freshly opened soil, settle down together with the soil crumbs in close contact, especially after some rain. This contact is not achieved if rain intervenes between land preparation and sowing. In fact, much of the effect of harrowing is lost if rain occurs before sowing, and germination will be delayed.

That compaction prior to sowing enhances germination is evident on most soils from the better establishment on the tracks in front of the sowing equipment. Rolling is advisable if the soil surface is loose, except on crust-forming soils and on slopy land liable to run-off. Rolling before sowing is better than after, but many farmers prefer to do both. Rolling is best done with a Cambridge-roller with alternate round and dented disks. An alternative device is a roller made up of partly water-filled oildrums. A common practice in Europe was to drive a tractor wheel-to-wheel over the land. This may sound excessive, but it shows how much importance is attached also in temperate grasses, either small- or largeseeded, to sowing grass seed on a firmly compacted seedbed.

7.1.4. Depth of sowing

No harrowing should be practised after sowing, not even a light raking. The chances are that raking buries the seed too deep which is a widespread reason of failure (Bogdan, 1961a; 1964a). A common practice among smallholders is to pull an Acacia thorn branch over the freshly sown land, which rubs the seed into close contact with the soil. Caryopses, *i.e.* the true seeds, are very small in relation to the spikelet. Placement at 1-cm depth may already prove fatal. It is, therefore, essential that seed is placed on top of freshly prepared and rolled soil so that the seed settles among and in close contact with the soil particles following some rain. If the topography of the terrain as well as the soil structure are conducive to sheet erosion, however, run-off may cause burial of the seed.

7.1.5. Broadcast sowing

For pasture purposes broadcasting is customary. Seeds of tropical grasses are very light, with 1000-seed (spikelet) weight of usually less than 500 mg. Thus, low seed rates are adequate.

Sowing can be done by hand which is the usual practice on small acreages. Scattering small quantities of seed evenly over the land requires considerable skill, but it helps if seed is mixed with sawdust or rough sand. Mixing with P-fertilizer (*7.1.8*) is an even better proposition, but the mixture should be broadcast immediately to prevent scorching of the seed. Scorching may also occur when the mixture is sown in concentrated bands in wide rows while the soil is neither wet nor dry but moist enough to let the fertilizer drain its acids over the seed. With broadcast sowing the risks of scorching in the soil are small. Mixing the seed with P-fertilizer is more laborious but it helps in obtaining a better scatter of the seed.

Tropical grass seed is often fluffy and does not flow freely in normal sowing equipment. Mixing with P-fertilizer is preferred when wheat drills are used. Essential is that the grass seed is not let down into the ground through the pipes; this would bury the seed too deep in the furrows made by the coulters. Either the coulters are lifted so that the seed drops through the pipes on the ground or the pipes are detached from the seed box at the top so that the seed drops freely. Light chains trailing behind the coulters will then rub the seed into the topsoil. This is the most common method of broadcast sowing. An alternative is to use fertilizer spinners. However, the fertilizer is swung over a wider range than the grass seed, so that care is needed to ensure that sufficient overlapping occurs.

7.1.6. Sowing in rows

Some farmers prefer to sow in 75–100 cm rows with maize drills. This used to be normal practice for seed production but even then such rows are too wide (Table 8.3). Stoloniferous species, *e.g.* Rhodesgrass, can be sown in wide rows and a sward will result eventually, but tillers which have to develop from nodes rooting between the rows will continue to lag behind in vigour. This is later observed after mowing and especially after a fire. Even in an old Rhodesgrass sward the original sowing lines may show up with better regrowth after a fierce fire. The heavier the seed- and P-rates within the rows the clearer this row effect, also because high seed rates suppress stolon development (*7.1.7*).

However, there can be no doubt that seeds establish faster when sown in rows. With maize drills it is possible to place seed and fertilizer close to each other. As a result more seedlings emerge and they do it faster. Sowing in rows gives quicker establishment in hoe-agriculture also, because:

1. When seed is placed in freshly drawn furrows, good contact between seed and soil is ensured.
2. P-fertilizer, mixed with the seed, remains in closer contact and is more effective than when seed and fertilizer are broadcast over the whole interrow space.

Sowing in rows offers the possibility of interrow cultivation, but this is only practised when the 1st crop is of special value, for instance, when intended for seed production. Preference is mostly given to broadcasting. This applies to both stoloniferous and tufted grasses.

7.1.7. Grass seed rate

Seed rates affect primary growth until compensatory tillering takes over the role of regulating sward density. After the 1st grazing or cut, effects of seed rates are noticed only when they have been too low or too high and this applies to stoloniferous grasses also (*6.6.2*).

The speed (vigour) with which grasses germinate depends on the species and is poorly correlated with 1000-seed weight (Table 7.3). Although Rhodes- and Molassesgrass have the lightest seed, they germinate as fast as or faster than Coloured Guinea or Congo Signalgrass. Under good Kitale conditions it takes Rhodesgrass normally 8–10 days to emerge. After some 3 weeks rapid growth begins. Needless to say that grass is vulnerable to competition from weeds, yet chemical weed control at the early seedling stage is hazardous as only few herbicides exist which are selective and effective. Unfavourable conditions can have a strong negative effect on establishment also. Although seed is best placed on or near the surface (*7.1.4*) it is inevitably exposed to temperature and moisture fluctuation. On sunny days topsoil temperatures rise and desiccation may occur. When it rains, temperatures drop and seeds get soaked. Part of the seed will perish if the conditions are too extreme or when they change too abruptly and too often.

It is, therefore, not surprising that not more than 25% of the seeds sown may succeed. Table 7.2 illustrates the relationship between the number of seeds sown and the number of seedlings which establish successfully.

Table 7.2. Density of seedlings (m^2) of Rhodesgrass (at 1000-seed weight of 275 mg) in relation to seed rate and failure rate.

Seed rate of PGS (kg/ha)	Failure rate of: 0%	50%	75%
0.5	180	90	45
1.0	360	180	90
1.5	540	270	135

Rate of failure may overrule the effect of seed rate. In fact, an estimated survival of 25% may even be over-optimistic. With a seed rate of 1 kg/ha it is not uncommon to find seedlings at far less than $90/m^2$. For a stoloniferous grass like Rhodesgrass this isnot necessarily too little; 50 seedlings is often adequate. However, as the seed rate or the rate of emergence declines, so does the evenness of distribution. The lower the seed rate the greater the risk that large gaps appear between seedlings. Some farmers, therefore, prefer to use heavier seed rates, of 2 kg PGS or even more. Especially those who harvest their own grass seed and who do not bother to check the % germination unwittingly use seed rates many times of what is required. High rates, however, have a directly negative effect on the number of tillers which grow effectively. Too much competition causes overcrowding and the production of weak tillers (Boonman, 1972b). Rhodesgrass when sown too densely fails to produce stolons and does not form a closed sward (*6.6.2*; *7.1.6*). Sowing Rhodesgrass too densely is more damaging than sowing too thinly.

Seed rate experiments are often inconclusive. Little difference was found between rates varying from 0.2 to 1.8 kg PGS in Nandi Setaria and Rhodesgrass, as far as herbage yield in the 1st year was concerned (Boonman, 1972a, b). In the 2 months after sowing differences were noted in seedling density only. However, the fewer the seedlings the more numerous the tillers per seedling were so that a sward was formed within 3 months. This time scale depends on the degree of weed control. Dicotyle weeds germinate faster than grass seedlings and cause shading. A low density of grass seedlings may perform well, eventually, only if weeds are controlled. Consequently, weed control can even out the effect of differences in seed rate.

Table 7.3 contains recommended seed rates which reflect empirical experience. It is evident that the seed rate is determined by the seedling vigour as well as by growth habit, rather than by seed (spikelet) weight. It should be noted that spikelet weight is independent of caropsis weight, while larger caryopses have not necessarily more vigorous embryo's (*8.7.2.1*).

The condition of the seedbed depends on the type of soil but soil-type effects are obscure, since

Table 7.3. Seed weights and recommended seed rates (kg/ha PGS) in broadcast sowing; in declining order of seedling vigour.

Species	1000-seed* weight (mg)	Seed rate (kg/ha PGS)
Coloured Guinea (*Pcol*)	450	1.0
Rhodesgrass (*Cgay*)	275	1.0
Congo Signalgrass (*Bruz*)	2000	1.5–2.0
Molassesgrass (*Mmin*)	150	1.0
Nandi Setaria (*Ssph*)	425	1.5–2.0
Common Guinea (*Pmax*)	550	2.5

* Seed = spikelet (commercial sample)

so much depends on its structure, OM-content and on the history of land preparation. However, 2 extreme soil-types are known to make pasture establishment a hard task. The light coastal sands develop high soil temperatures especially at the end of the dry season and the beginning of the rains. Direct sunshine and dry, windy conditions may kill seedlings. Sowing is more successful when followed by a number of overcast, rainy days. Black clays (Vertisols) have a bad reputation for pasture establishment also. In drought the soil shrinks leaving cracks and in rain swelling occurs. Young roots break off in the process of shrinking. These soils are difficult to work, irrespective of the crop grown.

Generally, it can be stated that where cereals can be grown successfully, grass seed will establish well also. Consequently, the question of the suitability of certain soil-types for sown pastures should be primarily a question whether they are suitable for arable farming. Where such is not the case and where good natural grassland occurs, the question of sown pasture is not likely to arise (*1*; *4*; *5*).

7.1.8. P-fertilizer in the seedbed

Soils in East Africa are commonly reported as deficient in P. This was confirmed by Le Mare (1958) on the basis of 200 field experiments, mostly with maize and groundnuts in Tanzania.

In tropical grasses, however, clear responses to P-fertilizer in terms of higher DM-yields are rare (Holme & Sherwood, 1954; *12.7.3*). Dougall (1954b) concluded that P-fertilizer promoted seedling establishment, a residual effect after the 1st cut being reflected in the P-content of the herbage, but rarely in yield. Dougall's data of the 1st cut are presented in Table 7.4.

Table 7.4. The effect of P-fertilizer on pasture establishment, in the presence of N; data of the first cut (after Dougall, 1954b).

P_2O_5 applied in the seedbed (kg/ha)	Yield of dry matter (kg/ha)	% in dry matter		
		Crude protein	P_2O_5	CaO
0	800	18.3	0.51	1.03
40	1920	16.7	0.57	0.91
80	2260	15.7	0.60	0.82
160	2370	15.5	0.67	0.88

It is evident from Table 7.4 that phosphate in the seedbed has a very large effect on establishment and on % P_2O_5 in the herbage. The decline in % crude protein and CaO is probably an age effect, because P accelerates early growth and development.

Dougall concluded that the initial optimum dressing would rarely exceed 60 kg/ha P_2O_5, as single superphosphate. This was later confirmed by Strange (1961a). When single superphosphate became cheaply available from the Tororo (U) factories in the early 1960s it was recommended to apply 30–60 kg P_2O_5 in the seedbed and the same recommendation was advanced to subsequent years (*7.2.4.3*).

Tororo single superphosphate contains 21–22% P_2O_5 in easily soluble form, 21–31% CaO, 13% sulphur, as well as micro-elements. Sulphur deficiency in legumes had been suspected in some soils (Strange, 1959b; Poultney, 1963b; Horrell & Court, 1965; Stephens, 1970) and in maize in the Kitale region (Allan, 1976); see also *10.2.1* & *17.3*. P-fertilizer imported from overseas is mostly of high purity, *i.e.* with little or no sulphur and micro-elements.

Although the P-effect on grass seedlings is large, especially when seed and fertilizer are applied in the row (*7.1.6*), little of this effect is carried through beyond the stage of the 1st cut or grazing, as Dougall (*op. cit.*) had shown. Much of the effect at emergence is gone within 2–3 months. Without applied-P seedlings may emerge slowly, but they catch up later on. A P-effect may also disappear early and not show up in any of the cuts (Boonman, 1972b). A lack of response may be more pronounced on regularly fertilized soils which are rotated in a grass-cereal rotation than on continuously cropped and P-deficient soils (*12.7.3*). Tropical grasses can draw their P-requirements from soils with low P-content, once their root system gets under way (*5.3.2*).

Circumstantial evidence suggests that the need for seedbed-P is reduced by early and effective weed control. P encourages the emergence of dicotyle weeds more than of sown grasses, so that grass seedlings suffer from competition and shade. Weeds such as Chinese Lantern (*Nicandra physalodes*) are highly responsive to P-fertilizer. Again this is most evident when the grass seed is applied in the same row as the fertilizer. In the interrow space weed seedlings develop slower and also in a smaller number. Interrow cultivation is then not an adequate form of weed control since it misses the more vigorous weeds in the sowing line.

From this it should be clear that weed control should be applied as early as possible. Unfortunately, cheap auxin-type herbicides such as 2,4-D and MCPA cannot be applied selectively at an early stage (*7.1.10.2*).

P-fertilizer is more critical on soils on which grass is cut and removed (zero-grazing). The subject of P-nutrition is, therefore, dealt with in more detail in Part II (*12.7.3*). From all evidence available it appears as if in sown and grazed grasses in crop rotations 20–30 kg/ha P_2O_5 in the seedbed is adequate for most low and medium content soils, respectively. The more effective the weed control in the early stages of establishment the lower the P-response. Subsequent top dressings do not seem necessary under grazing (*7.2.4.3*).

7.1.9. N-fertilizer in the seedbed

As a result of mineralization soil-N availability is usually high in the year of sowing. Grass yields are often as high as achievable only with 100–200 kg/ha N of fertilizer in later years. The 1st-year is of such great economic importance that every effort should be made to promote a rapid establishment after sowing.

A small dressing with N-fertilizer in the seedbed helps (Table 7.5) but the combined results of N and P are better than when any of the two is applied alone. Dougall (1954b) concluded that 20–30 kg/ha N in the seedbed should be adequate. The same precautions should be taken as with phosphate. Dicotyle weed growth is promoted by N-fertilizer also and the risk of seed scorching is even greater than with phosphate.

Table 7.5. The effect of N-fertilizer applied in the seedbed on pasture establishment, in the presence of P; data of the first cut (after Dougall, 1954b).

N applied in the seedbed (kg/ha)	Yield of dry matter (kg/ha)	% in dry matter		
		Crude protein	P_2O_5	CaO
0	1430	15.9	0.59	1.04
33	1750	15.9	0.57	0.88
66	1990	14.3	0.57	0.87
132	2260	17.9	0.60	0.83

7.1.10. Weed control

Arable weeds in East Africa have more vigorous seedlings than sown tropical grasses. Dicotyle weeds often emerge in very large numbers. Unless precautions are taken, grass seedlings can be completely overtopped by weeds. This is a disadvantage compared with grasses in temperate regions. However, the weed picture is reversed in subsequent years. After the 1st dry season very few weeds remain. Most of them are annuals, typical of arable cropping, which are rarely found in permanent grassland or among sown tropical grasses, after the 1st dry season. By contrast, temperate pastures continue to be invaded by dicotyle weeds which necessitate chemical weed control repeatedly. Allelopathic effects are found in tropical grasses such as Rhodes-grass and Setaria, which retard associated weed species but also legumes (*10.5*; Baker, 1978; Young & Bartholomew, 1981; Crowder & Chheda, 1982; Chheda *et al.*, 1983; Takahashi *et al.*, 1985).

First-year's tropical pasture can be rendered useless by weed growth. Weeds compete for light, water, nutrients and space, but they also reduce soil temperature. In times past when land was abundant and little awareness existed of the potential pasture productivity in the 1st year, many farmers did not bother about the pasture until the 2nd year after the weeds had gone (*7.1*).

Some weeds such as Chinese Lantern (*Nicandra physalodes*) and Mexican Marigold (*Tagetes minuta*) can grow up to 100 cm in less than 2 months. The opinion has been held for many years that mowing ("topping") was the best answer; this was thought not only to prevent these weeds from seeding and to tidy up the field but to make the botanical balance tip in favour of the sown grasses, when freed from shade. It was even believed that mowing would encourage frail grass seedlings to tiller more vigorously. Observations have shown, however, that vigorous weeds are hardly set back by early mowing, whereas mowing may further weaken the grass when the tops, *i.e.* the efficient part of the grass canopy, are removed. This can be quite damaging. Under a heavy weed cover grass seedlings tend to produce etiolated shoots, so that tillering and stolon development are reduced. Root growth is probably also retarded as this is usually one of the first plant parts to be affected by conditions of reduced light intensity.

A field scale experiment was conducted in 1971 at Kitale on a 10-week old sward of Nandi Setaria. Part of the field was topped in the traditional manner at about 30 cm height, a second part was sprayed with 2,4-D and the last part was left untreated, as control. Only short- and medium-height broad-leaved weeds occurred, not growing taller than 50 cm, mostly *Bidens pilosa* (Blackjack), *Commelina benghalensis* and annual grasses (Table 7.6). The best control was achieved by the chemical method. However, a much more vigorous stand of grass was found 2 months later in the control than in the part which had been topped. In the untreated part most of the annual weeds had completed their growth cycle early and had died early (Boonman, 1979b). Similar experiences were reported by Sands *et al.* (1970) and Wigg (1973), both in semiarid areas.

Consequently, weedy stands should not be topped unless there is sufficient clearance between the grass seedlings and the weeds. In these circumstances, however, weed-wipers ("rope-wick") soaked with Glyphosate herbicide are more appropriate (Matthews, 1984).

7.1.10.1. Weed species

The weed species which are important in the pasture establishment phase are the same as those found in arable crops. Relatively little is known about the ecology of important weed species in East Africa (Tiley, 1970b; Ivens, 1971; Popay, 1974; Popay & Ivens, 1982; Terry, 1984). Most weed expertise available was directed towards controlling weeds rather than to studying their ecology.

The most important weeds, apart from the perennial grass weeds (Table 4.4) are presented in Table 7.6 (see also Bogdan, 1950).

Some weeds such as Mexican Marigold can be found all over East Africa above 1000 m, not only in arable land, but also where vegetation is cleared and burnt. Most of the important weeds are in fact cosmopolitan and were introduced adventitiously as seed contaminants, especially from Latin America. Certain weeds side with certain crops. Maize and wheat, when grown on the same land repeatedly, develop distict weed vegetations. Mexican Marigold is favoured by maize growing and by the burning of maize stover. Another typical maize weed is Chinese Lantern. Annual Setaria weeds, on the other hand, are characteristic of wheat cultivation. Changes in agricultural practices can be held responsible for the spread of certain weeds. For instance, wild oats (*Avena fatua*) used to be confined to a few cereal growing areas but the contracting-out of combine-harvesters (which replaced farm-bound combines) has caused the spread of this serious weed into the Molo and Uasin Gishu areas of Kenya. Changes in the weed population have probably taken place also due to the increased use of auxin-type herbicides (2,4-D; MCPA; CMPP) and, more lately, of the triazines (Atrazine, Simazine and Cyanazine).

7.1.10.2. Chemical control

Chinese Lantern is extremely sensitive to MCPA. For the control of Commelina and Galinsoga treatment with 2,4-D is needed; its effect on germinating annual grasses can also be considerable. From experience with pre-emergence application of 2,4-D in maize it is known that 2,4-D acts against seedlings of Wimbigrass. This is a serious annual grass weed, since it grows fast and spreads out its tillers in a fan-like manner. It also has very tough stems which are rejected by cattle.

The use of MCPA and 2,4-D on freshly sown grasses is not without problems. These chemicals cannot be applied until the grass is in the 3–4 leaf stage, which is not early enough to be selective and effective against Commelina and Galinsoga. A search was made for new herbicides at Kitale (Table 7.7; Veenstra & Boonman, 1974). Good results were obtained with Ioxynil-Bromoxynil mixtures, Bentazon and Faneron. Ioxynil-Bromoxynil was very effective against most broad-leaved weeds and selective even when applied as early as 2 weeks after emergence of the sown grass. They had also a considerable effect against Wimbigrass. Bentazon gave an excellent control of Chinese Lantern and suppressed sedges (*e.g. Cyperus rigidifolius*) to some extent. Befenox/CMPP combinations (Verigal) can also effectively be applied at an early stage in Rhodesgrass (personal observation, 1991). Unfortunately, most of these herbicides are rather expensive compared with 2,4-D, MCPA and CMPP. Some of them are also used in wheat and this may help to bring the market price down to a more attractive level.

Table 7.6. Common weeds of arable crops and newly sown pastures of the Kitale and other regions.

Species	Common name	Characteristics
Galinsoga parviflora	Macdonaldi	Moderately resistant to auxin-type regulators
Bidens pilosa	Blackjack	Barbed seeds
Tagetes minuta	Mexican Marigold	Especially common in recently cleared land
Commelina benghalensis	Commelina	Perennial, moderately resistant to auxin-type regulators
Amaranthus spp.	Pigweed	Prefers good soils
Ageratum conyzoides	Ageratum	Common in Kikuyu- and Stargrass pasture
Nicandra physalodes	Chinese Lantern	Very susceptible to MCPA and Bentazon
Dichondra repens	Kidneyweed	Common in lawns
Oxalis latifolia	Oxalis	Perennial with tubers, common in frequently worked land
Digitaria velutina	Fingergrass	
Eleusine indica	Wimbigrass	Fan-shaped canopy
Setaria spp.	Annual Setaria	
Cyperus rigidifolius	Nutgrass	Susceptible to Bentazon
Rottboelia exaltata	Guineafowlgrass	Brittle, irritating hairs

Setaria and Rhodesgrass are generally more susceptible to Atrazine than Signalgrass and Common Guinea, which is important when establishment under maize is considered (*6.11*). Where annual Setaria is a problem, Endothal and Chlortoluron give some control. Terbutryne and Methabenzthiazuron are effective against Fingergrass and Wimbigrass.

Herbicides should be no substitute for poor husbandry. More benefit, in the short run, is to be expected from preparing the best conditions for grass seed establishment in the field:

1. Sowing as early as possible at the onset of the rainy season.
2. Harrowing, rolling and sowing organized in tandem.
3. Seed should be sprinkled on top of freshly prepared land and not be harrowed or raked in.

Seed removal by ants and other insects can be a hazard in some circumstances. Insecticides can be employed (Anslow, 1958) but these do not necessarily repel insects and may not prevent removal because of the time interval needed before they take effect. Pelleting seed with a dusty substance repels insects or causes them to drop the seed.

7.1.11. Early pasture husbandry after direct sowing

Grazing may be possible in 2 months after sowing. This is common in Rhodesgrass but Nandi Setaria takes considerably longer to establish. The newly bred varieties of Rhodesgrass, Elmba and Boma Rhodes, have stronger seedling vigour. The best field establishment ever recorded at Kitale was a Rhodesgrass pasture achieving a sward, 75 cm high, representing 4000 kg/ha DM in 2 months. The field had, in fact, been ready for grazing within 6 weeks after sowing.

Primary growth differs from subsequent regrowth in that the sward is thin and the tillers are few and large (Table 7.1). Some risk exists that cattle pull out the young shoots when these are not yet firmly anchored. Grazing should, however, not be delayed; delay results in spoilage due to treading, fouling and poaching. Poaching of primary growth can be serious as there is no dense sward to counteract the hoof impact on wet, partly bare ground.

The best use of primary growth is hay or silage, also for other reasons. First, it is the cheapest sourcc of herbage to be conserved. Second, when the pasture has been sown at the onset of the rains it may produce its bulk at the same time as most of the pasture on the farm and may not be

Table 7.7. Herbicide effects in seed crops at Kitale (after Veenstra & Boonman, 1974).

Chemical; Trade name	Application pre- or post emergence	Seed crops				Broad leaved weeds					Annual grass weeds	Recommended rate (kg ha^{-1})	Additional susceptible plants
		Rhodes	Setaria	Brachiaria	Desmodium	General	Nicandra	Amaranthus	Galinsoga	Commelina			
Atrazine; Gesaprim 80%	post	4	4	4	–	4	4	4	5	2	2	2.0–2.5	
Bentazon; Bas 3517H, 48%	post	5	5	–	5	3	5	2	2	4Y	1	1.6–2.0	*Cyperus* spp.
Bromacil; Hyvar 80%	post	4	3	3	–	3	5	2	5Y	2	3	0.8	annual Setaria
Bromophenoxim; Faneron 50%	post	5	5	–	1	4	5	3	5	4	1	2.0	Galinsoga
Bromoxynil; Brominil 33%	post	5	5	–	–	3	5	2	4	2	1	0.6	
„ +MCP ester; Buctril M(20/20%)	post	5	5	–	–	4	4	4	5	3	1	0.6	
Buturon; Eptapur 50%	post	4	4	–	–	3	4	3	3	2	2	2.0	
Butylate; Sutan 6-E-78, 25%	pre, inc.	–	–	3	4	2	3	–	3	2	3	4.0–5.0	
Chlortoluron; Dicurane 80%	post	4	–	4	–	3	5	1	–	–	3	1.0–5.0	annual Setaria
2,4-D amine; Phordene 72%	post	4	4	–	4Y	3	3	4	2	2	1	1.0	
„ +dicamba	post	4	4	–	–	5	5	5	4	4	1	0.7–1.0/0.2	
2,4-D ester; U46, 72%	post	4	–	–	4Y	3	3	4	2	–	1	1.0	
Di-allate; Avadex 40%	pre, inc.	–	–	1	4	2	2	–	2	2	3	2.5	
Dinoseb amine; Dynotox 18%	post	–	–	–	4Y	5	5	–	–	5	4S	0.7–1.5	young seedlings
EPTC; Eptam 75%	pre, inc.	–	–	2	4	2	3	–	2	2	3	4.5	annual Setaria
Endothal; Endothall 19%	post	4	4	4	3	2	2	2	1	1	3	1.2	annual Setaria
Ioxynil/bromoxynil; ARD 12/13, 40%	post	5	5	–	–	5	5	5	5	4	1	0.6–1.0	
Ioxynil octanoate; Fr 304/5, 25%	post	5	5	–	–	5	5	5	5	3	1	0.6–1.0	
MCPA salt; Phenoxylene 40%	post	4	4	–	4	3	4	3	2	2	1	1.5–2.0	
„ +dicamba	post	4	4	–	–	4	5	5	3	4	1	1.2/0.2	
Methabenzthiazuron; Tribunil 70%	post	4	4	2	1	4	5	3	4	3	3	4.0	
Terbacil; Sinbar 80%	post	3	4	2	–	3	5	3	5Y	3	4	0.8	
Terbutryne; Igran 80%	post	4	4	3	1	5	5	5	4	5	4	2.0	

Abbreviations: inc. = incorporated in soil; – = no information; Y = in young stage; 1–5: 1 = worst; 5 = best in agricultural sense; S = on seedling stage.

immediately needed for grazing. Understandably, first-year grass is often locked up for seed harvesting (*8.2*).

7.1.12. Undersowing

Thus far we have discussed the practice of direct sowing. The land is ploughed and prepared for the sole purpose of direct sowing of pasture. Direct sowing is usually preferred in more advanced systems of farming, since it yields more grass and confers a greater degree of control over operations. Undersowing has been reported on by many authors (Edwards, 1941c; Bouwer & Strange, 1948; Harker, 1954; Peers & Jones, 1962; Poultney, 1963a; Goldson, 1967; Thomas, 1975). Molassesgrass used to be mainly undersown to maize. In semiarid areas undersowing may not be a good proposition (Sands *et al.*, 1970).

Undersowing is done one year or one crop-season earlier and does not require separate land preparation. The seed is sown under maize or wheat. The only cost incurred is that of the seed and of the actual sowing. These are arguments of persuasion towards a beginner, who has never sown grass seed before. Goldson (1967) has emphasized that the smallholder is reluctant to set land aside for the establishment of grass when his first need is to grow food crops for his family. In Goldson's view undersowing presents to the smallholder with limited capital and equipment a suitable means of establishing grass.

7.1.12.1. Undersowing to maize

There has been some concern that undersowing may reduce maize yield, but this risk is small under low-technology farming. Only under regimes of complete weed control and of early sowing of grass seed may crop yield reduction be substantial, as data due to Goldson (*op. cit.*) illustrate (Table 7.8).

Table 7.8 shows also that undersowing after 3 weedings, when the maize is 90 cm high, generally led to a poor grass cover at the time of the maize harvest. Although the differences in effect between the early and medium sowing dates in the trials of Goldson were small, general field experience shows that the grass seed should be sown as early as current methods of weed control permit. This is usually after the 2nd weeding. When the grass grows up in close relay with the cereal it has the chance to be competitive for light, space, nutrients and water. By the time maize begins to ripen, with sunlight reaching the ground through a senescent leaf canopy, active growth will be assumed by the grass. The earlier the grass seed is sown, the more grass will be available after the harvest of the cereal, which is usually at the start of the dry season. Maize stover is a valuable bye-product in the dry season and the more grass there is the better the nutritive balance is between the fibrous stover and the protein-rich grass. When grass seed is sown at a maize height of 75–100 cm, less grass will be available after the maize harvest. In the following year the performance is then usually disappointing also.

What the negative effect of undersowing may be on the yield of maize depends, therefore, entirely on the circumstances. The higher the level of maize husbandry the greater the reduction in maize yield will be if the grass seed is sown early and under optimum conditions. Obviously, grass acts then in place of weeds which could have been controlled chemically or mechanically. Thus, at a high level of maize husbandry undersown grass competes strongly with maize, be it in an elegant way compared with weeds. Undersowing to maize is difficult to reconcile with mechanized maize husbandry. For instance, chemical weed control is only possible with herbicides which are selective towards both the cereal and the grass seedling.

Undersowing is, therefore, more appropriate in systems of low-technology farming, with incomplete weed control, since the effect of undersowing on maize yield is very small under such circumstances. When grass seed is sown after the 1st or 2nd weeding, grass grows up among weeds which are in any case more competitive and more harmful to the maize than the grass is. Consequently, there will be little or no reduction in the yield of maize caused by the sown grass itself. On the other hand, African Couchgrass is very competitive in the year of sowing both towards maize and the undersown grass, but will eventually be overgrown by the sown grasses in the following year as it is susceptible to grazing (*4.5.1*).

Table 7.8. The effect of height of maize at time of undersowing on the vigour of Rhodesgrass (visual scale, 1–10), at maize harvest in western Kenya, and mean maize yield (kg/ha); after Goldson (1967).

Height of maize (cm) at weeding and sowing	Grass vigour				Mean maize yield
	Kakamega		Busia	Bukura	
	1963	1964	1963	1963	
15 cm	9.2	2.0	6.0	4.5	4480
40 cm	7.6	2.5	7.0	6.3	4710
90 cm	3.2	2.0	6.5	1.0	4870
Control (weeded 3 ×)	–	–	–	–	4890

There are other aspects to be considered also:

1. Farmers, who have never sown pastures before, are more easily persuaded to undersow than to set land aside for the purpose of pasture establishment only. Farmers are impressed by the grazing value of the grass-maize stover combination in the dry season. Any campaign of introducing sown pasture into low-input agriculture should begin with undersowing.
2. When undersown to a cereal crop, there is less risk of the seed being inadvertently buried too deep mechanically.
3. Undersowing reduces the weed population following harvest as compared with crops not undersown (Peers & Jones, 1962).
4. The best legume available (*Desmodium uncinatum*) grows well in shady conditions and establishes most successfully when undersown to maize (*10.3*).

In smallscale farming, maize seed and P-fertilizer are usually planted by hand in hills. It is then not advisable to broadcast grass seed between the rows. A better method is to wait until after the 2nd weeding and to sprinkle grass seed around the maize plants. By placing the seed close to the maize plants, the grass seedling still benefits from the phosphate applied in the maize seedbed while further cultivations can still be carried out without interfering with the grass. Rhodesgrass stolons will have covered the interplant space by the end of the cropping cycle. Rhodesgrass is the best grass for undersowing as it is vigorous and stoloniferous but Molassesgrass may be more satisfactory on overcropped soils (*5.6*; *6.10*). It goes without saying that when grass is undersown stooking of maize stalks in the field at harvest time should be avoided.

7.1.12.2. Undersowing to wheat

Wheat has a shorter growing period than maize. However, the risk that undersown Rhodesgrass interferes with the wheat crop at harvest time is real. In fact, when Boma and Elmba Rhodes, which have stronger seedlings than the older varieties, came on the market in the late 1970s, farmers who were used to undersowing wheat with Rhodesgrass were taken by surprise. Grass seed rates should be reduced to the minimum, so as to ensure a stoloniferous habit while delaying upright growth in Rhodesgrass (*6.6.2*). On the estates of the Agricultural Development Corporations (ADC) at Kitale and Endebess it is now standard practice to sow all wheat in the maize-wheat-pasture rotation to Boma Rhodes. The grass seed is sown together with the wheat seed.

7.2. The following years: pasture husbandry

For a description of grazing management on small farms the reader is referred to the description by Stobbs in the Pasture Handbook of Uganda (Wendt *et al.*, 1970). As Stobbs put it, management is a compromise between plant and animal needs and there is skill in knowing when to sacrifice pasture for the sake of the animals and when to accept immediate detrimental effects on livestock for later benefits from the pasture. Stobbs advocated the 3-paddock system of 2-weeks grazing, followed by 4-weeks rest, and tethering where stock numbers are small. To achieve optimum production night grazing is necessary (Stobbs, *op. cit.*).

7.2.1. *Mixed-farming and rotational grazing*

Cultivated pastures are normally grown in rotation with, and consequently, side by side with crops so that subdivision of land and fencing of paddocks are inevitable. Rotational grazing and rotational farming often go hand in hand. Rotational grazing is applied not because it is better from the pasture utilization point of view, but for practical reasons, on natural grassland also (*4.3.1.1*). Farm operations such as topdressing, haymaking, ploughing and sowing are also sequential, necessitating subdivisions and, consequently, rotational grazing. It is good practice also to take the dairy herd around the paddocks frequently and offer them the best grass available at all times. Other stock will then follow directly afterwards before a rest break is introduced. A grazing spell also helps to break the life-cycle of internal parasites. A more intensive form of rotational grazing is strip grazing, with the aid of an electric fence or by tethering. Paddock grazing is generally better than strip grazing, especially when quality is more limiting than quantity (Joblin, 1963).

Under practical farming conditions, theories that continuous grazing should be preferred are largely academic and one may find both continuous and rotational grazing being applied on the same farm and at the same time. In public opinion continuous grazing has a bad reputation because it is often equated with overgrazing and with poor natural grassland (Table 4.1).

7.2.2. *Seasonality in pasture production*

The main challenge in grassland husbandry is how to cope with the seasonality of production. Before alternative inputs to deal with the dry season are considered which may be expensive and may even have an opposite effect, it is imperative to consider whether seasonality itself can be manipulated and be put to advantage.

The ultimate aim of husbandry is to regulate growth in a way that herbage is available throughout the year. It is basically a matter of matching supply and demand. This endeavour is not confined to the semiarid areas or to the dry seasons only. In subhumid zones such as Kitale the seasonality in grass growth is relevant also in the wet season (Fig. 6.1). Growth rates are not constant in the rainy season, but follow recognizable patterns.

7.2.2.1. *Dry season*

The dry season is the main bottleneck if sole reliance is placed on direct grazing throughout the year and if no provision is made for some relief in the dry season. Cattle may well be out at pasture in the dry season but grazing is little and its value is even less if seen against the damage that overgrazing can do to that pasture. In temperate climates with winter rainfall cattle must be taken off pasture during winter to prevent damage by poaching of the frosty or muddy swards. Also, there would be very little for the animals to eat, not even in mild winters without frost. In the tropics this is different. If poaching does inflict damage to certain soils it is in the rainy, *i.e.* the growing season. Secondly, grazing can be carried over as "standing hay", which in a consistently dry season can be of a relatively high quality (*4.3.4*; *6.3.2*). Thirdly, grass growth in the dry season is less uncertain than in winters in temperate countries. Every farmer in Europe knows that if no special provision is made for roughage for the next winter his animals will have nothing at all to eat. In the tropics, every so many years the "dry" season is not dry and grass may in fact remain green and active throughout. This happens often enough to prevent the farmer-optimist from making provisions as he believes "that the next dry season will again not be so dry" (*19.1*).

We should not simply think of setting special areas aside for the purpose of growing fodder crops, which are expensive, but first of all of the manipulation of existing pasture (Fig. 7.4):

1. Systems designed to shift grass growth potential from the rainy to the dry season.
2. Strategic N-application.
3. Mowing for conservation.
4. Fodder catch crops (oats or Grass Sorghum) at the end of the productive life of the pasture.

7.2.2.2. *Seasonality within the rainy season*

In spite of uniform patterns of rainfall and of mean temperatures within the rainy season at or near the equator (Fig. 3.3), a relatively large variation occurs in the supply of herbage. This variation is of a systematic nature (Fig. 6.1):

1. The early part of the rainy season has a much better potential for grass growth and this is largely due to the high soil temperatures following the dry season (Fig. 3.4).
2. Later, in the middle of the rainy season, growth rates are reduced, especially in Elephantgrass (mid-season depression).
3. The earlier N-fertilizer is applied, the higher the response per kg applied. One month of delay can reduce the annual DM-yield by up to 25%.

The examples quoted here are derived mainly from areas with uni-modal rainfall. No data are available on growth rates in areas with bi-modal rainfall but it may be assumed that in each rainy period, be it in the "long" or "short" rains, the growth rate rises to a peak within the first month.

Seasonality of grass growth is well known from temperate regions with large differences in temperature and radiation regimes between seasons. The further away from the equator the more pronounced these differences are. Seasonality in growth rate is associated with seasonal variation in climate but also with the transition from the vegetative to the generative phases as the specific daylength requirements are met. For instance, Perennial Ryegrass heads in May or June, well before the longest day. Spring vigour is more pronounced the more dormant the grass has been over winter (M.L. 't Hart, private communication).

However, also in subhumid, equatorial regions with relatively constant daylength and atmospheric temperatures, seasonality in grass growth occurs during the growing season, in many ways comparable with temperate climates. For instance at Kitale, with patterns of temperatures and rainfall being relatively constant over the first 7–8 months of the rainy season (Fig. 3.3), the bulk of grass is produced in the first 3 months. More than half of the grass may be produced in only one-quarter of the year (Fig. 6.1). The factors underlying this unevenness of grass supply are not fully understood. The peak in the early part of the rainy season is only to a limited extent associated with the tendency of the grass to become generative. It may even be suggested that this association is irrelevant on the equator as stem elongation and subsequent heading is independent of the time of the year but depends largely on moisture conditions (*8.7.1*).

When the rains set in, conditions for growth are at their best. Chudleigh *et al.* (1977) demonstrated that grass growth at Kitale is to a large extent a function of soil temperature, which in turn depends on the distribution of rainfall. In what may seem a uniform climate, interrupted only by a dry season, topsoil temperatures fluctuate considerably as a function of rainfall. Soil temperature reaches its maximum by the end of the dry season. At Kitale under bare soil, temperature at 5-cm depth drops from nearly 26 °C at the onset of the rains down to below 18 °C within 3 months (Law & Cooper, 1976; Fig. 3.4). This variation in soil temperature exceeds that of atmospheric temperatures and is, therefore, a major factor affecting plant growth. The growth rate of grass reaches an early peak, 4 to 6 weeks after the onset of the rains, and follows the decline in soil temperature at some distance in time. With soil temperatures already at their minimum, tropical grasses may suffer further from small drops in temperature and radiation levels at the peak of the rains, at Kitale in July-August, and may even face a temporary halt in growth, especially Elephantgrass (Fig. 6.1). In September temperature and radiation levels rise again and growth rates then display a second, but minor peak. No quantitative information on the seasonality of grass growth is available from other regions of East Africa. It may be inferred, however, that in the low-rainfall areas which are mostly at low altitudes with higher temperatures, seasonality is more affected by variation in rainfall than in temperature.

Seasonality in grass growth cannot be fully accounted for by meteorological data, however large their effect. Circumstantial evidence suggests that the drier the dry season and the more dormant the grass has been, the more pronounced is the peak in the early rainy-season. Apart from temperature- and dormancy effects, N-release ("Birch-effect") is an additional factor to be considered also when N-fertilizer is applied (*5.3.1*; *7.2.4.1*).

Although the seasonality of growth rate in the rainy season is very pronounced, it is not always so evident in practice. Many pastures are under such grazing pressure that they have no chance of showing excess herbage, not even at the best of times of peak growth. When the grazing pressure has been too severe in the dry season, all capacity for extra growth is first invested by the plant in the rebuilding of root mass. Also the requirements of the grazing animals for compensatory growth have to be met. Visible surplus growth will be rare in such circumstances. Only when pasture has been rationally grazed, without excessive use, will regrowth be very rapid with the onset of the rains and will the patterns displayed in Fig. 6.1 be followed.

High nutritive value accompanies the high growth rates in the early part of the rainy season (Fig. 6.5). Tropical grasses may excel at times in quantity rather than in quality. However, the capacity of tropical grasses to develop flowering heads at any time of the growing season should not automatically label them as stemmy (*6.3.4*). In pastures which are continuously under hard grazing pressure no heading may be seen at all. This is common on smaller farms. Stemminess is then not a liability. If rationally manipulated, *i.e.* when pastures are managed in agreement with their capacity, it is an asset. Unless this is appreciated wrong decisions may be taken:

1. Refuge may be sought in expensive complementary fodder crops, rather than in hay or silage, made to prevent the grass from growing stemmy.
2. Reliance may be placed on legumes to improve or maintain quality, with the grass itself allowed to go stemmy unhindered.
3. Refuge may be sought in purchased concentrates.

7.2.3. Manipulation of grass availability

From the above it should be evident that there is much that can be done to influence the grass growth curve with its peaks and its troughs. Assuming that the seasonal grass supply off sown and rationally managed pasture *on a farm basis* is represented by Fig. 6.1, we can push the curve upwards and enlarge the supply of herbage, with newly sown pastures and with N-fertilizer (Fig. 7.4). Peaks can be transferred through conservation in the form of hay or silage. These strategies imply extra inputs. However, at a constant level of inputs we should be able to influence the shape of the curve also. Peaks can be flattened, by delaying the timing of sowing, of N-application, and by earlier than normal grazing and mowing. An important decision on inputs is, therefore, their timing. Timing is often more crucial than the operation itself.

Farmers need strategies on both a long- and a short-term basis. Success of potential strategies depends largely on the thorough knowledge of the factors which affect supply and demand and their inter-relationships as explained above. The farmer can employ a number of tools with which he can regulate the supply side, even at short notice, in times of emergency. The most important of these tools are:

1. The variation in age of the sown pasture.
2. N-fertilizer.
3. Mowing for conservation.
4. Complementary fodder crops as catch- or break crops.

Fodder crops are expensive when brought in as substitutes for poor pasture. Elephantgrass is not *per se* higher yielding than sown grasses (*11.9*) and silage of maize is not *per se* better than that made of grass (*19.2*). The question is whether an artificial fodder surplus should be established on an extra piece of land, if a natural surplus can be utilized from existing pasture in the form of hay or silage. Within the farm system, fodder crops should complement rather than replace the role of sown pasture. For instance, fodder oats or Grass Sorghum are ideal catch- and break crops when an old pasture is due for ploughing and they can replace the latter to provide valuable fodder exactly in the dry season that follows. Also perennial fodder grasses such as Elephantgrass can be managed in such a way that they are complementary. The aspects of annual and perennial fodder crops are discussed in more detail in Part II.

7.2.3.1. The variation in age of the sown pasture

Quite apart from the inherent seasonality within established pasture, there are systematic differences between pastures of different age:

1. First-year pastures are more vigorous than old pastures and continue growing deeper into the dry season.
2. Late-sown pastures grow actively deep into the dry season.

First-year's Rhodesgrass pasture is usually very vigorous. It produces its bulk at a time of the growing season, the later part, when those of the 2nd and 3rd year tail off (Fig. 6.1). By sowing Rhodesgrass later in the year, after an early arable crop, but well in time to benefit by the last rains, a vigorous sward can develop before the dry season sets in. The late sowing of Rhodes pasture is a recommendable practice to supply a heavy hay cut in the dry season. Freshly sown pasture is much more drought tolerant than old pasture, and active growth will go on deep into the dry season. Various causes lie at the base of this phenomenon:

1. In freshly sown pasture soil-moisture has been saved and is stored in the soil.
2. Tillers are less numerous but more vigorous (Table 7.1).

Freshly sown grass behaves in many ways like annual cereal-fodder crops such as Grass Sorghum and oats.

In 2nd-year pasture, grass surplus is most likely to occur 2 months after the onset of the rains (Fig. 6.1). In older pastures, peak growth tends to occur at a later date, but also at a lower yield level (Sheldrick & Thairu, 1975). The most economic response to N-fertilizer can normally be expected in the 2nd and 3rd year (*12.7.2.1*). By an intelligent use of the age difference between these pastures, a farmer can influence the supply of grass over the seasons, corrected where necessary by N-fertilizer and by mowing for conservation.

Relatively little is said in the literature of tropical pastures in favour of grass conservation or N-fertilizer. These inputs were commonly regarded as uneconomic. Economics, however, are subject to variation due to location and time. Most of the negative reports came from regions with large properties and low-input management of beef cattle. Even then, high rates of N-fertilizer proved to be economic in Central America (Vicente-Chandler, 1974). In East Africa we are mostly concerned with small mixed-farms and with dairying with economic conditions favourable for N-fertilizer. Any input, however costly at first sight, should be explored.

7.2.4. Fertilizers on sown and grazed pastures

Fertilizer strategies depend on whether grass is grazed or cut or grown for seed. These subjects are best dealt with separately (see also *8.5*; *12.7*). Except on Stargrass (*4.2.2.2*), fertilizer is rarely applied on natural grassland and has little effect on improvement in botanical composition (*4.3.6*). In zero-grazing operations of foddergrasses such as Elephantgrass, emphasis is on ensuring that the soil is not depleted of P, K and other elements (*12.7*). This risk is small in grazed, sown pasture so that emphasis is here on N-fertilizer.

7.2.4.1. N-topdressing

Soil-N is a liberal supplier of N to grass in the establishment year due to mineralization (*7.1*). However, with the age of the pasture N-availability declines, although considerable N-flushes are released due to the "Birch-effect" when drought is followed by rain (*5.3.1*; Figs. 6.1 & 6.5). Large reserves of organic-N accumulate in living remains and in soil-OM making N less available as the age of the pasture increases (*5.3.2*). N-fixing legumes would be the obvious answer if these species could be relied upon as partners in grazed swards (*10.4*) and did not demand so much extra P- and other fertilizer instead (*10.2.1*).

In the intensive grass-based dairy economies of the world the most essential input to maintain or to increase herbage yield and quality is N-fertilizer. While the supply of other nutrients can be maintained at a sufficiently high level for long periods, the supply of available N which is sub-optimum on almost all soils should be constantly kept up. N is nevertheless the most powerful instrument to increase animal production per farm and to increase family income. It is, in fact, the most flexible tool with which the farmer can match herbage supply and demand.

In many tropical countries, however, N-fertilizer has been one of the most neglected inputs. In their survey of tropical grassland husbandry, mainly of Africa and Latin America, Crowder & Chheda (1982) quote no examples of N-experiments from East Africa, except those on grass seed production. Indeed, only a few published articles refer to the subject specifically, mainly from Uganda (Stephens, 1967a; Olsen, 1972; Tiharuhondi *et al.*, 1973) or those concerned with temperate pastures at high altitudes (Birch, 1959c; Morrison, 1966a). In Zimbabwe and Zambia, however, much more work has been done on N-fertilizer, especially on Star- and Rhodesgrass

(Weinmann, 1964; Brockington, 1964; Clatworthy, 1984a).

N-fertilizer had, in fact, been discouraged and dismissed as uneconomic in the days when N-fertilizer was expensive, while land and milk were cheap and grass was plentiful. N-fertilizer was believed to increase the bulk, not the quality, of herbage when not wanted. N may be expensive and is not always within easy reach, but the same applies to inputs that have been recommended with less reluctance but which are often less effective, if not more expensive, in the end. Examples that come to mind are P- and other fertilizers but also legume seed and inoculants (*7.2.4.3*; *10.2.1*). N-fertilizer is certainly cheaper than most forms of supplementary or substitution feeds.

Whilst it can be argued that it may be counter-productive to promote inputs which are not readily available to the farmer, the argument should not be selectively used against N-fertilizer alone. The erratic results obtained with P-fertilizer, although so commonly recommended (see below), may have had a negative bearing on the image of N-fertilizer also. The bias against N-fertilizer was continued by the legume approach (*10*) and more recently by preoccupation with animal manure (*5.2.2*; *12.7.1*). Current environmental concern in industrialized countries about N-emission and leaching may well provide an additional argument to be used against N-fertilizer, although similar objections apply to other N-sources and fertilizers (*10.2.1*).

Most of the early experience on field scale with N on grasses in East Africa was with grass seed production. Consistent responses averaging 65 kg DM/kg N of Nandi Setaria were recorded with 100 kg/ha/cut N at infrequent cutting (Boonman, 1972a). Its effect, also at low rates, can already be visible within 7 days. At Kongwa (T) increases of 3% crude protein were observed within 9 days of application of 45 kg/ha N (Owen, 1964). Unlike other fertilizers, N-fertilizer rarely fails to produce immediate and large effects. Residual effects are commonly observed (Owen, *op. cit.*), also in humid areas (Stephens, 1967a).

Experience with N-fertilizer on pasture on farm scale in East Africa is growing rapidly but is still not common. Larger farms usually have the capital but not the urge to apply much N-fertilizer regularly to boost herbage supply when there are alternatives of different kinds on the farm. Smaller farms are faced with exactly the opposite.

The quantity of N-fertilizer applied to pasture must be justified economically by the returns from livestock that feed on those pastures. In fresh-milk market situations with the high milk prices prevailing, cost-benefit analysis can give a rapid answer as to which fertilizer rates are economical. Also in intensive beef production economic returns seem possible with rates exceeding even 200 kg/ha N, as work at Makerere University has indicated (Tiharuhondi *et al.*, 1973). Conversely, stocking rates should be high enough to match the potential generated by N-fertilizer (Fig. 10.6). The residual effect of N-fertilizer on subsequent arable crops should also be taken into consideration. Stobbs (1969e) reported positive effects under grazing regimes but Tiley (1970a) observed that the residual effects were often not significant under regimes of cutting and removal.

The best response in terms of yield of DM/kg of N applied is obtained when N is given as early as possible at the onset of the rains; soil temperatures are high and growth rates of the grass are then at their best (Fig. 6.1; see also *8.5.1*). This extra growth, when surplus to grazing requirements, may imply a conservation cut lest it is wasted. Timing and intensity of grazing may need to be adapted also. The importance of early N-topdressing should be emphasized since there has been a tradition of delaying application unnecessarily (Tiley, 1970a), apparently based on the assumption that some regrowth must have taken place before N-topdressing can be effective (see also *8.5.1*).

Extra growth provided is in fact a growth-accelerating effect, also under conditions of low rainfall and temperature (Minderhoud *et al.*, 1974; see below). This accelerating effect of N is shown to full advantage when yield is determined at the appropriate time. It is not a matter of producing more herbage per defoliation at a given date but rather one of more defoliations per season.

Apart from the work mentioned, most of the research evidence available in East Africa on N-fertilizer is from work done with cut grasses, especially Elephantgrass. This is presented separately (Part II; *12.7.2*) and in more detail because:

1. The data of cutting experiments have more relevance under cutting- than under grazing regimes.
2. Much higher N-rates are required under cutting than under grazing (*13.3.2*).
3. When grass is cut and removed, other elements such as P and K become limiting, especially at high N-rates.
4. N-deficiency is more acute on small farms practising cutting.

However, even in the studies on cut grasses, only little research work was directed at N-fertilizer as a factor on its own. N-fertilizer has commonly been used as a blanket application in trials not aimed principally at N-fertilizer responses. In experiments of this kind, N-rates are often very high in relation to the low frequency of cutting adopted. N-fertilizer is often employed as a yardstick, to measure:

1. The effect of animal manure on grass yield and quality.
2. The amount of N fixed by legumes.
3. The amount of other elements (P, K) removed.
4. The interaction with other factors, *e.g.* harvest interval.

Under grazing a large part of N is returned through the urine. Losses by volatilization are considerable, but the same applies to stall-feeding operations, quite apart from the additional losses incurred in the latter system (*12.7.1*). Experience with night paddocks and close folding has shown that the fertility build-up under grazing is formidable (*5.2.2.3*).

Although most of the research work involving N-fertilizer was in experiments under cutting regimes, a notable exception has been the work done in Uganda at Kawanda by Stephens (*op. cit.*) and at Makerere by Tiharuhondi (*op. cit.*). Stephens observed that, under grazing, N-fertilizer at the rate of 250 kg/ha/year N increased herbage yield 3 to 4-fold and protein yield 5-fold, on average over a 3-year period. N increased the % crude protein in the edible leaf from 7 to 10. Similar increases were obtained by Tiharuhondi (*op. cit.*), under supplementary irrigation, in this case not of Elephantgrass but of a mixed sward of Setaria, Guinea- and Rhodesgrass (Fig. 7.2). The effect of N as % of the mean value of all treatments increased as time progressed, but this was observed only under grazing- and not under cutting regimes (Fig. 7.1; Stephens, *op. cit.*). A similarly reduced N-response under cutting regimes, when related to increasing N-rates, was observed by Simpson & Stobbs (1981).

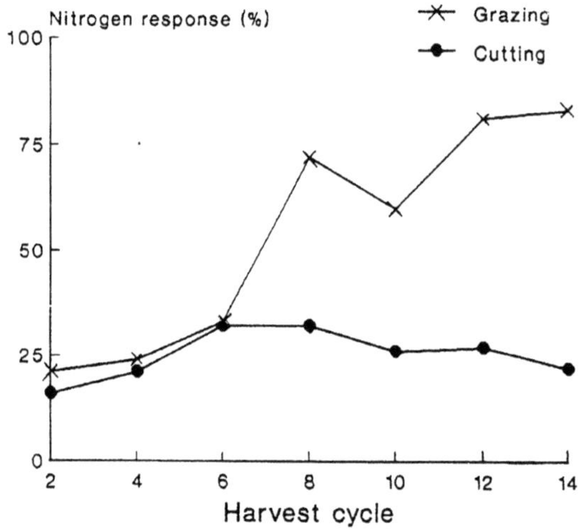

Fig. 7.1. Nitrogen-fertilizer response (%) as affected by grazing or cutting of Elephantgrass (after Stephens, 1967).

Higher grass root contents in the soil are obtained with N-fertilizer (Tiley, 1970a; Weinmann, 1964) and this may help to explain the widespread observation that N-fertilization increases water use efficiency if accompanied by increased yields (*5.3.4*).

Also in semiarid areas such as Kongwa (Wigg *et al.*, 1973) and Ukiriguru (Walker, 1969) yields of Buffelgrass were markedly increased by topdressing with N-fertilizer, while P had little or no effect (Wigg, 1973). At high altitudes, an application of 130 kg N trebled the yield of Perennial Ryegrass and markedly increased the % crude protein (Birch, 1959c; Suttie, 1967). As a general rule, by applying N at 100 kg/ha/year stocking rates can be doubled in the 2nd-3rd year. In the 3rd-5th year the rate may need to be increased if the same level of herbage production is to be maintained (Fig. 12.3); see also Fig. 10.6.

Tiharuhondi's (*op. cit.*) work has been remarkable in the liveweight gains recorded under rapid rotational grazing and the insight it provided in N-effects when supplementary irrigation is applied in a humid, warm area (Zone II). Under no-N, yields were even depressed by irrigation (Fig. 7.2). This was attributed to a suppression of the Birch-effect (*5.3.1*). N-fertilizer markedly increased % crude protein, from 4.3% to 11.5%, and decreased % crude fibre, from 32.1 to 29.5. Mean daily liveweight gains of about 2 kg/ha were achieved at the high N-rates.

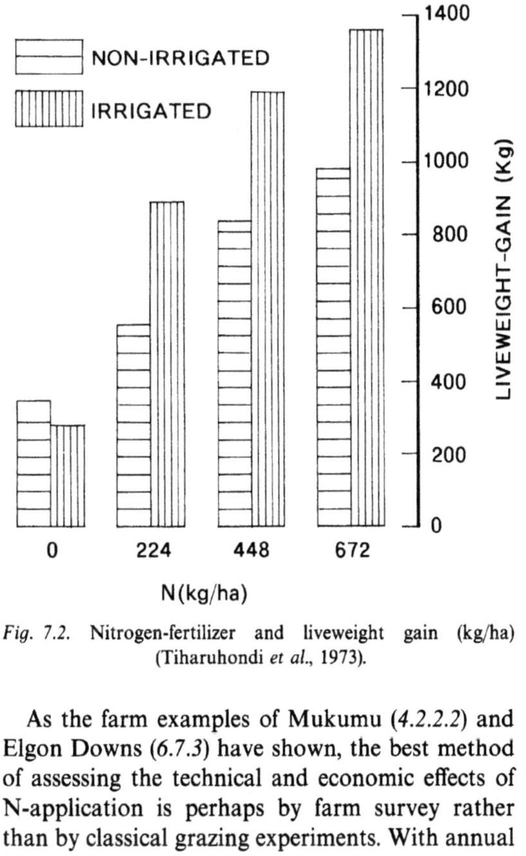

Fig. 7.2. Nitrogen-fertilizer and liveweight gain (kg/ha) (Tiharuhondi *et al.*, 1973).

As the farm examples of Mukumu (*4.2.2.2*) and Elgon Downs (*6.7.3*) have shown, the best method of assessing the technical and economic effects of N-application is perhaps by farm survey rather than by classical grazing experiments. With annual N-rates of 100–150 kg/ha it is possible to more than double stocking rates and milk yields. All available evidence suggests that much of what is known about the effects of N-fertilizer on pasture elsewhere in the world is directly applicable to conditions in East Africa.

Best results have been obtained with calcium ammonium nitrate (CAN) and ammonium sulphate nitrate (ASN). Urea is little used. In the early years sulphate of ammonia (SA) was preferred because it was effective and cheap but it increased soil acidity. This affects subsequent crops rather than the grass itself (see also *12.7.5*).

7.2.4.2. Strategic N-application

Over 80 ton DM have been obtained with high N-rates in the warm and humid tropics (Crowder & Chheda, 1982). However, our interest here lies in low but strategic applications, to get the maximum quantity and quality of herbage with as little N as possible.

Obviously, no farmer would act wisely by suddenly beginning to topdress all his pastures after every grazing. He will not be primarily interested in maximizing yields throughout the year, but rather in filling the gaps in the year-round fodder supply. Not the amount, but the timing should be a primary consideration. This is called strategic N-application.

Because of its powerful effect and because of the flexibility in choice of date, site and rate of application, N has become the major instrument with which to manipulate grass quantity and quality. Its rapid action makes it possible to adjust stocking policies at short notice and vice versa, as a function of both supply and demand of herbage. If a herbage shortage is expected, a gift of 50 kg/ha N applied 2 months ahead can work wonders. N is effective in prolonging grass production late in the rainy season and early in the dry season, and to increase quality within a matter of days (Owen, *op. cit.*). It is also used to speed up growth of pasture, weakened by the dry season (see also *12.7.2*).

N-fertilizer policies should begin with a strategic approach. It can move on later to blanket applications, when the aim is not simply to prevent temporary shortages but to increase long-term, year-round stocking rates as well as the production per animal. N is then applied regularly, in "split" applications, *i.e.* in a sequence of small doses spread over the year, directly after grazing. When rates of N exceed 50 kg/ha split applications are recommended for several reasons: better distribution over the year of yield and nutritive value; decreased luxury consumption and (at high rates) less loss from leaching and volatilization. Since the residual effect of N can be quite remarkable, as illustrated by the work done at Kawanda (Fig. 7.1), dressings after alternate grazings are also effective. Leaching losses under active-growing moderately fertilized swards are small and this applies to a wide range of climates (Mays, 1974).

A management model based on 100 N in the 2nd year and 50 N in the 3rd year is presented in Table 7.9.

7.2.4.3. P-topdressing

The legume approach may explain the benign neglect towards N-fertilizer in research circles but also why the cost-argument was rarely raised with regards to P-fertilizer. P-topdressing at 40–50 kg/ha P_2O_5 annually was recommended as a matter of routine whether legume or grass and whether grass was grazed or cut (Odhiambo, 1974).

In contrast with legumes (*10.2.1*), however, tropical grasses rarely respond to P-topdressing, except to initial dressings on acid soils (*12.7.3*) or when grasses are repeatedly cut and removed (Stephens, 1969). The lack of consistent responses was first reported by the Kenya Highland Fertilizer Scheme (Holme & Sherwood, 1954) in a survey of the P-requirements of natural pasture in various districts; deep placement produced no results. On cultivated pasture, Dougall (1954b) observed that the main effect of P was to promote seedling establishment, a residual effect after the 1st cut being reflected in the P-content of the herbage, but rarely in DM-yield (*7.1.8*). The general lack of response of pure grass swards to P, except when N was applied also, was noted by Whyte *et al.* (1959).

A later experiment in which no less than 16–20 ton of DM were harvested and removed annually over 3 years failed to reveal a response to rates varying from 0 to 80 kg per ha, applied in the 2nd year of a Nandi Setaria field (Boonman, 1972c). It appears that tropical grasses, once established, have the ability to draw their P-requirements even from soils with low P-content (*5.3.2*). Grasses of temperate origin do not seem to share this ability.

Recommendations on P-fertilizer have traditionally been on the "safe" side, with little regard to the cost-benefit ratio. Phosphate does not leach and is not lost, residual effects may be apparent and legumes may be more P-responsive, but now that P-prices have trebled internationally over the past 20 years (Fig. 7.3) it is worth considering the circumstances under which topdressing can be reduced, without undue consequencies for the grass, for livestock or for the crop-grass rotation. Fuller consideration will be given in Part II to fertilization under cutting regimes, since this has been the subject of considerable experimentation (*12.7.3*). Topdressing with P-fertilizer is indeed necessary under cutting regimes, as these involve the removal of large quantities of P in cut herbage. Fortunately, little P-fertilizer is required under direct grazing and it would seem justifiable then to restrict P-application to the cropping phase as much as is necessary for maximum P-response of the crop in question.

Many soils that produce poorly as they are found in nature respond enormously to good husbandry. The easily corrected deficiency of P in arable crops is a common example. P is the most important single nutrient in cereals, especially in wheat. Dressings up to 80 kg/ha P_2O_5 seem to be justified in the 1st year of wheat after grass. In subsequent years dressings beyond 50–60 kg/ha P_2O_5 very seldom give economic returns. Soil analysis data do not correlate closely with crop response to applied P, except at extreme values: low analysis: high response; high analysis: low response (Seitzer *et al.*, 1970). Long-term effects, however, should be taken into account also. P-fertilizer strategies should aim at maintaining the soil-P status at the level required to sollicit a high response in the cropping phase. It is becoming appreciated on a worldwide scale that P-application is generally more efficient on soils with a well-maintained P-status.

The only reliable test for fertilizer requirements is still the field experiment. A method to assess whether grass is in need of a P-topdressing is by means of a trial-strip application, although it may take time before a response becomes visible. This method can be extended to other fertilizers, *e.g.* potash and lime. However, no evidence exists that grazed grass in East Africa benefits from K- or Ca-fertilizer, except perhaps on the poorest soils (*5.2.2*). This is in sharp contrast with conditions of zero-grazing of Elephantgrass, requiring a replenishment with high levels of most nutrients on many soils (*12.7*).

7.2.5. Mowing as a husbandry tool

The value of mowing as a husbandry tool to improve pasture was recognized at an early stage (Halcrow, 1938). In addition to creating a fodder

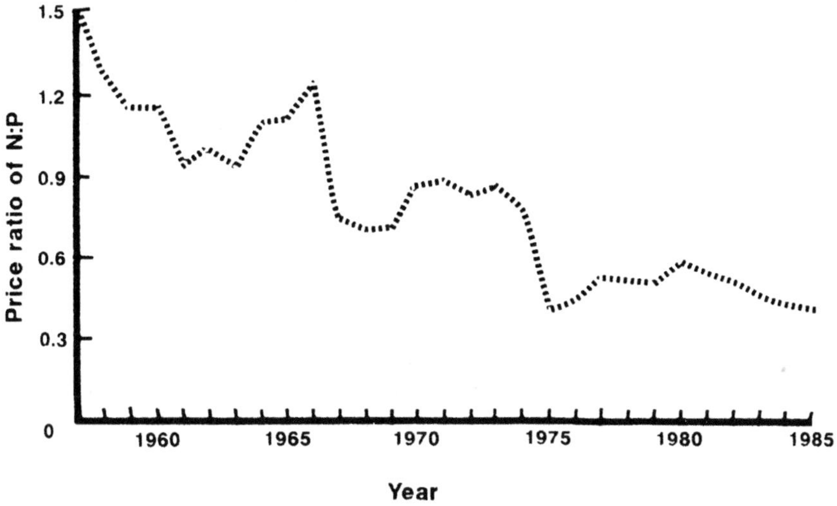

Fig. 7.3. Changes in the ratio of the price of nitrogen (as urea) to the price of phosphorus (as single superphosphate) between 1957 and 1985 (Harrison, 1986).

reserve, mowing lends itself as a husbandry tool in order to manipulate the distribution of grass growth. After a period of overgrazing, for instance in the dry season, a spell is advisable to let the grass regrow to a stage whereby a cut of some 3–4 ton/ha can be harvested. Thus, grass refills its reserves as an insurance for better regrowth in the nearby future. Conversely, mowing at a relatively early stage may help the farmer to put the brake on some pasture or to have especially nutritious young hay. These alternatives provide an illustration of how mowing regimes can affect and, in fact, shift the availability of herbage over the seasons. Mowing may also serve a purpose as a means of weed control, but as we have seen this is rarely necessary in the years after sowing (*7.1.10*).

7.2.5.1. Mowing for conservation

The spelling of pasture from grazing so that a conservation cut can be taken gives the pasture some rest. The longer such a rest the better the subsequent regrowth will be. Thus, the date of utilization of the grass is shifted to later in the season. Haymaking is more common, but excellent silage can be made of sown grasses without difficulty and this has been confirmed in feeding trials (Thomas, 1977a). Hay and silage making are discussed in more detail in Part II (*19*).

7.2.5.2. Mowing as "topping"

Topping is commonly practised in pastures in temperate regions to remove unevenness in the sward. Weed growth and dung pats can cause selective grazing, whereby cattle will leave such sections of the pasture ungrazed, unless they are topped.

It is uncommon to find dicotyle weeds in sown pastures in East Africa, after the 1st dry season (*7.1.10*). Only Ageratum and Dichondra (Table 7.6) may appear in overgrazed pasture. Mowing is, therefore, rarely necessary as a means of control. Dung pats are not a problem either in the rainy season, since faeces are rapidly buried by insects (*5.2.2.3*). In old pastures, however, it may at times be necessary to deal with bush regrowth, but also with weed grasses such as *Sporobolus pyramidalis* and *Eragrostis tenuifolia.* Mowing is rarely effective against these grasses as their tough stems are pushed over rather than cut. Such pastures are better ploughed in time and cropped.

Mowing also helps to promote evenness in sward structure. In tropical pastures animal selectivity in grazing can be very pronounced (Bredon *et al.*, 1967; Dradu & Harrington, 1972).

Finally, N-fertilizer and mowing can be made to join forces to regulate grass supply. Whereas N is applied to prevent possible shortage, conservation cuts should be taken when a surplus

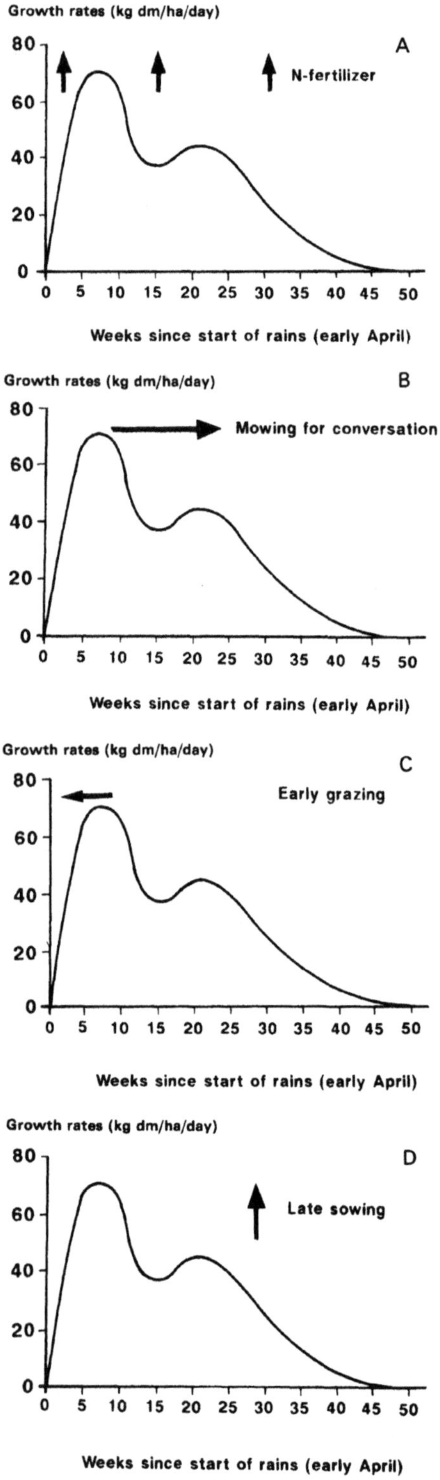

is anticipated. These strategies together with late-sown pastures and complementary fodder crops can be employed to even out the seasonality of herbage availability of the farm (Fig. 7.4).

7.2.6. Last-year pasture as holding ground

It is inevitable that pastures are grazed short in the (later part of the) dry season, but genuine damage to the pasture occurs when the grazing pressure is such that shoots are pulled out of the ground. This is when supplementary feeding with hay, silage or fodder crops should have begun. Part of this will be supplied in the farmyard at milking time, or by direct grazing of fodder crops, but part of it will be fed back in the pasture. The latter is then in fact not much more than a holding ground. The poorest and oldest paddocks which are due to be ploughed anyway should be assigned to function for this purpose. Younger pasture is better rested in the dry season, so that it can regrow vigorously when the rains start again. When livestock is concentrated on paddocks due to be ploughed and cropped, direct beneficial effect can be expected from dung and urine (*5.2.2.3*).

7.2.7. Other practical management aspects

In the previous paragraphs it has been shown that many natural and managerial factors interact to help the farmer regulate a more even supply of herbage throughout the year. Variation in age of pastures and seasonal fluctuations can be evened out so that periods of shortage are compensated for by periods of peak growth. It was shown that these peaks can be shifted by varying the date of sowing, of N-application, of grazing and mowing and that they can be transferred through conservation.

It is a feature of mixed-farming that the grazing pressure is high also in the rainy season because year-round stocking rates are not determined solely by the dry season as is the case in ranching. In addition to bye-products of cropping and to feeds acquired from outside more use is made of fodder conservation.

Fig. 7.4. Strategies to even out the seasonality of grass availability on a farm-basis.

It is important to realize that grazing policies on a farm can never focus on *one* paddock alone. What a farmer may decide to do with a paddock on one day may differ totally from what he chooses to do at a different date. This is the radical difference between normal farm practice and grazing experiments. The latter can offer only limited guidance to on-farm husbandry, as their management regime is by necessity too rigid (*4.3.1*). The grazing situation on the farm is one of constant "put-and-take" decisions, for instance due to unexpected changes in the weather.

The dry season is the principal bottleneck in the year-round system of pasture management. Many farmers are faced with the vicious circle: overgrazing – no fodder provision – overgrazing. This can be broken by having, at least temporarily, additional sources of herbage. The pasture can then be rested over the dry season, good regrowth occurs and conservation cuts can be taken for the next dry season. For temporary relief one may rely on fodder crops, grown specifically for the dry season (Part II).

Handbooks describing general aspects of cultivated grassland in various parts of Africa include: Bille *et al.* (1969), Boudet (1978; West Africa), Savory & Thomas (1977; Malawi), Thorp (1979; Zambia). In addition to those mentioned in this book, smaller country or area reports have been given by: Catford (1953; south Sudan), Timberlake & Dionisio (1984; Mozambique), Haile (1984; central Ethiopia).

7.2.8. Recommendations

To ensure an even supply of herbage from grazed pasture on a farm basis, it is recommended that:

1. New pastures are sown preferably right at the start of the rains; alternatively, sowing can take place after an early arable crop to provide hay in the dry season.
2. First-year's pasture is rested from the 2nd month of the dry season onwards.
3. Hay or silage is made of peak growth of 2nd year pasture in the 2nd or 3rd month of the rainy season.
4. N is applied when needed to encourage peak growth.
5. N is applied 2 months before the end of the rainy season.
6. Old pasture due to be ploughed serves as holding ground in the dry season to feed back hay, silage or fodder crops.
7. Pasture due to be ploughed is sown to a fodder catch crop, before the next cycle of arable cropping.

In Table 7.9 an example of pasture management is given which is valid for areas of Zone II and III (*3.4*), also for those with a bi-modal pattern of rainfall. The assumptions are that N is applied in preparation of either dry season grazing or of a hay crop, in the 2nd year.

7.3. Summary

A rapid establishment is necessary to have a maximum benefit of good and cheap 1st-year pasture, which can also be very drought-tolerant in the dry season. Direct sowing, at the earliest possible date in the rainy season, is preferred, but undersowing is a good practice to get acquainted with sown pastures. Late (direct) sowing maximizes productivity in the dry season. Grass seed should be sown on the surface of freshly prepared and preferably compacted soil. Seed rates depend mainly on the seedling vigour of the species and less so on its 1000-seed weight. Weed control is important but (low) mowing weakens the vigour of the grass seedling more than that of the weeds. Phosphate promotes seedling establishment but is not effective as a topdressing, except under cutting regimes. In systems of rotational farming, rotational grazing is convenient and allows best for conservation cuts which are needed to even out peaks and troughs (dry season) in the annual herbage supply. Grass growth rate reaches a peak soon after the onset of the rains and decreases with soil temperature in the rainy season. Measures must be taken to even out the supply of herbage. Within one paddock this can be achieved by a combination of N-fertilizer with conservation cuts. On a farm basis, herbage supply can be manipulated by making use of the age-difference between pastures. N-fertilizer, though traditionally stigmatized as uneconomic, is a valuable strategic aid in regulating herbage

Table 7.9. A management model of Rhodes pasture according to age. (Example: Kitale, uni-modal rainfall from April–November).

Age (year)	January/ March	April/ June	July/ September	October/ December
1st	—	Sowing I	Sowing II	Hay
2nd	Rest + 50 N	Hay	Grazing + 50 N	Grazing
3rd	Resting	Grazing	Grazing + 50 N	Grazing
4th	Holding	Ploughing	Catch crop	Catch crop

Legenda		
1st year:	Sowing I	: Sown at the onset of the rains
	Sowing II	: Following an early arable crop
	Hay	: Hay or silage made in November–December
2nd year:	Rest + N	: No grazing until a hay crop is taken in May/June, preceded by 50 kg/ha N at the onset of the rains
	Grazing + N	: The grazing in September is followed by 50 kg/ha N
3rd year	Resting	: No grazing in the dry season
4th year	Holding	: In the dry season animals are restricted to 4th year pasture only and are fed hay, silage or fodder
	Ploughing	: In preparation of arable or fodder-catch crops such as oats or Grass Sorghum (*e.g.* Columbus)
	Catch crop	: Strip grazed or cut green

supply and quality throughout the year. Optimum N-levels are lower under grazing than under cutting regimes. N-deficiency is the main factor limiting herbage production.

CHAPTER 8

Grass seed production in Kenya

The identification and introduction into farming in the early part of this century of the major cultivated tropical grasses (Table 6.8) was a farmers' achievement (*2.2.2*; *6.1.2*). One of the consequences was that on-farm seed production and seed trade developed alongside. Subsequently, government research added new varieties and an organized seed industry developed to take care of specialized multiplication and marketing. This has been of crucial importance in helping the improved varieties spread in and outside East Africa. Without seed multiplication and marketing, varieties are still-born, however excellent in performance.

Correct seed multiplication implies maintenance breeding in order to preserve varietal integrity. It is only a small step from there to positive selection, *i.e.* breeding. Rapid growing grasses are needed which are quick to cover and protect the soil. Seedling vigour is largely determined by genetic properties. It is, therefore, essential to treat seed production and breeding in conjunction (*9*).

Since Kenya has been leading in organized grass seed production, most of the following will refer to Kenya. Special mention should be made, however, also of Serere (U). Here the strategy from the 1930s onwards had been to collect grass seed of local stands and to sell it to farmers. This indicates that grass seed production has been meeting with a demand from the (small) farmer for over 50 years. Also in other countries in the region, *e.g.* Malawi, commercial grass seed production has developed in the past 2 decades to the point of export.

Like Kenya, Zimbabwe has had a flourishing seed production and export of Rhodesgrass for many decades. Major seed producing areas outside the African continent are Queensland, Brazil, India and Thailand.

8.1. The Kenya grass seed sector

In Kenya seed production and trading of Molasses- and Rhodesgrass was carried out by individual farmers and some trading companies until a group of Kitale grass seed growers joined hands in 1956 to form the Kenya Seed Company at Kitale (Combes & Verburgt, 1971). Only later came hybrid maize, sunflower, cereals and horticultural seeds. Legume seed production was undertaken also, of clovers and of Desmodium (Silverleaf), but production was suspended in the early 1970s due to declining demand (*10.10*). The area, however, under grass seed production has consistently risen and stands now annually at around 4000 ha (Fig. 8.1), mostly in the Kitale region.

Although Kitale had assumed a leading role as a seed region, seed continued to be produced elsewhere, *e.g.* of Rongai Rhodes in the Nakuru region (*6.6.3*) and of Nandi Setaria near Mount Kenya. Right up to 1970 Buffelgrass was grown for seed in water spreading schemes in Turkana. Production of Buffel (*6.9*) but also of Common Guineagrass (*6.8*) came, however, to a standstill due to lack of demand for these particular species.

The Kenya Seed Company sells some 400 ton of tropical grass seed annually, with 80% consisting of Rhodesgrass, but excluding Sorghum and oats. This does, of course, not include home-grown seed or seed exchanged between farmers, *e.g.* of Rongai Rhodes. Exports may exceed 50% of the sales in some years. Official statistics frequently refer to some 100.000 ha in Kenya of temporary pasture, but it is estimated that, on average, 40.000–50.000 ha are sown every year. These include some 2000 ha sown to temperate grasses, based on imported seed.

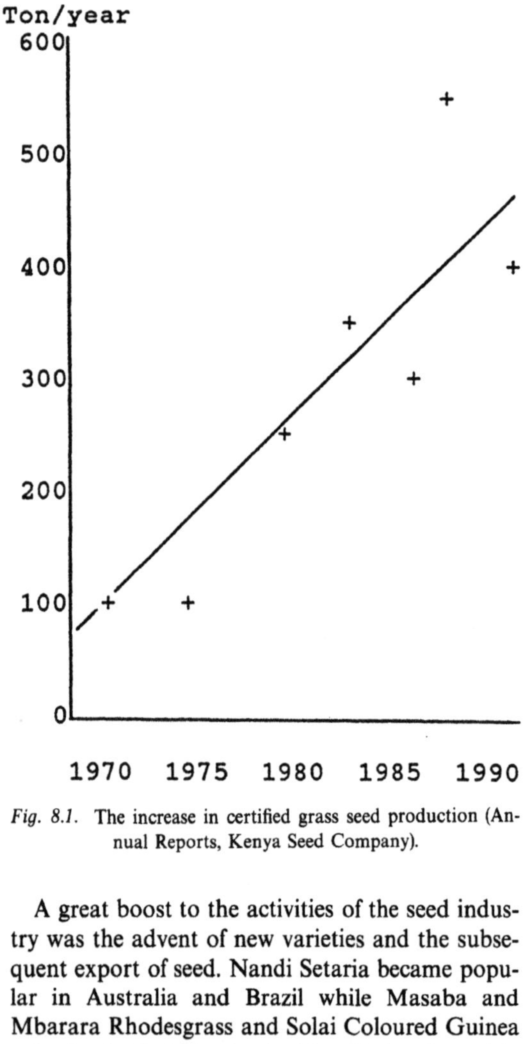

Fig. 8.1. The increase in certified grass seed production (Annual Reports, Kenya Seed Company).

A great boost to the activities of the seed industry was the advent of new varieties and the subsequent export of seed. Nandi Setaria became popular in Australia and Brazil while Masaba and Mbarara Rhodesgrass and Solai Coloured Guinea gained a foothold in the Arabian Peninsula and in Japan.

Although opportunist home-production of seed has continued up till the present day, commercial seed production began to be tackled in a more proficient manner with the advent of the Kenya Seed Company. Seed agronomy research but also breeding followed in the 1960s, which led to the release in 1975 of Boma and Elmba Rhodes (*6.6.4*) and in 1977 of Nasiwa Setaria (*6.7*).

The adoption of pre-set standards of % PGS (pure germinating seeds) has been a key factor in the success of grass seed marketing in Kenya. Since % purity and % germination are low and variable (Table 8.1) in tropical grasses, the Kenya PGS-standards enable the buyer to adjust seed rates and they provide the seed grower and the seed company with a realistic basis to conclude contracts. Seed growers are paid per kg PGS of seed cleaned after delivery.

The PGS test (ISTA, 1985) is based on % pure seed (by weight) and on % germination (by count) of pure seed. Pure-seed spikelet weight is an indicator of mature seed content but, since purity is difficult to delineate, "pure" seed may not contain a caryopsis and the latter may not be viable. Although germinability is affected by post-harvest conditions, such as hardseededness and physiological dormancy (*8.4.1*), % PGS is the most useful single figure of a seed test on which to adjust seed rates. Reproducibility of seed testing is the main ISTA criterion and is based on the presence of caryopses and maximum germination. This may be confusing in the evaluation of dormancy and hardseededness but viability assessments based, for instance, on tetrazolium tests also lack the practical reproducibility needed to assess the economic (commercial) value of seed.

Although seed yields and % PGS are low and variable by temperate grass seed standards, 1000-seed weights and seed rates are low also (Table 7.3). Some general characteristics of variation in seed quality are presented in Table 8.1.

Differences not only between but also within species, are large and systematic (Table 6.7; Figs. 8.5 & 9.1) and have formed the basis of the Kitale breeding programme (*9*).

Table 8.1. Variation in seed quality at Kitale.

	Clean seed		Purity	Germination
	kg/ha/crop	%	%	%
Nandi Setaria	100	60–80	25–65	20–60
Mbarara Rhodes	100	40–60	50–80	48–80

Certification schemes in the classical style of field- and seedlot inspection have been instigated on a number of occasions but a meaningful, cost-effective control is hard to put into effect without a close involvement of the seed industry and the seed growers. Field inspection may place undue emphasis on criteria which are easy to recognize in the field, such as isolation distances, although the risk of pollen contamination, if any, is small. More important criteria, for instance whether a seedlot is correctly named, can be assessed only when inspection on the spot is backed up by growing-out tests in comparison with the standard referee of that variety. This implies maintenance breeding which is, however, often accorded low priority (*9.1*). To be effective, regulatory schemes should be tuned to existing commercial standards of seed production with the aim of improving them. Process control, rather than routine field- or seedlot inspection, is cost-effective and has a powerful educational effect. It will, therefore, assume more prominence in the future.

8.2. Systems of contract seed production

In the year of sowing the number of harvests depends on the date of sowing and on the degree of weed control. In subsequent years, the Kitale climate (Fig. 3.3) permits 2 seed crops to be taken. The intensive 2-harvest system (June–July and December–January), developed in the 1960s at the Kitale Station by C.M. Moore, has subsequently been refined (Table 8.2) and has been adopted by specialist growers. A less intensive system of seed production is also found, quite apart from the practice of locking-up pasture not needed for grazing:

1. Seed fields are sown in June-July, when most of the arable operations of other crops have been completed. After the seed harvests a hay crop is taken in the dry season.
2. Also in subsequent years only one seed crop is taken, in November-December, following a N-topdressing in July. Hay is made in the dry season. Burning is rarely seen as beneficial.
3. Grazing is carried out between seed crops to benefit by the residual effect of the N-fertilizer.

The latter system, being more opportunist, suits the grower when grazing is plentiful although seed may then not be in much demand and not fetch a high price. The 2-harvest system is a more reliable basis of a sustained seed industry but developmental research was needed to answer some of the technical questions associated with this system. Harvesting grass seed in the early season, in July, when the rains are at their peak, was seen as inconvenient but research established with convincing evidence that seed harvests were as good in July as in the beginning of the dry season and that seed is in fact easier to thresh when harvested under humid conditions. June-July harvests of Setaria also have less bunt (*Tilletia echinosperma*) than progressively later-harvested crops (*8.7.2.5*). The aftermath of the June–July seed crop can be made into hay or silage, which is perhaps not a high-quality, but a useful bye-product. Alternatively, it can be slashed and chopped to act as a mulch. Precautions must be taken for adequate seed drying before storage, but this also applies when seed is harvested at the beginning of the dry season.

8.3. Home-grown seed

Home-grown seed has been the rule rather than the exception in many parts of the world where seed is otherwise too difficult or too expensive to obtain. It has been practised by large- and smallscale farmers in Kenya, since it is often easier to obtain a small quantity of the best seed available and multiply it to meet the home requirements. This is in fact how the early varieties of Kitale spread. Home-grown seed can be in the interest of both seed industry and farmers, since more high-quality seed may move into markets which are otherwise inaccessible. More of what are now grass weed fallows in the crop rotation (*5.1.2*) may then be turned over to productive fertility-restoring pasture. Where vegetatively propagated fodder-type grasses are not the answer, husbandry is easier with sown grasses while home-grown seed is easy and simple to produce. In fact, reserving primary growth of freshly sown pasture for seed harvesting is a good method to get pasture off to a solid start and also to make hay of the aftermath, without the risk of grazing too early or too hard (*7.1.11*). Grass seed harvesting can be made an integral part of pasture husbandry. Primary growth, *i.e.* the 1st year, is a reliable and cheap source to harvest seed from.

8.4. Husbandry of grass seed production

The 2-harvest system has been the basis for the research work carried out by the author

(Boonman, 1973b). In this and the following paragraphs an overview of husbandry measures (Table 8.2) and of their impact is presented first, after which an account is given of the principles of seed production (*8.6*; *8.7*). Unless stated otherwise, reference is made to Kitale conditions and to the 2 major grass species in multiplication: Rhodesgrass and Setaria.

In the following, 2 husbandry aspects are discussed which are of major importance to every seed grower, including the home-grower:

1. How to get the most out of the 1st year.
2. How to harvest.

8.4.1. The first year's seed crop

How 1st year's grass in general should be managed has been discussed (*7.1*). Unless weeds are a problem, 1st year's seed crops are usually highly reliable as they are less vulnerable than subsequent crops to stress factors such as drought to which especially Setaria is susceptible. They are cheaper to grow than in subsequent years because of a ready supply of available soil-N. More moisture is saved in the soil when the land is freshly prepared. First year's grass crops behave, in fact, like annual cereal fodder crops. Growth is more robust since tillers are fewer but much larger than in subsequent crops (Table 7.1).

The 1st seed crop can be harvested within 4–5 months after sowing. When this crop is cut at flag-leaf level, many late-formed tillers are ready to give a new flush of heads; there is no need of cutting the whole stand back to ground level. A 2nd and even 3rd flush may be harvested, until deep into the dry season. Obviously, this is a very efficient way of harvesting seed so that growers should take care to have the best growing conditions in the 1st year. The capacity to produce aftermath flushes decreases in subsequent years as sward density increases, except in species with a very "flat" head emergence curve, *e.g.* Makueni Guinea (Fig. 8.5).

Grass seed should be sown as early as possible at the onset of the rains (*7.1.2*). Traditionally, growers postponed their grass sowing operations until the arable crops had been taken care of. Seed should be sown direct (*7.1.1*), in rows or broadcast, on the surface of freshly prepared soil (*7.1.3*). Herbicides are useful but mowing ("topping") weakens grass seedlings (*7.1.10*). P- and N-fertilizer in the seedbed promote seedling establishment but also weed growth. The more effective the dicotyle weed control has been, the lower are P-response and P-demand (*7.1.8*).

It is common practice to sow fields intended for seed production in rows, often together with the fertilizer. There are practical advantages: depth of seed placement; P-availability; interrow

Table 8.2. Traditional versus research-based recommendations in the seed production of Rhodesgrass and Setaria.

	Practice		Impact through PGS-yield increase
	Traditional	Improved	
Time of sowing	Delayed	Early	Proportional
Seed rate	1–1½ kg PGS	Same	–
Superphosphate			
Seed bed	50 kg P_2O_5	Same	–
Topdressing	50 kg P_2O_5	Maintenance	–
Row-cultivation	Yes	No	–
Weed control	1 kg 2,4-D	Same	–
Rowwidth	90 cm	30–60 cm	30% (Setaria)
Nitrogen (N)			
Rate	50–60 kg N	100 kg N	50–100%
Date	Delayed	Early	100%
Seed harvest	At start of shedding	At 30% shedding*	50%
Burning	No	No	–

*Relative to spikelets present at time of harvest.

cultivation. The main benefit is a physiological one, in post-establishment crops, of regulating tiller density which should be high enough to promote synchrony in head emergence and flowering, but not too high to cause overcrowding of tillers. Seed growers used to employ the 75–90 cm rowwidth of maize drills. Best yields and % PGS are obtained, however, in rows of 30 cm or less in the 1st year (Table 8.3), although 45–60 cm is appropriate in subsequent years in tufted grasses at low N-levels (Fig. 8.3). Herbicides lessen the dependence on equipment tied to a standard rowwidth. In stoloniferous grasses such as Rhodesgrass rowwidth is important to achieve an early cover in the year of establishment, but some positive effect of narrow rowwidth may still be noticed in later crops.

Table 8.3. The effect of rowwidth on head properties and PGS-yield in the establishment crop of Nandi Setaria (after Boonman, 1972a).

Rowwidth (cm)	Heads		PGS	
	Density (No/m^2)	Length (cm)	%	Yield (kg/ha)
30	157	16.9	16.4	16.4
60	133	18.5	11.5	12.0
90	93	18.4	11.6	8.9

High seed rates should be avoided in Rhodesgrass since they impair stolon development (*7.1.7*). Seed rate depends on the seedling vigour of the species and less so on 1000-seed weight. The standard seed rates have been given in Table 7.3. Rates can be reduced when seed is sown in rows. A narrow rowwidth, *e.g.* 50 cm, is also of practical convenience. A closed canopy is formed earlier so that mechanical cultivation is unnecessary. The land surface is more even, without "ridges and beds", so that the sward is easier to mow eventually. Specialized seed growers prefer to grow their seed crops in rows but growers, intent on surrendering the field to pasture after the 1st seed and hay crop, sow broadcast.

8.4.2. Seed harvesting

As weeds are a problem in the establishment year, consideration should be given to the risk of contamination in the harvested seed since penalties can be unduly heavy in commerce. Methods of hand harvesting can cope better with weedy stands than mechanized harvesting; mechanical seed cleaning is often not very effective to remove weed seeds either and may inevitably imply the loss of good seed. Some weed seeds attach themselves to or inside the (clusters of) spikelets of fluffy grass seeds (Bogdan, 1966e).

Seed harvesting is usually carried out some 6–7 weeks after IHE (Initial Head Emergence; 5–10 $heads/m^2$), when "shedding" is already well on its way (Fig. 8.9). Inexperienced seed growers watch shedding with alarm as they equate early shedding with loss of valuable seed. The early shedding in tropical grasses, however, is false alarm because it consists largely of empty spikelets (*8.7.2.4*). In fact, shedding should be in progress for 2 weeks before net losses of PGS are incurred. Shedding may increase to 30% before the net loss of true seed (filled spikelets) outweighs the compensatory maturation of later formed seed (Fig. 8.2). It should be noted that the x-axis of Fig. 8.2 refers to the increase in % shedding as heading proceeds (Fig. 8.9). More valuable seed is "lost" by harvesting too early than too late.

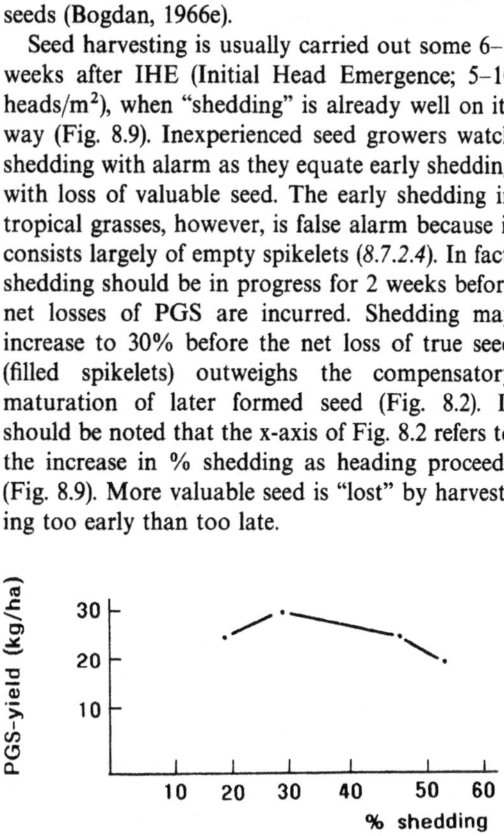

Fig. 8.2. The relationship between shedding and PGS-yield (derived from Fig. 8.9) (after Boonman, 1973a).

The interval of 6–7 weeks between IHE and optimum harvest date (OHD) is relatively independent of season or species or variety. Dry, bright seasons may seem to hasten OHD, but weather effects are generally difficult to translate in grass seed crops. The harvest safety margin is about 2 weeks, which is a comfortable moratorium for hand-harvesting procedures. Now that new varieties are available that mature more uniformly it may become feasible to employ more objective OHD criteria such as % seed moisture.

Hand harvesting is simple and weather-proof. Heads can be reaped whether wet or dry, but threshing is more difficult when reaping and stooking are carried out in a dry period. The labour demand may seem high but experienced reapers, binders and threshers need 15–20 mandays/ha all-in. The seed heads are cut with saw-edged sickles (reaping hooks or "ringa") just above the leaf canopy and are bound into sheaves. Sheaves are stood upright to form pyramid-shaped stooks, with one sheaf placed upside-down over the top, bent over in the middle and tied around the stook with its inside out. This is done to hold the maturing seeds and to protect them from direct exposure to the elements and to birds. The stooks should not be larger than 5 or 6 sheaves to prevent excessive heating. Some growers stook the sheaves *in situ*, others prefer to cart them to a corner of the field or to the farm compound. The latter method makes inspection of the stooks easier and also vacates the field so that the herbage residue can be mown. Essential is that stooks are placed in the open and are well-ventilated. They should not be in contact with bare soil, because insects easily contaminate the stooks with soil. A layer of grass under the stooks is effective and this is why many growers prefer to leave the stooks *in situ* on the stubble, although recovery of the stooks at threshing time is then less easy. The stooks may remain wet throughout and may get mouldy inside but this is no cause of concern. Threshing can be carried out after 2 weeks.

In the above method all heads that have emerged are cut in bulk and subsequent after-ripening in the stook helps to even out some of the age differences between spikelets. Plucking only the old heads, in successive rounds, is ineffective since late heads contribute little if any viable seed (Table 8.8). Also combine-harvesting of tropical grass seed is impractical, expensive and unnecessary and has thus far led to lower yields and % PGS under Kitale conditions. Reaper-binder machines are more useful and are the first to be considered in the event of mechanization.

Makueni Guinea is the least synchrone in heading of all grasses (Fig. 8.5) at Kitale and used to be harvested by bundling the heads *in situ* with a wisp of leaf over the top, followed by cutting and threshing 2 weeks later (Bogdan, 1966a). The increase in % PGS was at times considerable, yet not consistent enough for general recommendation as the costs in terms of labour were prohibitive.

Threshing can be a source of loss unless seed is collected properly. Some hand-rubbing of the heads or a beating of the sheaves with sticks may be necessary to free the seed. A 2nd threshing, after further maturation, has rarely proved worthwhile. Threshing with a combine-harvester is no more effective than when done by hand.

Threshing leaves a bulk of spikelets, leaf- or stem particles and live insects. The % moisture can be more than 60 and measures should be taken at once to prevent excessive heating. Leaf and stem parts are removed by scalping over a wire-meshed tray. The seed is then spread out in thin layers, regularly turned over, and exposed gradually to direct sunshine. Essential is that the natural drying process proceeds slowly and that seedlot temperatures above 40 °C are avoided. Hopkinson & English (1988) have shown how PGS can be damaged by the rapid, artificial drying of freshly harvested seed and how important it is to dry seed as slowly as possible. Seed should subsequently be kept under dry conditions to ensure long storage life (see below).

Low % germination is often attributed to seed dormancy but critical data are frequently lacking. Dormancy can be a problem in some grasses but is of minor importance in Rhodesgrass. Germination studies were carried out at Kitale by C.M. Moore in the early 1960s which showed the following:

1. Increase in germination with time in the first 6 months was quite marked in Solai Coloured Guinea, moderate in Nandi Setaria and negligible in Rhodes- and Molassesgrass.
2. The safe storage period before % germination dropped from 60 to 30 ranged from 2.5 years in Nandi Setaria, to 4 years in Rhodesgrass and Coloured Guinea, and 6 years in Molassesgrass. This factor allows plenty of leeway at Kitale for carry-over stocks.
3. There is no clear relationship between [1] and [2].

Physical constriction by encasing lemma and palea, impermeable seed coats (hard seeds) and chemical inhibitors in these structures may result in delayed germination. Hardseededness was recorded in Coloured Guinea in Kenya already by Edwards

(1932) but can be removed by scarification. To assess the relevance of apparent physiological dormancy, germination studies should be carried out with large numbers of samples per variety; per harvest season; at various levels of purity and germination (before and after OHD) and at various seed moisture levels. Pre-drying, pre-chilling, KNO_3- or mechanical treatment can be applied if an early germination test is required. Seed is, however, rarely sown straight after harvest, so that even if dormancy of a few months' duration is prevalent, little difficulty should be encountered in actual practice if seed is harvested, dried and stored properly.

In contrast with dormancy, the importance of dry storage and its effect on seed quality is often underrated (Behaeghe, 1960). Seed longevity is determined mainly by prevailing humidity. That longevity is controlled by humidity rather than by ambient temperature is slow to find acceptance by seedsmen. For temporary storage, after proper drying, a relative humidity of 40% and a temperature of 17 °C, which is close to the mean daily average at Kitale (Fig. 3.3), is a favourable combination. Artificial drying to 8–10% and packaging in moisture-proof material is necessary if the seed is to be prepared for shipment overseas or for long-term storage.

Seed cleaning usually gets more attention than drying but is not so essential; its relevance should be viewed in terms of product presentation (marketing) rather than as genuine improvement. To remove inert matter and soil, simple sieves and hand winnowing may suffice. Mechanical cleaning is often not very effective in removing weed seeds (see above) without losing good seed also and the same applies to the removal of empty spikelets. Conversely, "clean" seed (Table 8.1) is not necessarily of a high purity (*i.e.* high % caryopses), and "pure" seed may not be highly viable either (Fig. 8.11). Differences in physical measurements (weight, volume, surface) between filled and empty spikelets are often too small for the available equipment to sort out effectively.

It is a peculiarity of some species, including most tropical grasses, that the pure seed fraction, determined in accordance with ISTA rules, germinates to only a limited extent. The standard regulations for seed quality of Queensland stipulate a minimum of 10% germination and 30% pure seed for some grasses. This shows how important it is that the % germination of a seedlot is declared.

8.5. Post-establishment crops

These crops differ from the establishment crop in that they:

1. Have more numerous but finer tillers (Table 7.1).
2. Require substantial quantities of N-fertilizer.
3. Do not respond to P-fertilizer (*7.2.4.3*).
4. Have few, if any, weed control problems (*7.1.10*)
5. Are, in the case of Setaria, more sensitive to the effect of weather on the ultimate seed viability.

The start-to-finish cycle is shorter in post-establishment crops. Whereas no harvest can be taken until August-September in the year of establishment, a ripe crop may be ready as early as late June in the following years, within 90 days after the end of the dry season. This leaves plenty of time to deal with the grass residue after the seed harvest and to topdress in order to have a complete 2nd cycle within the same year (Table 8.2). The 2nd crop season usually lasts 100-120 days.

Aftermath flushes of heads are rare, except in species with a very "flat" emergence curve, *e.g.* Makueni Guinea (Fig. 8.5). Little problem is experienced with weeds or volunteer seedlings, after the 1st dry season (*7.1.10*). Unlike pasture (*6.3.1*), seed fields maintain yields at the same level of N-fertilizer for 3–4 years or longer (Tables 8.9 & 8.10), there being no evidence that optimum N-rates increase with age, either for herbage or for seed yield.

N-fertilizer is of crucial importance in post-establishment crops. Not only the rate but especially the time of application has an overriding effect (Table 8.4). Season is a major source of variation in post-establishment crops, even when well-fertilized, of Nandi Setaria at Kitale. High plant density is less crucial than in the 1st year's crop (Table 8.3), but is still important in order to improve % PGS (Fig. 8.3).

Table 8.4. The principal husbandry and environmental effects on seed yield and its components in post-establishment crops of Nandi Setaria (Boonman, 1973b); data relative to 100.

Negative effect	PGS		Head density	1000-seed weight	Crop index
	Yield	%			
Zero-Nitrogen (N)	5	60	15	100	30
N applied 4 weeks late	35	55	80	100	70
Season with dry spells	10	30	80	100	40
Optimum conditions	100	100	100	100	100

8.5.1. *Rate and time of N-application*

Without N-fertilizer there is little chance of having a good seed crop after the 1st year of harvesting (Table 8.4). Seed yield is even more sensitive to N than herbage yield, which is reflected in a low crop index, especially in Setaria (see also Table 8.5).

N is essential to bring on many heads and effective spikelets, while high plant density is necessary to promote the synchrony of their action. Consequently, N is the factor most responsible for the quantity of seed, while narrow rowwidth is instrumental in raising the % PGS. In non-stoloniferous grasses such as Setaria it is necessary to narrow the rowwidth to make maximum use of N, and vice versa (Fig. 8.3).

This interaction between N-fertilizer and plant density is of overriding importance, not only to increase yield but also PGS-capacity of tropical grasses. The response to N in terms of PGS-yield and its determinants is presented in Table 8.5.

By taking the N_0 levels as the base points it is apparent that 100 kg/ha N, regarded as the optimum in an average season, can increase PGS-yields 6–7 fold. Head density is the yield component most affected in the absence of N (Table 8.4). Top N-levels bring on more heads but not necessarily higher PGS-yields, except in favourable seasons. To the contrary, especially in Nandi Setaria, high N leads to a decrease in the % PGS at wide rowwidth (Fig. 8.12) since it makes late tillers develop heads and spikelets which do not reach maturity. This is probably the reason also why split application of N is deleterious and

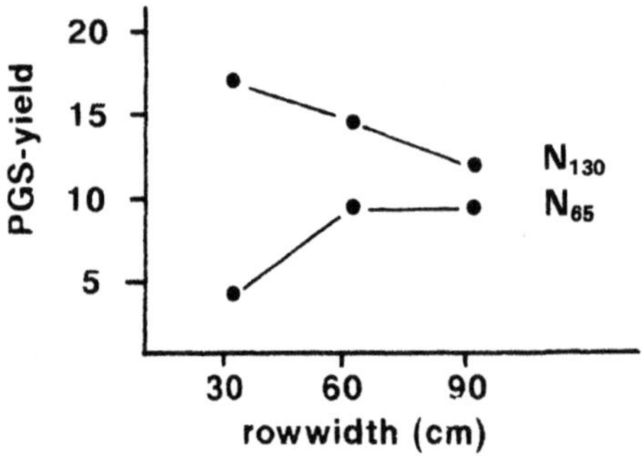

Fig. 8.3. The influence of N-rate on PGS-yield of Nandi Setaria in relation to rowwidth (after Boonman, 1972a; means of 7 harvests).

Table 8.5. The effect of N-fertilizer on post-establishment seed crops of Nandi II Setaria (Ndi) and Mbarara Rhodesgrass (Mba) over 5 and 6 crops, respectively (after Boonman, 1972a,b).

N (kg/ha)	PGS-yield (kg/ha)		Head density (m^2)		Heading tillers (%)		Crop index (%)	
	Ndi	Mba	Ndi	Mba	Ndi	Mba	Ndi	Mba
0	2.4	6.2	40	67	11	15	0.48	0.80
50	9.6	19.1	125	138	19	23	0.91	1.00
100	16.5	39.5	185	237	26	33	1.01	1.35
150	16.1	42.2	200	267	28	44	0.91	1.45
200	16.3	42.9	240	293	32	47	0.92	1.50

why N should be applied as early as possible in the life of a seed crop (Boonman, 1972c).

When N is applied late or when the season is unfavourable, the % PGS is the component most hit (Table 8.4). The 9-months growing season of Kitale can easily accomodate 2 crop cycles since the solar and temperature regimes are relatively uniform, but the timing of operations has practical complications:

1. The onset of the rains is a busy time on the farm so that topdressing of grass seed crops is easily postponed.
2. A late 2nd-season brings on a hay crop, after seed harvest, at a more convenient time, *i.e.* in the dry season.
3. A moratorium in July is needed to dispose of the residue of the 1st seed crop before the 2nd crop is embarked upon.

However, the underlying principles are of a seasonal nature and also apply when only 1 seed crop is taken per year. N should be applied as early as possible at the start of growth after the onset of the rainy season. N applied 1 month later reduces PGS-yields to only one-third, mainly due to a reduced % PGS (Table 8.4). This "early-season" effect is due to growing conditions, especially soil temperatures, being more favourable at the start of the rainy season than later on (Fig. 3.4; *7.2.2.2*). In the 2nd half of the year, timing of N-application is not so critical but, when the weather is cool, better growing conditions should be awaited until the end of August at the latest. Growing conditions are then again the determining factor. Aftermath herbage can be cut early and be followed by a grazing but N-topdressing should await warm weather. A combination of an early crop in the 1st half and a relatively late crop in the

2nd half of the year gives the grower a moratorium of 1–1.5 months in July to deal with the herbage residue of the 1st seed crop.

The importance of N-topdressing as early as possible at the onset of the rains cannot be overstressed. There has been a mistaken but persistent belief, perhaps borrowed from maize-topdressing practices, that N is better applied after some growth has accumulated (*7.2.4.1*).

8.5.2. The effect of season

Grass seed crops, the world over, are known to be inconsistent in yield, but in tropical grasses the variation in PGS-yield due to season may be as high as 8-fold (Table 8.6). This applies especially to Setaria, although nothing unruly may be noticed from the outside. Total DM-yields of seed crops are relatively unaffected by seasonal variation in rainfall. What may appear, from the head density present, to be a heavy seed crop of Nandi Setaria may yield no more than 5 kg/ha PGS at a % PGS of less than 10 (Fig. 8.11; Table 8.9) when dry spells, however short, intervene in the early phases of the seed crop; weather alone may reduce the average seed-set from about 20 to 3 mg PGS per head. Seasons with very low yields of clean seed of Nandi Setaria (< 50 kg/ha) tend to have a low % PGS. Seasons, however, with high yields of clean seed can have wildly fluctuating % PGS (Table 8.9).

Table 8.6. Species and season in relation to PGS-yield (kg/ha) at Kitale.

Species	1967	1968		1969		1970		1971	
Season	1	1	2	1	2	1	2	1	2
Nandi Setaria	16	10	5	13	6	9	25	40	35
Mbarara Rhodes	—	56	—	25	42	30	36	50	67
Coloured Guinea	—	46	—	46	—	78	30	62	51

Conversely, sparsely heading crops of Mbarara Rhodes can give a surprisingly high yield of clean seed and high % PGS (Table 8.10). Rhodesgrass and Coloured Guinea can produce high yields in seasons in which Nandi Setaria utterly fails.

When it was found that Setaria produced better in rainy seasons without too many dry spells seed production was concentrated south west of Kitale around Saboti in the foothills of Mount Elgon. This is a major coffee growing area, situated in Zone II (*3.4.2*). By contrast, seed yields of Masaba and Boma Rhodesgrass have been particularly good for many years at Chepchoina, north west of Kitale, on the edge of the Pokot Escarpment, close to Zone IV. Dry areas with poor soils should, however, be avoided. Jones & Roe (1976) reported a wide seasonal variation in seed yield in Queensland also.

8.6. Causes of low seed yields in tropical grasses

Even though seed production is taking place in various countries on commercial scale, tropical grasses are labelled as poor seed producers. It is, however, essential to distinguish:

1. The species. Some, *e.g.* Pangolagrass (*Ddec*), are virtually sterile. Others, such as Elephantgrass (*Ppur*) do produce seed but heading is shy and seedlings are weak and not "true-to-type". Vegetative propagation is the only alternative (*12.1*).
2. Seed yield in relation to seed rate. Seed yields are low but so are the 1000-seed weights and the seed rates (Table 7.3). More than 1 crop can be taken per year (*8.2*). Coefficients of multiplication of 50/year are not unusual so that the comparison with temperate grasses, with 1 seed crop not until the 2nd year, following vernalization, is not so unfavourable.

Where only small patches of amenity (*e.g.* lawn) or fodder-type grasses need to be established, vegetative propagation can be a practical proposition. The best lawn species are, in fact, those which combine the capacity to produce densely knit swards of an either stoloniferous or rhizomatous nature with the reduced ability to produce stems. These species belong to the genera of Cynodon (*3.4.2.2*; *4.2.2.1*), Digitaria and Paspalum (*6.4.1*; *6.16*) and originate from higher latitudes (*6.4.1*). Fodder-type grasses, *e.g.* Elephantgrass, are multiplied vegetatively also (*12.1.1*). Vegetatively propagated species require intensive plant breeding programmes to keep abreast of diseases which inevitably develop as the acreage of single clones increases. Control measures of strictly enforced plant hygiene are needed, also to keep out weeds, which are easily transplanted together with the target cultivated grass: African Couchgrass (*Daby*), sedges (*Cyperus* spp.) and Oxalis (*11.8.4*).

Although vegetative propagation may be practised on some scale, it is a very poor alternative for grasses which can be sown from seed (see also *4.3.5* & *6.11*). Apart from the labour involved in vegetative transplanting, establishment is slower than from seed and more weather-dependent. Acquisition of recommended material for initial vegetative propagation is not without cost, at least of transport, either and the argument that grass seed can only be obtained by purchase has been refuted by the standard practice of home-grown seed (*8.2.1*). On the other extreme, arguments to develop a seeding Elephantgrass can be heard (*9.7*; *12.1*). Such attempts have been widely publicized but have thus far produced no tangible results.

Those East African species, which are normally vegetatively propagated, are hard to eradicate by ploughing and present a weed problem in arable rotations. Thus, vegetative reproduction assumes both physiological and ecological significance. The indigenous species which are multiplied vegetatively are poor seed producers and they mostly originate in Zone II. Conversely, seeding species originate mainly in Zone III and are more difficult to multiply vegetatively (Table 6.6; *3.4.2*; *4.5.2*; *6.4*).

8.6.1. Low seed yield in vegetatively propagated grasses

This inverse relationship between the generative and vegetative abilities is of great consequence. Consistent, abundant and normal head production is uncommon in vegetatively propagated grasses. At Kitale many varieties of Elephantgrass and amenity grasses fail to produce flowering heads. Kikuyu- and Stargrass are shy headers also. Elsewhere in the world, species such as Pangolagrass do produce flowering heads but fail to set viable seed; it is thought to be a sterile inter-specific hybrid. Polyploidy and aneuploidy are common causes of low seed-set in "giant" grasses (Hacker, 1966). However, their very hybrid nature or their ploidy level made their vegetative characters stand out so prominently. Hybrids display heterosis while polyploidy often makes plants more robust. Their ease of vegetative reproduction has been of great adaptive significance in the absence of an adequate mechanism for generative reproduction (Boonman, 1971a).

8.7. Field studies on seed formation processes

Early literature on seed production of tropical grasses used to stress the prevalence of cytogenetical barriers and apomixis as the principal causes of low seed yield, also in seed-multiplied grasses. This had a delaying and negative effect on a practical approach being taken towards agronomy and breeding. Suspected "sterility" is at times due to auto-incompatibility and can be rapidly remedied by adding other genotypes as pollen sources. Apomixis is not an impediment either since Molassesgrass, one of the most prolific seeders, has been grown commercially for many decades; the same is confirmed by apomictic temperate grasses such as *Poa pratensis*. In fact, apomixis is the very instrument of a plant to overcome cytogenetical barriers that would otherwise impair seed-set. Finally, lack of domestication is often brought to the fore to explain the low seed yield of tropical grasses. It is, however, difficult to see what the difference in domestication has been between varieties of world repute, such as Nandi Setaria or a temperate grass such as Melle Perennial Ryegrass (*6.1.2*). Both were collected from natural vegetation in the mid 1930s and both came from areas with heavy grazing pressures.

Casual observers are commonly bewildered by, what seem to them, disastrous seeding characteristics of tropical grasses: [1] the rate and earliness of shedding; [2] flocks of seed eating birds; [3] debilitating spikelet diseases; [4] the bulk of herbage to be dealt with before the seed can be reached. These impressions have not failed to have strongly negative impacts on nascent seed production efforts.

The growing commercial demand for seed of the Kitale varieties in the 1960s and their relatively high prices made it necessary to find out what the actual determinants of seed yield were and to increase the efficiency of seed production in terms of cultural and genetic inputs. The work concentrated on field experiments, the results of which could be tested immediately on a commercial scale in the multiplication fields at the Kitale Station. Most of the work was carried out with Nandi Setaria and Rhodesgrass but was found applicable to other seed-multiplied grasses. The main impediments to high seed yield were identified by Boonman (1971a) as follows:

1. Prolonged head emergence within plants.
2. Prolonged flowering within heads.
3. Decreased duration of flowering in late-emerged heads.
4. Low seed setting.
5. Low density of head producing tillers.
6. Extended spread in heading date between plants in a variety.
7. Other factors including low seed retention, spikelet diseases and bird damage.

Most grasses show a combination of these determinants which are also interdependent. An increase in one is commonly offset by a decrease in another, and vice versa, so that the same seed yield may be achieved via different pathways.

8.7.1. Tillering and heading

Head emergence within plants, and within crops, is extended over months. Tillers may reach their maximum numbers within 6 weeks after the cleaning cut, but head emergence goes on for 3–4 months (Fig. 8.4).

In temperate grasses in temperate climates the situation is almost the opposite. Daylength effects which cause tillers, differing many months in age, to elongate after vernalization and to exsert heads within a matter of 2–3 weeks at most in temperate climates are absent on the equator. Also at 2750 m in upper Molo heading in Perennial Ryegrass is negligible (Birch, 1958). In tropical grasses heading may start at any time of the rainy season and continue its ascent provided enough vegetative mass is present. Heading date (IHE) is highly variety-specific (Figs. 8.5 & 9.1)) and is related to the length of the growing season at the site of origin (*6.6.3*). Seed is grown over a broad range of latitudes, from the equator to over 30°N & S in the subtropics. Further from the equator, seasonal effects may be apparent and date of IHE may be advanced or delayed relative to other varieties. Masaba Rhodes heads 2 months later than Katambora in autumn in southern Japan (33°N) so that cold weather and frost interfere with its maturation. In regions below 30°, with adequate moisture and without frost, it is normally possible to obtain 2 seed crops per year, one in early summer and one in late autumn. For some equatorial grasses such as Nandi Setaria and Masaba Rhodes short-daylength effects have been reported but contrasting daylength responses have been found also (*6.6.1*; *6.7.1*; Nada, 1985). The issue is complex and seems confounded with temperature- and light-energy regimes (Tainton, 1969). Dirven *et al.* (1979) postulated that short-day grasses are better multiplied at higher latitudes. This would be justified if not only heading date but also heading synchrony were truly sensitive to short-days. In Australia, commercial seed production has, in fact, moved in the opposite direction, from the border area between New South Wales (30°S) and Queensland to northern Queensland (17°S) (Humphreys & Riveros, 1986).

Date of IHE (5–10 heads/m^2, with the base of the head clear from the subtending leaf sheath) is relatively easy to estimate visually under field conditions. At IHE the elongating culms have reached 50–75% of the final culm length of the crop (Boonman, 1971b). Head counts at later stages require destructive sampling or counting *in situ*; tiller counting *in situ* is only possible in the early stages. IHE was found to be the best yardstick to mark the beginning of the linear ascent in the head emergence curve.

IHE-date is dictated by variety (Figs. 8.5 & 9.1) and is accelerated by favourable (warm) weather and growing conditions, for instance after the dry season (Fig. 3.4). Guineagrasses are invariably the first to exsert heads after the dry season at Kitale, but their subsequent heading curves can follow any extreme form, from very steep (*Pcol*) to relatively flat (*Pmax*). If the IHE-ranking between varieties is related to the length of the growing season at the site of origin, within varieties plants differ greatly in date of IHE because of differences in vigour (Fig. 9.2). They can be selected on that basis to constitute new early-heading, but more vigorous, varieties (*9.4.1*). Thus, Elmba and Boma Rhodes are earlier-heading and more vigorous than their parent varieties, Mbarara and Masaba Rhodes, respectively (*6.6.4*), while Nandi II is later-heading than Nandi I from which it was selected for apparent leafiness (*6.7*).

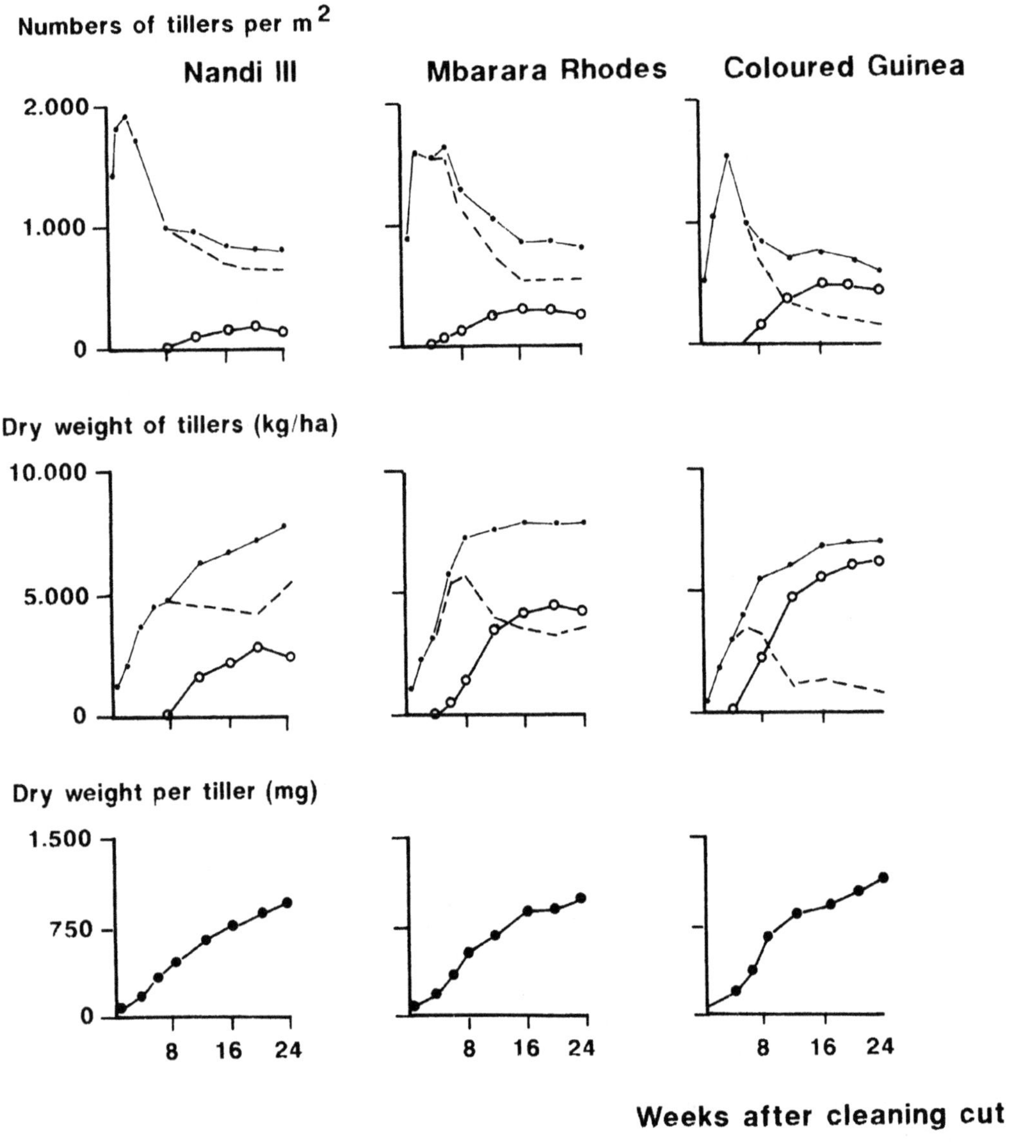

Fig. 8.4. The tillering patterns of three varieties: ●—● total of all tillers; --- non-heading tillers; ○—○ heading tillers (Boonman, 1971b).

Species and varieties differ markedly in final head numbers (Fig. 8.5). These may reach 500/m^2 in Coloured Guinea and Congo Signal. In normal seed crops of Rhodesgrass head numbers rarely exceed 350/m^2 at seed harvest and out of the tillers then present 40–50% may be heading. In Nandi Setaria the figures are 250/m^2 and 30%, respectively. These refer to the final numbers (Fig. 8.4) but since seed-set efficiency diminishes rapidly in later-emerging heads, OHD takes place when head emergence is still only half-way (Fig. 8.9). Species and varieties differ in tillering patterns (Fig. 8.4; Tables 6.7 & 7.1) but, apart from the % heading tillers and the resulting number of heads, no simple correlation with seed yield exists. Primary growth in the 1st year is characterized by a low

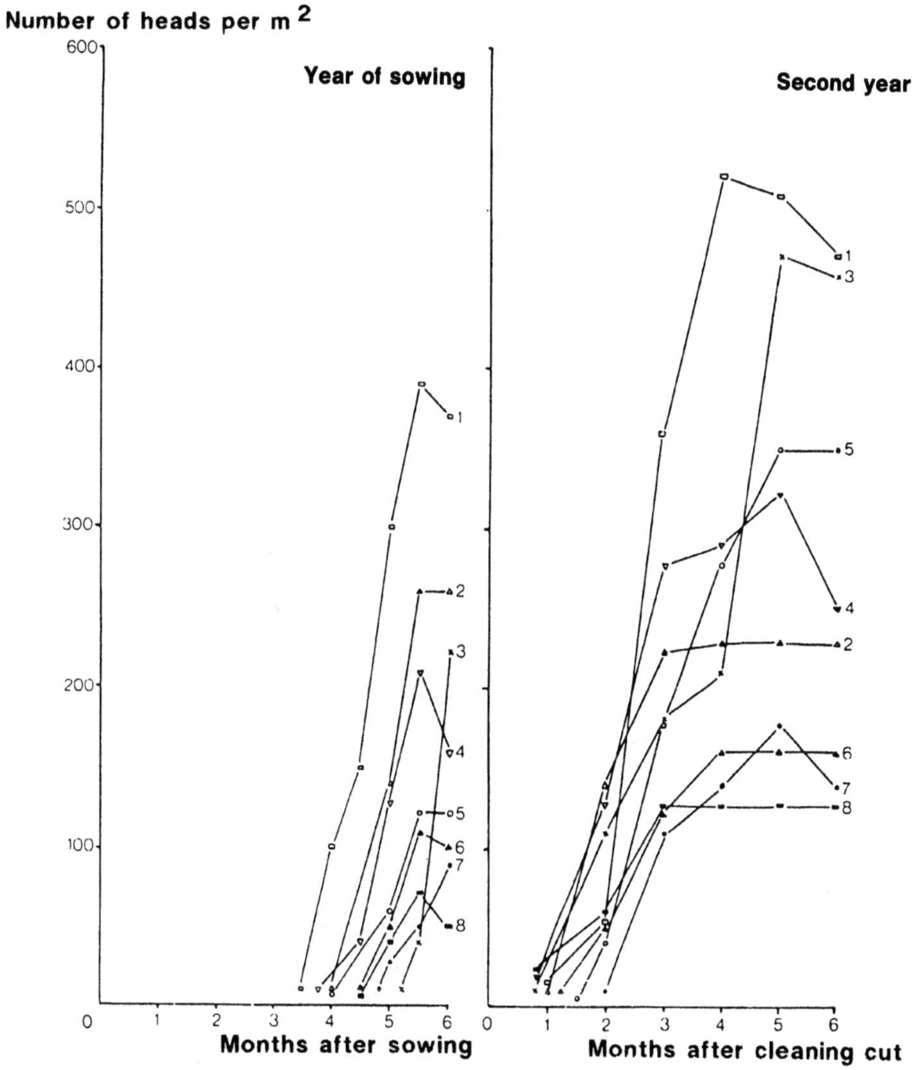

Fig. 8.5. Pattens of head emergence in the year of sowing (left) and the 2nd year (right): 1 = Coloured Guinea; 2 = Masaba Rhodes; 3 = Congo Signal; 4 = Mbarara Rhodes; 5 = Nandi I Setaria; 6 = Pokot Rhodes; 7 = Nandi III Setaria; 8 = Makueni Guinea (Boonman, 1971b).

density of heavy tillers (Table 7.1) and this may well bring on the high and reliable seed yields invariably obtained in the 1st year, although head numbers are lower than in subsequent years (Fig. 8.5).

At high density (narrow rowwidth) and high N-nutrition, tiller numbers rise rapidly to a peak of 800-1000/m^2 in the year of sowing and to 2000–2750 in subsequent years (*7.1*; Fig. 8.6). The subsequent fall is rapid also and begins around IHE (Fig. 8.4). Tiller numbers tail off 1–2 months after peak tillering. The average tiller weight increases in an almost linear fashion.

The tillering hierarchy of Fig. 8.6 is found back in the heading patterns of Fig. 8.7. Head numbers reach higher maxima at narrow rowwidth, at medium to high N-levels, than at wide rowwidth, whereas at zero-N head and tiller numbers are low and remain virtually constant.

High plant density and high N-nutrition accelerate the heading process after IHE. N has a strong

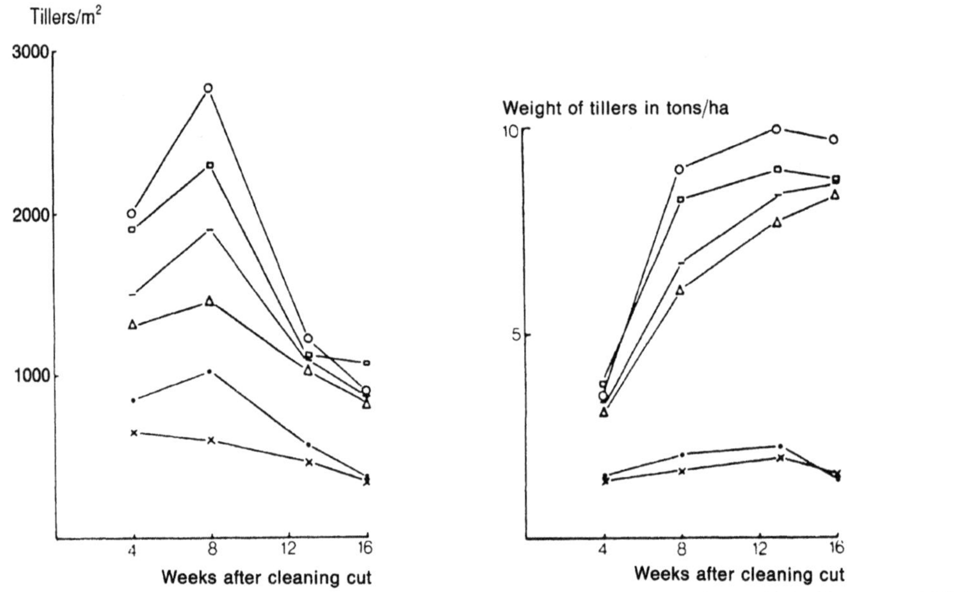

Fig. 8.6. Tillering in Nandi II Setaria, seed crop July 1969, at different nitrogen levels and rowwidths (Boonman, 1972a) ○—○ $N_{260}R_{30}$; □—□ $N_{130}R_{30}$; — $N_{260}R_{90}$; △—△ $N_{130}R_{90}$; ●—● N_0R_{30}; ×—× N_0R_{90}

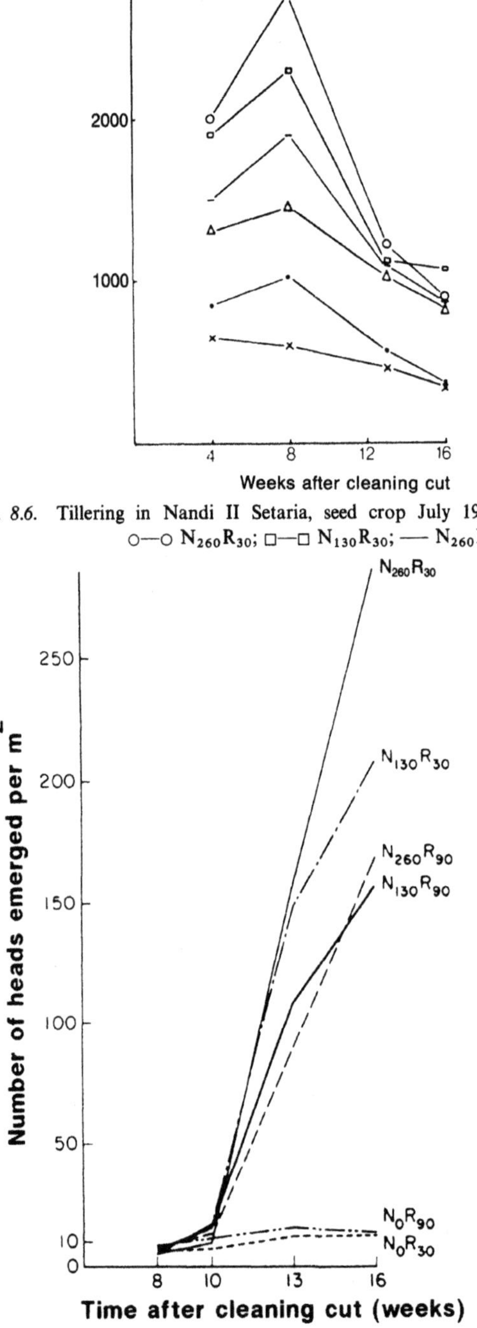

Fig. 8.7. Patterns of head emergence at 3 levels of nitrogen (0, 130 and 260 kg/ha) and 2 levels of rowwidth (30 and 90 cm) in Nandi Setaria (after Boonman, 1972a; 1980).

effect on % heading tillers (Table 8.5; Fig. 8.8) while high density ensures that they are emerging more rapidly or synchrone (Fig. 8.7); in wide rows tillers are too densely packed in narrow bands. At both zero-N and narrow rowwidth IHE is delayed (Fig. 8.8). This adds further to the positive effect of high-N and density on the synchrony of seed maturation.

Uniformity in head emergence within plants can be increased by husbandry measures as indicated, but the large intra-variety variation in IHE (Fig. 9.2) can only be tackled effectively by breeding (*9.4*). High seed yields will not be possible unless there is maximum uniformity in heading.

Head length may be seen as a measure of the number of potential spikelets, but it may equally be argued that shorter heads are more efficient due to more uniform flowering and seed-set. Heads are much longer in the season of sowing and the potential number of spikelet sites is much higher in Rhodesgrass (Table 8.7). Head length is increased by low plant density (Table 8.3) and high-N (Fig. 8.8).

Within a crop, head (and culm) length is determined by the hierarchy of emergence (Table 8.8). Heads in age-groups emerging prior to IHE are smaller than at IHE, remain constant in length for 4 weeks but decrease rapidly after peak emergence. Head length has, therefore, no direct relationship with seed yield characters. Table 8.8 refers to a

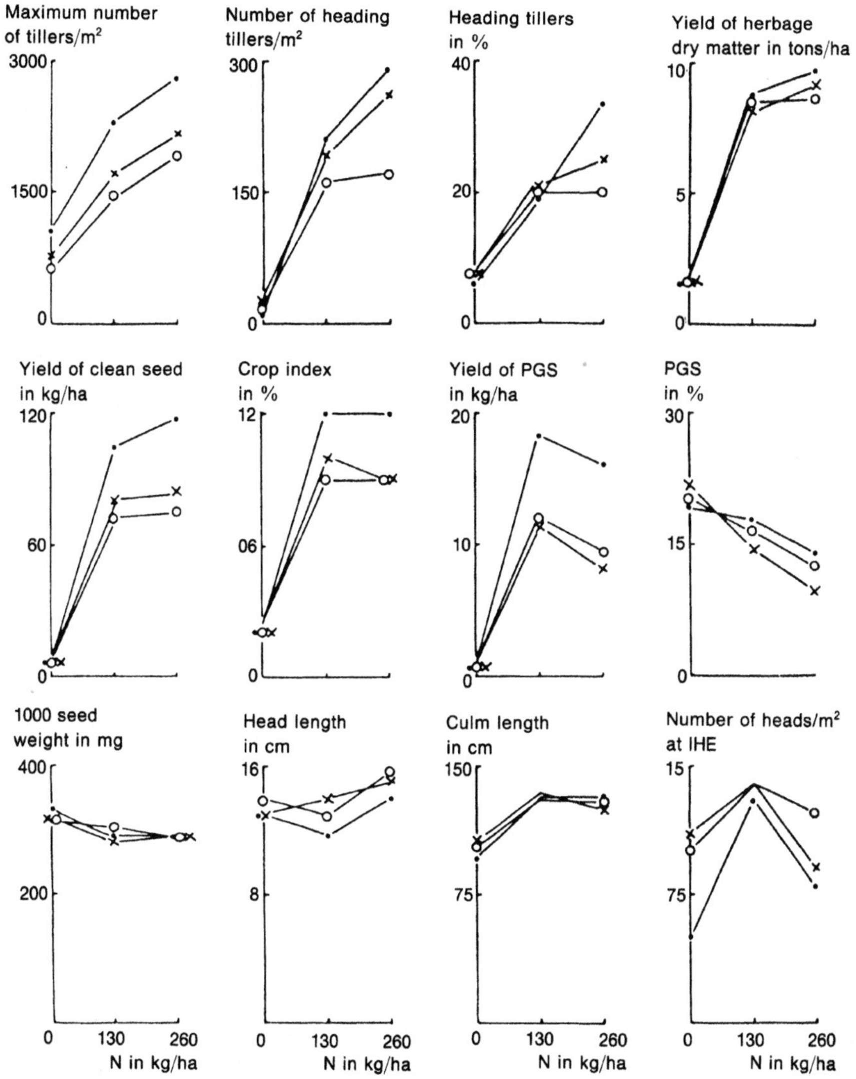

Fig. 8.8. Seed crop characteristics of Nandi Setaria (after Boonman, 1972a): Rowwidths •—• 30 cm; ×—× 60 cm; ○—○ 90 cm.

study in which the heads of successive age groups were tagged weekly with coloured rings as they emerged, over a period of 14 weeks.

8.7.2. Spikelet development

The asynchrony in heading is further complicated by unevenness in flowering (Table 8.8). Anthesis and stigma exsertion in Setaria occur at dawn, beginning in the middle of the head, often within a few days of the head becoming visible. Early-emerging heads of Nandi Setaria have the capacity to display flowering for 7 weeks, although it can be interrupted for days in succession. Dry spells, even of a few days, preclude flowering. Flowering is most intense after a previous bright day in otherwise rainy periods. In Rhodesgrass flowering occurs in the afternoon, is less sensitive to dry weather and is more restricted in duration.

Spikelets are difficult to monitor in the field since they are small and their stage of development cannot be judged by their age. When flowering is observed at OHD, it may well consist mainly of "late" spikelets in early-emerged heads, whereas the shedding that is recorded at the same time may refer to empty spikelets of recently emerged heads. What we know of spikelet dynamics stems, therefore, mostly from circumstantial evidence.

Flowering duration decreases as heads emerge later (Table 8.8) and it can be assumed that seed-set is similarly reduced. After peak emergence, some 5 weeks after IHE, the decline becomes very pronounced. Since late-emerging heads are very small and flower over a short period if at all, they contribute very little to the effective seed yield. Flowering is much more prolonged but, ironically, seed setting is more efficient in the early age groups, simply because the late groups are hardly capable of flowering. This explains why OHD is situated so early in the heading process, in fact when head numbers are only half-way (Fig. 8.9). Equally dramatic is the reduction of the interval between head emergence and spikelet shedding (Table 8.8). Early-emerged heads take long before their mostly well-filled spikelets are shed, whereas late-emerged heads shed their mostly empty spikelets almost instantly. After some 12 weeks, as the old crop begins to collapse, a new cycle of tillering and heading begins. By then all heads of the early- but also of the late-emerging categories have lost their spikelets. Empty spikelets, *i.e.* which have not flowered or which have not been fertilized, shed early. This should explain that the behaviour of a spikelet is no indication of the age of its heading category.

Table 8.7. Head length at seed harvest in the year of sowing and in subsequent years (after Boonman, 1972a,b).

	Head length (cm)	
	Setaria (Nandi II)	Rhodesgrass* (Mbarara)
First year	17–19	70–90
Subsequent years	13–16	50–80

*Number × length of racemes.

Table 8.8. Decrease in duration of flowering and shedding and head length in successive age-groups of head emergence in a seed crop of Nandi Setaria (after Boonman, 1971a).

	Successive heading age groups (in weeks)													
	1	2	3	4	5	6	7	8	9	10	11	12	13	14
			(IHE)			(peak)								
Head length (cm)	18	17	19	19	19	19	19	18	18	17	16	13	13	13
Flowering duration (weeks)	7	7	5	3	2	2	1	1	0	0	0	0	2*	2*
Interval (weeks) to 100% shed	9	8	10	10	10	9	8	7	6	4	2	1	—	—

*Aftermath heads.

8.7.2.1. Spikelet weight (1000-seed weight)

Data on 1000-seed weight usually refer to cleaned seed lots. Weights vary with season (Table 8.9 & 8.10), harvest date and species (Fig. 8.10), but are largely unaffected by N, rowwidth or other cultural techniques. Seasonal effects are strong but are not related to the seed-yield level of a particular season. In Nandi Setaria 1000-seed weights can vary from 270 to 400 mg and in Rhodes from 220 to 320, occasionally 370 mg (Tables 8.9 & 8.10). The 1000-seed weight is often at its highest in the year of sowing, especially in Rhodesgrass (Table 8.10). In Nandi Setaria and Coloured Guineagrass seed weight increases with harvest date. In Rhodesgrass the effect is small and sometimes even negative (Fig. 8.10). Seasonal variation in 1000-seed weight is not related to % PGS either. The same applies to varietal differences. Mbarara Rhodes has lighter seeds than Pokot (Table 6.7), but its field establishment is usually more reliable. Spikelet weight is independent of caryopsis weight, while larger caryopses have not necessarily more vigorous embryos (*7.1.7*).

It is notable that 1000-seed weight (together with head and culm length) consistently had low coefficients of variation, as compared with head numbers and yields themselves (Boonman, 1972a, b). Thus, statistical significance was observed, although differences were often negligible from the practical viewpoint except between seasons (Tables 8.9 & 8.10) and, in some species, between harvest dates (Fig. 8.10).

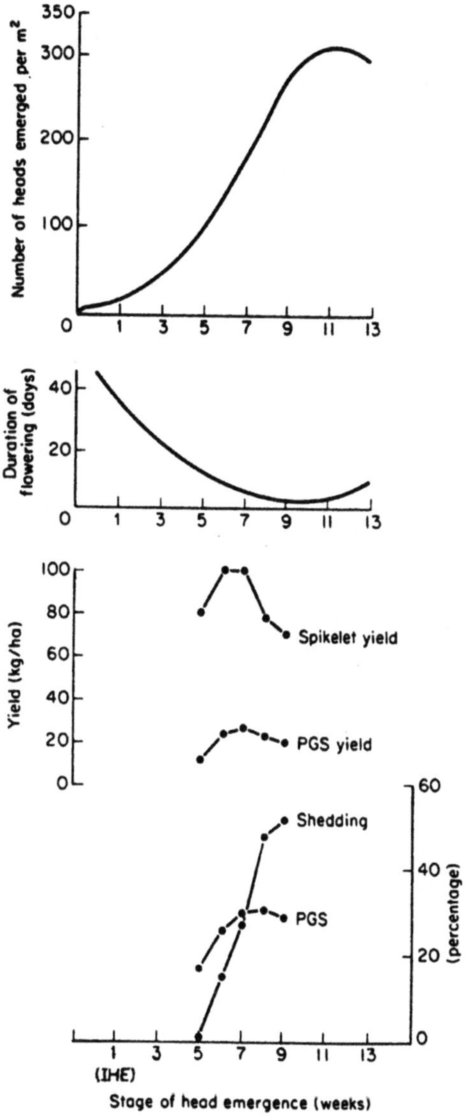

Fig. 8.9. Heading, flowering duration, seed yield, % PGS and shedding at successive stages of a seed crop (Boonman, 1980).

8.7.2.2. The Purity and Germination Capacity (PGS)

In the previous, attention was drawn to the fact that PGS values of tropical grasses are low, unpredictable and inconsistent from season to season. Seed yield without reference to % PGS is not appropriate (Table 8.9). Seasons with very low yields of clean seed of Nandi Setaria (<50 kg/ha) tend to have a low % PGS. Seasons, however, with high yields of clean seed can have widely different % PGS (*8.5.2*). Expressing viability in terms of % purity (*i.e.* % caryopses) without referring to its % germination is not adequate either since both components of PGS vary independently (Fig. 8.11).

Purity (P; as % of the clean seed)) and germination (G; as % of pure seed) each averaged around 40% in Nandi Setaria, ranging independently over 40 %-units (20–60 or 25–65) over seasons (Table 8.1). Although P and G were both high in season 8 (late 1970), the rest of the data shows no correlation and certainly not a positive one. Seasons 4 (late 1968) and 6 (late 1969) were the lowest in % G and also in yield of PGS.

From all evidence available it appears that % purity is no adequate predictor of % PGS. The caryopsis of Setaria is enclosed by the crustaceous palea and lemma making it difficult to ascertain whether a "seed" contains a caryopsis or not. Filled and empty spikelets are difficult to separate in the seed cleaning process, during which losses of filled spikelets can also be substantial. Even in official purity analysis we cannot separate the caryopses that are likely to germinate from those that will not. The only proper method to assess seed quality is the germination test itself. Seed should, therefore, always be expressed in terms of germination and this has come to be known as PGS (Table 7.3; *8.1*; *9.5.1*), even if the % purity has not been determined. In the numerous routine tests at Kitale more practical relevance was attached to germination results of a larger sample of summarily cleaned spikelets (say 500 spikelets of 40% purity) than to those of 200 "pure" seeds (*9.5.1*).

High plant density (narrow rows) has a positive effect on both purity and germination. In Nandi Setaria, both N-fertilizer and (wide) rowwidth lowered % purity by about 10 %-units. This effect was found to be significant in 6 out of 8 seasons. In these same seasons, N lowered % germination to an even greater extent, whereas wide rowwidth reduced germination significantly in only 1 out of 9 seasons. The combined effect of high N and wide rowwidth on % PGS can, therefore, be considerable (Fig. 8.12).

Fortunately, the negative effect of N on % PGS is limited to high rates, above 100–150 kg/ha N, which do not add to yield anyway (Table 8.5), both in Rhodesgrass and in Setaria, and particularly so at wide rowwidths of the latter. Obviously, when

Table 8.9. Nandi II Setaria seed crop characteristics at the central point of the experimental design (Boonman, 1972a); yields in kg/ha.

Crop characteristics	Sept 1968	July 1969	Nov 1969	June 1970	Oct 1970	July 1971
Yield clean seed	73	101	42	75	120	113
% PGS	12	12	16	12	33	21
Yield of PGS	9	12	7	9	40	23
Yield of dry matter	8000	9050	8950	9600	7050	8720
Crop Index (%)	0.9	1.1	0.5	0.8	1.7	1.3
Number of heads per m^2	170	260	250	170	160	200
% Heading tillers	34	36	27	23	—	35
Average head length (cm)	17.6	14.9	13.4	15.4	13.4	16.2
1000-Seed weight (mg)	400	280	360	270	320	—
Seed retention (1–5)	—	—	4.3	4.9	4.4	4.5
% Bunt (in unclean seed)	23	0	23	0	18	11
Culm length (cm)	167	146	146	135	—	—
Lodging (1–5)	—	—	3.1	—	—	3.7

Table 8.10. Mbarara Rhodes seed crop characteristics at the central point of the experimental design (Boonman, 1972b); yields in kg/ha.

Crop characteristics	Dec 1968	July 1969	Oct 1969	July 1970	Oct 1970	July 1971	Nov 1971
Yield of clean seed	125	53	107	74	81	137	128
% PGS	45	63	39	37	40	37	47
Yield of PGS	57	34	42	27	32	51	60
Yield of dry matter	7800	6500	8400	7900	7500	6700	4200
Crop Index (%)	1.6	0.8	1.3	0.9	1.1	2.1	3.0
Number of heads per m^2	350	320	110	250	120	330	310
% Heading tillers	51	38	20	36	-	51	46
Number of racemes per head	11.1	8.2	10.9	11.5	10.0	11.1	9.4
Average raceme length (cm)	7.1	6.5	7.0	7.0	7.1	7.0	6.9
1000-Seed weight (mg)	320	290	270	220	250	290	300
Seed retention (1–5)	-	-	4.5	3.9	4.7	4.1	3.3
Culm length (cm)	130	129	128	123	-	-	100
Lodging (1–5)	-	3.4	2.0	4.9	3.0	4.1	3.5

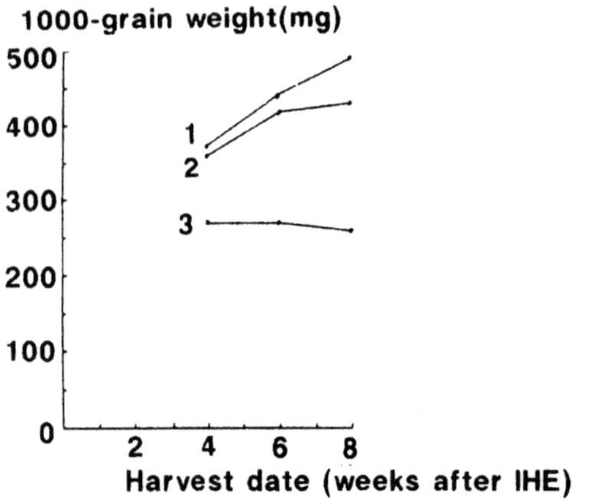

Fig. 8.10. The effect of harvest date and species on 1000-seed weight. 1 = Coloured Guinea; 2 = Nandi (I) Setaria; 3 = Mbarara Rhodes (Boonman, 1973a).

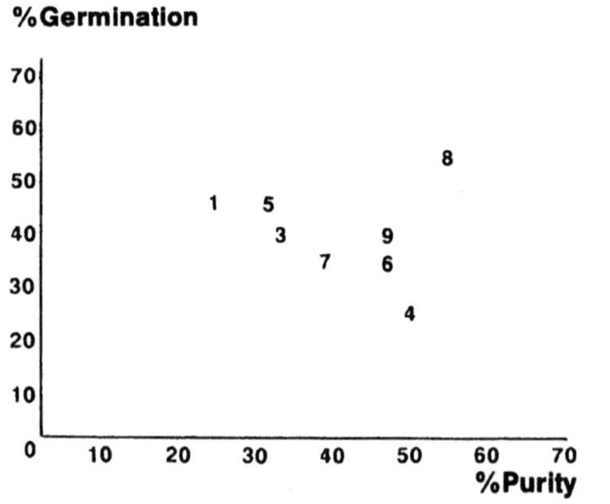

Fig. 8.11. The relation between % purity and % germination according to season in Nandi Setaria (9 consecutive seasons, 1967–1971).

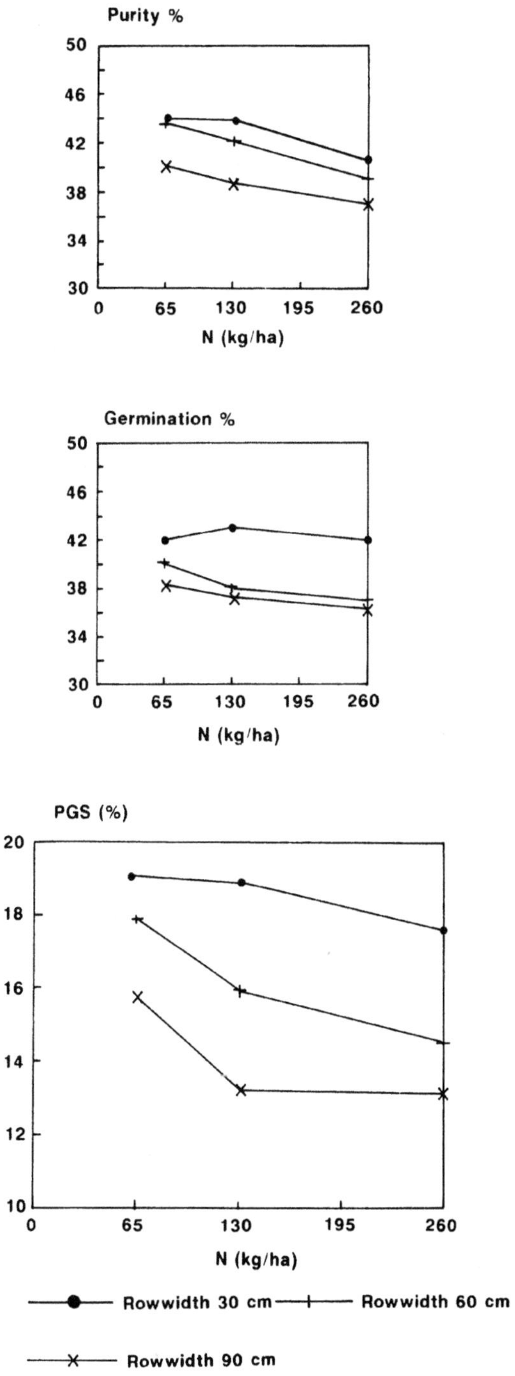

Fig. 8.12. The effect of nitrogen and rowwidth on % purity, % germination and % PGS in Nandi Setaria (9 seasons).

higher N levels decrease % PGS but increase PGS-yields, the net effects can still be positive.

There is little evidence to attribute the decrease in % PGS at very high N-levels to, for instance, drought stress or to lodging. Lodging is generally not a yield handicap, top yields often being associated with some degree of lodging. A more likely reason to explain the negative effect of high N on % PGS is the stimulation of late-emerging heads and late-developing spikelets, failing to reach maturity in time before harvest (*8.5.1*).

8.7.2.3. PGS and harvest date

The relation between % PGS and harvest date is a linear one, extending well beyond OHD (Fig. 8.9). The % germination remains fairly constant with harvest date, so that it is mainly % purity which varies, along with % PGS. While clean-seed yields are already decreasing, % PGS continues to rise and so does 1000-seed weight (Fig. 8.10). Thus, OHD differs according to the character measured. OHD of clean-seed yield precedes OHD of % PGS and PGS-yield by 2 and 1 week, respectively (Fig. 8.9). PGS-yields increase over a longer period with harvest date than clean-seed yields, apparently because full spikelets are retained longer and do not shed so readily. These spikelets have a higher proportion filled with PGS and heavy caryopses.

8.7.2.4. Seed shedding

The seed population is continuously changing due to maturation and shedding. The early shedding consists predominantly of empty spikelets of heads emerging from peak emergence onwards (Table 8.8). At some stage however, mature spikelets fall off also and these losses will be noticed when they are no longer compensated by later-maturing spikelets in later-emerging heads. Since seed setting decreases rapidly as heading proceeds, harvest date can suddenly become critical although the heading process is still only half-way.

It should be recalled that shedding entails the whole spikelet and that abscission layers are formed below the glumes. Differences in form, size and weight between full and empty spikelets of grass seeds are small. It is no simple matter to distinguish filled from empty spikelets with the naked eye. Seeds of tropical grasses shed rapidly when mature, but even earlier when spikelets are empty. Late-emerging heads which do not flower are void of spikelets within a week of emergence (Table 8.8). Also Hopkinson & English (1982) found that among the first-shed spikelets of Guineagrass the empty ones predominate.

At Kitale, we rated at each harvest date a sample of heads on a scale from "all spikelets present" to "no spikelets present". As head emergence continued linearly, each new sample referred to an expanding population. The pertinent observations, some of which have already been referred to, are listed here together:

1. As heading proceeds, shedding increases jointly with PGS-yield and % PGS until some 30% is shed (Figs. 8.2 & 8.9).
2. The later the heads emerge, the faster the spikelets are shed.
3. With the method of recording followed, shedding is negligible in the first 4 weeks after IHE. Its subsequent rise is linear but faster than rate of heading.
4. High N-levels tend to encourage shedding.

Figs. 8.2 & 8.9 refer to shedding in successive samplings of a numerically increasing head population. Early, probably modest, shedding may go unnoticed in such samples, but may be visible in the field. Spikelets shedding early may involve overmature seeds, but are more likely to consist of empty spikelets of more recently emerged heads. Wind and bird damage are often overrated in their effects but they may promote the loss of overmature seed rather than of empty spikelets, so that the distinction is even further complicated. Since the decrease in efficiency of flowering and the increase of shedding are equally rapid as heading proceeds, the quantity of empty spikelets increases markedly as harvesting is delayed. After 12 weeks no spikelets are left on any of the heads, not even in those emerged in the previous week. Late heads are so inefficient that they cannot compensate for losses incurred in the early ones. Hence, the still relatively critical role of harvest date in the later stages of the harvest safety margin (*8.4.1*). From all evidence available it appears that more harm is done in harvesting too early than too late.

Growth regulators such as CCC and α-NAA have been tested at Kitale but none of the seed crops appeared responsive.

8.7.2.5. Spikelet diseases and pests

Bunt (*Tilletia echinosperma*) commonly attacks all *Setaria* spp. (Bogdan, 1955) and Guineagrass. A head of Setaria can be converted into a string of swelling bodies (spikelets) already within 2 weeks after head emergence. It appears that unfertilized spikelets are infested more readily since late-emerging heads show the symptoms more rapidly while high PGS-yields are in some seasons associated with high levels of bunt. Seed crops should, therefore, never be abandoned for reasons of bunt. The fungus is not seed-borne (Bogdan, 1971), which agrees with the hypothesis that mainly unfertilized spikelets are affected. Crop hygiene is important as the disease spreads from infected plants left over from the previous cropping cycle. At the Kitale Station bunt is more common in fields near the station's grass nurseries. When bunted heads in the immediate vicinity are destroyed over the dry season, crops remain free from bunt until the middle of July which underlines the importance of having an early seed crop. The disease can spread rapidly from July onwards in wet seasons with a well distributed rainfall. N-fertilizer seems to reduce % bunt. This fungus may be the same as *Sphacelotheca setariae*, reported from Zaïre by Behaeghe & Blouard (1962), and as the ergot reported from Serere (U) by Martin & Badcock (1937).

Seed establishment of Buffelgrass was affected negatively by ergot (*Claviceps* spp.) (Brzostowski, 1961). Ergot is commonly found on Buffel-, Guineagrass, Paspalum, Stargrass and many other species under dry conditions. In humid weather, symptoms of sticky exudates and later of bodies of grey, pink or black colour appear. The fungi involved are *Sphacelia* (the imperfect stage of ergot), *Cerebellum* and *Fusarium* and are related to Sorghum honeydew disease (*15.7*; Ryley, 1981). Attacks are rare in Rhodesgrass, but when they do occur they appear serious. Due to the stickiness spikelets are glued together and are lost in the seed cleaning process.

Seed crops of Guinea-, Signalgrass and Setaria attract a multitude of small seed-eating birds of *Quelea* and other species. No day passes without a number of (small) visiting flocks. Their damage is perhaps more mechanical than one of actual consumption and has been impossible to assess thus far. No method was found effective in fending off birds, but more recently Mesurol (methiocarb) has been found to repel birds. Rhodesgrass seed fields are rarely visited, probably because the awns on the spikelets act as a deterrent. Setaria is well liked by birds but the favourite hosts are Guinea- and Signalgrass.

8.7.3. Seed recovery

When crops are liable to the early shedding of viable seed, it can be argued that recovery of the seed already present is a more realistic target than the production of more seed, part of which is likely to be lost. When this "recovery theory" is applied to tropical grasses, the spikelet population is pictured in time-relay, together with the head population, in consecutive age-groups: [1] early, overmature and shed; [2] semi-early and mature; [3] medium-late and unripe; [4] late and not-developed. Still according to this theory, recovery would be best if fraction [2] could be maximized at the expense of [1] and [3]. Fraction [1] was traditionally seen as the principal source of shedding losses, *i.e.* as the culprit of low seed yield.

In tropical grasses, however, there is no such a sequence in time-relay. Most of the shedding comes from relatively late heads, whereas the early-emergence groups produce most of the viable seed. Early "shedding" should be equated with abortion rather than with overmaturity. Efforts should aim therefore at maximizing both the size and the recovery of the early effective (viable) fraction.

Recovery, in terms of harvesting, poses special problems in mechanized harvesting as found in Queensland. Queensland workers: J.M. Hopkinson, D.S. Loch, J.B. Hacker, R.J. Jones and L.R. Humphreys have studied seed production in tropical legumes and in grasses. The work has been reviewed by Jones & Roe (1976) and by Humphreys & Riveros (1986).

8.8. Conclusion

Tropical grasses are characterized by a low efficiency of the seed formation process. Heading and flowering go on for a long time so that synchrony is lacking. Besides, late-emerging heads contribute little. As heads emerge later less seed is formed and more empty spikelets are shed. However, neither shedding nor spikelet diseases are the threats they were thought to be. Most of the valuable seed is found in the early-heading category. Efforts to increase seed yield and PGS should aim at maximizing this group. Head numbers are controlled largely by N-nutrition, but their efficiency in terms of PGS depends on weather conditions and on the synchrony of maturation. The latter can be increased by proper husbandry techniques, *e.g.* plant density. The synchrony constraints are also partly genetic in origin. Uniformity in heading date is a primary requirement if improved husbandry is not to be wasted. This can be tackled only by breeding (*9.4*).

Unjustified worries about: [1] suspected daylength requirements; [2] pollen contamination; [3] apomixis; [4] spikelet diseases; [5] early shedding and; [6] other putative physiological constraints have hindered strategies of practical husbandry and breeding necessary to make seed production more efficient and seed prices more competitive.

8.9. Summary

Tropical grass seed has been home-grown and produced commercially since before World War II. A major boost in the mid 1950s has been the release of Nandi Setaria, Mbarara and Masaba Rhodes, and, concomitantly, the joint effort of grass seed growers to establish the Kenya Seed Company at Kitale. The multiplication rate is rapid, but yields and germination capacity (% PGS) are low and variable; seed is best expressed in terms of PGS. The most reliable and least expensive seed harvests are obtained in the year of sowing. In the following years, 2 seed crops can be obtained but early and adequate application of N-fertilizer is crucial to bring on a large number of heads capable of setting seed; weather also has a great effect on seed setting, in Setaria especially. Better yields and PGS result from more synchrony in heading. Tiller density should be sufficient to allow a high number of heads, brought on by N-fertilizer, to emerge and flower in synchrony. From peak emergence onwards, heads lose efficiency of seed-set and spikelets shed rapidly. Early shedding of empty spikelets should be no cause of alarm. Shedding should, in fact, be well in progress before the optimum harvest date is reached.

CHAPTER 9

Grass breeding at Kitale

Viewed from the general status of tropical grassland development grass breeding may seem a luxury. However, where varieties are being multiplied to move into trade they might just as well be made to be in their best form and be improved where possible.

Breeding is not equated here with hybridization or with breaking apomictic barriers, however dominant these topics may be in scientific literature. Donald (1975), working in Australia, has pointed out that many herbage breeding programmes have been unproductive because hybridization was regarded as purposeful *per se*. Sophisticated genetic designs can be useful when demanded by the circumstances, but there is no general relationship between degree of sophistication and practical success, measured in terms of superiority of the new varieties produced. Realistic objectives and practical approaches based on the true genetic variation at hand are generally more effective than the genetic designs chosen. A highly successful example has been the identification and release of aphid-resistant lucerne, based on a survival population, in California (Lehman *et al.*, 1985; *17.7*).

To identify the best ecotypes available and to work upon them by selection are the first but decisive steps in breeding. This is usually referred to as "ecotype breeding". The challenge then is to recognize variation that is genuine and useful, and to unravel it from a much broader variation, due to associated inter-dependent characters or due to environment. Intra-variety variation observed in leafiness, for instance, may look promising at first sight but may turn out to be negligible when truncated at levels (classes) of heading date or herbage yield in the course of the selection process. The ultimately correct choice of plants is more crucial than that of the genetic design into which they are fitted. Consequently, in the breeding concept based on intra-variety variation of cross-pollinating grasses, described below, no distinction is drawn between "breeding" or "selection". Breeding is here seen as the steering of useful variation through a judicious choice (selection) of plants constituting a new variety.

Breeding has maximum impact when conducted in unison with the multiplication of seed, in fact as the first step in the spiral of production of breeder's, basic and certified seed. In a similar fashion, it is best carried out in close contact with, if not in joint responsibility for, marketing so that the breeder knows what happens in the field and what will be needed in the future. Varieties in cultivation are the eyes and the ears of the breeder. This serves to explain why breeding, multiplication and marketing have produced the best results thus far there where they developed parallel within the same seed house. Where links between marketing and breeding are missing, breeding may slide off into an excercise of applied genetics and may move away from variety development (van Wijk *et al.*, 1993).

The necessity of breeding at Kitale arose from the results of earlier work. Under the influence of schools of thought in Europe in the 1950s and early 1960s, especially of Aberystwyth and Wageningen, grasses had come to be distinguished in early- and late-heading varieties with the latter considered superior. The same principle was also extended to tropical grasses, *e.g.* Rhodesgrass (*Cgay*) and Setaria (*Ssph*). Thus, although great success had been achieved with early-heading varieties such as Mbarara Rhodes and Nandi Setaria, it came to happen that Pokot Rhodes (*6.6.3*) and equally late-heading, *alias* leafy, selections out of Nandi Setaria (*6.7*) were brought to the fore. When seed yields turned out uneconomically low, research on the variety issue became necessary. It was found that these late varieties also produced less herbage and were inferior in many other aspects (Boonman & van Wijk, 1973). As a result, Nandi Setaria as well as Mbarara and Masaba Rhodes were rehabilitated. The latter two which are of early and medium-late maturity, respectively, serve as examples to show that variety superiority does not depend on its heading date. We will see below, however, that when ecotypes are subjected to intra-variety selection, early-heading sections are far superior in most of the agriculturally desirable aspects. This is how Elmba, Boma and Nasiwa came to exceed Mbarara, Masaba and Nandi, respectively. Elmba and Boma Rhodes were released in 1975, Nasiwa Setaria in 1977

The Kitale top varieties, most of which had already been identified and put to use by farmers, had been recognized by Bogdan and Strange at the very start of their exploration work in the early 1950s (*6.1.2*; Table 6.8). Subsequent exploration, totalling 4000 introductions (Boonman, 1979a), proved less successful, partly because the material collected was outside the range of direct interest to farmers and partly because it missed the same degree of domestication under seed multiplication. Consequently, in his final report of 1965, Bogdan advised to concentrate on breeding, rather than on exploration. Also Whyte *et al.* (1959) postulated that exploitation of already existing varieties would be more productive than ecotype collection. In the 1970s, however, renewed exploration campaigns were mounted by various agencies. Although more material was collected than ever before and more than could be evaluated or stored, preliminary screening within the commercially most important species did not bring any new ecotypes of comparable potential to light (*6.1*).

In addition to the work of Bogdan (1959a), grass breeding was undertaken in the region in Zaïre (Germain, 1954; Behaeghe & Blouard, 1962), in Tanzania and Zambia by van Rensburg (1956; 1969b) and in Zambia by Crauford (1978); see also *11.7*.

9.1. Variety maintenance

Genetic shift was also a hot item in the 1960s. Unless maintained in the state in which they had shown their original superiority, varieties under continuous seed multiplication might alter their botanical composition and, consequently, also their agricultural performance. The generally suspected risk of shift towards earliness, at that time equated with low nutritive value, led to steps at Kitale in 1968 to stabilize varieties-in-multiplication in the form of semi-permanent populations of spaced plants from which inferior plants could be eliminated and breeder's seed issued; these were named "foundation fields".

Was the original aim of these fields one of maintenance and to promote stability, the variation they displayed in heading date was so huge that it called for positive plant selection, *i.e.* breeding. Uniformity in heading date is a necessity if a variety is to express its seed yielding capacity (*8.8*).

It must be recognized that the main concern in the 1950s had been to multiply seed in quantity to meet the demand. Maintenance to ensure varietal integrity was not applied and priorities were poorly understood. In 1950 a government multiplication scheme was set up for Nzoia Rhodes at Thika and in 1953 at Eldoret. These sites were chosen because of the paucity of natural Rhodesgrass around so that contamination from foreign pollen could be excluded. Although contamination can be considerable, partly due to bee activity (Bogdan, 1962), we know now that this is not a priority consideration. Contamination-from-within, when varieties contain inferior genotypes, overrides the effect of foreign pollen, as we will see below. The Eldoret scheme was finally superseded by the establishment of the Kenya Seed Company at Kitale in 1956 (*2.5*; *8.1*). Basic (then called mother) seed was obtained from the Kitale Research Station, which itself remained also one of the main producers of certified seed. Specially isolated blocks were reserved for basic seed production for each variety. Seed certification in the form of field- and seedlot inspection was carried out, also by the Kitale Station, in cooperation with the Seed Testing Station in Nairobi. Although no maintenance breeding was undertaken, if we exclude the original plots that were part of the Kitale living-nursery collection, there was unanimity that the varieties in use could be distinguished by their main character, heading date. The system as a whole worked well in that it ensured that there was seed on the market to meet local and overseas demand.

The establishment of foundation fields was completed by 1970, also of Coloured Guinea (*Pcol*), Makueni Guinea (*Pmax*) and Congo Signal (*Bruz*).

Intra-variety uniformity (due to apomixis) was observed only in Buffelgrass (*Ccil*) and work was limited to removing a few of the obvious off-types, probably caused by admixture. Due to the low economic importance of the latter group of grasses, work was in the end confined to Rhodesgrass and Nandi Setaria. Elephantgrass, a vegetatively propagated grass, was embarked upon instead (*9.7*).

It is internationally recognized that the responsability for variety maintenance as well as for breeder's and basic seed lies with the breeder or the authority which produced the first seed for release. Maintenance is undertaken to ensure the integrity of the variety. A variety should be distinct, uniform and stable (D.U.S.) so that it can be recognized and registered. Heading date is an important criterion of distinctness, so that plants constituting a new variety should be uniform in this character. D.U.S. standards have reached their zenith in countries with Plant Variety Protection and Plant Breeders Rights (Mastenbroek, 1988) but should be applied everywhere in the interest of varietal integrity. Without strict maintenance, little meaning can be attached to variety names, let alone to legislation or to certification schemes since the variety may too easily lose its identity. Varieties occur that have become distinct, under multiplication elsewhere, but still carry the same name, while others occur inadvertently under several different names. A typical example of the latter in Elephantgrass is Gold Coast, *alias* Banagrass (*11.7.3.1*).

9.2. Source material for breeding

Intra-variety variation in Setaria but also in Rhodesgrass was reported by Bogdan (1959a; 1961c, d; 1964b) in contrast with Hutton (1960). Since most of the breeding research of that time was concerned with hybridization and other designs to increase variation *per se*, little importance was given to intra-variety variation, let alone to its exploitation. Very little was known about the consequences of certain selection procedures, for instance that selection for "leafiness" could have such a negative effect on herbage and seed yield (see also *6.6.3*; *6.7*).

This poor experience with Nandi Setaria made it clear that the course of events had to be reversed. Since the ecotypes had not been maintained (*9.1*), the dilemma arose whether to reselect from existing pastures or from seed fields. When Nandi Setaria pastures were sampled to compare botanical composition with that of seed fields, differences were observed which suggested that genetic shift had taken place in different directions depending on the husbandry given (Boonman, 1971a). It was assumed then that early-heading genotypes were eliminated under heavy grazing in favour of more competitive, but less productive types (Montgomery effect; Boonman & van Wijk, 1973). However, in more detailed studies on contrasting Nandi genotypes under cutting regimes, van Wijk (1980) found the erect types the most productive and competitive. In Australia, Hacker (CSIRO Ann Rep 1978) confirmed the early-flowering characteristic of Nandi Setaria populations obtained from heavily stocked and N-fertilized pastures. However, the preliminary findings had not encouraged us to base our breeding on survival populations of grazed pasture. Instead, the choice fell on the foundation fields (spaced-plant populations) derived from the seed fields of the Kitale Station. Additional reasons in support of this choice were:

1. The seed fields were by their very nature the most reliable representatives of the variety.
2. Advantage could be taken of natural selection, over the previous generations of continuous seed multiplication, towards better seeding properties.
3. More variation of general interest was expected to be found in seed fields than in pastures.

As the seed agronomy programme (*8*) was being concluded it became evident that little further progress would emanate from husbandry measures unless the heterogeneous heading and flowering structure of the varieties was altered. By choosing seed fields as source populations a better equilibrium might be achieved between the natural selection for herbage and seeding properties, if at all the two were as antagonistic as so often suggested at the time. The ecotypes had, obviously, been adopted originally for their herbage, not for their seeding characters. Interestingly, Birch (1964) had found a steady increase in % PGS in samples submitted to the Nairobi Seed Testing Station over the period of 1954–1964. However, the explanation given then did not refer to natural selection or to improved cultural techniques as the most likely causes.

The Kitale varieties such as Nandi Setaria, Mbarara and Masaba Rhodes had been multiplied continuously, without plant selection, for some 5–10 generations before the foundation fields were established. This continuous multiplication was likely to have helped their domestication.

9.3. Breeding methods

The characters sought in tropical grasses in East Africa have been outlined (*6.3*). Quality and yield aspects are not easy to separate, but yield measurements are useful and handy to register plant responses.

Since preliminary evidence had shown that most of the characters of interest had a relatively high heritability, procedures of positive mass selection were applied, followed by half-sib selection at later stages. Variation in characters was soon found to be inter-dependent so that selection for character Y (*e.g.* leafiness) was meaningful only when conducted at a given level or class of character X (*e.g.* heading date). Selection procedures, therefore, involved the vegetative transfer and grouping of classes of X and subsequent evaluation of target plants (with the desired character Y) within classes of X. This procedure, which proved powerful also to reduce environmental variation and location interactions, is referred to below as "class transfer". High value was attached to recording as many data as possible on target plants at consecutive stages of the transfers and to the final evaluation of target plants under sward conditions.

As it became increasingly clear that rapid progress could be made by exploiting the variation existing within the already popular ecotypes, purposeful crossing was deferred. This may be employed at later stages to add useful variability, if required and when located. Crossing programs are a time-demanding alternative to ecotype breeding and may be superseded by the latter as recent results with *Cynodon* spp. in the USA and with *Poa pratensis* in Europe have shown. Classical procedures of progeny testing should also be treated with caution. The polycross test, for instance, does no justice to specific combining ability and may fail in its objectives by preponderance given to dominance and epistasis and by lack of panmixia (Boonman *et al.*, 1979). For a discussion of different genetic designs that are possible in grass breeding, see Crowder & Chheda (1982). For details on ploidy levels, see Table 6.8.

The source populations (foundation fields) were established by direct vegetative transplanting but when this proved unreliable in Rhodesgrass (*6.6.2*), plants were raised from seed in polythene sleeves before transplanting. Plant density varied from 1 × 1 m in tufted grasses to 4 × 4 m in Rhodesgrass (stoloniferous). The original populations consisted of 4000 plants of Nandi I and Nandi II and 1000 of each of the other varieties.

This meant $1000 \times 16\,m^2 = 1.6$ ha per Rhodesgrass variety. In order to maintain uniformity in plot size in Rhodesgrass and to check stolon expansion, plots were separated by paths planted to Paspalum (*Pdil*) (Bogdan, 1964b) or sown to Congo Signalgrass. This also helped to reduce edge effects. At later stages, higher densities were introduced (2 × 2 m or 1.5 × 1.5 m) but there was no proof that this led to a more precise discrimination between target plants, to a more efficient spread of pollen or to a higher work efficiency, except in the amount of land required.

Plant-to-plant variation in heading date or in vigour of regrowth is probably more accentuated at wide spacing. Therefore, as soon as a pattern of variation emerged, target plants were transplanted and grouped in classes, according to heading date and, subsequently, in subclasses of other target characters, *e.g.* plant type or herbage yield level. This method of transfer showed its advantage in that the various classes when grown adjacently stood out clearly against each other while, within the relatively smaller-size classes, plants could be evaluated with greater precision. Early-heading Rhodesgrass plants, to name an example, displayed their erect growth-habit more clearly when grown adjacent to late-heading, prostrate plants (Table 9.2).

Classification was the first objective of selection when it was found that most of the variation in target characters, *e.g.* nutritive value or seed yield, depended on the variation in heading date (maturity) and vigour (*9.4.1*). Class transfers provided a handy tool to verify maturity and vigour classes prior to subsequent selection for the dependent variable. Thus, gross and net variation were separated (*9.5*) to make subsequent selection more realistic.

The initial class transfer from the source population took place after 4 seasons of plant observation, from late 1971 onwards, and the final transfers were in 1973 (Table 9.1). Seed collected from these transfers was used in 1974 to establish the first range of test varieties (Table 9.8) and to multiply seed. Interprovincial variety trials were laid out in 1975 over 9 locations throughout Kenya (Kenya District Trials, Table 6.2).

shed their seed by the time the last are coming into head.

It should be noted that the sigmoid curves of head emergence are representative of normaldistribution curves; there was no evidence that admixture from different ecotypes or contamination with foreign pollen played a major role. In fact, when ecotypes introduced straight from the wild flora are subjected to intra-variety screening, variation tends to be wider than after some generations of seed multiplication (Boonman, 1978a).

When, in the following, reference is made to maturity classes such as Early, Medium, Late or Very-Late these are meant to cover a range of heading date of 1–2 weeks each, tentatively.

Table 9.1. Class transfers out of 3000 plants of Mbarara, Masaba and Pokot Rhodes.

Transfer	Number of plants	Target of selection
1971	735 unreplicated	Maturity class
1973a	350 unreplicated	Seed- and herbage properties
1973b	148 clonal replicates	Seed properties

9.4. Heading date (maturity class)

As already mentioned, maturity is largely accountable for much of the variation found in agriculturally important characters of the population and is, therefore, a major tool of classifying the variation observed. Moreover, closely matched heading date of plants constituting a new variety is a prerequisite if breeding is to be successful. Heading date was defined as IHE (Initial Head Emergence; 5–10 heads/m^2). Since IHE is an easy to determine and an objective character it deserves utmost attention (*8.7.1*).

Plants were classified first on the basis of heading date. When IHE was observed in ± 5% of the population, these early plants were classified as "Week-1" (W_1); observations continued at weekly intervals.

Within the source populations head emergence was spread out over 5 weeks in Mbarara Rhodes and Nandi II, and 7 weeks in Masaba and Pokot Rhodes and Nandi I, if one considers only the time span between the first and the last 5% of the plants to come into head. If this period is extended from the first to the last 1%, the time span is 10 weeks or more (Fig. 9.1). It should be obvious that in such broad populations early-heading plants have

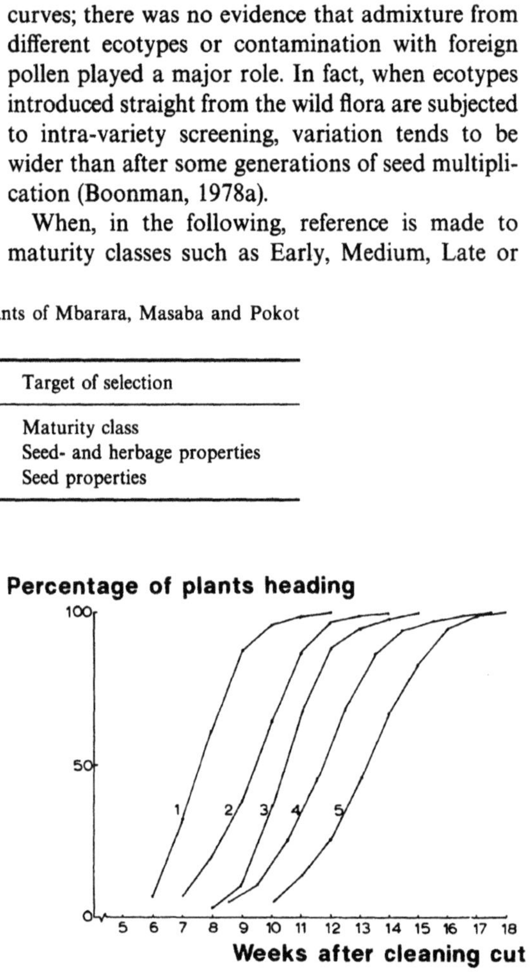

Fig. 9.1. Heading patterns in spaced-plant populations of Nandi Setaria and Rhodesgrass varieties. 1 = Mbarara Rhodes: 2 = Nandi I; 3 = Nandi II; 4 = Masaba Rhodes; 5 = Pokot Rhodes (Boonman & van Wijk, 1973).

9.4.1. *Heading date and plant vigour*

The heading sequence of Kitale varieties is related to the length of their growing season at the site of origin of collection (*6.6.3*; *8.7.1*). Within a variety, however, the heading sequence of individual plants is dictated by their vigour and vice versa. High-vigour plants tend to head early, whereas low-vigour plants are more frequent in the late-heading classes (Fig. 9.2). Conversely, late-heading classes consist mainly of low-vigour plants. Very few plants are found that are both late-heading and vigorous. "Vigour" is taken to mean the herbage mass, rated visually for every plant on the same date, at IHE of the W_1-class, and is thus a measure of growth rate. Because of "false-starts" in primary growth after vegetative establishment, data on vigour refer here to regrowth in post-establishment crops. The vigour variation found within the source populations was so great that visual scoring on a l-5 scale, whereby 5 represents high vigour, was found adequate.

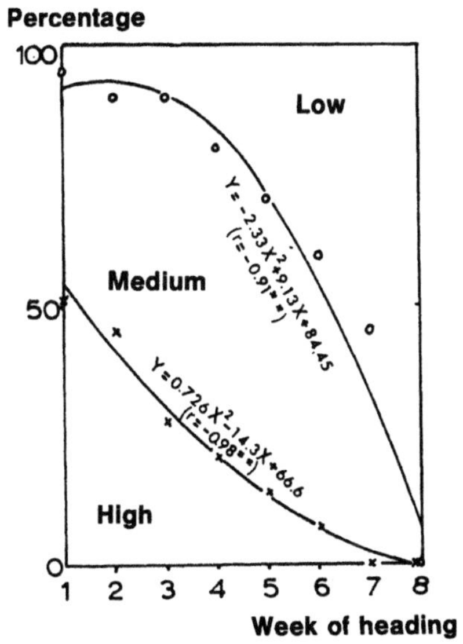

Fig. 9.2. Intra-variety variation for heading date (measured in weeks) and vigour: low, medium and high vigour (or growth rate) (Boonman & van Wijk, 1973).

It should be noted that the period prior to IHE is not analogue to what is termed in temperate grasses as the juvenile period which tillers require before they become inductive to vernalization.

9.4.1.1. Vigour and plant type

The plant type in the early stages of vegetative establishment was found to provide an early indication in what vigour class and, thus, in what heading class a plant would end up eventually. In Rhodesgrass, types with an erect rather than a stoloniferous habit in primary growth tend to head early; their stoloniferous development is delayed considerably. However, as soon as the individual swards have closed and differences in spreading ability have evened out, the same plants will again be heading early in subsequent seasons. Those with a stoloniferous habit in primary growth are late heading, also in subsequent seasons (Table 9.2).

Table 9.2. Heading date and plant type; the spreading ability, on a 1–5 scale, of Rhodesgrass plants following vegetative establishment (Boonman, 1978a).

Variety	Maturity class			
	Early	Medium	Late	Very late
Mbarara	3.0	3.2	3.8	not present
Masaba	2.3	2.9	3.1	3.6

Grave errors are made if this relationship is ignored in the establishment year. Stoloniferous, late-heading plants look very attractive in the 1st season. By contrast, erect early-heading types produce many heads before stolons are seen and before a sward is formed. They may easily be discarded as off-types. Were selection based only on 1st year's vegetative attractiveness of appearance, spreading but eventually late-heading plants would be in the majority, with less vigour etc. Conversely, what appear to be atypical, unattractive plants in the planting season, would be chosen more readily in the post-establishment years because of their high vigour. Apparently, vegetative propagation gives rise to a kind of primary growth, different from that of a seedling. Since early-heading plants are our primary target class, as we will see below, subclasses of the same maturity are better established from seed to ensure early stolon-development and to speed up the selection process.

Similarly in Nandi Setaria, erect/tall plants turned out to be more vigorous and early-heading, and vice versa.

9.5. Gross and net variation in dependent characters

Heading date and vigour are closely correlated. Heading date is correlated with many other important characters but is easy and reliable to measure, more so than vigour or plant type. Target plants may be of different heading dates but obviously only of one vigour class, *i.e.* the highest available. In the following, therefore, heading date will be regarded as the independent character. Since heading date is largely accountable for much of the variation in the dependent characters to be discussed we call such variation, when due to heading date, gross variation; when due to the dependent character on its own, it is called net variation. Other terms used for gross and net are primary and secondary variation, respectively (Boonman, 1978d).

When in source populations like the foundation fields described above, measurements are made of seeding or herbage properties, a lot of (gross) variation is found. High values of heritability may be established and selection will be considered promising. It will continue to stay "promising" without tangible results ever, when it concerns dependent characters. Seed shedding and leafiness are the most obvious examples of possibly elusive objectives. Seed shedding is common in tropical grasses (*8.7.2.4*) and it is often said that breeding aimed at improvement of seed retention or of seed recovery should lead to rapid increase in seed yield. In the above source populations it was no problem to find plants which had not shed their seed at all. Such plants were, however, merely late-heading (Fig. 9.2). The same applies to variation in leafiness. However logic all this may seem, this is exactly the trap into which many prospective breeders fall (see also 6.7). Variation in leafiness and shedding should, of course, be considered in relation to heading date first.

For breeding to be successful it is imperative that gross and net variation are separated. Gross variation, due to variation in maturity, is taken care of by splitting the population into maturity classes. Supplementary subdivision is often necessary as a refinement because, within a maturity class, variation in characters such as leafiness depends further on herbage-yield level (*9.5.2*). It should, therefore, be no surprise to find that subsequent upon the division of gross variation in maturity- and yield classes, very little net (*i.e.* genuine) variation is observed in leafiness or in other characters of nutritive value such as digestibility (*9.5.3*).

9.5.1. Heading date and seeding characters

In the foundation fields, plants with low vigour produced very few heads. Some plants had almost no heads at all. Out of the 1971 class-transfer (Table 9.1), 735 target plants (of high vigour) were arranged according to maturity class, covering no less than 14 days each, and were assessed in 4 m^2 plots for seed yield and head density (Table 9.3).

Table 9.3. Variation between target plants of different maturity classes of Rhodesgrass; relative to Early = 100 (after Boonman, 1978a).

Maturity class	PGS yield	Head density
Early	100	100
Medium	67	100
Late	58	81
Very late	49	79

Table 9.3 shows that the seed yield potential of target plants declines with maturity class. The later they head, the lower the head density, but especially PGS-yield. Head numbers, although reduced, are still relatively high in late-heading target plants (of high vigour), but PGS-yields are reduced markedly. Since 1000-seed (spikelet) weight varies little within a particular season (*8.7.2.1*), yield reduction is caused by reduced seed-set (Fig. 9.3c). Spikelet abortion is also more common, due to lower seed-set (*8.7.2.4*). Head numbers and seed-set are the two most important components determining seed yield (*8.7*). Genotypes with a better synchrony of heading, *i.e.* more heads produced per unit time, seem desirable but variation in this character, within subclasses of high head-numbers, was hard to detect.

The next step undertaken was to evaluate plants with a high vigour and high head density. 148 Target plants (class transfer 1973b, Table 9.1) were assessed in plots of 4 m^2 with 6–8 clonal replicates each, over 3 seasons. This clonal refinement was brought in to assess the correct IHE-ranking order, before top clones could be identified and assigned to a range of test varieties (Table 9.8). The results are given in Table 9.4, only for the maturity classes Early and Medium. Breeding work was not continued in the Late and Very-late classes due to generally poor performance, although they consisted of the most vigorous target plants available.

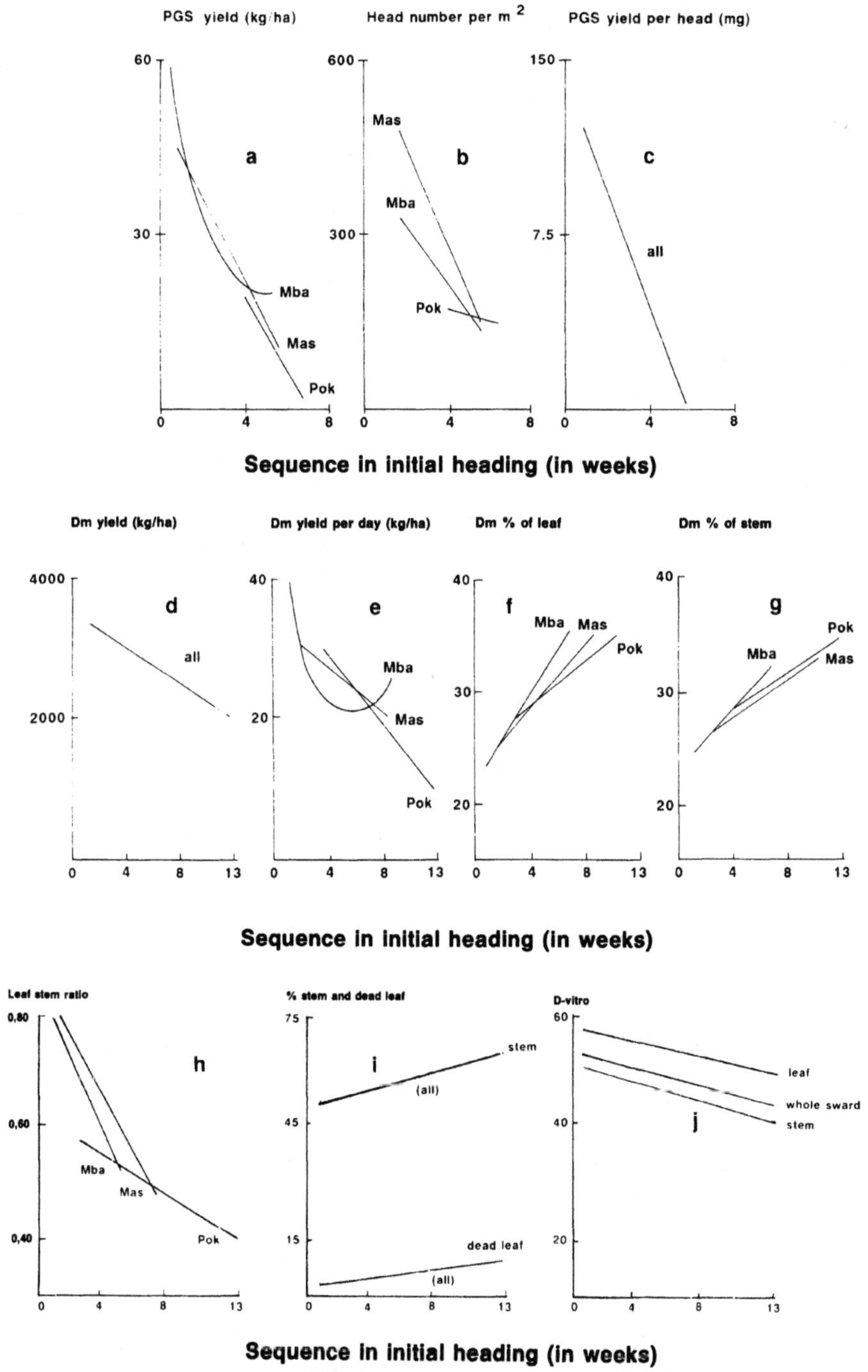

Fig. 9.3. Intra-variety variation in relation to date of initial heading; 1. ..., 13: plants heading in 1st., ..., 13th week; (a) & (b): data taken at initial heading stage of each plant in 1976; c: data taken at seed harvest, 1975; Mba = Mbarara, Mas = Masaba; Pok-Pokot; all = all varieties pooled (after Boonman, 1978d).

Table 9.4. Variation between target clones of Rhodesgrass within maturity classes; over 3 years, average = 100 (after Boonman, 1978b).

Source variety	Maturity class of target clones	PGS yield	Head density
Mbarara	Early	50–163	69–130
Mbarara	Medium	51–195	67–139
Masaba	Early	55–162	72–125
Masaba	Medium	59–130	77–119

This proved how difficult it is to find enough plants, within the late-maturity classes, with consistently high vigour and head density.

Table 9.4 illustrates that within maturity classes large differences remained in head numbers but even more in PGS-yield and, consequently, in seed-set (calculated as number of PGS per head). The following additional observations were made:

1. The range in IHE within the classes was on average still some 15 days, partly due to differences still prevalent in vigour, whereas between the class averages it was reduced to 4 days. In fact, some overlapping was found between classes.
2. Significant clone-season interactions were noted, probably due to one poor season (early 1974), although this did not affect the overall ranking order. The 8 best Mbarara-Early clones which were later assembled to form the Elmba variety were among the top-10 in PGS-yield in all 3 seasons.
3. High yield of clean seed was normally not accompanied by an equally high % PGS, which demonstrates the importance of determining % PGS of every target plant (*8.7.2.2*).

In order to cope with the large number of germination tests, a routine method was developed. Seed samples were first sharply cleaned over a "Brabant" clipper to achieve a high % purity (*i.e.* high % caryopses). The seed was then germinated on pats under bell jars on zinc trays in a greenhouse. The latter was heated with charcoal stoves when necessary to maintain a sufficiently high temperature to speed up the germination process (Boonman, 1972b). For the same purpose, water in the trays was replaced by warm water once a day. With this method 3000 samples could be tested between 2 planting seasons. At later stages of the breeding programme special attention was given also to seeds giving a fast development of radicles and coleoptiles. Seedling vigour is the first indication of a productive genotype. When the radicles are less than 1 cm long, the seedlings can be lifted without damage from the germination pats and be transplanted for subsequent selection. This proved to be an effective tool in half-sib family selection procedures.

Perhaps the most difficult but nevertheless most desirable component to improve is seed-set, *i.e.* the number of PGS per head. This was measured in the target clones of Masaba of Table 9.4, out of which the top clones were subsequently assembled to constitute Boma Rhodes (Table 9.5).

It is evident from Table 9.5 that the 25% increase in PGS-yield was accompanied by a 6% increase in head number and a 17% increase in seed set. The same was found in test varieties (Table 9.8). Differences in 1000-seed weight were statistically significant but very small (see also *8.7.2.1*). An effect of spikelet weight on seedling vigour has never been established, whether based on intra- or on inter-variety variation. Mbarara Rhodes has smaller seeds than Pokot (Table 6.7) but its field establishment is usually more reliable. It is concluded that within target plants of the same maturity class, of high vigour and head density, selection for seed-set and PGS-yield is rewarding.

9.5.2. Heading date and herbage characters

Similar attempts were made to establish whether in such target plants useful variation remained in herbage characters. It was observed that drought tolerance and persistence were markedly better in early-heading classes, although selection had not been directed at those characters (Boonman, 1978a). Attention had concentrated on exploiting variation that was found in leafiness and digestibility.

The 1973a class transfer plants (Table 9.1) were grown in 2.5 m^2 swards, separated by 30 cm paths sown to Congo Signal and studied for herbage characters in 1976, following seed yield studies in 1975. Every plant was sampled at its own IHE. True genetic differences may thus emerge, because of the negligible differences in climate within the rainy season at Kitale (Fig. 3.3). DM-yield,

Table 9.5. Seed yield components in target clones of Masaba Rhodes (after Boonman, 1978b).

	PGS yield (kg/ha)	1000-seed* weight (mg)	Number of heads/m^2	Number of PGS per head
Top-5 clones	44.6	261.7	366	48.2
All clones	35.7	253.5	345	41.2
Gain (%)	+25	+4	+6	+17

*Seed = spikelet.

leaf:stem ratio and *in-vitro* organic matter digestibility (D-*vitro*) were determined. It should be remembered that these target plants had already passed through 2 selection cycles for maturity class and, therein, for plants with high vigour. Yet, plant-to-plant variation was still such that DM-yields dropped by one-third in the last plants coming into head, although the duration of growth of the latter had been twice as long (Fig. 9.3d *etc.*). The daily growth rate of late-heading plants was, therefore, reduced by about one-half. Leaves had a higher % DM than stems and this difference increased as plants came into head later. Leaf:stem ratios and D-*vitro*, both determined at the same growth stage (IHE), dropped as plants came into head later (see also Fig. 9.4), whereas % dead leaf increased. D-*vitro* was on average 7.6 units higher in early plants than in plants heading 8 weeks later, but the extremes varied as as much as 17 units between some early and late. In the previous year, even more marked declines had been observed in terms of seed yield and its components, especially seed-set (Fig. 9.3c). Late heading has a more drastic effect on seed yield than on herbage characters.

Late-heading results in low digestibility, at the same stage (IHE), which is accompanied by a progressive decline in leaf:stem ratios and an increase in % dead leaf. From the nutritive value point of view, late-heading plants have obviously little to commend them. As far as the variety populations as a whole were concerned, Pokot Rhodes was late-heading, yet the lowest in % leaf, at the same physiological stages (Fig. 9.3h). At Makerere (U) Reid *et al.* (1973) had also observed a considerably lower digestibility in Pokot than in Mbarara or in Masaba Rhodes. That digestibility is high in early-heading Rhodesgrass had previously been shown for Pioneer Rhodes in Queensland by Milford (1960).

When varietal comparisons are made on the same calendar date, *i.e.* not necessarily at the same growth stage, differences observed in nutritive value are confounded with stage of growth. Because of the variation in IHE between and within the varieties in the populations under study (Fig. 9.3), the unique opportunity arose to compare subsets of the 3 variety populations not only at the same growth stage, IHE, but also on the same day. In this way, the confounding effects of age and heading date were eliminated (Table 9.6). These subsets consisted of 20–30 plants per variety.

It is evident from Table 9.6 that Masaba had a much higher leaf:stem ratio and a higher D-*vitro* of stem and of "whole sward" (including dead leaf). This shows that Masaba is medium-late heading but genetically more nutritious than the other varieties, either early or late, independent of growth stage.

Table 9.6. A comparison of plants, heading on the same day, out of different Rhodes varieties (after Boonman, 1978d).

	Mbarara	Masaba	Pokot
Leaf:stem ratio	0.54	0.63	0.48
D-*vitro*:			
–Leaf	50.9	50.2	51.8
–Stem	46.9	48.2	45.2
–Whole sward	49.7	50.5	48.8

In a supplementary study the same analyses were carried out but now of 20 clones of Masaba-Early (Table 9.4), at 3 stages of regrowth prior to IHE. Significant differences of D-*vitro* of no less than 5 units were found especially in the stem fraction, but they could be accounted for by variation in DM-yield, since the two are negatively correlated: the 7 clones highest in D-*vitro* were among the 9 lowest in DM-yield. Conversely, clones of similar yield level had a closely similar D-*vitro*. The latter was not correlated with leaf:stem ratio. Thus, by eliminating differences due to heading date and yield level very little remains of the variation in D-*vitro* that may be observed in a randomly sampled population with a wide range in heading date or vigour. Any potential gain in D-*vitro* or in leaf:stem ratio is offset by a corresponding decline in herbage yield (Boonman, 1978e).

It is concluded that true genetic differences in D-*vitro* and leaf:stem ratios may be found between varieties, for instance between Masaba and Pokot Rhodes, and that early-heading plants within a variety are superior in this respect. But when plants of the same maturity class and yield level are subjected to further screening, very little net variation remains, which is in contrast with what is found in seed set (*9.5.1*).

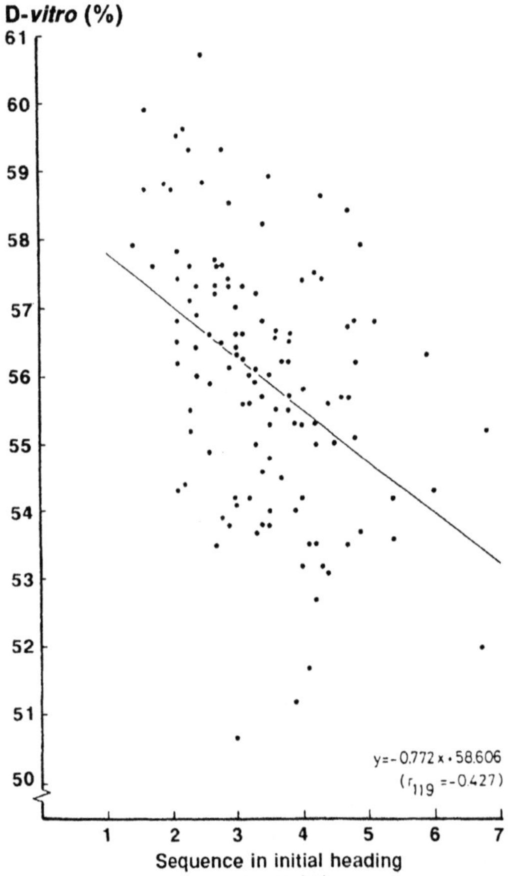

Fig. 9.4. Digestibility (D-*vitro*) and heading date in 121 random plants of Nandi Setaria (van Wijk, 1980).

9.5.3. Genetic correlations in Nandi Setaria

Van Wijk (1980) has decribed in detail his breeding research with Nandi Setaria at Kitale. His work has contributed greatly to the Kitale breeding programme by analyzing genetic backgrounds and by demonstrating that breeding for high herbage and seed yield can be conducted simultaneously. This refuted the negative experience obtained earlier with Nandi II and Nandi III (*6.7*).

Setaria is a tufted grass and easy to propagate vegetatively (Table 6.6). Precision of measurement may be smaller when dealing with tufts than with swards, but this can be corrected by using clonal rows in semi-continuous drills. Van Wijk has shown that assembling and evaluating target plants in test varieties are time-efficient methods which are hard to beat by genetically more sophisticated designs of progeny testing. A refinement was his testing of Setaria clones within a sward of Rhodesgrass.

Van Wijk's work was geared to unravel herbage quality relationships, based on intra-variety variation. His work has demonstrated how difficult it is to improve D-*vitro* by breeding without losing DM-yield (Fig. 9.5). The heading date × vigour relationships mentioned earlier (Fig. 9.2) were refined by van Wijk by measuring herbage yield at (seed) harvest. Differences in D-*vitro* are easy to demonstrate in spaced-plant populations but this is gross variation (*9.5.2*), depending on heading date (Fig. 9.4) and on yield level (Fig. 9.5). When digestibility data are classified according to maturity and herbage yield, little net (genuine) genetic variation remains within such classes.

When target plants were assembled in test varieties according to maturity class and cut at 2-weekly intervals, differences in D-*vitro* hardly exceeded 1%-unit and were not significant at any stage of regrowth. The ultimate selection gain achieved in yield of digestible organic matter was due to an increase in yield rather than to improved digestibility. Earlier work (van Wijk, 1976) had shown that late-heading selections out of Nandi were lower in digestibility.

In addition to heading date, other plant characters pertaining to D-*vitro* were examined (Table 9.7). Low repeatability and heritability, as found in tiller angle, indicated that results as observed at times may be difficult to reproduce.

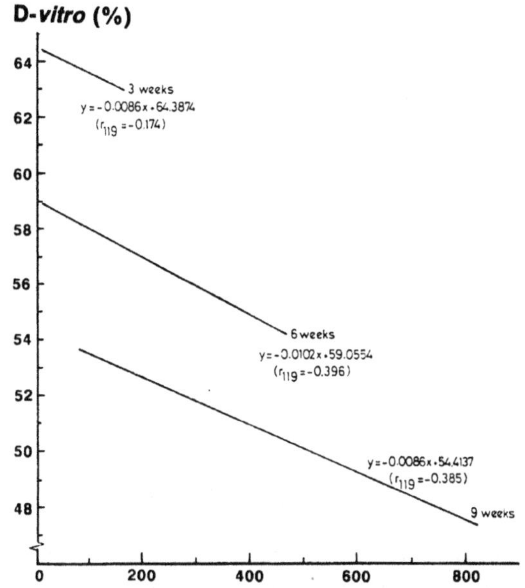

Fig. 9.5. Digestibility (D-*vitro*) and yield relationships in 121 randomly selected clones of Nandi Setaria, according to stage of regrowth (van Wijk, 1980).

Table 9.7. Plant characters, relevant to herbage quality (after van Wijk, 1980).

	Repeatability	Heritability (s.l.)
Tiller numbers	0.62	0.56
Tiller weight	0.54	0.70
Tiller length	−0.12	0.70
Tiller angle	0.13	0.41
Leaf number	0.30	0.51
Leaf length	0.53	0.67
Leaf width	0.64	0.67
Stem diameter	0.43	0.45
Fresh weight	0.16	0.52
Heading date	0.21	0.64
D-*vitro*	0.19	0.75

Erect, early-heading plants with a large number of heads proved to result in the most productive varieties, from the point of view of both herbage and seed. A high genotypic correlation was found between seed- and herbage yield.

9.6. Variety trials

The example of Rhodesgrass will be presented in brief to show the progress that has been made by breeding (Table 9.8). Test varieties, constituted by assembling seed from top target-plants, were compared with the original (standard) varieties. Data were grouped into 2 maturity classes only: Early and Late. Early embraced the classes already mentioned (Table 9.3) of early plus medium. Late included late plus very-late classes.

Table 9.8 demonstrates the dramatic effect that intra-variety selection, primarily directed at high vigour and uniformity of heading date, has had in the resulting varieties, especially in heading date and seed yield.

The range in heading date which was 19 days in the standard varieties was increased to 37 days in the test varieties. The standards were generally intermediate between "Early" and "Late". This shows that breeding for (uniformity in) heading date was very effective.

Selection for high vigour within maturity classes was efficient, since all selections had a higher DM-yield, although "Late" could not keep up with "Early", probably because of the lower frequency at which target plants occurred in the late-heading sections (Fig. 9.2). Also in Pokot Rhodes selections great improvements in DM-yields could be achieved. High seed yield is, however, hard to combine with late-heading. Seed yield is much more affected by heading date than herbage yield is.

Seedling vigour was usually much higher in selections than in standard varieties, so that DM-yields in the first herbage cut were often 50% higher (Boonman, 1978c). The consistency of this observation over a range of variety trials and farming situations (*7.1.12.2*) suggests a genetic effect. It should be stressed, however, that in variety trials genetic differences can only be established with certainty if variation in seedling vigour due to environmental factors, *e.g.* harvesting or storage conditions, can be excluded.

For more details on variety trials see Boonman (1978c, d; 1980). For performance data of new selections in the Kenya District Trials, see Table 6.2 (*6.3.1*).

9.7. Breeding Elephantgrass

Elephantgrass is a foddergrass grown to complement the fodder resources in the dry season and

Table 9.8. Test-variety trials at Kitale; yield data relative to Mbarara (after Boonman, 1978c).

Variety	Selection/ Standard variety	Heading sequence (in days)	Yield (3-year means)	
			Dry matter	PGS
Mbarara	Early (Elmba)	0	113	194
Mbarara	Standard variety	11	100	100
Mbarara	Late	16	109	109
Masaba	Early (Boma)	12	109	122
Masaba	Standard variety	16	100	100
Masaba	Late	23	106	75
Pokot	Early	28	108	38
Pokot	Standard variety	30	93	27
Pokot	Late	37	108	21

for stall-feeding (Part II). It is high-yielding in the warm and humid parts of Zone II, but its dry-season greenness in Zone II or III should not be taken to mean that this grass can survive in semiarid areas (Zone IV).

Breeding of Elephantgrass was prompted by the rapid spread in the early 1970s of diseases in and among the few productive clones available and by complaints about the hairiness and sharp leaf edges which make manual handling unpleasant. Attempts at Makerere (U) of producing hybrids between Elephantgrass and pearl millet had not produced the results that had been anticipated (*11.7.3*; *12.1*).

Elephantgrass is cross-pollinating and tetraploid. Although a shy-heading grass, caryopses are formed which, when freed from their spikelets, germinate very rapidly. Like most species which are normally transplanted vegetatively it is highly heterozygous, giving rise to a very heterogeneous population of seedlings, which are not "true-to-type". This and the slow development after germination explain why attempts at sowing Elephantgrass have failed. Seedling progenies offer opportunity for selection and this was how most of the now still famous varieties emerged in the 1930s. Uganda Hairless was developed in Uganda by A.S. Thomas. Cameroons and Gold Coast were developed in South Africa from seed of West African origin (*11.7*).

In August 1973 at Kitale, caryopses of French Cameroons were germinated and 300 seedlings were transferred in April 1974 to 1.5 × 1.5 m hills with French Cameroons as control and perimeter. 200 Plants were transplanted to clonal rows in August 1974, while 100 plants were screened for seeding properties. Seedlings give rise to thin stems with abundant heading, already at a height of 1.5–2 m, but they do so only in primary growth. In 1975–1977, 15 clones out of the 200 and 6 clones of the "seeding" type were yield-tested in replicated clonal rows. The "seeding" clones were disappointing in herbage yield, but out of the group of 200 clonal rows 1 clone emerged that outyielded French Cameroons (van Wijk, 1977). This clone came to be known as "Clone 13" and has remained disease resistant ever since (*11.7.4*).

Work was started in 1975 on X-ray irradiated seed and this approach was intensified by van Gastel (1978). Many thousands of seedlings were treated. The optimum dosis was found to be 8 krad for germination on filter paper and 5 krad on soil substrates. Some seedlings with smooth-edged leaves and with chimerism were observed, but as plants grew older no differences remained. Attempts were renewed to develop a seeding Elephantgrass, because of the high % germination, already mentioned, and the abundance of heading and variation in seedling progenies. From the 2nd year onwards, however, very few plants came into head and they did this only after a long delay and at a greater plant height. Plant-to-plant variability was by then also greatly reduced. The contrasting performance as displayed by seedling populations has been observed consistently. This programme was finally abandoned.

It is not known whether clones other than French Cameroons would have given the same promising results upon selection within seed progenies. To keep abreast of diseases, which have been known to assume devastating proportions in Elephantgrass clones, breeding is the best answer and the more seedlings are evaluated the better the progress is likely to be.

Clone 13, obtained from no more than 300 seedlings, has proved an outstanding success, both in yield and in disease resistance. It is yet another example that simple breeding techniques can be effective. Stems of Clone 13 are thin, so that establishment from root splits is better than from cane cuttings. Whether digestibility is related to stem diameter or not, recent work (Kitale Ann Rep 1986) has demonstrated that D-*vitro* of Clone 13 is 2.5 units higher than of Gold Coast. It should be noted, however, that digestibility was not an objective of selection in this particular breeding programme.

9.8. Apomixis

Although the most important cultivated species in East Africa are cross-pollinating (*Cgay*, *Ssph*, *Pcol*, *Bruz*, *Ppur*), facultative apomixis is common in species such as Common Guineagrass (*Pmax*) (*6.8*) so that methods of ecotype breeding as enumerated above can be applied effectively. Congo Signal and Coloured Guinea had earlier been believed to be largely apomictic (*6.11*; *6.12*). In our studies of the intra-variety variation, however, a normal distribution of quantitative characters, similar to that of cross-pollinated varieties, was found also in these 2 species. Subsequent selection produced botanically distinct test-varieties.

Apomixis has commonly been associated with polyploidy. Diploid sexual types have been found to occur naturally, however, in and among tetraploid apomictic types of Common Guineagrass (*6.8*), Signalgrass (*6.11*) and *Paspalum notatum* (Bogdan, 1977).

In highly apomictic species such as Buffelgrass, continued reliance will have to be placed on ecotype collection if useful genetic variation is to be added (*6.9*). Procedures of bypassing the apomictic barrier by searching for or inducing sexual plants have been attempted in Buffelgrass (*6.9*) but on a larger and more commercial scale in temperate *Poa pratensis* (Boonman *et al.*, 1979). In common with vegetatively propagated species, apomicts are highly heterozygous and a wide variability is released if the apomictic barrier is broken. Conversely, the apomictic mechanism circumvents problems of sterility and has been an effective evolutionary force in genome preservation, polyploidy build-up and permanent fixation of heterosis. Understandably, the prospect of being able to fix new hybrid combinations by seed aroused a great interest in apomicts. However, when commercially superior apomictic varieties are used as parents or source material, their sexual progenies often contain no genotypes of comparable value. The trouble needed to identify superior material is then in fact even less rewarding than that of ecotype screening. Secondly, it has proved very difficult to fix superior genotypes again in strictly apomictic form in subsequent generations, because sexuality recurs (A.J.P. van Wijk & J.J. Hintzen, private communication).

9.9. Conclusion

What began as an attempt to stabilize the commercial varieties (ecotypes) of Kitale in 1968 developed into breeding in order to exploit the wealth of variation found within these mostly cross-pollinating varieties.

This variation-on-offer was matched with objectives, some of which had been considered either antagonistic, *e.g.* seed versus herbage, or unorthodox, *e.g.* early- versus late-heading. Breeding research had generally been pre-occupied with genetics rather than with plant relationships. Knowledge of the latter is required to unravel relationships such as those between heading date and dependent characters. This unravelling of useful variation is a crop-physiological rather than a genetic exercise. Without such knowledge breeding rarely reaches beyond demonstrating "promising variation" which may turn out to be deceptive and even counter-productive. In the Kitale breeding approach, simultaneous improvement in seed- and herbage characters proved possible, especially within the early-heading section of an ecotype. Within already existing popular varieties, early-heading plants were found to have higher seed- and DM-yields, persistence, drought tolerance and, at the same growth stage, leaf:stem ratio and digestibility.

Uniformity, especially in heading date, is essential in new varieties moving into seed trade so as to ensure distinctness, uniformity and stability (D.U.S), and to make registration and recognition possible. Unless varieties are maintained at the point of first release, seed lots in circulation may deviate not only in botanical characters but also in performance, so that variety trials lose their meaning. Synchrony in heading date is also a prerequisite to achieve higher seed yields of high PGS.

The huge variation displayed at Kitale by varieties of tropical grasses, mostly cross-pollinating to a high degree, represents a normal distibution curve, although the earliest and latest plants to come into head seem rather far apart to intercross. How this wide variation, which is larger within than between ecotypes, came about and developed under continuous seed multiplication, has been discussed by Boonman (1978a). It would be too simple to attribute the causes of heterogeneity to admixture or to contamination with foreign pollen. Actually, comparisons with more recently introduced ecotypes such as Pokot Rhodes (Fig. 9.1; *6.6.3*) or with varieties such as Mpwapwa and Kapedo, which have undergone fewer cycles of multiplication, show that variation in heading date may be reduced under continuous multiplication. Whether this implies that the mean of the population has moved also, probably in the early direction, cannot be confirmed. It is unlikely that the section of late-heading plants with low vigour (Fig. 9.2) has much chance of survival under seed multiplication, except when helped by their prostrate growth habit and by late seed-harvesting. All evidence available suggests that a broad variation in heading date and in dependent characters has somehow been maintained.

A wide botanical variation may have served the original ecotype in its adaptation. However, botanical variation is no insurance of broad adaptation under cultivation. To the contrary, Boma Rhodes has been found to have a wider adaptation than Masaba, its parent variety. That within-variety variation has little to do with wide adaptation is corroborated by apomictic and, therefore, uniform ecotypes of species such as Buffelgrass. East African ecotypes of the latter are just as widespread in agricultural use in other parts of the world as Rhodesgrass.

Heading date is spread out over more than 2 months in spaced-plant populations of ecotypes, because low-vigour plants take a progressively longer time to amass enough herbage before heading sets in. This explains why, in the absence of daylength effects in the tropics, vigour and heading date are so interdependent. Vigorous plants are very rare in the late-heading class, which is reflected in a poorer performance in terms of DM-yield, head numbers and especially PGS-yield (Table 9.3: Fig. 9.3c). Selection for late-heading (*alias* leafiness) with no regard for vigour had already led to the poor performance of Nandi II (*6.7*). Conversely, selection for high seed and herbage yield goes hand in hand in the early-heading sections. Simply by grouping early-heading plants of high vigour together, one is assured of highly improved seed yield of high % PGS. Heading date has more effect on seed than on herbage yield.

Although lower DM- and seed yields are produced by late-heading ecotypes of tropical grasses, the above observations refer essentially to intra-variety (ecotype) variation. Masaba Rhodes which is of medium-late maturity serves as an outstanding example to show that principles governing intra-ecotype variation cannot simply be translated in terms of inter-ecotype variation. In fact, Boma Rhodes has become more popular than Elmba, probably because of the wide adaptation it inherited from Masaba (*6.6.4*). Pokot, by contrast, has little to commend it (*6.6.3*; Table 9.6). Pokot is an example of a grass that is late-heading but not leafy (see also Fig. 9.3h). Conversely, Elmba and Boma are the new exponents of early and medium-late Rhodes varieties, but it must be recognized that both were actually chosen from the early-heading sections of Mbarara and Masaba Rhodes, respectively.

Seed yield (including its main components: head density and seed-set) is positively correlated with early-heading, but the same applies to digestibility and leafiness, at the same stage of growth. However, the latter are negatively correlated with DM-yield within maturity classes and are major but less obvious examples of dependent or secondary variation. A range of 17%-units in D-*vitro* between early and late, as observed in the source populations, may be reduced to 5 between target plants, *i.e.* within classes of similar heading date, and to even less when compared at similar DM-yield levels. Very little net variation is left, because of the negative correlation between D-*vitro* and DM-yield.

Obviously, chances of breeding high-yielding varieties with higher digestibility by direct selection within maturity classes are slim. However, the early-heading sections of a population tend to be higher due to the positive correlation between digestibility and earliness. Thus, it is generally more efficient to screen test varieties for D-*vitro* at the end of the selection programme (as with Clone 13 of Elephantgrass). The same has been experienced in temperate grasses (van Wijk *et al.*, 1993). Despite high expectations of breeding for higher digestibility (Wilson & Minson, 1980; Humphreys, 1991; see also *4.2.2.1* & *6.3.4*), practical results have been limited. Van Wijk (1980) suggested that if higher quality is the aim better results may be obtained in farming practice by developing higher yielding varieties and exploiting them at a younger stage of growth.

Mass selection and ecotype breeding are simple conventional techniques but they are rarely highlighted in the scientific literature in this day and age of genetic engineering, however remote the practical application of the latter. Mass-selection methods as described (although modified via the system of class transfers) were effective to manipulate the variation under study, but it may be argued that progeny tests are needed to unmask the variability remaining. The polycross test had been employed to develop Nandi II and Nandi III (Bogdan, 1959a), but the results obtained serve to indicate that genetic designs are only of secondary importance. Mass selection is adequate to handle broad variation within source populations. Half-sib family testing, however, with progenies being retained of the best-yielding mother plants may be effective for continued population improvement (*9.5.1*).

It may equally be argued that variety- or species crosses may add useful variability (*9.3*). Evidence from successful commercial grass breeding programmes in Europe suggests, however, that as long as ecotype variability for the characters in question is readily available, no time should be lost on sophisticated methods which are not only very slow but can even be counter-productive.

Variation in spaced-plant populations increases as time goes by. Low-vigour and late-heading plants have difficulty in fending off weeds which is reflected in low persistence: poor plants get poorer and poorer. Short-cuts should always be pursued and there is no better way of assessing their effectiveness than in test-variety trials, under sward conditions. Fast progress is achieved by assembling open-pollinated seed of target plants into test varieties. When target plants are assembled in various combinations, some indication of ecological combining ability may be found. We have, on more than one occasion, observed that simply by assembling target plants in different combinations into test varieties, higher yields are found in some than in others, although no obvious reason could be established. The procedure can be repeated with the best test-variety as source population. Recent evidence has shown that Elmba and Boma Rhodes can be selected upon still further.

Plant evaluation under sward conditions is almost automatic in spaced-plant populations of Rhodesgrass, which is an advantage over tufted grasses, but the final proof is evaluation in test-varieties. Grazing trials are welcome as additional information, but unless they are followed-up by test-cropping they provide no information on the complete role of a new variety in the broader context of grass-crop rotations, for instance on nematode control as effected by Katambora Rhodes (*6.3.1*; *6.6.4*).

The classical example of Nandi II and Nandi III (*6.7*) shows that when breeding objectives are proposed without awareness of the prevalence of dependent relationships, *e.g.* in the case of leafiness, progress may be obliterated by losses in other important characters. Conversely, relationships have often been proposed as antagonistic, without adequate justification. Progress is hampered when relationships are assumed, *a priori*, to be in mutual conflict, with one taking precedence over the other. The Kitale breeding work has demonstrated that herbage characters and seed yield can be improved at the same time and that the range of maturity from early to late heading can be maintained, if required.

Finally, the example of Clone 13 of Elephantgrass demonstrates that in vegetatively propagated grasses progress with selection from seedlings can be rapid and effective for yield and disease resistance but also for digestibility.

9.10. Summary

Grass breeding is described as ancillary to seed multiplication. Cross-pollinating ecotypes of tropical grasses contain a wide intra-variety variation in vigour of regrowth and, therefore, in heading date. Among plants selected for high vigour, heading date is the main independent character to which variation in most of the remaining characters is related. Early-heading plants are higher in DM- and PGS-yield but also, at the same growth stage, in leafiness and digestibility. Very few plants are found which are late-heading but of the same potential; their seed yield is particularly low. Within classes of similar heading date and herbage yield, useful variation remains for PGS-yield but not for digestibility. Test varieties, consisting of target plants, showed an increase in DM-yield, also in the late-heading classes although to a lesser extent. By contrast, PGS-yields dropped very sharply with heading date. New varieties, Elmba and Boma Rhodes, were constituted from the early heading classes of Mbarara and Masaba Rhodes, respectively. The same applied to Nasiwa Setaria (out of Nandi). In Elephantgrass, 300 seedlings of French Cameroons produced one outstanding clone, Clone 13, which is disease resistant and of high yield and digestibility. For improvement in truly apomictic species reliance is to be placed on additional ecotypes.

CHAPTER 10

Legumes in sown and grazed pastures

No subject has been described more extensively in the scientific literature of tropical grasslands than that of tropical legumes. This chapter will not seek to emulate the many accounts of their potential role. However, as Crowder & Chheda (1982) remarked: "in view of the benefits gained by growing legumes in combination with grasses it is surprising that they have not been more widely used".

Humphreys (1991) has argued that it would be foolish to claim that pasture legumes are widely used in tropical farming. Recent statistics of CIAT show that while tropical grass seed production stands at 5000 ton in Latin America, true pasture legumes do not exceed 100 ton (Ferguson, 1990). In Queensland, where much research has been carried out, legume seed production amounts to 250 ton/year (Walker & Weston, 1990) and the acreage sown to tropical grass-legume mixtures has not exceeded 100.000 ha in most years. In Kenya the production of grass seed alone has trebled to 400–500 ton/year in the past 15 years (Fig. 8.1) but the legume seed market has stayed at the 1970 levels of only a few ton. Although legume seed is available, lack of demand is the principal cause of stagnation (Griffiths, 1990).

This lack of interest in pasture legumes on the part of the farmer should be viewed against the rapid rate of adoption not only of cultivated grasses but also of hybrid maize and of new crop varieties in general, exactly among smallholders. Cereal legumes, by contrast, are important in cultivation and are mostly grown in mixture with maize or other food crops. Consequently, farmers are familiar both with legumes and with crop mixtures.

The focus in this chapter is on the possible reasons why pasture legumes are not in more demand. This is relevant, since some explanation is needed why the interest of the farmer is so much at variance with, if not opposite to, research policies which have emphasized legumes rather than grasses up to the present day. In particular, we will need to know whether pasture legumes, of either present or possibly future species, have any chance at all when faced with the high grazing pressures found on small farms. We will also have to examine what effects legumes may have in the case of dairying, which is the main livestock purpose for which herbage plants are cultivated in East Africa. Such an overall analysis may provide a better platform to arrive at practical solutions for the "legume problem" than when theories heralding an impending tropical pasture revolution, based on tropical legumes, are repeated once again. The role of legumes outside sown and grazed pastures is discussed elsewhere: green manures (*2.1.1*); natural grassland (*4.1*; *4.3.5*); in mixture with Elephant-grass (*12.8*); lucerne (*17*) and other leguminous fodder crops (*18.5*).

Theory and praxis are obviously in conflict. Theory centers mostly on the role which N, fixed in symbiosis with Rhizobium bacteria, could play. To the uninitiated, this vision of "free fertilizer" still has an enormous appeal. Praxis, however, is aware that extra inputs (phosphate and other fertilizers) are needed which may wipe out any such saving and is more concerned with how legumes can be grown and maintained in grazed pasture without sacrificing grass potential. Opinions differ widely. There are those who continue to subscribe to the view that no tropical pasture worthy of that name exists unless it is legume-based. Others treat legumes like other herbs, potentially useful or not.

The main assumptions on which the pasture legume theory has been based are that:

1. Legumes fix atmospheric N which is ultimately transferred to the companion grass.

2. Legumes have a high nutritive value.
3. Legumes are useful in the dry season.
4. Grasses are N-deficient.
5. N-fertilizer is uneconomic, rarely available and mostly imported at the cost of scarce foreign currency.

In some situations all 5 conditions prevail, very often only a few and sometimes none. All 5 assume that legumes are easy to be put to work. However, legume lability, defined here as lack of dynamic presence (persistence; *6.3.3*) or lack of resilience in the grazed sward, is the major obstacle to their sustained use. This weakness of adaptation applies to legumes, whether twining, scrambling or erect (Humphreys, 1980) but also, albeit for different reasons and to a lesser extent, to the clovers (*Trifolium* spp.). For the farmer there is probably no greater deterrent to the use of certain plants than to see them disappear from his field for no fault of his own.

In the following an analysis is given of how tropical pasture legumes came to be considered important and what their strengths and weaknesses are. Subsequently, some suggestions are offered whereby legumes may be retained in agricultural practice.

10.1. The clover model

The question is intriguing whether tropical pasture legumes would have drawn attention had their promoters not been familiar with clover. Would these legumes on their own strength have captured the imagination of the scientist, in environments where legumes are mostly no more conspicuous than other herbs?

Clover may have stood model, but it should be realized that clover is the exception rather than the rule among the legumes. Although Leguminosae originated in the tropics, clover is one of the highly specialized genera into which this order evolved (Norris, 1967). There are other points of difference to consider also. Clovers have been sown in Europe for centuries; they were no invention of plant exploration or plant breeding and were not borrowed from cover-, shade- or manure crops of plantation agriculture. Clovers left Europe in the wake of colonization and found their way, adventitiously or as assisted migrants, into other regions of the world where conditions for their spread were favourable and could be improved. White clover spread to New Zealand as well as to eastern and southern Australia. Here clovers assumed much more importance than in their areas of origin. Subclover (*10.9.3*) invaded southern Australia and became the cornerstone of the wheat rotations over most of the region, early in this century. Subclover is, however, insignificant agriculturally in its Mediterranean home region.

A similar history is found, within tropical legumes, in Townsville Stylo (*Shum*; *10.8.6*). It invaded Queensland at the turn of the century and became of great importance for livestock. This plant is insignificant in its area of origin in Central America (Roseveare, 1948). It was the natural spread of this legume around Townsville in Queensland which triggered off the interest of Australian science in tropical legumes as a whole. It was believed then that, given the appropriate genotype and fertilizer, other tropical legumes could be induced to emulate Townsville Stylo and even subclover.

In East Africa, however, and in most regions outside Australia and New Zealand, no examples can be found of such successful naturalizing legumes of either native or exotic origin. All the tropical legumes we deal with had to be purposefully introduced or imported and most of this was done by research institutions. This in itself was a sign on the wall. The choice of cultivated species in East Africa had already been fixed by farmers, from what they saw or obtained from elsewhere (*2.2.2*; *2.3*; *6.1.2*). With the exception of lucerne (*17*) and cereal/fodder legumes (*18.5*), pasture legumes were not among the chosen. Suttie (1970b) has shown how little the success in Kenya had been of official introduction of exotic plants in the previous 60 years. Since then the picture has not substantially changed.

Since N-fixation is the top-priority requirement for any potential legume, this will be dealt with here separately, with special reference to Desmodium (*D. uncinatum*; Silverleaf) which is still the best performing legume in East Africa. *D. intortum* (Greenleaf) can be equally or even more productive (Compère, 1961c; Tiley, 1970a; Keya, 1974) but its seed yield is far below that of *D. uncinatum* and its seedling vigour is weaker. Nearly all field scale experience is limited to *D. uncinatum*. Unless indicated otherwise, this species will be referred to below as Desmodium.

10.2. Nitrogen fixation

There is no doubt any longer that the more successful tropical legumes fix worthwhile quantities of N from the atmosphere, in symbiosis with Rhizobium (or Bradyrhizobium) bacteria. Most of them do so naturally, without special assistance. In fact, white clover can be more demanding on this point, as inoculation and lime are frequently required (Table 10.1). This specificity is related to their evolutionary specialization, already mentioned. Although many of the tropical species had a reputation of good nodulation in plantation agriculture, failures when subsequently tried under pasture conditions were equated, as in clover, with lack of inoculation and/or liming. These were the 1950s, when work on pasture legumes was resumed in East Africa after a lull of many years (*2.1*; *2.5*). Rhizobium studies were intensified (Bumpus, 1957). In the 1960s, bacteriologists were specifically assigned to pasture research teams to help develop Rhizobium strains as well as methods of lime-pelleting and of overcoming suspected mineral deficiencies (de Souza, 1969; Kitale Ann Rep 1968; 1969).

It was rapidly confirmed, however, that nodulation was not a handicap and that most soils contained promiscuous Rhizobium bacteria of the cowpea type which can inoculate many species. Sowing into existing swards proved possible, also without great difficulty (*4.3.5*).

Lime-pelleting, borrowed from subclover inoculation (*10.9.3*) and at one time recommended for all legumes, may assist inoculum survival should seed come into contact with superphosphate but has a negative effect on most tropical legumes. The cowpea-type of Rhizobium is alkali-producing, adapted to acid conditions and intolerant of the alkalinity of the lime coating of the pellet, in contrast with acid-producing Rhizobium of clover and lucerne (Norris, 1967). Most tropical legumes are tolerant of acid soils and can extract Ca from soils low in this element. Applications of lime are rarely needed by the plant itself, especially if phosphate dressings contain Ca. An exception should be made for *Leucaena leucocephala* (*18.5.6*), a specialized species of a distinct family (Mimosaceae), that tends to occur naturally on coral benches, volcanic soils of high base status or on limestone areas, and responds to application of lime when planted on acid soils. Leucaena should be inoculated with an alkali-producing Rhizobium (Table 10.1). Similarly, Kenya white clover and other clovers (*10.9*) inhabiting the East African

Table 10.1. A guide to inoculum and lime requirements of legumes used in tropical pastures (after Norris, 1967).

Inoculum	Lime response	Species	Common name
Cowpea	No	*Calopogonium mucunoides*	Calopo
Cowpea	No	*Dolichos* spp.	Dolichos
Cowpea	?	*Neonotonia wightii*	Glycine
Cowpea	No	*Lablab purpureus*	Lablab
Cowpea	No	*Macroptilium atropurpureum*	Siratro
Cowpea	No	*Pueraria phaseoloides*	Kudzu
Cowpea	No	*Stylosanthes guianensis*	Stylo
Cowpea	No	*Stylosanthes humilis*	Townsville Stylo
Cowpea	No	*Vigna* spp.	various
Specific	No	*Centrosema pubescens*	Centro
Specific (?*)	No	*Desmodium* spp.	Desmodium
Specific	No	*Lotononis bainsii*	Lotononis
Specific	Yes	*Leucaena leucocephala*	Leucaena
Specific	Yes	*Medicago sativa*	Lucerne
Specific	Yes	*Trifolium repens*	White clover
Specific (?*)	Yes	*Trifolium semipilosum*	Kenya white clover
Clover	Yes	*Trifolium subterraneum*	Subclover

?*Specificity doubtful in East Africa.

highlands show specificity at times, although inoculation studies have been far from conclusive and natural nodulation is commonly found (Strange, 1961a; Jones & Evans, 1961; Kitale Ann Rep 1968).

The cowpea type of Rhizobium and its adaptation to acid soils are primitive characteristics that have survived in the majority of the tropical legumes (family Papilionaceae). *Trifolium* and *Medicago* spp. (*17*), with a lime requirement and an acid-sensitive Rhizobium, have developed from the primitive type. Not all temperate legumes, however, behave in this fashion. Lupins (*18.5.3*) grow on acid soils and are intolerant of lime. *Lotononis bainesii* (*10.8.10*), a species from southern Africa, requires no lime, but is Rhizobium-specific.

In Kenya, in most of the tropical legumes tested on field scale, including Desmodium, no effect was found of specific inoculants; field establishment and natural nodulation were good (Kitale Ann Rep 1969). It is worth noting that although more optimistic results were reported with pasture legumes in Uganda than in Kenya, little mention was made there of the need of inoculation of any of the species used (Horrell & Tiley, 1970; Wendt, 1971; Wendt *et al.*, 1970). It has been the achievement of Norris (1967) to have saved us from unnecessary worry and expense.

10.2.1. Amounts of nitrogen fixation and transfer

Techniques of direct measurement of N-fixation do exist, *e.g.* lysimeters and acetylene reduction, but for practical reasons indirect measurements are mostly employed by agronomists. In mixed legume-grass swards, fixed N in tops can be estimated from the total N-yield of legume and grass tops in the sward minus the uptake of mineral N from sources other than fixation by the legume (*5.3.1*). This mineral N can be estimated from the N-yield of the grass growing alone in adjacent plots.

A modification of this method is to compare the N-yields of legume-grass mixtures with grass swards receiving a range of rates of N-fertilizer. When the N-yield of a legume-grass combination is the same as that of grass receiving 100 N (kg/ha), it is commonly concluded that the legume has "fixed" 100 N. A more correct interpretation, however, would be to conclude that the N-yield of the mixture is equivalent to that of grass fertilized with 100 N. At Kitale, Suttie (1968) and Thairu (1972) estimated that the effect of Desmodium on the total DM-yield of a Desmodium-Setaria mixture under a regime of cutting and removal was equivalent to Setaria receiving rates of about 125 N (Fig. 10.3). However, the amounts of apparent N-transfer, obtained by subtracting the N-yield of the Setaria-only sward from the N-yield of Setaria in the mixture, did not exceed 16 N in any of the 3 post-establishment years of Thairu's experiment. Obviously, for a significant contribution legume tissue must be returned either by grazing or by decay.

Very high levels of N-fixation have been reported by some authors, based on estimates of gross N-yields of undisturbed pure legume stands, including roots and stubble. More realistic data, however, also take into account the amount of N taken up from other sources. Based on gross N-yields of tropical legumes and taking into account the amount of total available mineral N (from soil-N, animal excreta and plant decay) Henzell (1968) calculated the amount of N available to grasses (Table 10.2) and concluded that grasses in legume-based pastures may show symptoms of N-deficiency.

Table 10.2. Estimate of nitrogen (kg/ha/year) made available by tropical legumes to companion grasses (after Henzell, 1968).

Average legume growth	55–115 N
Good legume growth	175 N

The amount of N fixed is linearly related to the amount of legume present in the sward (Crowder & Chheda, 1982). The proportion of Desmodium in the sward, however, frequently declines due to cutting (Fig. 6.1) and due to grazing. It is unlikely that a legume struggling to survive in a sward is capable to fix N, let alone to make it benefit the grass.

Yield increases of grass-legume mixtures over grass alone are usually made up of legume only, while the grass component may be hardly affected (Fig. 12.6) or even reduced (Tables 10.3 & 12.10), especially in the 1st year (Strange, 1961a; Smith, 1962; Horrell & Court, 1965; Suttie, 1968; Thomas, 1976b). Thus, the amounts of N transferred may be negligible. Grass rarely exhibits the green appearance that is expected in the company of legumes (Smith, 1962; Horrell, 1964). At Kitale, Elephantgrass-Desmodium mixtures were observed to show a visible N-transfer effect, as compared with adjacent Elephantgrass grown on its own, only when left undisturbed for some 6 months. Such transfer effects were most evident during and just after the dry season (*12.8.3*).

It should be emphasized that if 100 kg N were effectively contributed, this should be seen as a considerable achievement since it implies that about one-third or more of the sward is composed of legume (Figs. 10.3, 12.5 & 12.6). This could mean a financial saving on (N)-fertilizer, if it were not offset by extra P- and other fertilizer required by the legume. For many years it has been emphasized that the rate of N-fixation is linearly related to the rate of P-fertilizer (Whiteman, 1980). In Queensland maintenance rates of up to 125 kg/ha P_2O_5 were recommended (Jones & Jones, 1971) but according to Hutton (1970) responses to higher levels were commonly found. In Kenya, recommended rates were 40–60 kg/ha/year P_2O_5 (Odhiambo, 1974; Keya, 1975), but responses to much higher rates were found by Suttie (1970a). Kulich (1988) mentioned 100 kg P_2O_5 as well as 120 kg K_2O in Zambia. Double superphosphate applied at 150 kg/ha P_2O_5 doubled Desmodium yield at Serere, which was attributed to better nodulation (Wendt, 1971).

How much P to apply depends on soil type, environment but also on the legume yield level required. High requirements of P were observed by Olsen & Moe (1971) at Makerere (U). Nodulation was significantly improved with P-application. The authors concluded that "there was no advantage in applying more than 224 kg/ha P_2O_5 (1 ton of superphosphate) in the seedbed". Obviously, such rates are prohibitive and defeat the purpose when the N:P price ratios become unfavourable for P (Fig. 7.3). Rates of 50–100 kg/ha/year P_2O_5 in the form of single superphosphate were applied by Stobbs (1969f, g) which had a marked effect (Fig. 10.1).

P-fertilizer is often enriched with micro-elements such as Mo (molybdenum), which is essential to the N-fixation process. Single superphosphate contains 13% S (sulphur) and 21–31% CaO (*7.1.8*). Responses of legumes to S have been reported from Serere (Horrell & Court, 1965; Wendt, 1970;

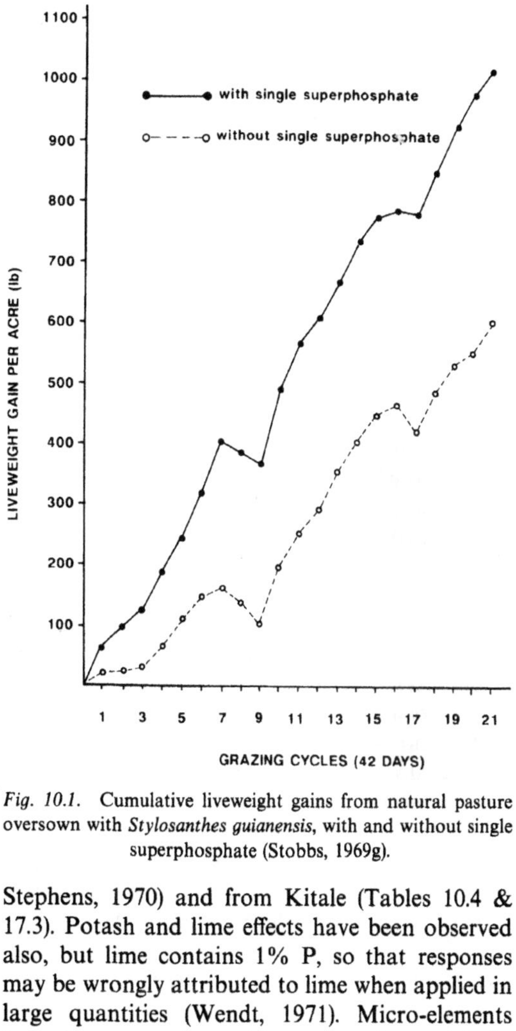

Fig. 10.1. Cumulative liveweight gains from natural pasture oversown with *Stylosanthes guianensis*, with and without single superphosphate (Stobbs, 1969g).

Stephens, 1970) and from Kitale (Tables 10.4 & 17.3). Potash and lime effects have been observed also, but lime contains 1% P, so that responses may be wrongly attributed to lime when applied in large quantities (Wendt, 1971). Micro-elements have been a major subject of study at Kitale, especially in lucerne (*17.3*), but the need for Ca and S has also been generally questioned on the basis of trials with groundnuts (Le Mare, 1958; Foster, 1972). In any case, single superphosphate seems to contain sufficient S and Ca to correct suspected shortages. For other deficiencies which may impair legume-Rhizobium symbiosis, such as Co (cobalt), see Whiteman (1980).

The combination of legumes and superphosphate has been seen for a long time as the key to pasture development and as the most economic alternative to N-fertilizer. The rising P-cost is now focusing attention on lower P-rates and direct P-supplementation of cattle. However, widespread legume-based pasture degradation occurred in Queensland when P-fertilizer was withheld. Legumes, however, produce far lower yields at high P-rates than do most of the grasses fertilized with N (Jones, 1990). It has been proposed to develop legumes with lower P-requirements, but it should be obvious that these are generally less productive.

Widespread soil acidity under P-, S- and Mo-fertilized legume-based pastures has been observed in Australia (Hochman & Helyar, 1989); see also *5.2.1*. Long-term use of legumes and such fertilizers has caused acidification resulting in poor establishment and persistence. The use of lime to overcome the acidity problem is hampered by high cost.

In contrast with legumes, grasses in East Africa can go without P-topdressing (*7.2.4.3*), while responding so markedly to N-fertilizer. Consequently, the more emphasis is placed on the need of P- and other fertilizer to prevent legume failures, the more acceptance should be taken of the fact that such extra costs wipe out the N-saving effect of the legume. When legumes designed to seek an alternative to N-fertilizer need equivalent or more costly amounts of other nutrients, for their maintenance alone, the fertilizer economy may on balance prove negative and defeat the purpose. Enriched sources of P-fertilizer are frequently harder to obtain than ordinary N-fertilizer. Increased freight costs have led to an increased use of high analysis P-fertilizers which contain little S.

10.2.2. The nutritive value of legumes

N-fixation cannot be divorced from its relationship with % protein which has been the conventional yardstick of nutritive value of legumes. Their N-economy as a whole and % crude protein in particular were the main criteria for which legumes were valued. Traditionally, nutritive value of grasses was viewed as low because they were regarded as stemmy (*6.3.4*) and of legumes as high because of their high % crude protein.

Legumes have a higher % crude protein than grasses, especially at advanced stages of growth. At such stages the decline with age in % protein is less pronounced than in grasses. This is often advanced as important for grazing systems in which the stocking rates are kept low in the rainy season in order to see the animals through the following dry season. Low grazing pressure in the wet season, accentuated by long grazing intervals, favours the legume content in the sward, especially if the grass is grazed in preference to the legume. Standing hay is then valued highly (*4.3.4*; *19.1*). The inevitable loss of nutritive value of older grass is partly rectified by the legume. Digestibility (also of protein), animal intake and utilization of the legume are lower at such late stages of growth but higher than in the grass companion. The benefit of legumes to foddergrasses is, therefore, largely a matter of compensating for the lower nutritive value in grasses defoliated infrequently (Fig. 10.6).

Digestibility over a wide range of tropical grasses and legumes at Makerere (U) was studied by Reid *et al.* (1973) who found that tropical legumes had a lower initial *in-vitro* DM digestibility than cultivated tropical grasses but the rate of decline was lower (Table 6.4; Fig. 10.2). Grasses, younger than 60–70 days of regrowth, were higher in digestibility than the legumes under investigation. The situation was reversed at 100 days. It is notable also that, although valuable in the dry season, many legumes such as Desmodium lose their dry leaf due to shedding or to rotting as the result of dew and the occasional shower of rain.

Voluntary intake is an additional, important criterion of nutritive value. At Kitale in 1969, 3- and 6-weeks old regrowth of Rhodesgrass and Desmodium were fed to young steers indoors over

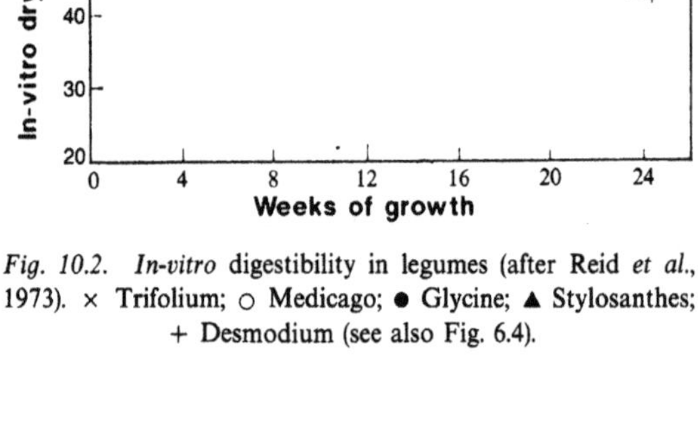

Fig. 10.2. *In-vitro* digestibility in legumes (after Reid *et al.*, 1973). × Trifolium; ○ Medicago; ● Glycine; ▲ Stylosanthes; + Desmodium (see also Fig. 6.4).

100 days at 10% refusal (G.H. Edwards, unpublished). Both intake and liveweight gain were very similar in both the grass and legume treatments. These results were considered anomalous at the time and were not published. It is now commonly known, however, that the *in-vitro* digestibility of *Desmodium* spp. is low, even of young leaves, because of a high content of tannins which protect the protein against microbial break-down. The nutritive value of proteins in forages is presently subject of reappraisal in many parts of the world. The level of digestible protein is being considered against how much of it can be converted efficiently and how much of it is persistent protein.

Legumes are valued also for their contents of minerals: Ca, and possibly S and P (Whiteman, 1980). Bloat and goitrogens, encountered with clovers, are rarely a problem in tropical legumes, but Leucaena contains a toxin, mimosine, which affects animal health and production adversely (*18.5.6*). Toxic compounds are very common in legumes (Bogdan, 1977), but are obviously rare in the cultivated forms. More serious is the depressing effect Desmodium and Siratro have, when fed to dairy cows, on milk yield (Stobbs, 1971). Less milk was obtained from cows grazing pure stands of either *Desmodium intortum* or Siratro (*10.8.4*) than from N-fertilized Pangolagrass. This was attributed to reduced intake of digestible energy. Grazing time on trailing legumes was shown to be excessively long due to low sward density. When milk yield is depressed due to a low intake of digestible energy, % butter fat increases and % protein decreases. This is, of course, of particular relevance to East African conditions, as most of the cultivated pastures are grown specifically for dairy purposes. There is also an increasing awareness that livestock faced with shortage of herbage suffers more from insufficient intake of energy than of protein so that protein may then be largely wasted.

It may be concluded that legumes have value in over-mature grass swards but to young grass under high grazing pressure their contribution to the diet is less obvious.

10.3. Husbandry of Desmodium-based pastures

In actual farming practice it has been established beyond all doubt that many of the well-known tropical legumes can grow well under East African conditions.

Desmodium has been tried extensively in Ecoclimatic Zones II and III in Kenya, Uganda and Zaïre between the altitudes of 1000 and 2000 m, *i.e.* in the principal agricultural regions. Towards the lower limit of this altitude range Siratro, Centro and Stylo (*10.8*) may perform better but Desmodium has the advantage of having a wider adaptation over both zones (for instance at both Kitale and Kampala) and of having been tested on a commercial farm-scale under grazing. It has also remained free of diseases, unlike Siratro and Stylo (*10.8*). Because of the evolutionary and ecological similarities of most legumes, much of the following applies equally to twining, scrambling or erect tropical species.

Many tropical legumes are slow to establish, especially Glycine and Centro. By contrast, Desmodium seedlings are vigorous and can establish generally without failure, also under cool conditions. Requirements of P, S, Ca and inoculation have been discussed (*10.2*). Seed rates of 2–3 kg are adequate. Chemical weed control is possible in Desmodium; auxin-type weed killers such as 2,4-D are tolerated by young Desmodium seedlings, and also after a grazing or mowing, provided the shoots are still very young. Elongating shoots are vulnerable. Very good results have been obtained by adding 1.6 kg Simazine to 0.7 kg 2,4-D amine (Veenstra & Boonman, 1974; Boonman, 1979a).

The seed is mostly broadcast, but sowing in 75–100 cm rows is also effective as the semi-decumbent plant habit of Desmodium makes the stems (stolons) root between the rows. Conditions of shade favour establishment (Compère, 1961c). Undersowing to maize is a very effective method. Field experience has shown that by the end of the 1st year Desmodium is established more firmly when undersown than when sown direct (Thomas, 1975). Undersowing to maize proved so efficient that a large seed production acreage of Desmodium was established in 1971 at the Kitale Station by mixing Desmodium seed with N-fertilizer at the time of topdressing. The method was unorthodox but cheap and simple. No extra P was added, as this had already been applied at the time of maize sowing. Seed and fertilizer should, however, not be mixed until the last moment, to prevent scorching (*7.1.9*).

The same benefit of shade is gained when Desmodium is grown with Elephantgrass or with other fodder-type grasses. Desmodium's vine-like growth habit allows it to elevate its young leaves above those of its companions and to overtop tall grass. Fodder grasses are best grazed or cut infrequently; they require long resting periods and so does Desmodium (*12.8*). These are the main reasons why Desmodium combines better under grazing with fodder grasses than with short, sown grasses (Fig. 10.6).

In mixture with grasses, Desmodium persists generally well under cutting regimes (Olsen & Moe, 1971; Wendt, 1971; Thairu, 1972). At Kitale, even with some 5 cuts per year Desmodium gained rather than lost strength to companion grasses over the years (Fig. 10.3; Thairu, 1972). High vigour can be maintained over many years in seed

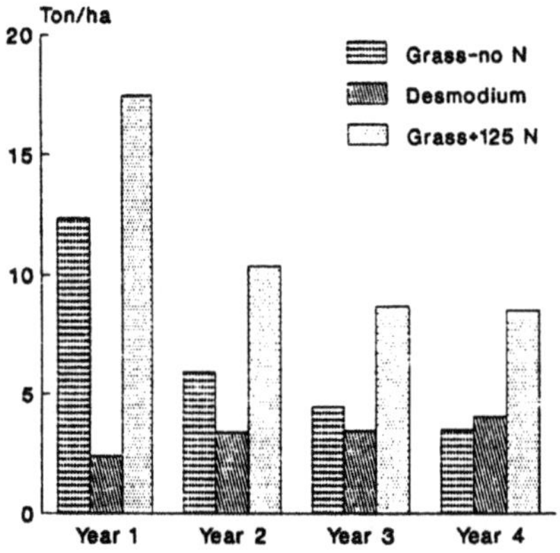

Fig. 10.3. The performance of grass and legume over time (after Thairu, 1972).

production fields also, except when overrun by African Couchgrass (*Daby*). By applying a grass-weed killer such as Dalapon at the beginning of the rainy season, when shoots are young, Desmodium can regain dominance (Boonman, 1979a; see also *17.4*). When protected from grazing temporarily (spelling), Desmodium's ability to recover is remarkable and is proportional to the length of the resting period. Under Kitale conditions it was observed on many occasions that a resting period of 4–6 months was enough to tip the balance in favour of Desmodium (*12.8.2*). Two practical examples stand out:

1. Fallen branches of Acacia thorn in a frequently grazed pasture show an increase of Desmodium underneath. Some effect of shade cannot be ignored, but the main benefit is thought to accrue from the protection from grazing.
2. To reduce seed production costs, Desmodium-based pastures were locked up to be harvested for seed at the Kitale Station in the mid 1970s. These pastures, with only little Desmodium left at the start, were rested during the rainy season. Legume dominance was achieved in 4–6 months. An additional advantage of this method was that the presence of a sown grass precluded the ingression of African Couchgrass that normally infests monocultures and seed fields of Desmodium.

These two examples are perhaps no practical propositions for general application but they show that a long resting period is effective in restoring the proportion of Desmodium in mixture with grasses. How often such resting periods need to be repeated is an open question, but a practical solution lies in having a spelling followed by a hay crop, whenever a rest is needed.

Desmodium has a remarkable drought tolerance. In Zone III, it survives long dry seasons vegetatively, unlike clovers. Older leaves, however, shed to the ground (*10.2.2*). Under grazing at Kitale, young Desmodium remains green throughout the dry season, but companion grasses suffer more from drought than when grown on their own (*12.8.1*).

In crop-pasture rotations, legume establishment is commonly regarded as too slow, while aggravating the weed burden, but where time is not a limiting factor, it can be concluded that there is little problem with getting Desmodium established properly, also without inoculation. With adequate application of P and possibly S in the seedbed a good stand of Desmodium can be obtained. Its proportion in mixture with grasses is strengthened by long resting periods.

10.3.1. Disappearance under rational grazing

By contrast, when such mixtures are grazed at the high frequency and intensity dictated by companion grasses grown on their own and rationally grazed in the rainy season, Desmodium disappears rapidly. Rational husbandry was defined as the intensive system aimed at leafy, dense swards without sacrificing grass potential (*6.3.4*). When a farmer decides to include a legume in his grazed pasture he rightly expects to maintain or to increase his stocking rate. Farmers know what stocking rates can be achieved with grasses when grown on their own.

At Kitale, soon after their successful passing of the nursery tests it became apparent that most tropical legumes lacked persistence and could not stand up to current grazing regimes, however moderate. It had been the policy of Bogdan to evaluate under farm-routine grazing at the Kitale Station before release was considered (*6.3.1*). This explains why Bogdan (1965a) was hesitant to recommend legumes, whatever the species (*10.6*). Yet, Desmodium had been known in the region since the 1940s (Compère, 1961c).

This release policy lapsed with time and from 1965 onwards Desmodium was sown in all new paddocks of the Kitale Station together with mostly late-heading grasses and less vigorous grasses, *e.g.* Nandi II Setaria and Pokot Rhodes (*6.6*; *6.7*; *9*). Elephantgrass was also intersown with Desmodium. The whole area planted amounted to about 300 ha, sown over 3 years. Most of the pastures were sown after maize as the previous crop. Desmodium established well and very impressive grass-Desmodium swards were obtained. Field days were held in 1966 and 1967 to demonstrate the benefit of Desmodium-based pastures to farmers. Demand for seed rose rapidly. The seed multiplication area under Desmodium increased from 20 to 100 ha in 1968, whereas seed production of both Kenya white clover and Louisiana white clover was phased out in favour of Desmodium. The Kenya Seed Company also began contracting out seed multiplication of Desmodium.

Early suspicion in 1967 grew into alarm in 1968 when it became clear that Desmodium was disappearing gradually but steadily from the grazed swards. As the stocking policy on the farm had not changed from the standard rate of about 1.5 LU/ha, and as the grass companion was not showing any sign of overgrazing, it was obvious that Desmodium had an unexpectedly low tolerance to grazing. The same was observed on a large number of farms, which had always been keen to try new varieties of the Kitale Station following the successful release of so many grass and cereal varieties. The news spread rapidly and local demand for seed dropped instantly. Donor-assisted livestock projects initiated at that time at Naivasha and Lanet had a strong fodder component but, apart from lucerne, legumes were not among them. Seed multiplication was, subsequently, confined again to the Kitale Station and changed to hand-harvesting from closed pastures in order to reduce costs.

The period of the mid 1960s saw the departure from Kenya of many research officers of long standing (*e.g.* Bogdan and Dougall) and the arrival of a new generation of officers, trained overseas in the principles of what was then believed to be a legume-based revolution of tropical pastures.

The nursery- and cutting trials at Kitale and subsidiary stations showed, however, no sign of legumes disappearing. In Elephantgrass, that was rested in the rainy season to provide a fodderbank for the dry season, as well as in seed production fields Desmodium continued to perform reliably. Unfortunately, observations under grazing on farm-scale were not continued due to the fact that the farm-routine sowing of Desmodium had been suspended. Had there been early research recognition of the inability of legumes to tolerate current grazing regimes, this would have helped to identify the problems involved, and solutions as suggested below (*10.7*) would have been tested and put into practice much earlier. Unfortunately, not the lack of grazing tolerance was advanced as the principal reason of legumes disappearing from the grazed swards, but rather: poor husbandry ("overgrazing"); poor establishment; poor inoculation; mineral deficiencies and diseases and pests. However, as already indicated, none of these factors was at a minimum as far as Desmodium was concerned.

It has been suggested that the East African highlands are too cool for tropical legumes (Wheeler & Jones, 1977). Obviously, above 2500 m tropical legumes are out of the question, but so are tropical grasses. In the present discussion we are concerned with the 1000–2000 m altitude belts where tropical grasses and legumes can both grow very well. Legumes such as *Desmodium uncinatum* are more productive at 1000 m altitude than at 2000 m, but so are Nandi Setaria and Rhodesgrass. In fact, tropical legumes have generally lower optimum temperatures than grasses (Fig. 10.4; Whiteman, 1980), although this should not

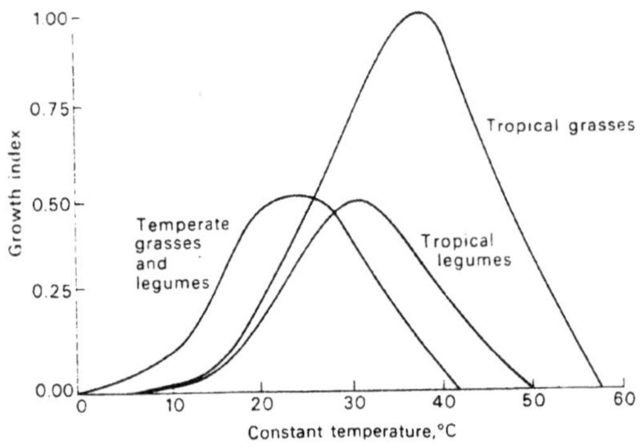

Fig. 10.4. Temperature responses of grasses and legumes (Whiteman, 1980).

necessarily apply to all combinations of species of each of the 2 groups, neither to low temperature regimes. If temperatures in general or night temperatures in particular are too low for Desmodium, then this should affect the legume in an equally negative fashion under cutting and in seed production fields. Other legumes such as Glycine, Siratro, Centro and Stylo may be more adapted to warmer areas. However, at lower altitudes their nursery performance is not necessarily better (*e.g.* Naveh & Anderson, 1967) and lability under close grazing is an equally serious problem, *e.g.* in Malawi (Thomas, 1976b).

Obviously, rapid establishment, good health and vigorous growth do not predispose a legume to tolerate rational grazing. As high animal production per unit area cannot be sustained unless stocking rates are at least as high as on grass alone, lability under grazing is a major drawback to the use of tropical legumes, whether of twining, scrambling or erect growth habit. Lability of legumes under grazing is a general phenomenon, also in clovers, but the plant attributes underlying persistence may differ (*10.4*). In fact, Kenya white clover is frequently found to take over from Desmodium or from Louisiana white clover. Slow recovery, a limited period over which a high leaf area is maintained and reduced stolon production are likely causative factors in the poor persistence of trailing, tropical legumes.

The disappearance under grazing of already well-established Desmodium should make us realize that solutions should be sought in adapting available husbandry techniques. These should be designed to strike a happy medium between productivity of the present and that of the following seasons. Rational grazing regimes are a varying continuum of grazing pressures in rhythm with the demand and can, therefore, not prevent that overgrazing occurs at some time. Sown species must be expected to have some tolerance to abuse.

10.4. Approaches to the problem of legume persistence

The inability of legumes to tolerate rational grazing policies is not confined to East Africa alone and is apparently independent of climate or latitude (Marten *et al.*, 1989).

Persistence was defined as the capacity of a species to maintain a dynamic presence in the sward (*6.3.3*). Plant attributes underlying persistence are varied and complex. For perennial legumes that exist as distinct plants (*e.g.* lucerne) persistence can be clearly defined in terms of plant density, since self-regenerating seedlings seldom survive. Defining pathways of persistence for legumes that propagate by stolons and/or seed is less obvious. Persistence of Desmodium relies to a very great extent on adventitious rooting from stolons (Jones & Carter, 1989). Where regular disturbance is experienced (*e.g.* drought) adapted species survive by developing a strong habit of reseeding, for instance in clovers, but differences between years and sites may still have a great effect. Persistence is usually less of a problem in woody legumes, *e.g.* Leucaena, once the upper canopy gets beyond the reach of livestock (*18.5.6*).

Some scenarios with which tropical, twining legumes could be maintained in the sward were suggested by C.R. Roberts, farmer and representative of the Australian seed industry (private communication, 1977):

1. Use of less aggressive grasses.
2. Lower seed rates of grasses.
3. No mowing, no slashing and no burning.
4. No management decisions based on the condition of the cattle, nor of grass, but on the amount of legume only.
5. A stubble height of 30–60 cm at all times.
6. Variable, yet continuous stocking.
7. “Sacrifice paddocks”, fertilized with N and kept in the ratio of 1:3 as stand-by in periods when the legume is endangered.

On Roberts' property legumes were indeed well maintained in the paddocks (personal observation, 1977). The system appeared simple, but illustrated at the same time to what kind of extremes one had to resort. When scenarios must be designed to weaken the grass in order to strengthen the legume, when the net gain is so uncertain, one may lose out on both eventually.

Kitale experience has shown that Roberts' recommendations [1] and [2] have little effect. Legumes declined in Rhodes-, Setaria- as well as in allegedly weaker Hyparrhenia grassland (*4.3.5*) but were, by contrast, easily maintained in vigorous Elephantgrass growing very tall (Fig. 10.6; *12.8*). In general, seed rate effects are important in the year of sowing only and cede to other dominant effects after the 1st grazing except when grasses tend to grow in isolated clumps. Experience at Serere has shown that when robust grasses such as Guineagrass are sown at low rates they do not form a continuous sward and may accomodate more legume to occupy the open spaces. An open sward, however, increases the risk of soil wash between the clumps and because Guineagrass itself is also vulnerable to grazing (*4.4*; *6.8*) weeds may invade.

The proposition to breed slower growing grasses to benefit the companion legume had been made by Hutton (1970) but this approach did not remain unchallenged (Boonman, 1973b). So long as it is uncertain what one is to gain with legumes in order to compensate for what one is sure to lose by inferior grasses, such strategies cannot be recommended. Not only was there no evidence at Kitale that slower growing Nandi II and Nandi III were more compatible with legumes than the original Nandi Setaria, but the use of slower growing grasses surely meant a sacrifice in grass (and seed) yield (*6.7*). The net photosynthetic rate and efficiency of light utilization of tropical grasses with their C-4 pathway (Hatch *et al.*, 1967) are greater than those of tropical legumes over a wide range of temperatures. These characteristics give grasses a considerable ecological advantage over legumes.

The relative merits of either continuous [6] or rotational grazing have been discussed (*4.3.1.1*; *7.2.1*). In East Africa continuous grazing is commonly equated with overgrazing, but in Australia it is often presented as a strategy conducive to variable and, more important still, light stocking to permit maximum forage allowance and, if anything, legume survival (Humphreys, 1991).

Experience at Serere in the pasture-evaluation systems developed by T.H. Stobbs showed that Roberts' proposals [3], [4] and [5] help to encourage the legume but the grass is inevitably penalized. Obviously, to leave the sward undisturbed is conducive to legume survival and this explains why mowing (for conservation) is often advised against (Stobbs, 1975). In the Serere grazing trials, grasses were allowed to reach highly advanced maturity. Long grazing intervals and elevated stubble height help to maintain the legume in the sward. In swards maintained at a height of 30 cm or higher it is difficult for new grass tillers to emerge because of shading. At Serere, in paddock cross-sections cut to ground level for yield measurements prior to each new grazing cycle, grass regrowth was more vigorous (personal observation, 1967).

Consequently, the main objection to the system proposed by Roberts is that the role of the grass companion is reduced to that of no more than a chaperon: if there were no sown grasses to accompany the legume, weed grasses would have taken their place.

It is not clear how much of the grazing pressure must be diverted to the N-fertilized "sacrifice" paddocks [7]. When these are laid out in a 1:3 ratio as far as area is concerned, they may well represent 1:2 or 1:1 in grazing-on-offer. It may in such cases be more economic to have the same total of N-fertilizer shared out to all paddocks.

The system, proposed by Roberts, may be applicable to large properties where livestock fattening or breeding is the main objective and where stocking rates are designed mainly to carry stock through the dry season so that grass inevitably runs to stem in the rainy season. However, for dairying in the context of small mixed-farms the applicability of this strategy is very limited. Besides, it has proved an illusion to assume that farmers are willing to moderate their stocking policies.

Periodic severe grazing appears to be more detrimental to legumes than cutting. Prostrate stems may escape under cutting but not under grazing, thereby reducing the regrowth potential. Legumes are generally also less tolerant to treading than grasses. Unlike clovers, most tropical legumes have small numbers of leaves and of growing points at ground level and display slow regrowth of these points. Hence the need for long resting intervals. Experience to date suggests that good legume husbandry must allow build up during the growing season by lenient grazing throughout or by occasional rests followed by a hay cut (*10.7.1*).

Very few tropical legumes excel in palatability which can, therefore, hardly suggest that legumes are "grazed out" in preference to grasses.

10.5. Other causes of legume disappearance

Numerous researchers have demonstrated that root-knot nematodes (*e.g. Meloidogyne* spp.) may be limiting legume production and persistence. Specific reference to *Desmodium uncinatum* and *D. intortum* has been given by Lenné & Stanton (1990). Galls of unspecified nematode species were found also at Kitale (Kitale Ann. Rep 1969). If nematodes were a factor in the persistence of Desmodium in grazed pasture, it would be difficult to explain the apparent health of Desmodium when not grazed, unless nematode damage is aggravated by grazing stress. This is unlikely, but the matter requires investigation because Meloidogyne affects many crops used in pasture-based rotations and extra host plants will not be welcome. In fact, much effort is made in Zimbabwe, Malawi and Zambia to reduce the infestation of Meloidogyne by sowing certain grasses in tobacco rotations (*6.6.4*; *6.16*). Lists of fungal and bacterial pathogens affecting various legumes in the region have been presented by van Rensburg (1967; 1969d) and Thomas (1973).

Effects of allelopathy (phytotoxicity) in grass-Desmodium associations should not be ruled out as Young & Bartholomew (1981) have demonstrated for *D. intortum.* Grass root residues were reported to have a strong negative effect on legume growth. Again, it would be hard to explain the differential effect, *i.e.* the healthy performance of grass-legume mixtures when not grazed, unless damage is accelerated by grazing stress or unless the damaging residues are more active under grazing. The issue of allelopathy possibly displayed by tropical grasses in general (Baker, 1978) would help to explain why land ploughed from pure grass is so free of germinating dicotyle weeds in contrast with land after Desmodium. In fact, this weed control effect is a major benefit of sown grasses to subsequent cropping (*5.6.1*; *7.1.10*). For the negative effect of Stylo on subsequent cropping, see *5.5* & *10.8.7*.

10.6. Legumes in early research in East Africa

The search for legumes has dominated research in East Africa off and on ever since its beginning (*2.1*) and there is no evidence to suggest that the absence of pastures legumes in farming has been due to lack of effort. East Africa has contributed greatly to the world's genetic resources not only of grasses but also of legumes (*e.g. Glycine, Lablab, Vigna, Stylosanthes fruticosa,* African clovers) and this is sufficient proof of the high level of interest and activity in this field. Many local legume species are unpalatable (Bogdan, 1949a) which stimulated the introduction of exotic species.

Most of the early work on legumes was done in Uganda, with smallscale farming only, so that claims that past research was geared to largescale farming only are unjustified. To mention an example: Leucaena was introduced into Uganda back in 1915. Its performance and that of other legumes used as green-manure crops had been disappointing (*2.1.1*; *5.2.1*; *5.3.3*) and interest in legumes in general subsided as a result. Summing up the results of research done in Uganda at the Colloquium on Grassland Research in Nairobi in 1949, Kerkham (1950) did not mention legumes at all.

At the same Colloquium, Edwards (1950a) concluded after 20 years of work that: "The search for legumes has so far produced disappointing results. Although several species have given promise at first, they have usually not succeeded in competition with sown grasses. Apart from lucerne, no other exotic pasture legume has proved successful". Earlier (1936) Edwards had said that the ultimate goal should be sown grasses and legumes. The same research objective had been expressed by Hosking & Stephens (1941) for Uganda.

Bogdan who was Edwards' colleague and successor spent a lot of his exploration and testing work on local and imported legumes, but was no less hesitant to recommend any in particular. In their joint publication (Edwards & Bogdan, 1951) little is pictured of a positive perspective of either natural or cultivated legumes. Numerous species had been recorded in natural grasslands, but they did not appear to contribute significantly to grazing. In an attempt of drawing up a list of cultivated herbage varieties of Kenya, Bogdan (1965a) referred, apart from the clovers, only to *Glycine* and *Desmodium uncinatum*. In the same year, Bogdan (1965c) reported that some farmers had come to regard Kenya white clover as a weed and sprayed their pastures against it with herbicides such as 2,4-D. In the introduction to his final work describing the most important species, Bogdan (1977) emphasized that legumes seldom increase the yield of companion grasses, but often decrease them. In the book "Rangeland Management and Ecology of East Africa" (Pratt & Gwynne, 1977) which is fully devoted to the improvement of natural grassland and which was pieced together by some 25 authors, very little is said about legumes. Describing grassland development in the Kampala area, Harker (1962a) who had collected creeping Vigna (*10.8.12*) commented that legumes grew well but contributed little. The same conclusion was reached by Mathieu (1960) in Rwanda-Burundi.

In (semi)arid areas results have been particularly poor. At Kongwa Ranch (T), following the closure of the Groundnut Scheme in 1951 and the introduction of beef cattle, considerable work was undertaken on sown pastures. Many legumes were tested but when some attractive property was observed, it was offset by another being unsatisfactory: Glycine (not persistent after 2 seasons); *Stylosanthes humilis* (failure); *Stylosanthes fruticosa* and *Clitoria ternatea* (no productivity). No legume produced any visible or measurable increase in grass yields (Wigg, 1973). A similar range of legumes, including *Dolichos* and *Indigofera* spp., was tested at Katumani but Sands *et al.* (1970) were even more adamant in their conclusion: "None of these legumes so far tried can be relied upon to persist either in pure stands or when grown in association with a grass". Although their conclusions were clear, the works of Sands and Wigg have rarely been quoted. It was concluded by French (quoted by Bumpus & Poultney, 1960) that pasture legumes in tropical grasslands rarely confer any benefit through N-fixation on the associated grasses and that they have no particular advantages (French, 1957). Legumes were not among the alternatives that Pereira and co-workers (1952–1967) considered useful for dryland farming.

It is significant that many East African workers had been so frank in stating that they had no practical success with pasture legumes. This was in no small measure due to the fact that most stations made use of the facilities available to evaluate new material under farm-routine grazing before any recommendations were issued. Moreover, farmers were usually involved at a very early stage to have new varieties tested under commercial farming conditions. This had worked particularly well with the grasses released in the 1950s and farmers were, therefore, keen to test new legumes also. This interest was, however, short-lived and the subsequent publicity had a negative impact on recommendations as a whole. As a result, all commercial production of legume seed ceased in 1970.

Serere workers were generally more optimistic (Horrell, 1963). The pasture evaluation methods study of T.H. Stobbs has been of singular importance, although his methodology has drawn less attention than the increases in liveweight gains achieved with legume-enforced pasture and with NPKS-fertilizer. A typical example of the general responses obtained is given in Fig. 10.5 (Stobbs, 1969i).

The species mixtures and treatments in Stobbs' works were often quite complex, but the increments in liveweight gain due to the inclusion of legumes averaged about 25% in most experiments, with generally low levels of performance of the control. His results with Centro (Fig. 10.5) are relevant because of the problems encountered with Stylo and Siratro (*10.8*).

It should be noted that some legumes were recommended in the past that were highly persistent under grazing when sown with grasses. The examples that come to mind are Kenya white clover at higher and Calopo at lower altitudes (*10.8; 10.9*). As mentioned already, Kenya white clover can assume weedlike proportions (Bogdan, 1965c). Calopo seed was issued to farmers in Uganda in the early 1960s for pasture purposes. It proved unpalatable and grew so vigorously with sown grasses under grazing that it became a weed.

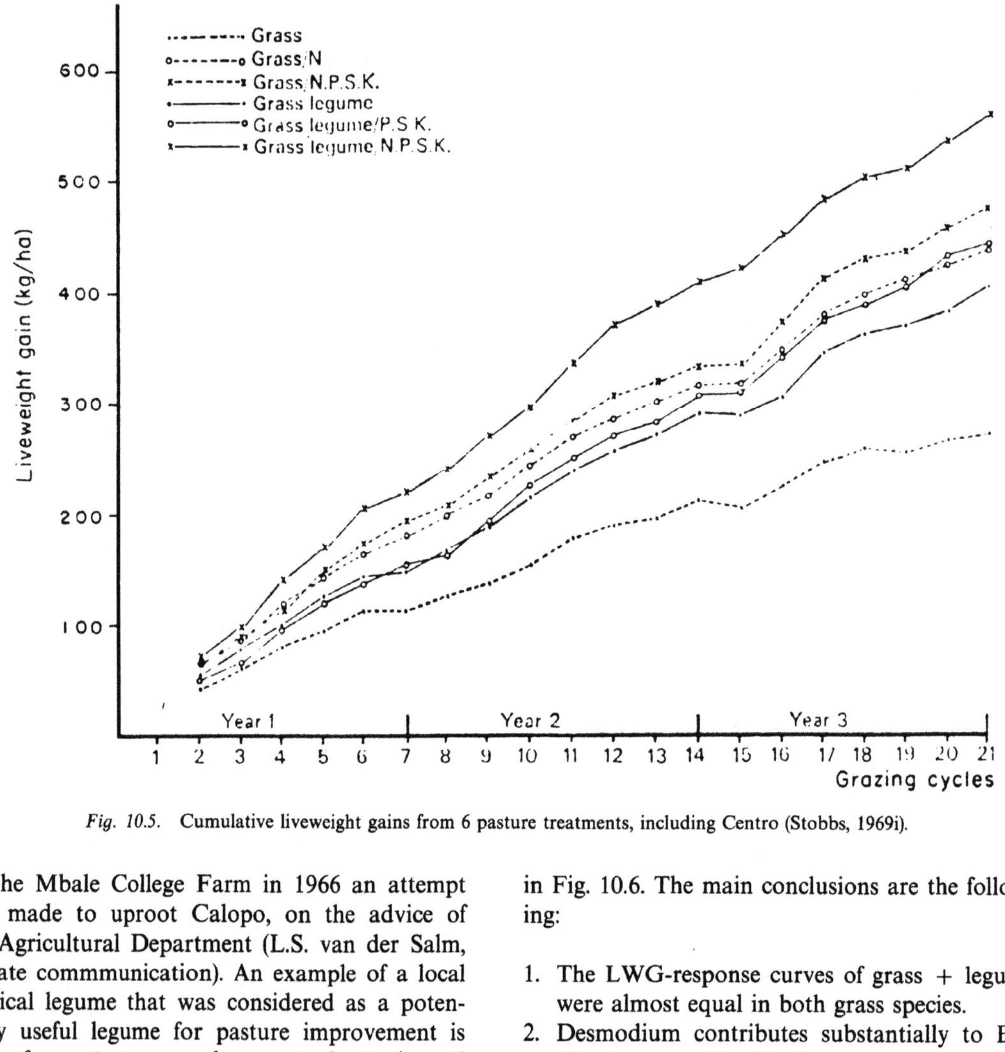

Fig. 10.5. Cumulative liveweight gains from 6 pasture treatments, including Centro (Stobbs, 1969i).

At the Mbale College Farm in 1966 an attempt was made to uproot Calopo, on the advice of the Agricultural Department (L.S. van der Salm, private commmunication). An example of a local tropical legume that was considered as a potentially useful legume for pasture improvement is *Indigofera spicata*, now known to be toxic and a weed, even of frequently mown lawns (Ivens, 1971).

In an effort to re-examine the role of Desmodium (*Dunc*) in grazed pastures, 2 replicated randomized grazing blocks were established in 1969 at Kitale to compare the effects of Desmodium and N-fertilizer on both Mbarara Rhodesgrass and French-Cameroons Elephantgrass. The plots were grazed by yearling Herefords for one week every month or every two months, respectively, over one rainy season of 8.5 months in 1971 (Odhiambo & Edwards, Kitale Ann Rep 1972). The data are presented in Fig. 10.6. The main conclusions are the following:

1. The LWG-response curves of grass + legume were almost equal in both grass species.
2. Desmodium contributes substantially to Elephantgrass, not to Rhodesgrass.
3. Desmodium contributes to Elephantgrass by compensating for the lower nutritive value found at wide harvest intervals and low stocking rates (12.5). LWG/head was highest in grass + legume in Elephantgrass, in contrast with Rhodesgrass.
4. Rhodesgrass produces much higher liveweight gains in the rainy season than Elephantgrass.
5. In Elephantgrass, much higher liveweight gains per ha are possible with 250 kg/ha N than with Desmodium, provided higher stocking rates are chosen to match the amount of grass on offer.

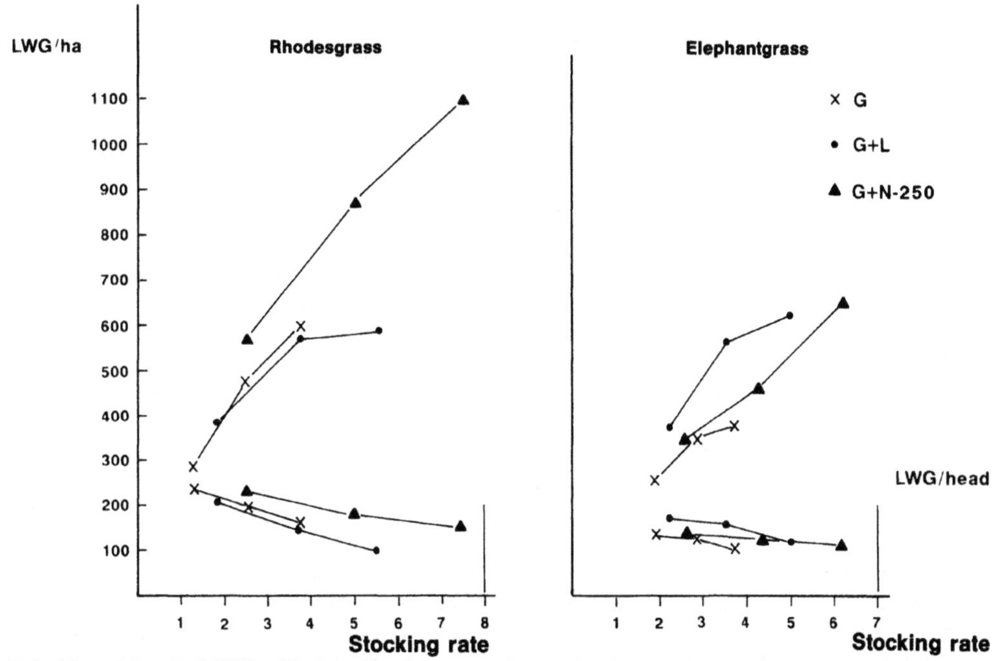

Fig. 10.6. Liveweight gain (LWG) at Kitale in the rainy season of 1971 (after Odhiambo & Edwards, Kitale Ann Rep 1971); ×—× G (Grass only), •—• G+L (Grass plus Desmodium), ▲—▲ G+N (Grass +250 kg/ha N).

6. The response to the range of stocking rates chosen shows that the potential of N-fertilizer had been under-estimated.

It should be noted that the differences observed between the responses of Rhodes- and Elephantgrass were due in part to the difference in grazing interval. Both Elephantgrass and Desmodium benefited by long resting intervals in contrast with Rhodesgrass. Although Desmodium was well established in both types of pastures at the start of the trial, Desmodium maintained itself well only in the Elephantgrass treatments at low and medium stocking rates.

10.7. How to maintain Desmodium in grazed pastures

Protein is a scarce commodity in tropical forage systems and legumes may help where no alternative N-sources are readily available. Legumes are a means to improve Elephantgrass and similar grasses requiring long rest intervals. The same applies to extensively grazed natural grasslands through over sowing (*4.3.5*), provided this can be done at little extra cost other than the seed. However, in intensively grazed sown pastures in East Africa the role of legumes is small because of their limited added value and because of their lack of persistence. When grass is young, legumes seem redundant.

In the following an attempt is made to design practical scenarios to preserve Desmodium or similar legumes in mixed-farming systems of East Africa. The main properties of Desmodium are summarized as follows (in Zones II & III):

1. Desmodium accumulates a large amount of slowly decomposing litter, important for improvement of soil structure (*5.3*).
2. Desmodium can fix considerable quantities of N when allowed to grow undisturbed for longer periods (*10.2.1*).
3. Young Desmodium retains greenness deep into the dry season.
4. Desmodium establishes easily when undersown to maize (*7.1.12.1*) and improves the nutritive value of the maize stover in the dry season.
5. Desmodium persists relatively well when cut or grazed infrequently, *e.g.* at 2 months intervals.

Husbandry option I
In the last year of cropping, Desmodium is undersown to maize, at 3 kg/ha with Rhodesgrass at 1 kg PGS, as soon as possible after maize emergence. This gives Desmodium a reliable establishment and a competitive advantage thanks to its tolerance to shade. The stover-legume-grass mixture is grazed in the dry season. Grazing is continued in the 1st half of the rainy season until the maize stover is disposed of. A hay crop is allowed to develop towards the end of the rainy season. In the next dry season grazing is withheld. In the 2nd and 3rd year the pasture is spelled from grazing, whenever the legume loses strength and drops to below 10% in the sward, and a hay crop is taken at an advanced stage. Phosphate is applied when trial strips (*7.2.4.3*) show that this is beneficial. For a comparison with the management model of pure-grass pasture, see Table 7.9.

This option is based on the principle that botanical changes are reversible and that Desmodium can restore its proportion in the grass-legume mixture following spelling if its density has not yet dropped to insignificance. A hay cut taken after 3–4 months is then effective to restore the grass-legume balance in favour of the legume and to utilize this herbage without wastage.

Husbandry option II
Desmodium frequently maintains its presence in the pasture better when topdressed with N-fertilizer. N-fertilized Desmodium gains a darker colour and grows more vigorously. This may have sounded heretic in the past, but a N-response is not at all unusual in legumes even when well nodulated (Russell, 1973). At Kitale, undersowing maize to a mixture of Desmodium seed and N-fertilizer at the usual time of maize-topdressing was found to be both practical and effective (*10.3*).

Positive effects of N on legume-based pasture were observed also in Uganda at Kawanda by Tiley (1970a) and on the Mbale College Farm (L.S. van der Salm, private communication). On this small dairy farm with semi-permanent pasture, Guineagrass and Brachiaria used to be sown together with Desmodium, Siratro and Centro or Stylo. When milk prices increased on the local market in the late 1960s, due to import restrictions on milk from Kenya, N-fertilizer was applied increasingly and silage was made also. The response by both grasses and legumes was impressive. The stocking rate was just over 2 grade cows per ha and the milk production was on average 2500 kg per lactation. The many paddocks, into which the farm was divided, lent themselves well to rotational grazing, N-topdressing and silage-making.

Husbandry option III
If these attempts at maintaining legumes in grazed pastures are not adequate, the only solution left is to grow legumes in monoculture, under infrequent cutting only, similar to lucerne (*17.5*). However, as with lucerne, such monocultures are liable to grass-weed invasion. They are better grown in association with cultivated grasses. This can be equally successful with Setaria, Rhodes- or Elephantgrass. At Makerere (U), Olsen & Moe (1971) found that Desmodium and Stylo yielded twice as much as lucerne in half as many cuts, in monocultures harvested at the early flowering stage. When grown with Nandi Setaria and cut to a stubble height of 20 cm, cutting at intervals of 6–9 weeks maintained an adequate proportion of Desmodium in the mixture. Lower stubble heights are generally permitted when the harvest intervals are widened (*12.6*).

10.8. Legumes in experimental grazing practice

The description below is restricted to those legumes which showed enough promise at some stage in the past that seed was available for experimental cultivation and grazing on field scale. It would be an impossible task in this chapter to refer to the numerous reports on nursery-only observations. The "promising" results reported at the nursery- or cutting trial stage are often in sharp contrast with those obtained later after grazing, if at all reported. For species being (re)screened in various countries in the region, see Dzowela (1988). For morphological and other details, see Bogdan (1977) and Skerman (1977).

10.8.1. Desmodium uncinatum (Silverleaf)

Trailing stems which root when in contact with the soil. From South America. Of all the tropical legumes tested, this has been one of the few with which, under conditions enumerated above, some degree of success seems possible at altitudes from 1000–2000 m. Desmodium produces seed relatively well and can be established without inoculation. It combines very well when sown with maize and fodder crops as these provide shade. Fodder crops also give some "lifting" of the Desmodium vines which then grow up along the elongating grass stems. In common with other trailing legumes but also with companion fodder crops (*12.8*), Desmodium benefits by long resting periods between defoliations (Fig. 10.6). Under cutting regimes Desmodium gives impressive yields and increases its proportion often with time (Fig. 10.3). Under frequent grazing, however, Desmodium disappears quickly from pasture, unless precautions are taken as specified (*10.7*). *In-vitro* digestibility is low due to tannins (*10.2.2*).

As far as pests and diseases are concerned, little damage has been encountered except from looper caterpillars, which though serious are only temporary and can easily be controlled. Pod-boring insects such as American Bollworm (*Heliothis armigera*) are, however, a problem in seed production. It is possible that, given the limited use of Desmodium, pests and diseases have been restrained. In spite of reports from elsewhere (Lenné & Stanton, 1990) no damage due to nematodes has been noticed.

Seed used to be harvested in the Kitale region with combine-harvesters. When the demand for seed declined, hand-harvesting of seed from locked-up pastures (*10.3*) proved more economic (Boonman, 1979a).

10.8.2. Desmodium intortum (Greenleaf)

Trailing stems. Differs little from *Desmodium uncinatum* in growth habit, ecological requirements and management. It may at times be more productive (Tiley, 1970a; Keya, 1974) but seed yields are much lower and its seedling vigour is weaker. This legume was introduced from Guatemala in 1942 as a cover crop and found to do well, particularly under conditions of shade (Compère, 1961c). It was also one of the major legumes in the programme at Makerere (Olsen, 1973).

10.8.3. Glycine (Neonotonia wightii)

The old name of this species was *Glycine javanica.* The newly proposed name of *Neonotonia wightii* may still be premature (Sue Edwards, private communication).

Glycine has trailing and twining stems. This is an indigenous legume, occurring in grassland and in bush from sealevel up to 2000 m. Glycine found some application as a cover crop in Kenya (Gethin Jones, 1942). For pasture purposes Glycine did not go beyond the experimental stage in Kenya (Bogdan, 1966b), but in Tanzania Naveh (1966a) reported a vigorous type (T2063) grown by a Moshi farmer, who had collected it near Tabora. In Australia, considerable success has been achieved with introductions from East Africa: Tinaroo (probably ex Nakuru), Cooper (ex Kongwa) and Malawi. Very impressive stands of Tinaroo-based pastures occur on the Atherton Tablelands in northern Queensland, in which Tinaroo is so aggressive that it nearly weighs down the wire fences (personal observation, 1977). Ecotypes (K51394) introduced from the Eldoret region by Strange (Bogdan, 1965a; 1966b) aroused some interest in Australia (Skerman, 1977).

From all evidence available in East Africa it appears that the rainfall should be moderate (1000 mm), the climate not too cool and the soil rich; such conditions prevail in the Arusha-Moshi area (Anderson & Naveh, 1968). However, even then establishment from seed is slow and failures are very common. Also when well established, Glycine lacks the vigour that is so characteristic of Desmodium. Seed production of Glycine is a major problem, even on trellises. Diseases are common, especially rust (*Synchytrium dolici*) (Bogdan, 1966b).

10.8.4. Siratro (Macroptilium atropurpureum)

Trailing and twining stems which root. Siratro met initially with great publicity. It was developed in Queensland from Mexican ecotypes (Hutton, 1962). Subsequent selection attempts remained fruitless. It would be worthwhile to study the distinctness of Siratro in more detail since Siratro would be the first example in the history of plant breeding whereby wild plants yielded a new cultivated plant simply by being crossed.

The acreage sown to Siratro in Queensland was 21.500 ha in the top-year 1972–73, with its nearest rival, Townsville Stylo, standing at 9.000 ha. Seed production dropped from 130 ton in 1972–73

(Hopkinson, 1977) to only 30 ton more recently (Walker & Weston, 1990). Siratro is not mentioned in the legume seed statistics of Latin America (Ferguson, 1990).

The interest in Siratro dwindled due to lack of persistence and later due to diseases. Siratro has been one of the major disappointments of tropical pasture development (Gramshaw *et al.*, 1989). It was this legume which had raised hopes that liveweight gains on P-fertilized Siratro-Buffelgrass could be increased 4–5 times as compared with unimproved natural pasture. In subsequent years, however, Siratro declined severely in density and productivity ('t Mannetje & Jones, 1990).

Common bean rust (*Uromyces appendiculatus*) appeared in Australia in 1978 and developed into the most important disease reducing herbage yield, quality and seed yield substantially; other diseases are *Rhizoctonia solani*, *Synchytrium phaseoli* and *Pseudomonas syringae* (Lenné & Sonoda, 1984; Irwin, 1989). Severe attacks due to Rhizoctonia were observed in Zambia as early as 1966 (van Rensburg, 1967). At Kitale, Siratro was introduced in 1961, in 1967 and many times since but it did not perform well, not even in nurseries (Suttie, 1968). Better results in East Africa were reported initially at the East African coast and at lower altitudes such as Serere (Stobbs, 1969c; *10.4*), but seed production was complicated by pests and weeds (Dradu & Ogwang, 1979). When fed pure, Siratro limits milk production because of low intake of digestible energy (Stobbs, 1971).

Intolerance to rational grazing regimes is a greater weakness than disease susceptibility, since it applies also there where disease is not prevalent.

10.8.5. Centro (Centrosema pubescens)

Trailing and twining stems. An old-timer among the legumes, of South American origin, but already established as a cover crop in plantation agriculture in south east Asia in the last century. Centro is essentially a legume for the humid lowlands, little of which is found in East Africa. It was tested as a green manure and cover crop at Bukalasa (U) in the 1920s but preference was later given to other methods (Nye, 1940; *5.2.1*). Centro repeatedly failed to perform well at Tengeru near Arusha (Naveh & Anderson, 1967). At Kitale, nursery performance was poor also (Strange, 1955b). It was tested at Serere in Uganda with good results, but preference was later given to Siratro and Desmodium. A drawback which it shares with Siratro it the tendency to shed leaves in the dry season. Centro is, however, hardier and more persistent under grazing than Siratro.

Of interest are the grazing trials carried out by Stobbs (1969h), in which remarkable increases in liveweight gains were obtained with Centro. (Fig. 10.5). Centro climbs up the grass stems and small Zebu, being tongue grazers, select the leafy part from the stems (Juko & Bredon, 1961). Drawbacks of Centro are problems with seed harvesting (best grown on trellises or fence-lines), poor seedling vigour, lack of stoloniferous spreading and the need for fertile soils. With Siratro it shares the mixed blessing of being ignored by livestock at times during the wet season. This allows the legume to increase its proportion in the sward.

10.8.6. Townsville Stylo (Stylosanthes humilis)

Ascending fine stems. This annual, of Central American origin, was recognized as a potential, cultivated pasture plant when it was found spreading naturally in the Townsville area of north east Australia at the turn of this century. As already mentioned (*10.1*), it had been the success with this legume and earlier with subclover, also accidentally introduced, which led to the concept developed in Queensland that other, possibly better, tropical legumes could be introduced artificially to perform the same duties. Its spread was, however, not as spectacular and as long-lasting as that of subclover.

Townsville Stylo's main contribution has been the improvement of native Australian pastures, mostly consisting of Speargrass (*Hcon*). P-fertilizer and hard grazing are essential for its performance, but perennial grasses often disappear as a result and annuals invade.

Townsville Stylo was tested repeatedly in East Africa, but with little result. At Serere Townsville Stylo did not perform well (Horrell, 1958b). Similar attempts which centered on Zone IV and V in Kenya and Tanzania (Sands *et al.*, 1970; Wigg, 1973) also proved abortive.

Interest in Townsville Stylo has dwindled further due to anthracnose (*Colletotrichum gloeosporioides*). Since 1974 this disease has spread rapidly causing enormous damage, also in countries to which Townsville Stylo has been exported (Jones *et al.*, 1984).

Anthracnose infects all Stylosanthes species causing severe damage in *S. humilis*, *S. guianensis*, *S. hamata*, *S. scabra*, *S. viscosa*, *S. capitata* and *S. fruticosa* (Irwin, 1989; Lenné, 1986). Anthracnose was recorded in Brazil more than 50 years ago and has been a major problem in Colombia since 1972. Extensive pathogenic specialization and variation for virulence have been identified. Previously resistant varieties have been overcome by new physios increasing in complexity over time as more genetically diverse host genotypes are deployed. Pathogenic specialization threatens tolerance found in Seca shrubby Stylo (*S. scabra*), in Oxley fine-stem Stylo (*S. guianensis* var *intermedia*) and in Verano Stylo (*S. hamata*) (Gramshaw *et al.*, 1989). Lenné has argued that lack of exposure to diseases during evaluation may result in selection of susceptible varieties and has also pointed out that pathogens on newly introduced material may cause serious damage to already cultivated pasture legumes and pulse crops. New outbreaks of anthracnose are not necessarily cases of resistance breakdown.

10.8.7. Stylo (Stylosanthes guianensis)

Erect, coarse branched perennial, from South America. Like Siratro, Centro and Townsville Stylo, (Perennial) Stylo is not adapted to high altitudes (above 1500–2000 m), neither to semiarid areas. It is basically a legume for warm and humid parts of Zones II and III in Zaïre, Uganda and western Tanzania. The most common type used to be Schofield (Skerman, 1977). Experience with the fine-stemmed and prostrate Oxley variety has been more recent, *e.g.* in Malawi and in Zimbabwe (Thomas, 1973; Clatworthy, 1984a). Oxley fine-stem Stylo has a narrow adaptation, is low yielding and is slow to start.

Schofield Stylo has been severely attacked by anthracnose in East Africa (Karachi, 1983; Gaboryaheze, 1984). Kulich (1988) and Lazier (1989) reported that the Cook variety was badly affected in Zambia and Ethiopia, respectively. Termites are also a problem; termite damage is often symptomatic of poor adaptation (Harris, 1949; *11.9.1*). At Serere, Otim (1975) observed an unexplained disappearance of Stylo in 3rd year pasture stocked at only 1 LU/ha.

Before the outbreak of diseases, a major setback in the use of Stylo had been the repeated finding that this legume reduces seed germination and yields by 30% in subsequent cotton crops (Table 5.9). The toxic effects were confirmed by pot tests and evidence was produced that leaf and stem exudates depressed germination also of pearl millet (Horrell & Newhouse, 1965; Brockington *et al.*, 1965; Tuley, 1968; van Rensburg 1969; Skerman, 1977; Bogdan, 1977). Also sisal crops were negatively affected, in Tanzania (Rijkenbusch, 1967). These observations effectively ruled out the use of this legume within mixed-farming rotations.

In semi-permanent pastures in the Mbale area of Uganda, Stylo was successful if grazing pressure was moderate and alleviated by an occasional topdressing with N-fertilizer (*10.7.2*). Stylo does not have a high nutritive value as the stems tend to become woody. On the positive side, Stylo was suitable for oversowing into natural pasture in warm and humid climates on relatively poor soils. It had the ability to persist rather well into the dry season. The ability of Stylo to enforce natural grassland was tested widely. Good results were obtained in Zaïre (Blouard & Thuriaux, 1962; Risopoulos, 1966) and in Uganda (Stobbs, 1969g; *4.3.5*). The experiments were continued in northern Uganda in the late 1960s (Maxwell-Davies, private communication). The dominant grass was *Hyparrhenia filipendula* (Fig. 3.6) and grazing pressure was low in the vast and lowly populated stretches in this region. Maxwell-Davies achieved the best results by sowing Stylo seed in the ash at the beginning of the rainy season, but later work showed that establishment is a problem when larger acreages are involved (H. van Arkel, private communication). Unfortunately, the programme came to an early end and no data are available on whether Stylo has been able to maintain itself. Stylo is easily killed by fire, which is a major risk in most areas of natural grassland.

10.8.8. Stylosanthes fruticosa

Shrub-like. This legume was mentioned by Walker (1969a) as the most common natural legume on hillsand soils in western Tanzania (*4.3.1*). Skerman (1977) observed that in areas protected from grazing and intermittently mown it forms a dense sward (Dodoma airstrip, Tanzania), and that it was grazed heavily by stock in the Sudan and Tanzania. More recently, however, Gaboryaheze (1984) reported that it was not well accepted by livestock in Burundi. At Kongwa (T), Wigg (1973) found that it was not very productive and the same was observed at the coast (*4.4*). Poor growth was reported from Zambia (van Rensburg, 1967). It was collected and tested also in Uganda in the late 1950s (*2.1.1*). Pests and diseases are common (Dzowela, 1988).

10.8.9. Other species of Stylosanthes

Verano Stylo (*S. hamata*) is an annual or short-lived perennial. Seca shrubby Stylo (*S. scabra*) is assured by strong perennation and seedling regeneration. These legumes came to the fore in northern Australia in the 1980s after the disappointing results with most of the older Stylo species mentioned above. Some of these legumes have been introduced to eastern Africa (Ibrahim & Orodho, 1983; Dzowela, 1988). Hardseededness is a widespread problem. Verano persisted poorly in Zone IV at Katumani (P.K. Kusewa, private communication, 1989). *S. scabra* gave slow initial growth and had low palatibility which make its use unlikely in intensive production systems (Lazier, 1989). Termites are a major problem in semiarid areas.

10.8.10. Lotononis bainesii

Indigenous to southern Africa. A dense sward is formed which becomes prostrate under grazing. Lotononis is unusual in that it needs a highly specific inoculant (Table 10.1). Seed production is difficult since the seeds are very small. Following earlier work (Bogdan, 1955; Strange, 1955b; 1958b; Richardson, 1965) an attempt to fit this "clover-type" legume into the pasture systems of the Kitale region was made by P. Bauer (Kitale Ann Rep 1975), in the hope that it would persist better than the trailing legumes. Lotononis produced attractive and dense swards, but grazing observations showed that it had little capacity to maintain itself. This confirmed the observation by Bogdan (1977) that Lotononis does not last long in mixture with grasses.

10.8.11. Calopo (Calopogonium mucunoides)

A very aggressive legume, from Central and South America, widely used throughout the tropics as a cover crop, also in Zaïre (van den Abeele & Vandenput, 1951). This legume is still the most widely sown for that purpose in South America. After it had been recommended and released in Uganda as a pasture legume in the early 1960s, an eradication campaign was launched in 1966 when it was found to be unpalatable and too aggressive (*10.6*).

10.8.12. Other tropical legumes

Virtually all the legumes listed by Bogdan (1977) and Skerman (1977) and many other indigenous legumes were tested in East Africa at various locations at some stage in the past 60 years. Research stations and equally many agricultural schools have in their files lists of entries which were tried on an experimental scale. Most of these were reported on earlier by Bogdan (1955), Strange (1955b; 1958a, b), Horrell (1958b) and Verboom (1965); see also *18.5*. The most important species, with their references, are:

Aeschynomene americana	Horrell, 1958b;
Clitoria ternatea (Kordofan pea)	Bunting & Lea, 1956; van Rensburg, 1967; Sands *et al.*, 1970; Wigg, 1973; Crowder, 1984
Dolichos sericeus (syn. *formosus*)	Strange, 1955c; van Rensburg, 1967; Anderson & Naveh, 1968; Bauer, Kitale Ann Rep, 1975
Indigofera hirsuta	Dougall & Bogdan, 1966; Sabiiti, 1985
Indigofera subulata	Rogerson, 1955; Naveh & Anderson, 1967
Macrotyloma axillare cv Archer	Bauer, *op. cit.*; Thomas, 1976b
Pueraria phaseoloides	Pereira *et al.*, 1954; Tiley, 1965d; Rijkenbusch, 1967; Hopkinson & Breitenstein, 1969
Vigna luteola	van Rensburg, 1967

A list containing 50 reputed legume species with a description of their growth habit and environmental adaptation was prepared by Kretschmer (1989). In addition to *Stylosanthes scabra* and *S. hamata* already mentioned (*10.8.9*) some species are being re-evaluated in Queensland. These include *Aeschynomene americana* and *Cassia rotundifolia*. However, under heavy grazing most of these did well at only 1 or 2 sites out of 9 (Cameron *et al.*, 1990). Among the old introductions being retested is *Vigna parkeri* (syn. *V. gracilis*; creeping

Vigna), known to provide excellent grazing (Bogdan, 1977). Creeping Vigna was collected by K.W. Harker (of Entebbe Veterinary Farm) in 1955 and sent to Australia. It tolerates grazing by growing close to the ground and by rooting from stolons, but is susceptible to moisture stress and slow to establish (see also Oram, 1990).

It has been said time and again that new legumes tolerant to grazing or to diseases are around the corner. However, as things stand as present, rather than gaining new ones we seem to be losing existing varieties and even species, due not only to lack of persistence but also to increasing susceptibility to diseases and pests, as observed in Stylo and Siratro. It goes without saying that negative experience with one species is often enough to prevent a farmer from trying another, possibly better, entry.

In view of the fact that so much effort over so many decades has already been spent it is unrealistic to expect that plant exploration or plant breeding will force a breakthrough in the foreseeable future. For the time being we are left with what we have. A continuous plant breeding effort will be required to keep abreast of diseases in species already available and to correct minor deficiencies, but it is unrealistic to expect breeding to lift species out of lethargy. It has already become increasingly apparent, due to the small commercial acreages under pasture legumes realized thus far, that it is hard to find the necessary funding for legume research programmes.

Whereas the future may seem bleak for the true tropical legumes, more emphasis on lucerne seems justified (*17*).

10.9. High-altitude pasture legumes

Of the tropical legumes Desmodium does relatively best, in the altitude zone of 1000–2000 m. Above 1500–2000 m many of the tropical legumes fail to perform well, even in nurseries, and indigenous (*e.g. Trifolium semipilosum* and *T. rueppellianum*) or exotic clovers (*T. repens* and *T. subterraneum*) do better. The latter two did not spread voluntarily to or in the highlands of East Africa (*2.3*); indigenous *Trifolium* spp. had already evolved before.

Symbiotic relationships of clovers with Rhizobium were found to be highly complex (Norris & 't Mannetje, 1964). Trifolium species are numerous (Gillet, 1953; Bogdan, 1956; 't Mannetje, 1966). Daylength responses have been reported ('t Mannetje & Pritchard, 1968), as in other tropical herbage species, but daylength effects are often confounded with temperature and light energy regimes (*8.7.1*).

Many introductions were made, beginning around the turn of the century. Like most of the temperate grasses, many of the exotic clovers do not flower well in equatorial regions. In general clovers survive better under frequent and intensive defoliation than the trailing types of tropical legumes. Unlike tropical legumes, clovers are capable of adapting their growth habit to the defoliation regime. There are, however, many circumstances in which also clovers have difficulty in maintaining themselves in the sward. Sunny, but moist climates are favoured most, while the companion grass should not grow tall nor overshadow them. Clovers produce highly nutritious grazing. Seed production of adapted varieties (Louisiana white clover) is relatively easy. Clovers are relatively free from pests and diseases. A virus causing bright yellow vein-banding and reduction of leaf size of Kenya white clover is related to pea mosaic virus (Bock *et al.*, 1973). The performance of clovers is often erratic, as they may come and go when and where unexpected.

10.9.1. Kenya white clover (Trifolium semipilosum)

This clover is found above 1500 m and can reach up to 3000 m altitude. It grows poorly at high altitudes such as upper Molo (2750 m), although Birch (1959b) found it doing better east of the Rift Valley. It is particularly at home in association with grasses in short swards, such as of Kikuyugrass (Edwards, 1936), or when the sward is regularly kept short by grazing or mowing. Kenya white clover (KWC) is a component of post-cultivation communities and can invade sown pasture (Jones & Evans, 1961; Suttie, 1970a). Inevitably, this indigenous clover aroused interest but has continued to be controversial. Seed was marketed in the 1960s and clover-based pastures won first prices in grassland contests. However, it has many drawbacks. Nodulation and establishment from seed are often poor and erratic (Strange, 1961a; Suttie, 1967). Little herbage is produced by KWC, but it has strong competitive power and can be found in almost pure stands. It is aggressive against associated short grasses, and can eliminate these without making up for the losses by growth of its own. Where Desmodium has disappeared from the sward, KWC may emerge. Pure-clover patches of 10-cm height are no exception. Its frequency is, however, often not meaningfully related to any management or environmental parameter tested. KWC is usually regarded as having a greater tolerance to moisture stress and acid soils than white clover.

Cattle bloat is not uncommon but KWC is not always relished by livestock (Bogdan, 1956). Some farmers came to regard it as a weed and sprayed their pastures against it with herbicides such as 2,4-D (Bogdan, 1965c). This was the situation as it arose in the 1950s, and although seed was produced at the Kitale Station of the Kabete variety (K5154; Bogdan, 1965a) up to 1968, interest waned as alternative and more bulky legumes such as Desmodium began to draw interest.

The effect of KWC on pasture yields in the 1st and 2nd year was investigated at Kitale by Strange (1961a). Although this applies to legumes in general, KWC caused a depression in total DM-yield in the 1st year (Table 10.3); it had a highly positive effect on the yield of DCP (digestible crude protein) in the 2nd year. See also Table 12.10.

Table 10.3. The effect of Kenya white clover (KWC) on yield of dry matter (DM) and digestible crude protein (DCP) in kg/ha; after Strange (1961a).

Year of pasture	Grass alone		Grass + KWC	
	DM	DCP	DM	DCP
1	6060	335	4930	321
2	3560	88	3660	170

Strange (*op. cit.*) further reported that in the absence of gypsum ($CaSO_4$) KWC failed to raise DM-yield, while the influence of gypsum on grasses in pure stands was negligible. P-fertilizer raised DM yield only in the presence of gypsum (Table 10.4), indicating a sulphur requirement (see also *10.2.1 & 17.3*).

10.9.2. *White clover (Trifolium repens)*

Louisiana white clover was introduced in 1957 from southern USA in the form of a short-daylength variety. Louisiana has been the most successful legume for sown pastures in the Kenya highlands from 2000–3000 m altitude. It can stand close and repeated grazing in association with Rhodesgrass and Setaria and at higher altitudes also with Cocksfoot and Perennial Ryegrass. It has a better productivity and less weed potential than Kenya white clover and rarely grows into a pure clover sward, although it is at times replaced by Kenya white clover. At high altitudes average temperatures may be too low (Birch, 1960).

Table 10.4. The effect of P-fertilizer on the combined yield of grass and Kenya white clover (kg/ha), in the presence of gypsum; after Strange (1961a).

Year of pasture	Dry matter yield of grass and clover	
	No P_2O_5	50 kg P_2O_5
3	2690	3760
4	2700	4310

The optimum P_2O_5-rate for grass/clover is 150–200 kg/ha in acid soils at high altitudes, but only 100 kg at Kitale (Birch, 1959b, c). High rates of N-fixation were recorded at Molo by Morrison (1966e), but in some dry seasons most of the clover died. Survival in the sward can then be effectuated only by seed. Such annual behaviour, although of great adaptive significance, has the disadvantage that clover occurs in wildly fluctuating proportions from year to year. In some seasons clover is found only sporadically. Much of the year-to-year variation of clover content may be principally associated with rainfall and its distribution. Strategies to significantly increase the white clover content in pasture through calculated hard grazings (for instance, by the end of the rainy season) are poorly understood by farmers. However, the seed persists well in the soil and it is commonly found volunteering after some years of a cereal break.

Morrison (1966d; 1969) listed 3 crucial factors influencing the establishment and subsequent persistence of clover mixtures:

1. Grass competition must be reduced by lowering the seed rate of the companion grass (*6.15*).
2. Adequate levels of P_2O_5 and effective nodulation, which can be ensured by the use of effective strains of Rhizobium (*10.2*).
3. Lime-pelleting (*10.2*).

Louisiana white clover attracted widespread attention and was grown on farm scale in Kenya in the early 1960s. At the Kitale Station, a mixed Louisiana-Nandi Setaria sward produced 3000 kg/ha/year milk in its best years (Kidner, Kitale Ann Rep 1963) and was kept in a productive state for nearly 10 years. Although this showed that Louisiana white clover was capable of high production, it fell out of favour in the mid 1960s and seed production ceased in 1968. Official interest had by then switched to Desmodium and other trailing tropical legumes which seemed to produce more bulk and to show less trouble in surviving the dry season (Suttie, 1965; Suttie & Moore, 1966). Although it took only a few years before doubts arose whether Desmodium was a good replacement, white clover had unduly been dealt a heavy blow, from which it has not recovered. An attempt at reviving the interest in clovers was made by Bauer (Kitale Ann Rep 1975).

Louisiana white clover was also tested in Kivu in Zaïre, but better and lasting results were achieved by Compère (1961a) with Ladino white clover, which is of a larger size than Louisiana. This Ladino type persists particularly well on fertile soils at 2250 m altitude with a high rainfall and a short dry season (e.g. at Ngungu), also when grown in association with relatively vigorous Kikuyugrass (*3.4.2.1; 4.2.1.2*). It is a poorer seeder than Louisiana (Bogdan, 1965c) but a common method of establishment in Kivu is to vegetatively transplant Ladino clover straight into existing or new pasture (personal observation, 1992). This is not always effective enough to achieve more than a 10% proportion of the sward but, although nodulation may be adequate, further advances should be possible with the aid of phosphate fertilizer.

Another white clover introduced via Cameroon and named Bamenda white clover (K53945) is occasionally mentioned in reports but it had no special merits of further testing.

10.9.3. Subclover (Trifolium subterraneum)

Subclover (subterranean clover), which had spread so voluntarily and successfully in the subtropical farming areas of southern Australia to become the cornerstone of wheat farming in the region (*10.1*), had not failed to draw the attention of farmers and scientists alike in the highlands of East Africa. A map of Kenya showing climatic regions where Australian varieties of different maturity rating were expected to do well was produced by Davidson (1958); this map was based purely on length of growing season and on mean temperatures. The idea was that this legume should do equally well in Kenya, contribute to soil improvement in wheat rotations and increase the productivity of pastures. Some success was achieved above 2250 m with the varieties Bacchus Marsh and Clare in upper Molo in the 1950s and 1960s (Birch, 1960; Morrison, 1969). Subclover is an annual regenerating from seed buried in the ground. This, however, does not do seem to do so well under Kenya highland conditions. Inoculation with effective Rhizobium and lime-pelleting (*10.2*) were necessary (Morrison, 1966b, d).

10.9.4. Miscellaneous annual clovers

Trifolium baccarinii, T. quartinianum, T. rueppellianum, T. steudneri, T. tembense and *T. usambarense* are the more important indigenous annual clovers distributed in natural grassland above 1500–2000 m altitude. Although these species are relatively common (Bogdan, 1955; 1956) and can be impressive, their potential as pasture species is small (Bogdan, 1977). Only *T. rueppellianum* was multiplied commercially, because it seeds well and develops quickly (Bogdan, 1966d); it was reported to do well in the Cameroon highlands (Borget, 1969). Chilean red clover (*T. pratense*) was imported in Kenya in the early 1960s. Contrasting aspects of nutritive value of different types and origins of clovers were observed by Dougall (1962a) (Table 10.5).

Renewed interest in some of these clovers has been shown recently by ILCA at Addis Ababa (Kahurananga & Tsehay, 1991). Between entries, date of heading was found to be related to the rainfall at the site of origin (see also *6.6.3*; Fig. 9.1) and, contrary to earlier expectations, no nodulation problems were encountered.

10.10. Conclusion

Like elsewhere in the tropics, legume research in East Africa has been characterized by a coming and going of species. With the passing of time,

Table 10.5. Nutritive value of indigenous and exotic clovers (after Dougall, 1962a).

	Indigenous clovers		Exotic clovers	
	Annual	Perennial	Annual	Perennial
Crude protein (%)	21.1	17.7	16.1	19.1
Crude fibre (%)	20.2	22.5	17.4	16.1

disadvantages experienced in farming practice gained the upper hand over the benefits so widely acclaimed in the early days of research observations. This may explain why legumes have been accepted so poorly by farmers, large- or smallscale, in such sharp contrast with cultivated grasses. As enumerated above, it has been common but incorrect to attribute this lack of interest to inadequate inoculation, to poor management or to non-availability of seed.

The examples, quoted above, serve to show that farmers cannot be expected to invest in plants that require extra care and fertilizers for their own upkeep and then to see them disappear because of lack of persistence. The benefit that legumes, when present in the sward, pass on to companion grasses is less than commonly believed and legumes do not prevent the decline in grass productivity in later years (Figs. 10.3 & 12.7). Legumes seem redundant when grass is grazed relatively young (Fig. 10.6).

The "free N-fertilizer" concept of legumes may still have a great appeal but is not valid since the specific fertilizer (N) to be saved upon has to be replaced by one or more others (P, K, S, Ca, Mo) without which legumes cannot function but which are equally expensive (Fig. 7.3) and difficult to obtain. Moreover, legumes which are slow to establish or which affect subsequent crop yields adversely due to weeds or other problems (*5.2*; *5.3*; *10.8.7*) are not welcome in crop-pasture rotations.

Severe attacks by diseases in some of the most important species (*Stylosanthes* spp.) have aggravated the problem. Intolerance to rational grazing regimes is a greater weakness than disease susceptibility, since it applies also there where diseases are not prevalent. The acid test of legume adaptation is its persistence. It may be better to have a low-yielding, persistent legume than a high yielding, non-persistent legume. Whatever the choice, legumes are lower yielding than grasses and they require fertilizers that grasses can do without.

The contrasting performance under nursery- and grazing regimes is frustrating. Since Edwards (1950a) had drawn attention to this point, legumes were tested under farm-routine grazing as a matter of policy (Strange, 1958b). This explains perhaps why Bogdan (1965a), Pratt & Gwynne (1977) and so many others (*10.6*) were so hesitant to lay much emphasis on the feasibility of pasture legumes. The same tendency was found in southern Africa (Scott, 1967) and elsewhere (Tempany & Grist, 1958; Whyte, 1962). In the world-wide wave of legume publicity of the past decades their advice was not taken seriously. Farmers, however, when unhappy with what they sow, respond by not buying a second time. And a seed industry, forced to suspend production, may be reluctant to resume activities when new and perhaps better varieties come along.

The need of timely evaluation under farm-routine grazing regimes is not yet accepted universally. The emphasis is still too often on the introduction and screening of large numbers of accessions. Timely multiplication of seed to be issued for farm-evaluation is then a major bottleneck since there are always new and allegedly more promising varieties in the offing that capture priority attention. In spite of the massive searches that have been mounted by CSIRO and by international research institutes, mainly CIAT in Colombia, there are still no new species or varieties capable of replacing the old-timers which had, incidentally, been borrowed from plantation agriculture. As a result, research resources have come under mounting financial pressure, which makes it even more unlikely that the "elusive tropical pasture legume" (Whyte, 1962; Crowder & Chheda, 1982) will soon be discovered. On the other hand, breeding has produced only limited results also (Clements, 1989) and is certainly unlikely to lift potential, still uncultivated, species out of lethargy. Jones *et al.* (1984) have stressed that some legumes with potential in Australia have been unsuitable elsewhere and that the same technologies may work out quite negatively when transferred to other regions with different farming systems.

Visiting overseas scientists have often queried the apparent lack of concern in many parts of Africa about the absence of cultivated pasture legumes in practical farming. Absence of certain species in farming is, however, no proof of them having been ignored. Legumes were studied in East Africa at an earlier time and in more detail than elsewhere because of the interest both in green-manure and cover crops (*2.1.1*; *5.2.1*; Table 5.4) and this was extended to their possible role as pasture legumes (*10.6*).

However, it has become increasingly clear in recent years that in many other parts of the world farmers have responded in much the same way. The lack of persistence of legumes has now come to be acknowledged as the major weakness (Marten *et al.*, 1989). A farmer may accept disease risks but he is less prepared to sow legumes that simply disappear for no fault of his own. On the other hand, persistence alone is no guarantee to success either, e.g. Calopo (*10.6*) or tall Leucaena (*18.5.6*).

This may explain the low volume of pasture legume seed production, less than 500 ton/year, which is little on a global scale compared with the 4400 ton of white clover seed exported by New Zealand alone (Frame & Newbould, 1986). The acreage of tropical legume-based pasture sown annually does, therefore, not exceed 150–200.000 ha world-wide, which is very small even when compared with acreages sown to minor pulses (FAO, 1982). For Zimbabwe, Clatworthy (1984a) gave a cumulative total of pasture legumes of only 6500 ha. For Kenya the figures are no better, in sharp contrast with the 400 ton of certified tropical grass seed produced annually (Fig. 8.1).

Not unexpectedly, the low rate of return on the resources spent has revived old debates. Can tropical legumes be expected to function adequately within the intended frameworks? Is it reasonable to expect tropical legumes to form stable components in mixtures with sown grasses, especially when the latter are grazed to their potential. The white clover of the tropics has not been found (Whyte, 1962). Unlike in clovers, naturalization of tropical legumes is rarely spontaneous and is not easy to induce artificially in stable swards. Nowhere in the world did tropical legumes form substantial components of the natural grasslands. Nearly all the legumes that were "introduced with success" in the past were borrowed from plantation agriculture. None of the old tropical civilizations developed legumes for forage purposes, in contrast with so many cereal legumes. Ironically, these are commonly grown in mixture with other crops, in densities that can be regulated. In grazed swards, however, mixtures of species cannot easily be manipulated. To obtain legume-N via decay of protein of plants grown for that very purpose may not be an efficient system, because proteins claim a lot of the plant's energy.

Although legumes have not lived up to the early hopes for their use, many scientists find it hard to admit that legumes in sown and grazed pastures are after all no practical proposition. The concept of tropical legumes for grazing was developed in the belief that farmers needed *and* wanted legumes, but it was not realized that the choice of species had been the farmer's and still is. This choice had been fixed long before in agricultural practice, with pasture legumes playing only a minor role (*2.2.2*; *2.3*; *6.1.2*). Over-optimistic assessments of their potential role were not balanced by realistic analysis of the constraints or of the associated indirect costs:

1. Tropical legumes fix N; the saving made on N-fertilizer may be wiped out when large amounts of extra P- and other fertilizers are required by legumes, only to start them off or for their own upkeep.
2. Tropical legumes transfer N to companion grasses; this effect becomes visible most often when plants are already mature.
3. Tropical legumes add protein, but its true value for livestock production may have been overestimated in relation to intake of energy, especially in dairy cows.
4. Legumes increase the herbage mass. The grass companion itself is often unaffected or even reduced; this limits the added value of a legume component in the sward.

It is postulated that legumes have application, provided they can be oversown and maintained cheaply, to improve the dietary value of extensively grazed and mature natural grassland (*4.3.5*), and also when grown as fodder crops ("protein banks" or in association with fodder grasses (Fig. 10.6; *12.8*; *17*). Under most other conditions, legumes are useful when present but they cannot be relied upon as productive partners in stable associations.

However, legumes have been regarded as ideal, exactly for the smallholder because of the "free N". Legumes fix N but they are usually not in a sufficiently strong position to display this ability. It should be obvious that they stand no chance under the grazing or cutting regimes commonly found on smallholdings. Their potential benefits are also difficult to reconcile with those of rationally grazed grass swards. Frequent defoliation is needed to lift the quality of the grass but the legume is then likely to disappear. On smallholdings low milk yields are due to the lack of quantity rather than of quality of feeds (Potter, 1987).

The question may be asked, what alternatives the smallholder has if legumes cannot be maintained reliably, except as fodder crops. Possible answers are explained in some detail in the chapters as indicated:

1. Higher 1st-year herbage yields are achieved with grass alone (*10.3*; *12.8*).
2. Weed control is easier with grass alone (*7.1*; *12.2*).
3. Resources spent on P and other inputs, only to encourage legumes, can be diverted directly to N-fertilizer (*7.2*).
4. Sown pasture, when becoming N-deficient, is better ploughed, cropped and sown to grass again after 2–3 years, in order to recycle soil-N more effectively (*5.3*; *7.2*; *12.7*).
5. Rather than have mature legumes improve the value of old grass, the latter should be grazed or conserved when young (*7.2*; *19*).
6. Mixed-farming based on crop-grass rotations enables efficient use to be made of N-rich bye-products. Crowder & Chheda (1982) quoted experiments conducted in Malawi on supplementation of Rhodesgrass pasture in the dry season. Liveweight gains were the same whether animals were fed Rhodesgrass plus 0.34 kg/head of cottonseed cake or pasture grown in strips and consisting of two-thirds of Rhodesgrass and one-third of Stylo and Siratro. Similarly at Kitale, when animals were supplemented with 0.45–0.90 kg/head of an oil cake with unrestricted access to pasture, liveweight gain continued to progress in the dry season (Kidner, 1966).

Research to solve the problems of legume-based pastures is becoming increasingly difficult to justify in terms of priorities because feasible higher production levels have already been reached with alternative farm systems, especially in dairy production. To the farmer, grass plus N-fertilizer has more appeal than grass-legume mixtures plus P-fertilizer. It is hard to persuade the farmer to put more phosphate on his legumes in order to save nitrate on his grasses. Similarly, farmers are rarely prepared to relax their grazing regime. Productive grasses are a prerequisite. Under no circumstances should "aggressive" grasses be withheld from recommendation on grounds of incompatibility with legumes.

While it is desirable that some resources should always be devoted to the search for new and better pasture species, we should not lose sight of those we already have and we should continue to work on them for their betterment. It is astonishing that searches are still being made for legumes to be compatible with grass varieties that are already more than 50 years in cultivation. (*6.1.2*).

We cannot miss what we have not had, and we had been warned. Perhaps white clover was the wrong model (*10.1*) for tropical legumes. Not white clover but lucerne is world's most widely sown legume (Marten *et al.*, 1989). Lucerne, however, is mostly grown as a fodder crop (*17.5*). Lucerne is an example to demonstrate that legumes do not need to be persistent under hard grazing to be useful. If tropical legumes are wanted but are unable to accompany, let alone encourage, grasses under current grazing regimes, for whatever reason, it is postulated that they also should be managed as fodder crops (*10.7.3*; *12.8*).

10.11. Summary

Unlike tropical grasses, legumes are little in demand by farmers, in spite of the fact that a large part of research resources has been devoted to legumes, especially to the search for suitable species. Although some legumes require special treatment, *e.g.* specific Rhizobium inoculation, additional phosphate and other fertilizer and although diseases can cause havoc, the overriding problem is that legumes are unreliable partners in grazed swards. In young swards legumes are redundant. Most legumes are capable of fixing nitrogen and of growing vigorously when relatively undisturbed. Desmodium is discussed as an example to show that tropical legumes can perform very well both in monoculture and in mixture with grasses under cutting regimes, but fail to persist under even moderate grazing regimes. In addition to suffering from lack of persistence, Stylo and Siratro have been further curtailed seriously by disease. If legumes are indispensable, they may be better managed as fodder crops. If N-deficiency is the main limiting factor in grasses in crop rotations, shorter rotations are the obvious answer to recycle soil-N more effectively.

CHAPTER 11

Elephantgrass as fodder crop

11.1. General introduction

Fodder crops and fodders are best defined as sources of herbage or feeds grown or utilized for special purposes, in contrast with sown and grazed perennial pasture for more general use. Certain fodders may have a place in a given farm situation but elsewhere or at some other time they may be less desirable, for want of other important characters. This may explain why their use has not been as universal and of long standing as the sown and grazed grasses described in Part I of this book.

Fodder crops often have something spectacular not found in sown grasses. A massive stand of Elephantgrass, whatever its age, is more impressive than a short sward of a sown grass. Sorghum, green at the peak of the dry season, is more spectacular than a stack of hay. Their excellence in such virtues has often made them to be heralded as miracle species, especially when newly (re)introduced and associated with "hybrid" characters. Conversely, weaknesses in other characters halted their advance or made them less popular than they actually deserved. The history of many a fodder crop has been one of wildly fluctuating interest.

Fodder crops are neither superior nor inferior. Their role should be seen as complementary, or as an alternative in special technical, economic or ecological circumstances such as when:

1. Farming is situated in a high-price milk market.
2. A protein-rich crop such as lucerne can be grown successfully.
3. Complementary feeds are needed in dry areas or dry seasons.
4. Break- or catch crops like oats or Sorghum are needed.
5. Green-feeding (stall-feeding) is preferable or inevitable.

Fodders are often associated with certain practices: lucerne with hay; Sorghum with dry-season grazing; Elephantgrass with dry-season grazing or with stall-feeding. However, Elephantgrass has also been used for all-year grazing in East Africa and elsewhere (Vicente-Chandler *et al.*, 1974). Conversely, Rhodesgrass is normally grazed but it has been the major grass irrigated and cut for intensive dairy stations in the Arabian peninsula (Farnworth, 1976). Rhodesgrass was the grass tested in the first published experiment on zero-grazing in East Africa (Ogborn, 1965) and has been used in stall-feeding in Tanzania (Naveh & Anderson, 1966) and in Malawi (Addy & Thomas, 1977; Agyemang *et al.*, 1988). Very few, however, would describe sown perennial grasses as fodder grasses (*13.4.3*).

Table 11.1 lists the main fodder species in present use but is not meant to be exhaustive. Various additional species have been described at some point as useful fodders in East Africa: mangolds, Brassica's; sweet potatoes; cereal legumes etc. (*18*). These crops are discussed not so much for their actual economic importance but because the active interest in these crops returns every so many years, often unaware of previous work. The list could be completed with the bye-products of arable crops which are major animal feeds in mixed-farming systems (*5.1*; *10.10*); these have been described in more detail by French (1940; 1943b; 1946a); Kategile & Said (1981), Kevelenge *et al.* (1983), Kiflewahid *et al.* (1983), Potter (1987), Preston & Leng (1987), Reed *et al.* (1988), Rogerson (1956), Thomas (1977b).

When perennial fodder crops are established primarily to cater for the dry season, their purpose

Table 11.1. Main species of fodder crops in East Africa.

Species	Main zone of use	Main advantage or purpose	Disadvantage
Perennial:			
Elephantgrass	Humid lowland (Zone II)	Yield	Labour requirements
	Zone III	Dry-season greenness	
Lucerne	Alluvial soils	Nutritive value	Soil requirements
Annual:			
Maize (green)	Zone II/III	Yield	Costs
Grass Sorghum	Zone IV	Drought resistance	—
	Zone III	Break crop	
Oats	Highlands	Break crop	—

of use in the main growing season is less clear. Conservation is often the only solution. However, the surplus growth of the sown and grazed pasture is also best converted into hay or silage (*7.2.5*; *19*). Fodder crops have, therefore, no monopoly on the dry season. The need for and role of certain fodder crops should be viewed against year-round livestock feeding systems.

Elephantgrass is discussed first, but much of it applies to other fodder-type or giant grasses. These last terms are often used to distinguish them from sward-type grasses which are sown and grazed.

Separate chapters on fodder grasses are justified by the fact that the responses are quite different under cutting than under grazing regimes (*1.4*; Vicente-Chandler, 1974). Since most of the research data available are obtained from cutting experiments, those relevant to fodder crops are directly applicable to the farm situation: hay or silage, green-feeding or stall-feeding.

11.2. Name and natural distribution

Elephantgrass is the common name when referring to the grass occurring naturally (*3.4.2.3*; *4.2.3*). In its cultivated state it is also known as Napiergrass, especially in Kenya.

Its history is relatively recent (Melle, 1918; Kennan, 1952). The first published note on the value of this grass dates back to 1905 after Mynhardt, a Hungarian missionary in Barume in what was then north west Rhodesia, had sent material to Zurich Botanical Gardens. It was named, however, after Colonel Napier of Bulawayo who called the attention of the Agricultural Department to this grass. In Kenya, Maher (1936a) referred to "Napier's fodder". Many now internationally used names of African flora and fauna alike were named first in southern Africa. East African vernacular names of Elephantgrass are Mabua (Kiswahili), Ebisagazi (Luganda), Agada (Luo), Thara (Kikuyu), Lussi (Kiluhya), Morongiot (Kinandi).

The natural distribution and ecology have been described (*3.4.2.3*; *4.2.3*). It is the major grass in the Elephantgrass zone in Uganda, in a fertile belt of forest potential (Zone II), where sugarcane thrives also. In the rest of East Africa the grass is restricted to riverbeds, since there are few areas outside Uganda with forest potential at the altitude of < 1250 m. Reports that it does not tolerate seasonal waterlogging are, therefore, surprising. Material has been brought into cultivation from nearby streambanks and its cultivation in vleiland was propagated for mulch production (Robinson & Brooke, 1959; Gosnell, 1963). Gosnell & Weiss (1965) reported that, on camber beds in vleiland near Turbo in west Kenya, it suffered from waterlogging during exceptionally wet periods only. Large areas of low-lying land in west Kenya have now been developed for sugarcane growing, so that conditions for Elephantgrass should be equally favourable.

The ecological zonation of its area of origin indicates that Elephantgrass is essentially a grass for the hot and humid low- and medium highlands, preferably below 1500 m. It has difficulty in surviving the long dry seasons of central and northern Uganda, *e.g.* at Serere in Zone III, not to mention Zone IV. Consequently, its dry-season greenness in Zone II should not be mistaken for drought tolerance in less favourable zones, *e.g.* the East African coast (*11.10*).

The upper altitude limit of cultivation is 2250 m (*6.5*), but attempts at cultivation can be witnessed in central Kenya at higher altitudes.

11.3. Importance as a cultivated plant

The massive growth of ever-green foliage displayed by undisturbed Elephantgrass in riverbeds could not have escaped the attention of the early agriculturalists. It is significant, however, that usages other than as a ruminant feed aroused the first interest, namely in soil regeneration and mulching in plantation crops, especially coffee. This was pioneered in Uganda (Tothill, 1940).

11.3.1. Soil regeneration

The soil regenerating effect of grass had been appreciated by local farmers in Buganda for a long time. Only when green and animal manures were found not to be adequate in restoring the productive capacity of the soil, it was officially realized that the indigenous custom of natural grass regeneration was a much better method (*2.1.2*; *4.2.3*; *5.1.2*; *5.2*). Direct planting was often necessary to speed up the process, so that the traditional rotation of 6-year fallow could be reduced to 3 (Maher, 1936a). Giant Setaria is preferred as a bund grass and for erosion stops because of the marked edge effect Elephantgrass may cause to adjacent crops (van Rensburg, 1969a).

11.3.2. Dry-season grazing

Maher (1936a) reported from Kitale that an increasing number of farmers grew "Napier's fodder" as a reserve feed for their working oxen in the dry season. Maher's was the first report of the grazing value of this grass in practical farming conditions. Maher also had the foresight that "dairy farmers in districts where there is a dry season lasting from December to March should take the last cut in September or October for silage, mulching or compost". Elephantgrass thus established itself as a dry-season fodder resource which is where its main strength still lies, in spite of all expense of time and labour. It is a common farmers' experience that milk yields in the wet season are lower than on sown pastures, which was confirmed by Odhiambo (Kitale Ann Rep 1967), who observed that "milk production fell at a much faster rate on Elephantgrass than on ley pasture". It is not a good source of fodder in the wet season (Fig. 10.6). This, however, can be turned to its advantage. When this grass is to be utilized in the dry season, the rainy-season's growth potential can be used to rehabilitate the grass from the damage inflicted by the previous dry season and the field can be rested (spelled) in order to maximize the next dry-season's production (*13.2.1*).

11.3.3. Mulching

From Uganda, Tothill (1940) reported on an experiment carried out in 1926 which proved the value of Elephantgrass mulch: "Mulching saved the Bugishu coffee industry from a rapid decline; in areas where Elephantgrass is plentiful the department now strongly recommends the alternate line mulching of coffee on the contour as the best treatment yet devised for the maintenance of coffee in good condition". The main beneficial effects of mulching are considered to be: improved water- and soil conservation, weed control and the addition of organic matter and nutrients, especially in the drier areas.

Pereira & Jones (1954) found mulching more effective than animal manure because of the moisture it conserves; rainfall penetrates to a greater depth (*5.2.2.1*). Producing enough mulch is, however, the main obstacle to realizing these benefits as acreages of Elephantgrass far above those of the coffee itself are needed to supply the 25 ton/ha required for alternate interrow mulching (Jones & Robinson, 1965). Tiley (1959) quotes file reports of a 3:1 ratio, *i.e.* 3 ha of Elephantgrass to 1 ha of coffee in Uganda. This explains the use of vlei for mulch production. Another disadvantage of mulching with Elephantgrass canes is the risk of cation imbalance. Robinson & Chenery (1958) noted that the high K-content of this grass (Table 12.5) induced Mg-deficiency, affecting the quality of the coffee berry.

Elephantgrass was actively promoted in central Kenya (KDA Ann Rep 1935). In spite of government campaigns in the 1950s very few smallholders mulched their coffee, but found it more profitable to sell manure and grass to coffee estates or to feed the cut grass to their livestock (Acland, 1971).

11.3.4. Stall-feeding and zero-grazing

The mulching campaigns helped inadvertently to get Elephantgrass established as a major fodder for dairy animals in the small-scale coffee growing areas with a rapidly growing demand for fresh milk. Stall-feeding often had its origin in circumstances, different from those of efficiency of fodder production or utilization. Even with nearby land available for grazing, it had been a customary practice to shut animals up at night in kraals. When the kraal period is too long animals need to be fed. Traditional reasons for keeping and feeding animals both day and night inside have been:

1. Security (against the risk of theft and predators).
2. Tick-borne diseases.
3. Tse-tse and Stomoxys biting fly.
4. Land tenure systems and land fragmentation (no fencing).
5. General livestock control.

The need for security has a bearing also on the use of some other crops in stall-feeding. The risk of theft of maize when ripe from the field may explain the extensive use of green maize (Stotz, 1977), a practice which can hardly be considered as efficient maize- or fodder production (*13.4.1*; *19.2*).

Bringing in fodder from outside was a common practice among the Wakara in Lake Victoria and the Wachagga of Mount Kilimanjaro (Staples, 1934; French, 1938c; 1940). Among the Wachagga, stall-feeding had been an established practice since time immemorial, because of cattle theft and tse-tse (Swynnerton, 1949). Fodder brought in from lowland areas at great distance included grasses, crop weeds, cereal stover, bean straw but use was also made of tree loppings and woodash; see also Naveh (1966a). Stalks, leaves and surplus bunches of bananas were the principal home-grown fodder and drinking-water supply of the livestock. All manure and refuse were applied to the banana. Government propaganda for Elephantgrass to be grown on the contour banks met, however, with little success. Around Arusha, van Rensburg (1969a) noted an increased demand for Setaria (*6.7*; *14.1*) for stall-feeding. An earlier report (van Rensburg, 1948a) referred to stall-feeding in Mahenge district with Elephantgrass collected from stream banks. In Rwanda stall-feeding was advocated by Deschuytener (1967).

As already mentioned, Elephantgrass had been promoted officially in Uganda for the purpose of soil regeneration and in Kenya for the purpose of mulching in coffee. Smallholders were quick to learn that this grass was more profitable to them when sold or when fed to animals than when used as a mulch in their coffee. Tiley (1959) reported the use of Elephantgrass for stall-feeding around Kampala. The spread of tse-tsefly right up to Kampala by the early 1940s (Robertson & Bernacca, 1958) and the nuisance of Stomoxys biting fly (Nye, 1940; Kerkham *et al.*, 1947) induced farmers to keep cattle indoors. The lower risk of tick-borne disease also helped to establish stall-feeding practices around Kampala when exotic dairy animals were first introduced around 1960 (Stobbs, 1962; Hall, 1966).

Malawi was the first country to have government-initiated stall-feeding schemes, in 1957 (Royle, 1961), to fatten stall-fed steers on smallholdings. This involved about 1000 animals in 1971 (Spurling & Spurling, 1972), but fattening on Rhodesgrass pasture turned out to be more efficient (Addy & Thomas, 1975).

In Kenya, cutting and carting of road-side grass, even of short swards of Kikuyugrass, were a common sight near Nairobi in the late 1960s (personal observation). At that time stall-feeding was vigorously promoted by NGO's (non-government organizations), Volunteers Corps and Farmers' Training Centres to help employment in the densely populated areas of central Kenya where land was severely limiting. Stall-feeding became an officially endorsed policy in the 1970s and donor-supported campaigns were initiated to extend it to other regions, including areas where direct grazing had been the normal practice. Its image was in no small measure enhanced by the beef feedlots which were set up around the country, mostly based on maize silage (*19.2*). These were started at the Beef Research Station of Lanet and subsequently, as a commercial proposition, on large farms of ADC (Agricultural Development Corporation) and on some private farms (Creek & Squire, 1976). A feedlot system, based on bye-products and concentrates, was extended to smallscale farming through experiments carried out at Farmers' Training Centres (*13.4.2*; Mulder *et al.*, 1973).

The change-over from largescale to smallscale dairying was very fast (Chema, 1984). Milk prices rose sharply after the milk quota system of the Kenya Cooperative Creameries was abolished in the late 1960s which led to milk shortages in the dry season. This helps to explain why Elephantgrass with its dry-season greenness became the dominant grass in zero-grazing systems, even in areas above 2250 m, which are too cold for this grass.

Although Elephantgrass may be very much in evidence as the most conspicuous grass in zero-grazing systems, because a lot of it is grown along road-sides, other grasses and sources of roughage do make up more than half of the feeding on many farms: green maize; bye-products such as reject grain, maize stover, cereal straw, sweet potato vines and banana residues; purchased materials such as hay, brewery waste and poultry manure. The data of Stotz (1977) suggested that 30% of the fodder in stall-feeding systems consisted of green maize, whereas Elephantgrass contributed 45%, the remainder being made up of maize stover, sweet potato vines etc. This was investigated in more detail in a survey by Wouters (1987a). Only farms which had Elephantgrass were included in the survey (Table 11.2).

Table 11.2. The contribution (%) of Elephantgrass to the dry matter (DM) available per livestock unit (LU) per day; results of a survey on stall-feeding in Kenya (after Wouters, 1987a).

Kg DM/LU/day of Elephantgrass	1983–1984	1984–1985	1985–1986
>10	35	27	33
5–10	30	27	29
0–5	35	46	38

Table 11.2 demonstrates that only a third of the farms had more than 10 kg DM/LU/day available in the form of Elephantgrass, and 38% less than 5 kg. In some districts such as Kiambu and Meru, where most of the stall-feeding is concentrated, the average contribution by Elephantgrass was not more than 4–5 kg DM/LU/day. Since the survey covered only farms with this grass, it can be estimated that Elephantgrass provides about 25% of the roughage used in stall-feeding, on a national basis.

Potter (1987) confirmed that the on-farm availability of forage in smallholding areas is very low, typically being less than 5 kg/head/day DM. At the farm level the reason for low milk yields is much more the lack of quantity than of quality of feed.

11.4. A re-appraisal of Elephantgrass

Elephantgrass has been subject to irregular waves of interest on the part of officials in both extension and research. Its giant appearance and dry-season greenness are easily interpreted as a capacity for high yield and drought tolerance. Others, however, object to the labour requirements of planting and weeding, to the problems of eradication, to the low yields in dry or cool environments and to the mineral depletion of the soil in zero-grazing systems. Elephant- and Rhodesgrass seem to take turns in the priority they receive, although their roles should be seen as complementary, as epitomized in this book.

It is worth noting that, despite the publicity it received in soil conservation and mulching campaings (*11.3*), Elephantgrass was hardly mentioned at the Agricultural Research Conference (1947) or at the Colloquium on Grassland Research (1950), in both of which meetings all experts concerned with pasture development in East Africa were convened. Significantly also, Henderson & Preston's (1959) "Fodder Farming In Kenya", a 149-page book, devotes only 1.5 page to Elephantgrass, even less so than to fodder sunflower. Of the many articles of Bogdan, none has Elephantgrass in the title. It has also received only limited attention from the International Research Institutes, although it is one of the few cultivated grasses compatible with legumes (*12.8*).

Elephantgrass has been credited with a high yield potential, but we need to confirm whether this is confined to the humid zones; we need to consider what relevant proportion of this yield (edible leaf) could be produced when it is wanted most (in the dry season). Choice of clone, plant density, fertilizer and cutting regimes are of consequence in realizing this potential. On the other hand, full attention has to be given to the costs of planting, maintenance and replacement, in order to arrive at an assessment in comparison with the alternatives at the farmer's disposal, such as surplus conservation off sown pastures and annual fodder crops. Last but not least we should consider the implications of zero-grazing on mineral depletion and the productive capacity of the soil, compared with direct grazing. There has been a tendency recently to underrate: [1] the dry-season potential following spelling in the rainy season; [2] the expenses in terms of labour and time which the various operations imply; [3] the mineral depletion of the soil against the low efficiency of recycled manure (*5.2.2*).

It is essential to distinguish stall-feeding and zero-grazing. Stall-feeding is essentially a livestock system and may imply any kind of fodder, as exemplified above by the Wachagga in Tanzania. Zero-grazing denotes that a field of grass is being utilized in a cut-and-carry system and is not grazed directly. The arguments that plead for stall-feeding (*e.g.* security; tse-tse) may be quite different from those that plead for zero-grazing, when fodder can only be obtained from outside.

11.5. Growth habit

Mature Elephantgrass reaches plant heights of 5–6 m and up to 20 nodes per stem; we came across escapee cultivated plants in over-grown bush in riverbeds at the Kitale Station with a height of 10.5 m, having 105 nodes (Table 11.3). It is also common to find short, thin stems with numerous nodes growing from the same basis as normal, usually thicker stems. Combinations of 3.4 m height and 60 nodes have also been encountered. The stems are used in building for roofing or to form a wall-frame which is plastered with loam; as fencing material for poultry pens and as marker pegs in hand-planted row crops. Elephantgrass is used for the manufacture of paper pulp in some east Asian countries.

Table 11.3. Some maximum heights and node numbers of mature stems of Elephantgrass at Kitale (after Boonman, 1979b).

Habitat (undisturbed)	Maximum height (m)	Maximum number of nodes
Open field	5–6	40–60
Riverbeds	7–9	80–100
Riverbeds, with bush undergrowth	10.5	105

Underground there is a woody rootstock consisting of short, horizontal branched rhizomes which are thickly intertwined; at the base of the aerial stem the rhizome is swollen and oval in cross section (Tiley, 1969). Aerial tillering from axial buds, usually from the top half of the stem, is common in French Cameroons, but uncommon though rarely completely absent in Gold Coast (Tables 11.4 & 11.6). Smoothness of leaf refers only to the absence of a dense cover of short hairs on the leaf blades. Gold Coast is very hairy. Density of clumps denotes the absence of bare centres, with no shoots in the middle of the plant, as often found in Uganda Hairless, which makes the plant more susceptible to weed invasion.

Varieties such as Gold Coast rarely produce inflorescences (heads), except under the most favourable conditions. It may take up to 12 months from the onset of regrowth before heads emerge. Heads are more readily produced in primary growth grown from seed, which is characterized by short and thin stems (*9.7*; *12.1*).

While it may take a year before an individual stem reaches the flowering stage, greenness is retained for considerably longer and it may take 2–3 years before such a stem will collapse. The evergreenness of Elephantgrass in Zone III is a major property and is at the base of its popularity as a ruminant feed.

11.6. Tillering and leaf characteristics

Most reports on management aspects deal with yield or chemical composition in response to plant density, cutting regime or fertilizer application. Information on tillering dynamics and on the partition of dry matter over plant components is lacking although this grass has been studied for more than half a century. This is important because attention should concentrate on maximizing the yield of leaf in the dry season (*13.2*).

Fig 11.1 presents data on tiller and leaf characteristics of a crop of French Cameroons over the dry season from November until April at Kitale (for climatic data, see Fig. 3.3). Regrowth followed a mature crop grown in 200-cm rows which was cut and topdressed with 50 kg/ha N on 1st November.

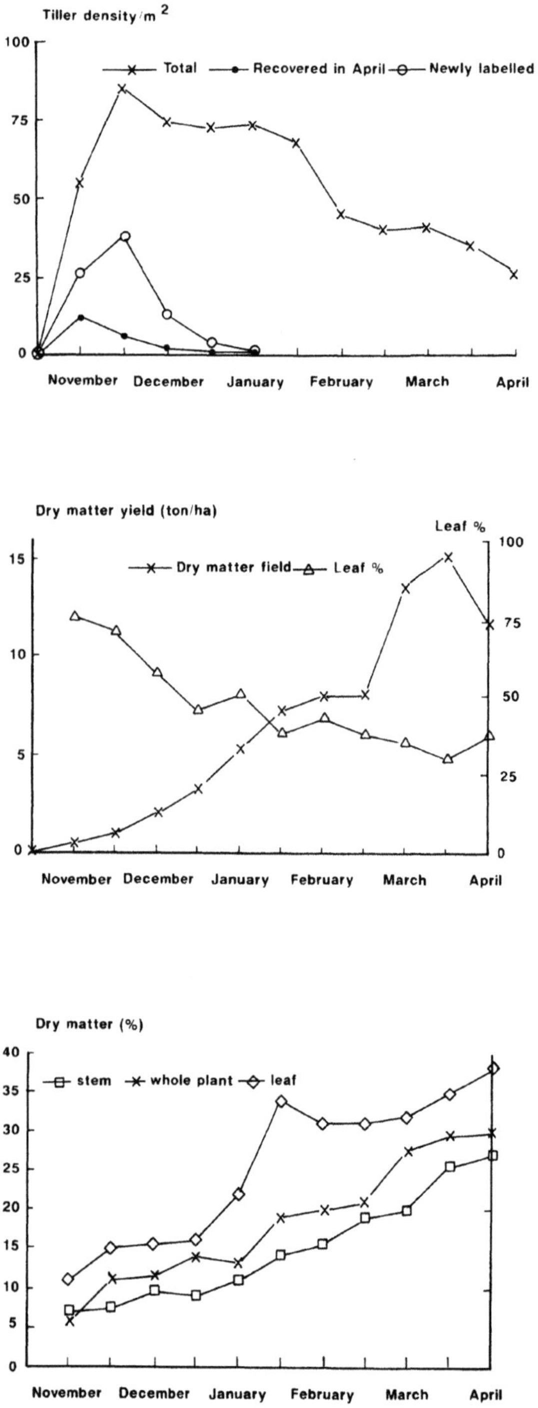

Fig. 11.1. Tillering in Elephantgrass (Boonman, 1979b).

Row sections of 3-m length were cut to ground level at weekly intervals and analyzed. Tillering was monitored with the aid of coloured rings.

Tiller density rose rapidly in the first 3 weeks to $85/m^2$ and then declined to only 25 at the end of the dry season by which time the stems had reached a height of just over 2 m. DM yield increased at an almost constant rate to 12–14 ton by early April, which included 4 ton of leaf. The % leaf blade dropped from 75% in early November to 30% in early April. Although some rain was received during this dry season, the high yield was attributed mainly to the long rest to which the grass had been subjected during the rainy season. The graphs on the % DM of the leaf and stem components illustrate the high moisture content of Elephantgrass and the large difference in this respect between leaf and stem. In young herbage the % DM in the leaf blade was twice that of the stem. Stems of 2 months of age still had less than 10% DM. The top leaves had the highest % DM.

Peak emergence of tillers occurred 3 weeks after the cleaning cut, with 36 new tillers. No new tillers appeared after early January. In early April all surviving tillers were counted and measured for length. Mortality was lowest (53%) in early-emerged tillers and increased progressively. Tillers having appeared at peak emergence had a mortality of 80% but the survivers of that group reached a greater average length (210 cm).

At every sampling date 50 tillers were examined for leaf number and leaf length. By the end of the sampling period, early April, 20 leaves were recorded on most tillers and these had appeared fairly constantly over the 20 weeks of sampling, at the average rate of about 1 leaf/week. Leaf appearance was close to the rate of 2 leaves/week in November, but these were only a few cm long. Average leaf length increased from less than 1 cm at the start to a maximum of about 90 cm in the 8th leaf appearing early January. Subsequently emerging leaves remained at this length, although there was a tendency of leaf length decreasing past the 10-15th leaf. Some individual leaves reached a length of 125 cm.

11.7. Elephantgrass varieties

Although they are truly clones, reference is normally made to them as varieties.

Edwards & Bogdan (1951) reported that the variety used generally in Kenya at that time was said to have been introduced from southern Rhodesia (Napiergrass) and to be more productive than types obtained locally in Kenya and Uganda. Apparently, local varieties were not highly regarded. At Bukura in western Kenya, Bradford (1946) recommended contour planting with "Uganda Napiergrass, not to be confused with the rank, coarse Kenya or Sudan type of Napiergrass". However, local varieties were still used in the late 1960s. Kusewa (1974) reported on the Limuru and Meru varieties, named after towns in central Kenya (Table 11.5). In Uganda farmers were advised to choose from leafy material, growing locally (Tiley, 1969). In Rwanda (Deschuytener, Ann Rep 1965) work focused on IAR 252, Keyberg and Minnasmall, the latter probably being the same low-yielding dwarf type referred to in Tables 11.4 & 11.5.

Table 11.4. Variety characteristics; on a 1–5 scale (5 = agriculturally desirable) (after Strange, 1959a).

Variety (clone)	Palatability	Smoothness of leaf	Tillering from nodes	Density of clumps
Gold Coast	3	1	2	3
Cameroons	2	1	3	5
French Cameroons	5	5	5	4
Uganda Hairless	5	5	3	1
Babala hybrid	3	4	1	1
Minna (dwarf-type)	5	5	3	2

Two events have been important as regards the variety issue:

1. In the late 1930s in Uganda, A.S. Thomas (quoted by van Rensburg, 1950; Tiley, 1959) selected from seed collected from various parts of Uganda and the clones retained were numbered Kw (Kawanda) 1-15. Although generally lower yielding, Kw-4 (Uganda Hairless, registered at Kitale as K5182) became well known since, as Strange (1959a) put it: "hairs on Elephantgrass can cause intense skin irritation" (Tables 11.4 & 11.6). Bogdan (1977) reported that varieties with stem nodes and leaf sheaths covered with stiff, easily breaking hairs are avoided by animals because eye damage can be inflicted by the hairs. However, Uganda Hairless is not completely hairless and still has sharp, awned leaf edges. Tiley (1969) reported that all known varieties had "razorlike edges" which cause irritation in the mouth of the animal. Hairless and awnless material is more pleasant to handle manually. Uganda Hairless (known as "Uganda") reached some fame in Central America, where it was said to be tolerant to eye-spot disease (*11.8*).
2. In the 1940s and 1950s a number of West African varieties became popular all over southern and eastern Africa. Kennan (1952) reported that these had been selected by Dr Cobb at Prinshof (South Africa) from seed collected by I.B. Pole Evans in West Africa. Cameroons (K52502) and Gold Coast (K52503) were brought to Kitale. This was followed later by French Cameroons (K53377); the original colonial names have proved persistent up to this date and the attempted renaming of Gold Coast to Ghana has led to great confusion (*11.7.3.1*).

In Table 11.5 a list is given of the main varieties which have been studied. The superior yielding ability of French Cameroons and Gold Coast over the local varieties is quite evident. In Malawi, Jackson (1951) found Gold Coast and Cameroons leafier than the indigenous varieties. For eastern Uganda, Horrell (1963) recommended Cameroons. This variety is, however, unlike French Cameroons, not thin- but thick-stemmed. Tiley (1970a) found the differences between varieties not large and since leaf yield was considered of paramount importance his recommended variety was Kw-16 rather than Cameroons (Kw 107) or Gold Coast (Kw 108). Tiley stressed the importance of equal initial stands, to avoid "false starts", so common in clonal comparisons. When new introductions are screened it is important to multiply them beforehand together with local standard varieties so that propagation

Table 11.5. Variety trials of Elephantgrass (ton/ha).

Variety	Dry season		Whole season		
	1956/1959[a] Kitale		1968/1970[b] Embu	1977–1978[c] Kitale	Kakamega
	Total	Leaf			
French Cameroons	7.2	5.7	21.2	16.6	19.6
Gold Coast ("Bana")	7.0	6.1	18.9	21.9	25.0
Cameroons	5.7	5.1	17.8	—	—
Meru	—	—	14.1	—	—
Uganda Hairless	5.3	4.2	11.8	—	—
Limuru	—	—	11.1	—	—
Babala hybrid (K5487b)	4.6	3.8	—	—	—
Minna (dwarf)	3.6	3.3	—	—	—
Clone 13 (Artona)	—	—	—	23.0	27.6
Elmba Rhodes	—	—	—	17.0	26.0
Nasiwa Setaria	—	—	—	18.8	20.0

[a]Strange, 1959a
[b]Kusewa, 1974
[c]Van Gastel, 1978.

material can be used that is of the same initial vigour. It often happens that entries with weak planting material continue to perform poorly, and to remain behind in subsequent years due to inter-plot effects.

11.7.1. Varieties and varietal identification

Confusion of variety names has become very common. Varieties are in circulation with more than one name. Examples will be quoted below of old, almost abandoned varieties which were inadvertently reintroduced and which became popular under new names. Also in demonstrations and variety trials, well-known varieties may be present under the old name (mostly denoting the site of origin) and a different name, normally the location from which it was last collected. An example of this is the Baraton variety. This is in fact Uganda Hairless, but reintroduced from Baraton. The confusion around Gold Coast, at one time renamed Ghana, has been even more dramatic as will be decribed below ("Bana").

No keys to distinguish the better known varieties have ever been published, but Bogdan's description (1965a) helps to recognize Gold Coast, Cameroons, French Cameroons and Uganda Hairless. The listing by Strange (1959a) of differences in aerial tillering, clump density and relative palatability (Table 11.4) is a useful addition. However, little use has been made of these descriptions so that name confusion has persisted. The parameters of Bogdan and Strange are listed in Table 11.6; some other characters and a description of Clone 13 are added.

Clones can be distinguished by quantitative and qualitative characters. They differ markedly in density and diameter of stems; degree of hairiness, of aerial tillering and of head production; commonness of a "black comb" on the upper edge of the leaf sheaths in mature stems. These characteristics are useful when comparing a range of varieties grown side by side.

However, qualitative differences are needed to distinguish individual plants and varieties grown by themselves. The easiest parameter is the pattern of hairiness of the leaf blade. Gold Coast has very dense short hairs, the Babala-Napier hybrid (K5487B) has long hairs which are not densely arranged. French Cameroons and Clone 13 have usually only a few short hairs which are scattered. Clone 13 carries a fringe of long hairs on the leaf blade near the ligule. Uganda Hairless often has leaf blades with almost no hairs at all.

In addition, habitus differences can be most useful. Tops of mature stems of Gold Coast carry a fan-shaped leaf arrangement and the lower leaves do not fall off early when dead; the leaf sheaths diverge slightly away from the stem which gives the combined stem and leaf sheath an oval diameter. At maturity Clone 13 has almost no leaves on

Table 11.6. Botanical characters of some important elephantgrass varieties (Boonman, 1979b)

	Number of stems at maturity (m^2)	Stem diameter	Hairs on leaf sheath	Black "comb" of hairs on upper fringe of leaf sheath	Long hairs on leaf blade near ligule	Hairiness upper surface of leaf blade	Aerial tillering from nodes	Flowering at Kitale	Special characters
Gold Coast (Ghana; *alias* "Bana") K52503	10–15	Thick	2–3 mm dense	Very common	Common	Hairy to the touch; very dense short hairs	Uncommon	Very uncommon	Palish colour; old leaves remain attached to the stems; top leaves give stem top fan-shape appearance; leafsheaths diverging which "flattens" the stem
French Cameroons K53377	15–20	Thin	2–3 mm not dense	Uncommon	Uncommon	Not hairy to the touch; scattered short hairs	Common	Common	Very fast establishment from cane cuttings
Clone 13	30–40	Very thin	3–4 mm not dense	Uncommon	Very common	Not hairy to the touch; scattered short hairs	Very common	Very common	Erect shoots; very poor establishment from cane cuttings
Uganda Hairless K5182	20–25	Thin	2–3 mm not dense	Uncommon	Uncommon; short hairs	Not hairy to the touch; hairless to scattered short hairs	Uncommon	Uncommon	Old leaves purplish; open clumps
Babala-Napier (K5487B)	15–20	Thin	4–5 mm	Very common	Very common	Hairy to the touch; long hairs not dense	Uncommon	Common	Not cultivated; very hairy

the main stem, but only small leaves on the tillers which spring up abundantly from the nodes. The general plant habitus is often sufficient for easy recognition, but in case of doubt, a comparison should be made both at a young and at a mature stage in adjacent plots.

11.7.2. French Cameroons

French Cameroons is still the standard variety in most countries of the region, although of late it has been attacked by white mould disease (*11.8.1*), in isolated pockets. The reasons for its popularity were the following:

1. French Cameroons gave the highest total yields (Table 11.5).
2. French Cameroons gave the highest dry season yields, being second only to Gold Coast in leaf yield (Table 11.5).
3. Unlike Gold Coast, French Cameroons is relatively free from hairs on the leaf blade and leaf sheaths (Tables 11.4 & 11.6) and is less unpleasant in manual handling.
4. French Cameroons has a high clump density, high palatibility and the capacity of aerial tillering to develop what Strange (1959a) characterized as a "leafy mop of regrowth" (Table 11.4). It also has a remarkable speed of establishment from cane cuttings (Table 11.6).

French Cameroons resembles the Merkeron type known in South America.

11.7.3. Gold Coast (alias Bana)

The variety widely known since the mid 1970s as "Bana" or "hybrid Bana", is in fact a misnomer of Ghana or Gold Coast (Boonman, 1979b; Karanja, 1981). The story is worth reporting in order to show how easily variety names get mixed up and what good or harm publicity campaigns can bring about.

The facts date back to 1974. It was reported then that a hybrid Napiergrass of South African origin, called "Bana" grass, had exceptional drought resistance, yield and leafiness. The origin of the claims and the planting material being distributed could be traced back to Delamere Estates in Elmenteita near Nakuru and from thereon to the Potha Farm of Mr Percival near Machakos, both locations being in Zone IV. Indeed, the green fields of this grass could be spotted from far in the middle of the dry country around. Word had long gone out and Rossiter (1974) reported that considerable areas were being planted in Machakos; planting material was leaving Potha Farm in lorry loads for different parts of the country. Stem cuttings were locally sold forward at what would now be the equivalent of 1 Kenya shilling per piece. "Bana" was said to be a hybrid between *Ba*bala pearl millet (*Pgla*; 2n = 14 chromosomes) and *Na*piergrass (2n = 28). It should be recalled that since the advent of hybrid maize, the word "hybrid" had come to mean something in Kenya.

Officials were reprimanded by higher authorities for having lagged behind farmers in their quest for productive grasses and for not having introduced the grass to Kenya any earlier. Most research stations subsequently joined in the promotion campaign and experiments were set up, inadvertently including comparisons of "Bana" and Gold Coast. Even the commercial beef feedlots (*11.3.4*) around Nakuru and the Beef Research Station at Lanet considered "Bana" as a possible alternative to maize silage (van Arkel, 1977). Research stations became an important centre of distribution of plant material.

The "Bana" of Potha differed from other known Napier × millet hybrids. Material was planted for identification at Kitale in 1974 and roots were sent for chromosome counts. If "Bana" were a hybrid of the above formula, it should have 21 chromosomes. Analysis showed, however, 28. Chromosomes were no different in pattern or number. "Bana" was, therefore, identical to Napiergrass (Kitale Ann Rep 1975).

Comparison in the meantime with the Kitale collection had shown that the "Bana" of Potha was markedly different from Bogdan's introduction of Bana, *i.e.* from the true Babala-Napier hybrid introduced from South Africa in 1954 (K5487b) and tested (Table 11.5). This material had been developed in South Africa by Gildenhuys (1950). Chippindall (1955) wrote that the *Ba*bala-*Na*pier cross was called Banagrass. The name "Bana" was referred to also by Lansbury (1959), working in Ghana, by Whyte *et al.* (1959) and by Borget (1969). Tiley (1969) and Bogdan (1977) referred to the hybrids as "Babala-Napier hybrids or Bana". Both authors and Horrell (1963) were rather firm in their conclusion that this hybrid was poor in productivity compared with Elephantgrass in East Africa, which confirmed the observation of Strange (1959a; Table 11.5). His Babala-Napier yielded only 4.6 ton as against 7.2 and 7.0 ton of French Cameroons and Gold Coast, respectively, over the dry seasons of the years 1956–1959. Bogdan (1965c) reported that hybrids are not recommended as they go easily to stem and are not as persistent. No further interest was shown in this material and no further references can be found in the Kitale files. Dr Käthe Brix, who continued Gildenhuys's work after the latter's death in 1965 writes that "nobody in South Africa seemed interested and the material was discarded" (private communication, 1979).

Another early reference is found in the July 1958 edition of the "East African Farmer and Planter" entitled: "New hybrid from South Africa shows promise". This article refers to roots imported in 1955 by air from Pretoria by Mr. R. Ballantyne of Kiambu of the "hybrid Napier/Bana*l* grass" and goes on to describe the grass as a useful coffee mulch or cattle food. As the name "Banal" given to this material indicates, difficulty with the correct spelling was apparent right from the start.

Comparison with the Kitale collection, as already mentioned, demonstrated a striking similarity between the "Bana" of Potha and Gold Coast (K52503). Gold Coast had been renamed Ghana (Bogdan, 1965a) following Ghana's independence in 1956. It was, therefore, deduced that the names Ghana and Bana had been confused along the way (Boonman, 1979b). Two very similar names had somewhere along the line been mixed up: gh*ANA* and b*ANA*. A spelling error with greater consequencies than the "Banal" already reported. Thus Gold Coast, *alias* Ghana, had staged a come-back under the guise both of "Bana" and of a "hybrid". It should be recalled that Gold Coast was no longer well known at that time, since French Cameroons had become the officially recommended variety. This explains why "Bana" of Potha was not recognized earlier as Gold Coast.

Why then could this grass be believed to be a hybrid? I assume that in the nearby coffee plantations in upper Machakos, not far from Potha Farm, "hybrid Napier/Banal grass" had been introduced around 1958 for mulching and animal feed, following the report in the "East African Farmer and Planter", already quoted. The grass itself was perhaps abandoned but the hybrid association must have lingered on for some time. The names Bana and Ghana were both adopted at the same time (late 1950s) and were, therefore, liable to confusion. What Potha had planted was Ghana (Gold Coast) but inadvertently named Bana and assumed to be hybrid Napier. It should be emphasized that the South African connection and the "hybrid" association figured highly at all critical moments of the story. The grass was, so it was claimed, of South African origin and had been imported. And indeed, the true Bana had originated from South Africa and was a Napier × millet hybrid.

However, also the commonly called West African varieties, Gold Coast and Cameroons, had been developed in South Africa (see above). Gold Coast used to be grown in South Africa and Rhodesia (Whyte *et al.*, 1959; van Zijl, 1970). The name's confusion and the hybrid associations may well have arisen there and then, so that the "Bana" of Potha was perhaps imported into Kenya directly. Ironically, both the correctly and the falsely named Bana-Napier hybrid were in Kenya already, but very few people recognized that the latter was in fact Gold Coast.

The probable reason why the "Bana" fields at Elmenteita and Machakos had attracted so much attention in the 1973/1974 drought was that they had developed a fodder bank in a period of high rainfall. They had, therefore, not been utilized heavily, and they stood on strategic places in the middle of dry country. And why 15 years later? Although Elephantgrass was reliable only at the upper limit of the semiarid zone in Machakos, Ghana (Gold Coast) had been performing relatively well (Table 14.1; Sands *et al.*, 1970). This was also the period during which hybrid Napier programmes, designed at Makerere University to develop a seeding Napier, attracted publicity (Fernandes, 1971; Mugerwa & Ogwang, 1976; Ogwang & Mugerwa, 1976). Most important of all, however, was that interest in Elephantgrass was beginning to stir nation-wide in the early 1970s. A new drought-tolerant variety, and certainly a hybrid, could not fail to draw attention. Table 11.7 gives a chronology of events which led to the confusion of Ghana and Bana.

As Bogdan (1977) reports, claims of miracle grass hybrids had previously appeared with

Table 11.7. The chronological events that led to the confusion of the varieties Ghana (Gold Coast) and Bana (Babala-Napier hybrid); after Boonman (1979).

Period	Gold Coast	Babala-Napier ex South Africa
1952–1954 Kitale (K)-introduction	K52503	K5487b, named Bana
1955–1960	Renamed Ghana and tested at Katumani	Discarded at Kitale (Strange, 1959a), Kawanda (Tiley, 1969), Serere (Horrell, 1963) and in South Africa (Brix, personal communication 1979)
		Publicity in Kiambu of a "Hybrid Napier/Banal" (*East Afr. Farmer & Planter*, 1958)
1960–1970	Superseded by French Cameroons	No records; hybrid Napier program started at Makerere University
1970–	Ghana (Gold Coast) confused with Bana	Bana confused with Ghana

Kavirondo Perennial Sorghum and Sorgrass (*15.2.4*) as well as with Ronphagrass, a Phalaris hybrid (*6.16*). Press reports and confusion of names can trigger a response, especially when the initiative comes from respected farmers.

The name "Bana" became so popular that it has almost become synonymous with Napiergrass. However, the lorries which had travelled all the way from research stations in west Kenya to Machakos could have stayed at home to collect planting material of Gold Coast from their own nurseries. And farmers would not have paid exorbitant prices to obtain a simple cane cutting. A positive aspect of the "Bana story" is that farmers played a major role, inadvertently, in having an old variety (Gold Coast) revived. It also shows what an effort taken up by officials of extension- and research services can do to promote a grass.

Bogdan (1965a) gave the following morphological description (see also *11.7.1*; Tables 11.4 & 11.6).

"GHANA (GOLD COAST) K52503. Shoots not numerous. Stems thick. Leaf sheaths with numerous stiff hairs which fall off early. The edges of the upper portion of the sheath are fringed with numerous hairs. Well developed leaf-blades are up to 1 m long and 45–60 mm wide, arching near the top; numerous soft hairs cover the upper surface. Flowering heads appear only very occasionally".

It is of interest to note that Bogdan and Strange had earlier considered Gold Coast to be less recommendable, as the hairs on the leaf sheaths are a deterrent to manual handling. Gold Coast produces more leaf (Strange, 1959a; Table 11.5), but Strange found French Cameroons superior because of its palatability, smoothness of the leaf and growth form (Table 11.4).

In the Kenya District Trials conducted in the mid 1970s (Table 6.2) no evidence of any yield superiority of Gold Coast was observed. The overall average yields of Gold Coast and French Cameroons were remarkably similar. In a subsequent trial at Kitale and Kakamega, van Gastel (Table 11.5) showed that Gold Coast was outyielded by Clone 13 and by sown grasses. Neither was Gold Coast any better in lower-rainfall stations such as Naivasha or Lanet. Particular effort was undertaken to test this grass at Katumani, almost adjacent to the cradle of the "Bana" of Potha Farm. Plantings failed, however, in the drought of 1975 and again in 1976 which led to the trial being abandoned. More recently, Karanja (1988) reported that this grass is more sensitive to drought than French Cameroons.

Consequently, little if any superiority of "Bana" has been established experimentally.

11.7.4. Clone 13

Clone 13 is the only variety which has remained free thus far from the white mould disease (*11.8.1*). It was developed in the 1970s by the Grass Breeding Team at Kitale. A simple breeding procedure was adopted based on the variability within the sexual progeny of French Cameroons (9.7).

Clone 13 was evaluated by van Gastel (1978), the results of which are reported in Table 11.5. Clone 13 proved outstanding both at Kakamega (Zone II) and Kitale (Zone III) in herbage yield and in tolerance to white mould. Clone 13 is somewhat similar in appearance to French Cameroons (Table 11.6). The stems are even thinner, which together with the habit of profuse tillering from the nodes may explain the fact that cane cuttings show a poor rate of establishment, compared with root splits.

A large amount of work has been done with Clone 13 at Kitale since its release. Clone 13 had a higher digestibility than Gold Coast *alias* Bana (Kitale Ann Rep 1986); the D-values were 57.6 and 55.1, respectively.

As already mentioned, the giant size of Elephantgrass is often equated with yield and its vegetative propagation is seen as an financial advantage over sown grasses. Ironically, such periods are usually followed by increased interest in seeding types (*9.7*; *12.1*) and in dwarf types (*e.g.* Minna; Tables 11.4 & 11.5); interest has recently been shown in a dwarf type from USA, but no relevant comparative yield data are available (S.C. Schank, private communication, 1992).

11.8. Diseases and pests

Like most other vegetatively propagated grasses, Elephantgrass is susceptible to diseases which can be very destructive. Mention has been made of eye-spot (*Helminthosporium ocellum* Faris (syn *H. sacchari*), which wiped out Elephantgrass in the 1930s in the Caribbean (Paterson, 1938; Blaser *et al.*, 1955; Takahashi *et al.*, 1966; Vicente-Chandler *et al.*, 1974). This fungus is known to cause damage in sugarcane in East Africa. The eye-spot symptoms have been observed in Elephantgrass at Kitale, but no certainty exists as to whether this fungus is responsible for some of the many unidentified blights, rusts and blotches, with which this grass is inflicted. A leaf discoloration when the plants grow older at Kitale was identified as *Piricularia grisea.* A "sooty mould" commonly found is associated with aphids (*11.8.3*) and was identified as *Khuskia oryzae.*

The risk of disease is enhanced by the monogenotypic (clonal) build-up and by vegetative propagation. Two major diseases have been identified as of economic importance in East Africa:

11.8.1. White mould (Beniowskia sphaeroidea)

This disease was already reported by Maher (1936a) and was described by Nattrass (1941):

> "is frequently seen, especially during the rains, to be covered on the lower surface of the leaves, and also to lesser extent on the upper surface, with numbers of small raised white pustules of a snow or cottony-like texture. Individual pustules vary in size from 0.5 to 2 mm, in diameter, and may be up to 1 mm in height. When viewed with a pocket lens they appear to have a faint pinkish tinge. The pustules may occur either single or more often in line, each group lying parallel with the midrib. There may be as many as eight or ten, decreasing in size from the centre. On the opposite surface of the leaf position of a group of pustules is indicated by a lenticular-shaped spot, which, according to age, may be light green or yellow or consist of dry pale tissue surrounded by a purple margin and which superficially resembles spots caused by Helminthosporium. The white mould is the fructifying part. It is common throughout East Africa".

This is how the symptoms had been known for many years but no damage was believed to accrue from the disease. In fact, the Plant Disease Services in Nairobi continued up till the late 1970s to advise that the fungus was harmless. The same symptoms of white pustules are also found in Setaria and in Kikuyugrass.

The disease has certainly become a serious threat to Elephantgrass in recent decades. In August 1974 I was called to Meru district, north of Mount Kenya, by the Kaguru Farmers' Training Centre. A serious outbreak had occurred on many farms. In some plants leaves were only showing the early symptoms of the white pustules but there were also badly stunted plants with purple-coloured short leaves. In 1975 serious outbreaks were observed in Kiambu district and Nandi district and in 1976 at the Kitale Station. It was only at the latter station that the name of the affected variety, French Cameroons, could be established with certainty. This variety came under attack also at the Kakamega Station in 1977.

In the absence of expert phytopathological support no proper investigations into the epidemiology of this fungus could be conducted but the following general observations were made:

1. Uganda Hairless, French Cameroons, Gold Coast (*alias* Bana) and local ecotypes are affected.
2. The symptoms of white pustules are often only seen in the 1st season of the infestation. In subsequent years affected plants show stunting and a purple or brown discoloration of the leaf, but often without white pustules. Other pathogens may have become involved at that stage.
3. Plants, once affected, do not recover.
4. Symptoms are rarely found on the youngest 1–3 leaves.
5. The infestation is usually not uniform through the field and can remain strictly localized.
6. It has not been possible to relate the incidence of an attack to the source of planting material. Primary infection, from plant to plant in the field, seems of greater importance.
7. Repeatedly low cutting to impose stress while screening for tolerance close to a source of infection was not effective.
8. Breeding Elephantgrass from sexual progenies is the easiest approach thus far to produce new tolerant varieties (9.7).

The absence of white pustules in the later stages, when the plants are purple in colour and stunted, may be the reason why the symptoms of white pustules were regarded harmless. Other pathogens may also be involved. Badly affected fields have to be ploughed. How common this is can be learnt from the survey by Wouters (1987a), referred to in Table 12.6. Some of the farmers' fields had to be uprooted while the experiment was only half-way.

At Kitale, the fungus had previously been noted most prominently in nurseries in the Uganda Hairless variety. However, French Cameroons became the standard variety and, therefore, the disease has been associated mostly with this variety. Early hopes that Gold Coast (*alias* Bana) was resistant were shattered in mid 1978 when it simultaneously appeared in 2 plantings, in one of which with a vehemence as had not been witnessed at Kitale before. Clone 13 has been tolerant thus far, and has given very high yields also when compared with disease-free French Cameroons or Gold Coast (Table 11.5). It will be necessary to continue the relatively simple breeding technique used in the development of Clone 13 to keep ahead of the disease (9.7). Chemical control has not been attempted but is unlikely to be practical.

11.8.2. Stunting disease

Tiley (1969) reported that stunting disease, probably a virus, was more serious than white mould:

> "It becomes visible in the regrowth after cutting or grazing. Affected shoots are pale yellow green in colour and seriously dwarfed, with a densely bunched (fastigiated) appearance on account of repeated proliferation of the axillary buds. Individual leaf and shoot size are greatly diminished and the leaves show clearly off the veins. The leaf sheaths are often enlarged. Root development is severely impaired and is reduced to a weak fibrous growth with truncated tips which tend to grow in an upward direction. Symptoms in the rhizome include narrowing and lengthening, with a flattened and twisted growth. Infection at an early stage results in symptom expression in the whole stool, with complete loss of yield and eventual death at 12–24 months. Later infections are frequently restricted to one or two growing points in the stool, the remaining stools staying healthy and contributing normally to yield. The amount of infection in the field is usually small and spread is slow, but can occasionally rise to 90% or higher resulting in a serious reduction in herbage production. At low levels of infection there is compensatory smothering of the diseased plants, which frequently cannot make use of the applied nitrogen due to the faulty root system. The cause of the virus is not yet known, but its behaviour suggests an insect-borne virus. Infection can be carried from place to place with the vegetative planting material, so that selection of healthy canes for planting is one method of control. However, the disease also arises spontaneously in healthy stems after periods of regular cutting or grazing. It can also be seen developing at the bases of caning Napiergrass and in very old canes appears to travel up the stem".

The possibility that there is some relationship between this stunting disease and white mould cannot be ruled out.

11.8.3. Pests

In Kiambu in central Kenya, where Elephantgrass is grown more than in any other district, many insects are found. Shootfly larvae kill young shoots when cut at ground level (see also *15.6.2*). Aphids are also abundant and are allegedly associated with "purple midrib" and "sooty mould" (see above). Bogdan (1965c) reported that Elephantgrass is "occasionally attacked by mites and such attacks can be severe. Mites settle on the underside of the leaf; leaves then become yellow in colour and die out. Hairy varieties are more affected because the mites prefer nesting between the hairs" (see also *6.11*). Termite damage is serious when the plant is weakened by drought and when dead material is allowed to accumulate in the crowns. Tiley (1969) also recorded a variety of insects "including leaf hoppers, ants, aphids, mites and flies. Noticeable damage has been recorded from a dipterous stem borer larva which lives in and damages the base of the apical leaf and growing point, and from mites, which give rise to reddened blotches on the leaves in heavy infestation". It is likely that the dipterous stem borer corresponds with the shootfly referred to above. For a list of insects of which Elephantgrass is host see Le Pelley (1959).

11.8.4. Disease and quality control

Given the fact that varieties consist of one single clone, Elephantgrass is extremely vulnerable to diseases and pests, especially when these are transmitted by vegetative propagation. The devastating effect of the eye-spot disease in the Caribbean has already been mentioned. The next grass to be introduced, Pangolagrass (*Ddec*), again a single clone, fell to the Pangola stunt virus (*6.16*). It is a matter of priority that breeding of Elephantgrass in East Africa is kept up to keep possible new outbreaks at bay. Measures should be taken also to improve the quality control of material offered for planting. Weeds such as African Couch- and Stargrass, Oxalis and various sedges are easily transplanted with the root splits of Elephantgrass (*12.1.1*). Farmers should be able to obtain healthy, weed-free material of specified varieties.

11.9. Herbage yields of Elephantgrass

Perhaps due to the giant stature and the huge herbage yields it can attain when allowed to grow undisturbed, Elephantgrass has often been credited with high yields. In the early days of the publicity around zero-grazing it was quite common to find farm fodder plans being based on the assumption of 25–30 ton/ha DM, even in Zone III/IV, and in any case twice as much as any other grass. Very high yields are indeed obtained in the hot, humid tropics on the better soils. Although even higher yields have been recorded experimentally, Vicente-Chandler *et al.* (1974) reported yields of over 50 ton/ha in Puerto Rico. However, the yields of digestible dry matter were only just above those of other grasses such as Congo Signal (*Bruz*), Common Guineagrass (*Pmax*) and Stargrass (*Cnle*), so that the livestock-carrying capacities were in fact similar. DM-yields of species under cutting can be in sharp contrast with animal performance under grazing (Fig. 10.6).

Canes exceeding 10.5 m (Table 11.3) have been measured at Kitale, though with less than 5% of the leaf still attached. Elephantgrass has a low sward density (*13.1.2*) and the ground is rarely shaded. Height should, therefore, not be mistaken for high yield. The main factor, however, responsible for the yield gap between green- and dry matter yields is the low % DM (Tables 6.2 & 6.3; Fig. 11.1). It was found that stands of Elephantgrass (16% DM) had to be twice as tall as swards of Rhodesgrass (21% DM) to achieve the same yield of DM. Elephantgrass can accumulate a large quantity of DM in one single cut. In August 1978 an average of 29 ton/ha DM was harvested in one single cut, grown uninterruptedly for 3 years. D-*vitro* (digestibility) was only 19. Such high yields of stemmy material obtained in one single cut are obviously advantageous for the purpose of mulching for which Elephantgrass has been extensively used (*11.3.3*).

High grass yields are normally obtained, without N-fertilizer, in the 1st year of planting. By contrast, the yields fall sharply to about half in the 2nd year and to a third or less in the 3rd year (Tiley, 1970a; Keya & Kalangi, 1972). Horrell (1963) recorded at Serere a yield of green herbage of 15 ton/ha in the 1st year, dropping to 6 and 4 ton in the 2nd and 4th year, respectively. A drop to one-eighth in the 4th year was observed by Deschuytener (Rwanda Ann Rep 1965). Fig. 11.2 illustrates the effect of age on non-fertilized Elephantgrass as observed by Tiley (1970a).

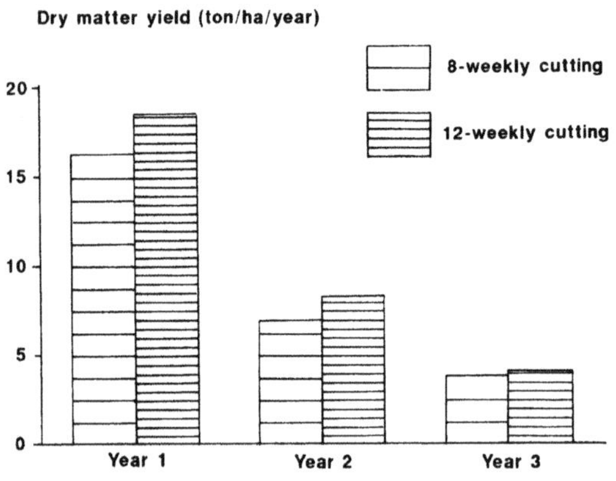

Fig. 11.2. The decline in yield with age of Elephantgrass (after Tiley, 1970a).

A decline in yield with age is typical of perennial grasses (Fig. 6.2) and is largely due to decreased N-availability common on almost all soils (*5.3.2*; *7.2.4.1*). Decline in yield is usually associated with lack of persistence and with weed ingression (*6.3.3*). Compared with sown grasses, there is some evidence to show that persistence is somewhat poorer in Elephantgrass under cutting regimes, but especially under frequent grazing (*4.2.3*; *12.5*; *13.3.5*).

11.9.1. Yield in relation to Eco-Climatic Zone

The highest experimental annual yields, on a 3-year basis, in Kenya (25.3 ton/ha) have been recorded at Kakamega (Table 6.2) in Zone II at an altitude of 1590 m, with 1845 mm annual rainfall. The rowwidth was 50 × 50 cm and an annual application of about 180 kg/ha N was given. With more N and at a lower altitude it should be possible to achieve nearly 40 ton in this zone as found experimentally at Kawanda by Tiley (1969). Elephantgrass has very high moisture requirements (Jackson, 1958), but it is rivalled in yield potential by sown grasses which was confirmed by van Gastel (Table 11.5). Also in the lowland humid tropics of Réunion, off the African coast, Fritz (1971) found Rhodesgrass and Elephantgrass equally productive.

In Zones III and IV, Elephantgrass is often outyielded by Rhodesgrass (Table 6.2). Zone IV is, in fact, too dry for Elephantgrass, even for its survival. Year-to-year rainfall varies greatly, so that Elephantgrass may do relatively well in some wet years and, once firmly established, perform equally well under lenient use in a subsequent dry year (*11.7.3*). Sands *et al.* (1970) working in Zone IV concluded that sown grasses were the best: "Elephantgrass is reliable only at the upper limit of the semiarid zone". Thomas (1960), working in the same area, had reported that Elephantgrass was attacked by termites when weakened by drought and unsuitable.

Zone III is where Elephantgrass should be put to its fullest possible use. Here it may play the complementary role in the prolonged dry seasons which are characteristic of this zone. Strange (1958a) suggested that the chief value of Elephantgrass from a nutritional point of view may lie in the "tonic" effect of the green leaf on milk production at a time when other grasses are relatively dry. Similarly, Scott (1967) referred to farmers' claims that a "green bite" in the dry season is important for dairy cows.

From the data reported in Tables 6.2 & 11.5 and from those of Wouters (1987a) of Naivasha, it may be concluded that the DM-yields as proposed in Table 11.8 are a realistic guideline.

The point should be stressed that these data refer to average yields obtained in the 1st and 2nd year after planting and when 100 kg/ha/year N is applied. In the warm, humid Lake Victoria region of Uganda yields are higher than in Kenya. Wendt *et al.* (1970) assumed 20–25 ton/ha. Allowing 70 kg/head/day fresh herbage they calculated a carrying capacity of exotic milking cows, under average conditions, of 2–3 head/ha producing 7 kg/head/day milk.

11.10. Drought tolerance

Elephantgrass and other giant fodder grasses have the impressive ability of staying green throughout a dry season lasting 3–4 months in Zone III. This phenomenon has mistakenly been interpreted as drought tolerance, *i.e.* as potential not only to survive but also to produce well in semiarid areas of Zone IV. However, the risks are too great. Elephantgrass will survive only in the wetter years and under lenient use (*11.7.3*; *11.9.1*). The same applies to the coastal areas, where Elephantgrass is often marginal (Goldson, 1968). Sorghum is a better proposition as a complementary fodder crop for this zone (*15.3*).

Table 11.8. Expected Elephantgrass yields in different regions (at levels of 100 N, 50 P_2O_5 and 50 K_2O in kg/ha/year).

Region	Zone	Typical location	Rainfall	Altitude	Expected dry matter yield
Central Uganda	II	Kampala	1600	1200	22.5 ton
West Kenya	II	Kakamega	1850	1600	20.0 ton
Central Kenya	III	Embu	1250	1450	15.0 ton
Central Kenya	III	Kiambu	1000	2000	12.0 ton
Coastal Kenya	IV	Kilifi	1000	0	7.5 ton

The ability to stay green in the dry season is not entirely confined to Elephantgrass or comparable giant-type grasses. It is rather a characteristic of grasses under conditions in which they have few tillers and large leaves. Also short grasses when sown in the 2nd half of the rainy season, have the property of staying green deep into the 1st dry season (*7.2.3.1*). In primary growth, after sowing, tillers are few and the leaves are large, relative to what is found in subsequent years.

11.11. Summary

Napiergrass is the name of the cultivated type of what is mostly known as Elephantgrass in its natural environment, *i.e.* the humid, warm areas north and west of Lake Victoria (U) or along riverbeds elsewhere. In (sub)humid areas it stays green in the dry season, but is not suitable for semiarid conditions. Elephantgrass was planted in the 1930s in Uganda to speed up its soil-regenerating capacity in the fallow period within the crop rotation. It was also recommended for mulching in coffee. Its value as a grazing grass or cut fodder for the dry season was rapidly appreciated. Stall-feeding had been a common practice in some regions. It developed steadily near the main towns where land and milk were high-priced but grazing scarce. Also security and insect-related problems (tse-tse, ticks, Stomoxys) encouraged the rearing of livestock indoors. Stall-feeding is commonly associated with Elephantgrass, but the bulk of the feeds used comes from other sources. Stall-feeding was officially promoted (*e.g.* beef feed-lots). The best Elephantgrass varieties are French Cameroons and Gold Coast (Ghana), of West African origin. The latter became also known as "hybrid Bana", because of a mistaken identification with a millet hybrid. Yield differences between varieties are small but tolerance to white mould and other diseases has become of paramount importance. Thus far, only the Kitale-bred "Clone 13" has remained unaffected. It is a fine-stemmed variety with a better digestibility. High yields of Elephantgrass are found in humid and warm areas (Eco-climatic Zone II). Yields are often lower than expected, however, due to a low % DM of the stem in particular, and due to a low sward density, in spite of the height Elephantgrass may attain when mature. Outside Zone II, annual yields are relatively low.

CHAPTER 12

Elephantgrass husbandry

12.1. Establishment

Many Elephantgrass introductions produce inflorescences (heads) only occasionally in favourable environments. For instance, no heads had been recorded from "Bana" (*11.7.3*), which was at one time believed to prove its hybrid nature, until heads were found in January 1979 both at Kitale and Embu.

Reports that no seed-set occurs in such heads are not correct. In the Kitale programme caryopses germinated as high as 90%, with little or no dormancy, and showed radicle emergence within 10 hours in germination tests (*9.7*). The seedlings' further development is, however, very slow, which is in support of the field experience that this grass is hard to establish from seed. Tiley (1959) referred to attempts made at sowing in 1939 in Uganda by A.J. Kerr, who would in the 1960s become a prominent farmer-promoter of Elephantgrass at Kitale (*12.2.1*).

Plants raised from seedlings are more variable in their primary growth than after subsequent cuts; tillers are also much more numerous and thinner and heads are produced readily, within 6 months after planting, already at a height of 1.5–2.0 m (*9.7*). After a cleaning cut, however, stems have to reach a height of 4–5 m and an age of 12 months or more before heads, if any, begin to appear at Kitale (van Wijk, 1977).

French Cameroons and, in particular, Clone 13 have been the best seeders so far, but the arguments militating against conventional seed multiplication can be summed up as follows (van Gastel, 1978):

1. Less than one crop is obtained annually of rarely more than 1–2 kg/ha PGS (pure germinating seeds).
2. The spikelets are fluffy, bulky and usually in clusters.
3. Field establishment from seed is very poor.
4. Sexual progenies are not "true-to-type": plants grown from seed harvested of the same parent are extremely variable, which leads to selective grazing with many plants left uneaten (Blaser *et al.*, 1955).

The recurrent statements about the desirability of a seeding Elephantgrass are an oddity of policy. It is often said that vegetatively propagated grasses suit the smallholder better than sown grasses because he does not need to spend cash on buying seed. However, when the labour burden concomitant with vegetative propagation begins to be realized, or when cash is demanded for the purchase and/or transport of vegetative material, the breeder or seedsman is asked whether they can make such a grass produce seed. This made Makerere University initiate hybridization with pearl millet to produce a seeding Elephantgrass (*9.7*; *11.7.3*), but the results were not encouraging.

12.1.1. Vegetative propagation

Vegetative propagation of Elephantgrass is a more reliable method. More vigorous plants are produced than with seed and they are true-to-type, so that a clone can be multiplied without losing, in practical terms, the specific characters for which it had been selected. For vegetative propagation a choice can be made between whole canes, cane cuttings or rooted tillers (splits). Each method has its advantages:

1. Whole canes. These are laid end-to-end in furrows and covered lightly. This is a quick but the least reliable method. Germination results

from the activation of dormant axillary buds, which can be erratic. It is advisable to remove leaves and top part of the stem, the buds of which are too young. Fields should be set aside for propagation at least 6 months ahead, so that the canes to be planted are fully grown, yet still green and with a waxy appearance. Thin canes are less viable.

2. Cane cuttings. This method allows for rejection of unsuitable canes, but inevitably requires a larger source area, 600 m^2 to plant 1 ha (Table 12.1). Germination is more even and faster, especially when 2 or more cuttings are used per hill. An average of 4 suitable cuttings with 3 nodes each can be obtained per cane, which should be about 75% mature, still actively growing, to give the best results; 1 cane is then needed to cover 100 cm length of row with 2 hills. Often the internodes are only 10 cm long so that cuttings of 5 nodes or more are needed to make pegging-in more practical. Bundles of cane cuttings are light and this method is recommended for long-distance transport. Cane cuttings can be stored for 2–3 weeks. At planting no furrows or holes need to be dug. The cuttings are stuck into the ground either vertically or at an angle with one-third above ground. It seems advantageous to plant the cuttings the right way up, so that the buds are above the nodes. Although pegging-in is a practical method of planting, cane cuttings establish better when buried completely.

Table 12.1. Requirements of planting material and manhours to establish 1 ha of Elephantgrass (100 × 50 cm) (adapted from Tiley, 1959; 1969).

	Requirements for planting 1 ha			
	Planting material			Manhours
	Number	Kg	Source area (m^2)	
Cane cuttings (2/hill)	40,000	4,000	600	285
Splits (rooted tillers)	20,000	10,000	1,000	445

3. Root splits (rooted tillers). This is the method most commonly applied when enough source material is nearby available. Field establishment is more reliable and faster. However, much more labour is required in collecting (digging) and preparing the material, as well as in transporting and planting. Although individual tufts can be split into single-shoot units and although even short rhizomes may germinate, the usual method is to plant the units as they are dug, with soil attached. Units of 2–3 shoots with soil may easily weigh 500 gram. Foliage is removed before digging. Table 12.1 illustrates that large source areas are needed, about 10% of the area to be planted, or more if it is not to be completely uprooted. Some 10 ton is needed to plant 1 ha at the spacing of 100 × 50 cm. Source plants recover poorly. When fields are ploughed mechanically, this is of great help to obtain planting material compared with when clumps are to be dug by hand. Old Elephantgrass fields are, however, commonly infested with weeds such as African Couch-, Stargrass, Oxalis and sedges. Weeds are easily transplanted with the actual planting material (*11.8.4*). Work is more amenable and better results are obtained when tufts are dug from plants, freshly established in a clean field, which have not yet developed a robust base clump and are, therefore, easier to split. New fields can be planted in stages, splits being taken from adjacent, already established strips.

The optimum time of planting is right at the onset of the rains (*7.1.2*). An even better start is made if the planting material has already been pre-rooted in nursery beds during the dry season. This is impractical with root splits since it implies digging in the dry season and transporting twice. Cane cuttings are more practical. Pre-rooting provides an efficient method of eliminating inferior canes and of overcoming the problem of slow field establishment. No use has been made of mechanical planters in East Africa.

12.1.2. Costs of planting

Tiley (1969) investigated the work load involved in the various methods of establishment (Table 12.1). Machine- and manhours were calculated. His calculations put the number of manhours required to establish 1 ha with 20,000 root splits at 445 hours; 2 tractor hours were required. Where transport is poorly organized labour requirements can be as high as 1000 manhours. Obviously, higher plant densities are prohibitive, but they are unnecessary (*12.4*).

12.1.3. Cost and labour efficiency of operations

Since there is no need to buy seed, vegetative propagation may be seen as a cheap alternative of establishing grasses. However, new planting material is rarely free of charge and the centre of distribution may be remote. Farmers paid the equivalent of what would now be 1 Kenya shilling per cane-cutting for "hybrid Bana" (*11.7.3*). On a hectare basis this was Kshs 40,000/-. Farmers may multiply their own stock, but this is costly in terms of time and acreage involved also.

Tiley (*op. cit.*) put the costs of cutting and transporting Elephantgrass for stall-feeding at 40 manhours/cut/ha. Stotz (1979) estimated the labour requirements of feeding and milking at 1 manyear per 2 cows + followers. If the labour needed for vegetative planting and weeding is also added, this may well amount to a total of 1 man per cow, on a year-round basis, for all the extra work involved, compared with direct grazing. It is crucial in stall-feeding systems that labour is always available. When human labour is translated in money equivalents, labour-intensive methods may be justified from the viewpoint of underemployment. Other farm operations may, however, come in conflict at times of peak labour demand. High requirements of labour may be sustainable on a family farm, but on larger farms they are prohibitive. The cumbersome work aspects of zero-grazing operations will certainly come to the fore.

12.2. Weed control

Elephantgrass has a more serious weed burden than sown pasture grasses, which has a negative effect also on subsequent arable crops. Even when harvested at the recommended stage (*12.5*) or when planted at high density (*12.4*), it leaves an open sward, liable to weed invasion, especially of African Couch-, Stargrass and various sedges. On the other hand, repeated cultivation encourages the spread of Oxalis (*11.8.4*; Table 7.6).

Imperata (*Icyl*) can be a major grass weed in Uganda, but most widespread trouble is encountered with African Couchgrass (*Daby*) which can effectively kill a crop (*4.5.1*). The longevity of a crop and its concomitant production are determined primarily by the measures taken to prevent the encroachment of Couchgrass. A clean seedbed and clean planting material are prerequisites, but repeated cultivation is needed until the leaf canopy is well established and, subsequently, after every cut or grazing. It has often been said that close spacing is desirable from the weed control point of view but even at a plant density of 50 × 50 cm weeds do invade. Good ecological weed control is achieved when the crop is allowed to grow to more than 200 cm height in anticipation of the dry season (*13.2.1*) and when N is applied to the individual plants rather than broadcast. Anything, apart from direct weed control, that is done to encourage crop competitiveness such as infrequent defoliation, wide row width, N-fertilization and a combination of all these will help in combating weeds.

12.2.1. Interrow cultivation

If frequent defoliation is the rule rather than the exception, hand weeding and interrow cultivation are necessary after every grazing or cut, irrespective of the rowwidth applied. Weeds are a problem also at narrow rowwidth because of the open-clump habit of Elephantgrass (Table 11.4). Wide rowwidths facilitate mechanical cultivation.

The system worked out by A.J. Kerr near Kitale is of interest. Kerr's method was geared to maximizing production in the dry season, but with limited N-fertilizer. A rowwidth of 300 cm was adopted. After the last grazing in August-September, some 3 months before the end of the rainy season, Kerr ploughed the interrow space with a disc plough in order to cope with the hard soil-surface. After such cultivations a much greener leaf colour was observed, probably the result of increased N-mineralization.

This N-effect due to cultivation (*5.3.2*) may offer some explanation for the relatively high yields of clean-cultivated Elephantgrass commonly encountered in the 4th or 5th year if no N is applied (Fig. 12.3). Bare ground and continuous cultivation are, however, not conducive to maintaining soil structure (*5.2*). Bare ground is vulnerable to the impact of raindrop, so that infiltration rates are impaired. Subsoiling, apart from being very expensive, produces little beneficial effect (Addison, 1958b) and dries out the soil more rapidly (Pereira & Beckley, 1952).

12.2.2. Herbicides

Triazine compounds have been used in the establishment stage at rates of 5 kg/ha Atrazine *a.i.*, which can control broad-leaved weeds and annual grasses while other herbicides such as 2,4-D can be applied also (post-emergence). Alachlor (Lasso) at 2 kg/ha *a.i.* was also effective (Kitale Ann Rep, 1984). See also *7.1.10.2.*

It has been found practical to use Dalapon at 100 g in 10 l water for spot application of weed grasses. Sufficiently wide rowwidth (200 cm) is necessary so as to prevent the effect of drift. Wallis (1962) observed that the hairless varieties were more susceptible to Dalapon than the normal hairy types. Dalapon application can be followed up with Paraquat (Gramoxone).

12.3. Eradication by ploughing

Elephantgrass stands, even when clean cultivated, drop in the 5th year to yields of less than one-third of the 2nd year, after the year of establishment, in the absence of N-fertilizer (Figs. 11.2 & 12.3). However, wide rowwidth and interrow cultivation, coupled with N-fertilizer and not too frequent defoliation can maintain a field in productive condition (Fig. 12.3). However, if grass weeds are not controlled from the start it is often better to plough the field at an early stage.

The great difficulty encountered in eradicating Elephantgrass by ploughing has militated against its wider usage in mixed-farming systems (Tiley, 1959). It can be cleared with heavy machinery, but not with ox-drawn implements. With ox-cultivation it is almost impossible to prevent the plough from jumping sideways whenever a heavy clump of grass is encountered, even when planted densely and spread evenly over the field. Hand-digging is often the only alternative to ox-implements. With hand-digging land was broken at a rate of less than 80 m^2/day, *i.e.* 125 mandays/ha (Tiley, 1959).

Horrell (1958a) reported that frequent hard grazing kills the grass eventually. This treatment has often been recommended to facilitate the reclaiming of Elephantgrass fields for cropping (Goldson, 1977). However, this is not a practical proposition in zero-grazing systems, without fencing, unless animals are tethered which is a practice found near Kitui in eastern Kenya. Secondly, while little benefit is obtained from grazing such unproductive stands, African Couchgrass and other grass weeds take advantage of the weakening of the Elephantgrass and aggravate the weed burden in the subsequent arable crops.

Even when ploughing has been successful it is common to find volunteering plants in subsequent crops. Where mechanical eradication is a problem, Elephantgrass can be effectively killed by herbicide treatment with Dalapon or Glyphosate (*4.2*).

12.4. Plant density

Whereas comparisons of Elephantgrass and other fodder grasses such as Guatemala (*Tand*), Giant Guinea (*Pmax*) and Giant Setaria (*Sspl*) are usually the first exercise of research programmes, trials with various plant densities are the second. Results are not always reported, probably because significant differences are normally found in the 1st year only. These are, however, not always taken seriously (*7.1*); weed growth and uneven start are commonly given as reasons. A clear exception is the work of Tiley (1970a).

In subsequent years extreme rowwidths (50–400 cm) and/or extreme weather conditions are needed if significant differences are to be found. Compensatory tillering evens out the differences in initial density, so that remarkably similar tiller densities and herbage yields are achieved. Under adequate irrigation or at the lower altitudes of Zone II, higher total yields may be observed under fertile conditions with high plant densities (50 × 50 cm) whereas in Zone IV it is commonly advantageous to plant at 300 × 100 cm, so as to reduce competition for moisture.

The discussion here will concentrate on Zone III, in which Elephantgrass can play the complementary role for which it is most useful (*11.9.1*). Not the total annual yield, but the yield in the dry season is of paramount importance, so that relatively low plant densities of 5000 plants/ha (200 × 100 cm) and less should be considered.

Table 12.2. The effect of dry-season rainfall and low plant density on yield of fresh herbage (ton/ha) of Cameroons Elephantgrass at Kitale; cleaning cut and 30 kg/ha N in August (after Strange, 1959a).

Year of dry season	Rainfall Dec–March (mm)	Plant density (cm)	
		300 × 100	600 × 100
1955–1956	320	22.5	17.5
1956–1957	130	7.5	5.0
1957–1958	360	35.0	25.0
Mean		21.7	15.8

Strange (1959a) examined extremely low densities of 1666 plants/ha (600 × 100 cm). The dry season yields were, however, lower than at 3333 plants (Table 12.2).

The advantages of low plant densities and wide rowwidths are obvious. It makes an enormous difference in costs of planting when only 5000 plants need to be planted (*12.1*). Wide rowwidth also allows for tractor- or ox-drawn cultivation to control weeds (*12.2 1*). At wide rowwidth, clumps expand and fill in the intra-row space so that a continuous line of growth is formed along the rows, which limits weed growth within the rows. In fact, at wide rowwidth, 250–300 cm, it is possible to widen the spacing within the rows to > 100 cm and still achieve the same number of shoots per length of row, once the crop is well established. At wide rowwidth, intra-row spacing becomes largely irrelevant. Kennan (1952) reported that Elephantgrass stayed productive longer at very wide rowwidths (400 cm).

Fig. 12.1 has been computed on the basis of numerous density and fertilizer experiments and presents in a general way the diverse effects of plant density, everything else being equal, on:

A. Yield in the establishment phase.
B. Yield in subsequent dry seasons.
C. Yield in subsequent wet seasons.
D. Total long-term yield over wet and dry seasons.

As Fig. 12.1 indicates, the effect of plant density is very marked in the year of establishment and also in the dry seasons, but in the opposite direction. In subsequent wet seasons plant density is of little significance. Consequently, the overall effect of plant density on long-term average annual yields is negligible. Most trials with plant densities

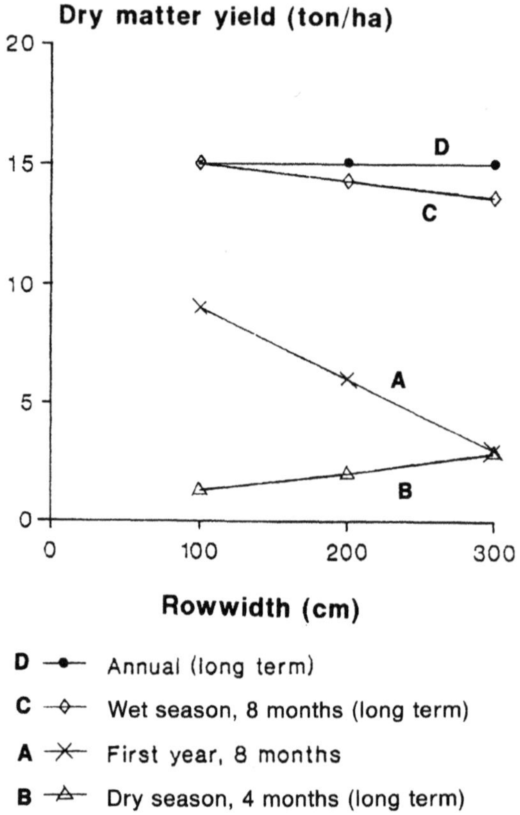

Fig. 12.1. Rowwidth and dry matter yield in Elephantgrass in relation to age and wet- or dry season (Boonman, 1979a).

varying from 3333 (300 × 100 cm) to 20,000 plants (100 × 50 cm) give almost identical herbage yields when totalled over wet and dry seasons for a number of years.

At the widest rowwidth, 300 cm, one-fifth of the total herbage may actually be produced in the dry season, when cuts are taken regularly during both the wet and dry season. Thus, by having wide rows more of the grass production potential can be shifted to the dry season.

In the 1st year intercropping with beans is often practised to recover the expenses of weed control and to compensate for low grass yields. Since weed control is so essential from the start, intercropping with beans is a better practice than having maize as a nurse crop. Plant density is relevant also when Elephantgrass is grown in mixture with perennial legumes, since legumes can provide the necessary soil cover (*12.8*).

No evidence has been found that plant density has any effect on the leaf:stem ratio, after the establishment year. It may further be assumed that higher plant densities require more N-fertilizer (see also Fig. 8.3).

It is concluded that for Zone III the recommended plant density is 5000 plants per ha (200 × 100 cm). Higher densities are unnecessarily costly at planting (Table 12.1), do not increase overall yield but decrease dry-season yields.

12.5. Harvest interval

Unlike sward-forming grasses, Elephantgrass and similar fodder-type grasses cannot survive under frequent harvesting, be it cutting or grazing. Its growth form dictates rigid resting periods. Much of the regrowth arises from stout, slowly-emerging basal shoots which sprout from the underground rhizome base. Being nutritious in their young stage and few in number these basal shoots are particularly vulnerable to grazing. Continuous defoliation, whether under cutting or grazing, deprives the crop from its main instrument of perennation. In the following some aspects of variation in harvest interval are discussed. The role of N-nutrition, although essential when discussing harvest frequency, will be discussed subsequently (*12.7.2*), but N is assumed here to be applied at the rate of 100 kg/ha/year.

Many experiments have been conducted on defining optimum harvest interval, also expressed in terms of target height, age of the crop and harvest frequency. Fig 12.2, due to Tiley (1970a), illustrates in a classical fashion the response of unfertilized Elephantgrass to increase in harvest interval from 8 to 16 weeks. The yield levels at a 12-week interval corresponded with those achieved at a foliage height of 125 cm, whereas those of a 16-week interval were well above the height of 200 cm. Fig. 12.2 shows that leaf yields were similar at the different harvest intervals. It is generally found that with increase in harvest interval there is an increase in total DM-yield, but this increase is largely composed of stem.

Maximum DM-yields are achieved with only 1 to 2 cuts annually (Table 12.3). The quality of such herbage is very low; only the tops of the 4–5 m long stems carry a tuft of green leaves.

Table 12.3 illustrates the effect of wide harvest intervals on the increase in total DM-yield, which is progressively made up of stem, and on the decline in leaf yield and quality. Widening the harvest interval from 2 to 8 months more than doubled the total yield (of leaf and stem combined) but reduced the yield of leaf protein by half. The leaf:stem ratio was reduced from 2.1 to 0.3. Mwakha's trial covered 16 months of growth after planting and his data referred to the 12 months after a cleaning cut in the 4th month; although 100 kg/ha N was applied, yields and/or quality parameters were unusually high; see also *13.1*.

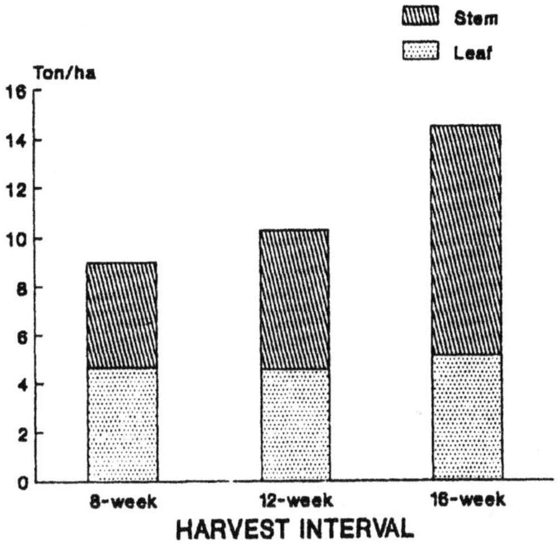

Fig. 12.2. Dry matter yield of leaf and stem in unfertilized Elephantgrass at Kawanda (after Tiley, 1970a).

Most studies have concentrated on 30–120 days intervals, kept constant over both wet and dry seasons. As a rule of thumb it can be said that, within this range, trebling the harvest interval doubles DM-yield but reduces the overall % crude protein and the leaf:stem ratio by half. Also, the nutritive value of leaves declines with increasing height of insertion. Newly expanded leaves have progressively lower nutritive values (Table 13.2). When the yield, attributable to a wider harvest interval, largely consists of more inedible stem, some compromise must be sought.

In areas with a relatively constant rainfall regime, such as the lower altitudes of Zone II, an interval of 8–10 weeks is the best compromise between yield and quality (Tiley, 1969), when N-fertilizer is applied. In the absence of N-fertilizer longer harvest intervals are required if vigour and longevity are to be maintained.

In less benign conditions, however, time-fixation

Table 12.3. Harvest interval and yield of dry matter, leaf:stem ratio and crude protein of French Cameroons Elephantgrass in the 1st year, with 100 kg/ha N (adapted from Mwakha, 1972).

Harvest interval (months)	Dry matter yield (ton/ha)		Leaf:stem ratio	Crude protein		
				(%)		Yield
	Leaf	Total		Leaf	Total	(leaf) (ton/ha)
2	10.6	16.8	2.1	13.8	12.1	1.46
4	10.3	23.6	0.8	11.0	7.7	1.13
8	7.9	35.9	0.3	8.7	4.1	0.68

of harvest intervals is not practical, since the optimum interval is much longer in the dry season. A more realistic approach is to fix the harvest date according to the physiological age, for which height of foliage could be taken as a criterion. In the latter approach both the wet- and dry season treatments are equally severe. Although foliage height is not a highly reliable indicator of yield, a height of 125–150 cm seems to offer a good compromise between yield and nutritive value. The available evidence suggests, however, that when the target is a % crude protein of 10 and a D-*vitro* of 55, harvesting should be at a lower height, of not more than 100 cm; the leaf:stem ratio is then about 1.0. This applies when no N is given. In actual practice in Kenya, Elephantgrass is cut at lower heights. Suttie (1965) reported that the grass is rich in protein (12–15%) when cut at the 60–90 cm stage. The dilemma "yield versus quality" can only be broken by applying N-fertilizer to the level required to meet both the quantity and quality demands (*12.7*; Fig. 12.3; Table 12.7).

Long harvest intervals are needed to achieve high yield, but the yield increment is progressively composed of inedible stem. To obviate this dilemma, management can be designed in such a way that the stems are rendered useful not as edible material, but rather as a framework to extend the tillering zone vertically on which more leaf can be produced and harvested (*13.2*; Fig. 13.1).

The consequence of a long harvest interval is a yield increase over the period itself, but also a growth acceleration over the subsequent interval. Regrowth is more vigorous the longer the previous harvest interval has been. This residual effect is most important. Dry-season productivity is maximized when long intervals are applied in the wet season. At Makerere, a rest of 9 months resulted in up to 50% more DM-yield in the subsequent 3-month period than that achievable from a crop regularly harvested at 3-months intervals (Boshoff, 1966). At Kitale a total yield of nearly 20 ton/ha/year was found when the number of cuts was reduced from 5 to only 2. If only 1 cut was taken, after the middle of the wet season, the dry season yield was as high as 7 ton (Table 11.5).

12.6. Stubble height

In East Africa opinion has been against cutting Elephantgrass to ground level, yet this is customary practice. With the chopping action of the "panga" (machete) there is no other way but to cut to ground level, in contrast with the "ringa" (reaping hook; *8.4.1*). Tiley (1959) quoted data of unpublished Kawanda work of W.H. Edwards, who cut monthly at ground level, beginning 2 months after planting, and noted higher yields. Working with Cameroons and other varieties in Brazil, Santana *et al.* (1987) observed an interaction between stubble height and harvest frequency. A low height (0–15 cm) was more effective with infrequent cutting, at 8–12 weeks. With a 4-weekly harvest interval, however, a stubble height of 15–30 cm had a favourable effect on yield.

The effect of stubble height on yield was investigated by Kamau & Kusewa (Table 12.4; Kitale Ann Rep, 1978). Three cutting heights, 0, 15 and 30 cm had been designed but stubble height increased inadvertently as the trial progressed. The ringa is mostly used in grass experiments but is unsuitable to deal with the tough stubble, which tends to be cut above the previous level; the stubble increases in height and in width with each subsequent cut. The target height of the 15 and

Table 12.4. Stubble height and dry matter yield (ton/ha) of Elephantgrass (adapted from Kamau & Kusewa, Kitale Ann Rep 1978).

Stubble height	Year			
	1976	1977	1978	Mean
Ground level	14.0	21.1	14.1	16.4
30–50 cm	12.0	24.0	16.0	17.3

30 cm treatments also reached the height of 30–50 cm, so that only 2 comparisons remained effectively: ground level and 30–50 cm. Cutting was applied every 2 months irrespective of amount of regrowth.

Cutting at ground level had a positive effect on yield in the year of planting but the effect was reversed in subsequent years.

The bias against a low stubble is, therefore, not justified and indeed the superiority of low cutting in the 1st year would suggest that low cutting is beneficial as long as the grass is vigorous. Continuous cutting to a height of 30–50 cm produces a stubble overcrowded with weak or dead shoots which may prevent new shoots from emerging. As the stubble height is raised, regrowth depends on terminal meristems, whereas basal and axillary tillering are promoted when the level of defoliation is lowered. However, overcrowding by old shoots hampers regrowth from basal meristems when these would otherwise be at a premium. Under such circumstances it is necessary to cut back to ground level. Consequently, there is no evidence to suggest that cutting to ground level is detrimental, provided the grass is tall and vigorous, *i.e.* cut infrequently. When Elephantgrass is, however, cut repeatedly to ground level, the effects are negative.

With the recommended system of wet-season spelling (*13.2.1*) low cutting should be of no harm. Frequent cutting is more damaging than cutting at ground level. However, many farmers faced with shortage of fodder cut too often and too low. The problem of cutting too low would be eliminated if farmers could be persuaded to employ the ringa rather than the panga. Much of the problem would be alleviated if adequate N were applied. As will be discussed later (*13.2.1*; Fig. 13.1), manipulation of stubble height may be employed as an additional tool to shift the productive capacity of Elephantgrass to times when it is wanted most and especially towards maximum leaf production.

Serious side-effects of Elephantgrass cut too low and too often are those on soil structure, mineral reserves and subsequent crop yields. The younger the grass is cut the higher the mineral contents are that are removed and the more damage is done to the soil structure. When the soil is bare – in the case of poor stands of Elephantgrass, repeatedly chopped to ground level, this is more than half of the time – the effect of sun and rain on the bare soil, on soil structure and on erosion are serious (*5.3.5*). Crops often perform poorly after zero-grazed Elephantgrass (Table 5.8); this may in part be due to mineral depletion (Table 12.5).

Table 12.5. Concentration (%) and removal of minerals in 12.5 ton/ha/year Elephantgrass (Boonman, 1979b).

	% in dry matter		Kg removed	
	Actual	Animal need*		
Nitrogen (N)	1.20	1.60	150	
Potassium (K)	3.00	0.44	375	(450 kg K_2O)
Phosphorus (P)	0.25	0.36	30	(70 kg P_2O_5)
Calcium (Ca)	0.30	0.43	37	
Magnesium (Mg)	0.25	0.12	31	
Sulphur (S)	0.30	0.10	37	

*In edible dry matter for dairy cows.

12.7. Manure and fertilizer

When considering plant nutrition and the fertilizer requirements for a particular system of harvesting it is necessary to have an estimate of what quantity of minerals is removed in a normal year's harvest and what the animal requirements are. Table 12.5 shows the percentage of some elements contained in the herbage and the actual quantities removed by cutting at an annual yield of 12.5 ton/ha, based on data of Dougall (1960) in Kenya, Foster (1969) in Uganda, supplemented with those of Vicente-Chandler *et al.* (1974). This table shows that the field is deprived of 150 kg N, 75 kg P_2O_5 and 450 kg K_2O annually. The yield level of 12.5 ton was chosen as an average yield achieved at recommended husbandry levels in Zone III (Table 11.8). Yields will be lower when harvests are taken frequently but since the grass is in a younger stage, containing a higher % of minerals, the nutrient removal will be the same if not larger. It should be

Table 12.6. The decline in soil-Phosphorus; soil analysis (ppm P) before (1983) and after 3 years (1986) of zero-grazed Elephantgrass in districts of Kenya (after Wouters, 1987a).

District and number of farms surveyed		Previous cropping	Stocking rate (LU/ha)	P ppm[a]		
				1983	1986	
					A[b]	B[c]
Kakamega	4	maize	5.0	30	13	7
Kericho	6	maize or grass	4.9	40	21	24
Kiambu	3	maize or arable	11.3	30	19	19
Meru	4	maize or arable	6.0	32	26	25
Kilifi	4	maize or "bush"	6.0	10	8	14
South Nyanza	3	maize or grass	9.3	7	12	10

P[a]: $P < 20$ ppm is considered deficient. The data refer to averages per district.
A[b]: Elephantgrass with animal manure.
B[c]: Elephantgrass was cut at either 100 cm height or at an age of 70 days. Fertilizer was applied at 100 N, 50 P_2O_5 and 50 K_2O, in kg/ha over 2 gifts/year, without animal manure.

noted that the level of removal of K and P under zero-grazed Elephantgrass is considerably higher than when a complete crop of maize grown under similar conditions is removed. As Drysdale (1965) has shown at Kitale, a good crop of maize (6000 kg/ha grain) deprives the field of 160 kg K_20, 55 kg P_2O_5, and 140 kg N, in grain, cob and stover.

Table 12.6 gives the results of a farm survey (Wouters, 1987a) designed to monitor the decline of soil-P under zero-grazed Elephantgrass receiving animal manure and/or P-fertilizer.

Apart from Kilifi (coastal sand) and South Nyanza, the soils of the locations mentioned in Table 12.6 rank among the best in Kenya. The data indicate that in spite of manure and fertilizer being applied such soils are rapidly being depleted. Extra nutrients have to be brought (back) to the field in order to restore the nutrient status of the soil to the level required for sustained production.

12.7.1. Animal manure on and off zero-grazed Elephantgrass

Animal manure is available for use there where animals are kept in relatively small enclosures and where it can be effectively collected. It played only a minor role in agriculture in East Africa, except in Chagga country near Mount Kilimanjaro, where livestock has been stall-fed since time immemorial (*11.3.4*). For general aspects of animal manure see *5.2.2*.

The value of manure is a function of its composition and the quantities involved, but especially of the circumstances under which it operates. As Goldson (1976a) has pointed out, the beneficial effects so often observed in kraals are due to the concentration of enormous quantities of faeces, equivalent to 170 ton/ha/year, not including urine (*5.2.2.3*). Opinions differ, therefore, greatly as regards the value of animal manure when available in only small quantities and when artificially applied. Already in 1944, Martin referred to the inefficient manner in which it increased crop yields. The effects of animal or green manure on soil structure have been regarded as very small. Any beneficial effect of animal manure observed in responsive crops can be attributed to its contents of K, occasionally of P, but rarely of N (*5.2.2.1*).

Zero-grazing and stall-feeding offer the opportunity to manure the same grass in return. It has been suggested that zero-grazing is an efficient recycling system, since N and other minerals in dung and urine would now not be "lost at pasture". It was thought that, under a stall-feeding regime, faeces and urine could be effectively accumulated, stored and returned at the optimum time to the Elephantgrass field, in an efficient manner.

Much of what we expect from animal manure is derived from experience in temperate countries. Part of the renewed interest is due to the fact that several countries in Europe are now faced with excess quantities of high-quality, but cheap animal manure. In temperate countries much of the manure is derived from cattle offered grass with luxuriant N-levels. N is subject to luxury consumption. As a result cattle slurry (10% DM and 6% OM) contains on average 4.5% N, as against

0.75% P and 4.5% K in the dry matter (for the composition of faeces, see Table 5.2). By contrast, slurry of 6.3% DM sampled at Kilifi at the Kenya coast (van der Noll & Janssen, 1983) contained only 1.14% N, against 0.42% P and 3.8% K in the dry matter (*5.2.2.1*).

In tropical grasses, and Elephantgrass is no exception, N is a major limiting factor in the nutrition of both the grass and the animal. Excreta can therefore not be rich in N. Also, in many stall-feeding systems the facilities for the collecting and storage of manure are poor. To begin with, concrete floors are required if the faeces and urine are not to be lost in the mud. Even then the efficiency is lower than anticipated. The losses due to volatilization of N from urine are larger when deposited on a hard floor than onto a pasture sward. Much is washed away also by rain. Additional losses occur in storage pits and during the process of spreading. The same applies to the faeces. By contrast, on pasture in the rainy season faeces are buried rapidly by coprophagous insects.

It should, therefore, not be surprising to find values as low as 1% N in manure dry-matter. When the grass that is fed is poor, excreta cannot be rich either. Secondly, losses in the animal feeding and manure collecting process may exceed 50%. Moreover, only about one-quarter of N that remains is as effective as N-fertilizer (*5.2.2.1*). Table 12.5 has indicated that 150 kg/ha N is removed annually in an average crop. Thus, the N-effect of animal manure produced off that one ha of grass cannot be larger than 20 kg/ha/year of N in fertilizer.

This low efficiency is confirmed by the findings of Hiddink & Wouters (1987) who observed at Naivasha that the N-effect of 27.5 ton of slurry, even when of high quality (not specified), was equivalent to not more than 25 kg of N-fertilizer. The grass yield was increased by only 370 kg DM from 2760 to 3130 kg/ha. On the assumption of 10% DM in the slurry it follows that 2750 kg of manure dry-matter produced only 370 kg of grass dry-matter. In the subsequent harvests no residual effects of N were observed. The authors reported further that a measurable N-effect, however small, cannot be expected unless the slurry is collected in a concrete-lined pit, distributed rapidly and incorporated into the soil. The effect of N was nil when the slurry was stored in a soil pit or in the open for 2 weeks. In a subsequent experiment, however, 60 ton/ha slurry with 2% N in the DM did not raise DM yields significantly, even when properly worked into the soil (Snijders *et al.*, 1992b). Under the same conditions, 120 ton of slurry was still no more effective than 100 kg/ha N applied as fertilizer.

At Kongwa (T), Wigg *et al.* (1973) applied dried manure, assumed to contain 0.5% N, at 50 ton/ha annually on grasses, which was found to be as effective as 50 kg N-fertilizer. The residual effects of manure and fertilizer were also about the same.

Rates equivalent to 100 kg /ha/year of N have to be applied if Elephantgrass is to benefit substantially. Only quantities of 100 ton of good slurry can be expected to produce such an N-effect. However, such situations rarely arise in practice as there is simply not enough manure produced within the farm circuit and the losses are great. Animal manure would have to be acquired from outside sources. Van Parijs (1959), in the highlands of Ituri in eastern Zaïre, found that 3 ha of grassland produced enough manure to fertilize about 1 ha of crops for an indefinite period. Our data suggest that some 5 ha of Elephantgrass have to be stall-fed to produce enough manure to maintain 1 ha of the same grass at the same level of productivity. On the many stall-feeding units which derive only one-third or one-quarter of their fodder resources from Elephantgrass (Table 11.2), it can be assumed that only when all the manure is collected and concentrated on the home-plot of Elephantgrass an N-effect will be noticed.

Ironically, where manure is plentiful, *e.g.* in the kraals on larger holdings, no manure is used effectively. Where manure is urgently needed, it is not available in quantity. Kraal manure is often so heavily contaminated with earth that it contains little more N than the soil it is supposed to improve (*5.2.2.1*).

It is concluded that recycling of faeces and urine through zero-grazed and stall-fed grass is no substitute for N-fertilizer. At best it may be adequate to replenish the removal of P and K, provided it is properly collected and stored. As we will see below, however, P and K are less critical than N is, except on the poorest soils. There is no alternative but to apply N-fertilizer.

The N-efficiency of *1 ton* of manure on grassland is not higher than of *1 kg* N applied as fertilizer. Although there are several barriers, mainly of a non-technical nature, to the use of fertilizer by farmers, they can be no excuse to promote a system of manuring of grassland, via zero-grazing, that is technically so inefficient and wasteful from the viewpoint of N-economy.

12.7.2. N-fertilizer

In the absence of N-fertilizer Elephantgrass yields drop rapidly from the 2nd year onwards (Figs. 11.2 & 12.3) and herbage quality is also negatively affected (Figs. 10.6 & 12.4). N-fertilizer has been under-exploited in East Africa, but Uganda workers have been more concerned (Stobbs, 1965; Tiley, 1970; Tiharuhondi, Olsen & Musangi, 1973). For some of the reasons why N-fertilizer was given so little attention, see *7.2.4.1*.

As a management tool, N-fertilizer is the most potent of all in raising and manipulating both herbage yield and quality, yet also the most flexible. With other inputs it is often difficult to anticipate an immediate positive effect, but with N one can greatly influence the fodder supply and quality that is needed within a span of 2 months. With rates of only 100 kg/ha/year N it is possible to double herbage production in frequent-cutting regimes from the 3rd year onwards (Fig. 12.4). In the absence of N yields decline to 25% in later years (Fig. 11.2). With N the decline is reduced to 65–75% (Fig. 12.3). At Kawanda, Stephens (1967a) concluded that 250 kg/ha/year N increased DM-yield 3 to 4-fold and yield of crude protein 5-fold. N-fertilizer is also effective in achieving a balanced ration and is in this respect more cost-effective than concentrates. N is needed if we are to achieve a stocking rate of 2 LU/ha (in Zone III) and milk yields of 6 kg/head/day. This level of performance presupposes an intake of 10 kg DM, 10% crude protein and a D-*vitro* of 55.

N-fertilizer is a powerful instrument in the hands of the farmer to make his fodder supply match demand. Because of these powerful effects, N-fertilizer must be viewed against year-round herbage supply as discussed for sown pastures (*7.2*). Although most studies have concentrated on the effects of increasing rates of N throughout the year, interest should primarily focus on strategic application (*7.2.4*), designed to alleviate temporary but acute N-shortages. Rather than at the rates used we should look at the timing of N-application in relation to season and to age of the field, to be more cost-effective. An important reason why the role of N on Elephantgrass is discussed here separately is the fact that most of the relevant fertilizer experiments were carried out in East Africa under cutting regimes. Deficiency in other essential nutrients is also more acute under cutting than under grazing. Some of the subfactors will now be described in more detail.

12.7.2.1. Rate and time of N-application

Keya & Kalangi (Kitale Ann Rep, 1971; 1972) used annual rates of up to 550 kg/ha N, divided over 4–5 cuts in the rainy season (Fig. 12.3). The experiment had been designed to assess the K-requirement of Elephantgrass (*12.7.4*). In contrast to K-, however, the response to N-fertilizer was quite marked and linear; even rates above 550 kg/ha N would have been effective, had the frequency of cutting been increased to match the high N-levels (*7.2.4.1*). In the 5th year, DM-yields dropped to one-third of those of the 2nd year in the absence of N, but only to about one-half at 200 kg/ha N. At the top levels the decrease was only 26–35%. Thus, N not only increases overall DM-yield, but also maintains a relatively high yield in later years.

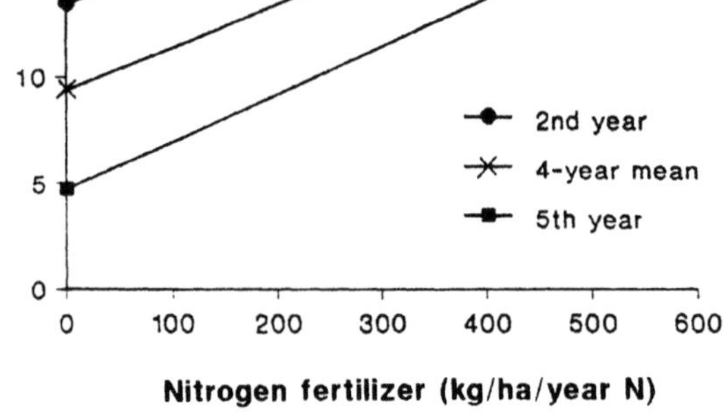

Fig. 12.3. Nitrogen fertilizer and dry matter yield in Elephantgrass in relation to age (after Keya & Kalangi, Kitale Ann Rep 1971; 1972).

Whether N has been part of husbandry or not, the question of "how much and when" is a permanent one. N is most effective in terms of kg DM/kg

N if applied at the beginning of the rains when the soil is still warm (*7.2.2.2*). Not only the growth rate (Fig. 6.1) but also nutritive value (% N) is then much higher than in the middle of the season (Fig. 6.5). N-application at the first rains may not be wanted so much from the livestock point of view but is a good proposition when the grass has been over-exploited in the dry season. Where, however, N is valued mainly from the viewpoint of strategic application (*7.2.4.2*), it is advisable to apply N some 2 months before the end of the rains. Wouters (1986) found at Naivasha that 100 kg N applied before the dry season produced 3000 kg/ha milk extra, compared with no-N, giving a return on purchased fertilizer of 680%.

Apart from the time of application within a particular year, the question also arises how to handle the age difference between the fields to be topdressed (*7.2.3.1*). Should a given quantity of N be applied to a 2nd year's field in its prime with a yield level, without N, already of 10 ton? Or is it better given to a 5-year old field producing only 4 ton? Fig. 12.3 suggests that the response per kg N was slightly higher in older fields. However, the yield level at no-N in the 2nd year was already very high and much depends also on the harvest interval (Table 12.8) and on whether the grass is cut or grazed. Under cutting regimes, the most economic response to N-fertilizer can normally be expected in the 2nd and 3rd year. N is usually not applied in the 1st year, the establishment year, since the supply of soil-N should then be adequate (*7.1*). Where the choice is to apply a given quantity to one field only or to divide it over more fields of the same age and condition, the latter should be preferred.

It is postulated that per individual cut the N-rate should not exceed 50 kg N but should not be lower than 25 kg. It is applied on or near the individual plants and directly after cutting. N is most effective when applied as early as possible in the growth cycle but there has been some tradition of delaying application unnecessarily (*7.2.4.1*; *8.5.1*).

The residual effect of N-fertilizer should be taken into account also. Stephens (1967a) has been one of the few to have investigated this aspect and he did this under both cutting and grazing regimes. N was applied at 75 kg/ha to alternate cuts or grazings. He observed that the residual effects were only slightly less than half their direct effects. The residual effect was much larger under grazing than under cutting. Higher N-rates had large and, as time progressed, increasing residual effects on grazed as compared with cut grass (Fig. 7.1; *13.3*).

12.7.2.2. Interaction of N-rate and harvest interval
Annual yields are increased not only by N-fertilizer but also by widening the harvest interval (Fig. 12.2). Within the 30–120 days range, trebling the harvest interval doubles DM-yield. Leaf:stem ratio is reduced by about half; leaf yield may not be affected but nutritive value is greatly reduced (*12.5*). It is important to know how these 2 factors may be combined to reconcile yield and nutritive value at a high level.

The interaction between N-level and harvest (cutting) interval on Elephantgrass has been studied repeatedly and is typified for Kitale conditions by Fig. 12.4. The effect of increasing levels of N on yield is roughly linear up to about 500 kg, but is more marked at the harvest interval of 12 weeks. Herbage quality, in terms of % crude protein, is positively affected, even far beyond the 500 kg level. However, the ranking of the harvest intervals is now reversed. Levels above 500 kg are shown to full advantage only under more frequent harvest regimes. The effect of N is to accelerate growth, so that the harvest frequency should be increased accordingly.

In Fig. 12.4, the relationship between age and quality, in terms of digestibility (D-*vitro*), is presented between brackets.

Table 12.7 is derived from Fig. 12.4 to demonstrate how the amount of N required to achieve a particular level of yield and quality, 15 ton or 10% crude protein respectively, depends on the harvest regime. Extrapolation of the data shows that at an interval of 8 weeks, the amount of N required to achieve both objectives is 300 kg. The foliage height will be about 125 cm.

Thus, much less N is required to produce a given total yield, when the harvest interval is increased, but much more N is required to give this quantity of herbage also the required % crude protein.

In addition to harvest interval, other factors must be considered which influence the response to N, in terms of yield increment per kg of N applied:

1. The yield level at no-N, hence the age of the sward (Fig. 12.3).

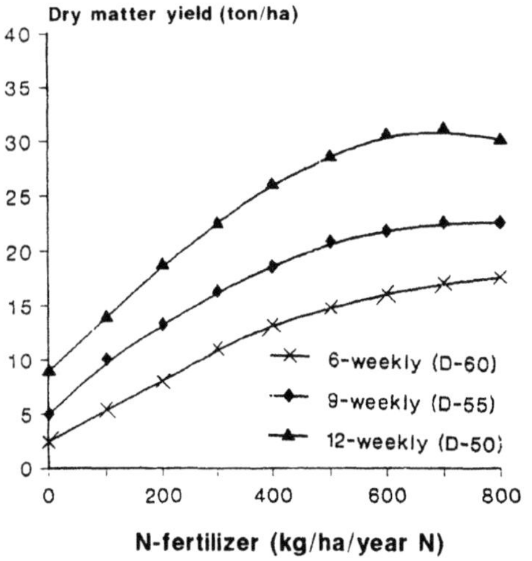

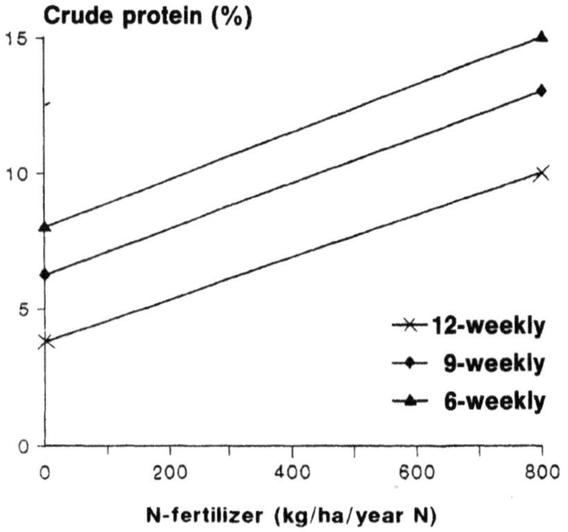

Fig. 12.4. Dry matter yield and % crude protein in Elephantgrass in relation to harvest interval and level of nitrogen fertilizer (Boonman, 1979b).

Table 12.7. Nitrogen requirement in relation to harvest interval in Elephantgrass (Boonman, 1979b).

Harvest interval	Nitrogen (kg/ha/year) required to produce:	
	15 ton dry matter	10% crude protein
6 weeks	500	200
9 weeks	250	400
12 weeks	100	800

2. The time of the year, early-applied N resulting in higher DM-yields and higher % N in the herbage (*7.2.4*; Figs. 6.1 & 6.5).
3. The harvest method, *i.e.* grazing or cutting (*13.3*).

Of the various factors, harvest interval has perhaps been the most spectacular, which explains why it has been frequently studied. Table 12.8 summarizes some of the responses found. The same was observed in Nandi Setaria and Rhodesgrass (Boonman, 1972a, b). When only 2 cuts were taken annually an average response of 65 and 47 kg DM/kg N was observed, respectively.

Table 12.8. The response of Elephantgrass to N-fertilizer under cutting regimes; increment in dry matter yield per kg of N applied in the 2nd and 3rd year (Boonman, 1979b).

Harvest interval	Kg dry matter/kg N applied
15 weeks	50–60
12 weeks	35–40
9 weeks	25–30

Harvest interval and N-fertilizer are, both separately and jointly, effective tools to manipulate the availability of herbage so that its supply is evened out over the year or shifted from the wet to the dry season (*7.2.3*; *12.5*). On the one hand, there is the obvious trough in the dry season, but there is also the enigmatic peak in the early part of the rainy season, followed by a mid-season depression (Fig. 6.1; *7.2.2*). The peak in yield manifests itself provided that the field has not been over-exploited in the dry season. In actual practice, many Elephantgrass fields never have a chance of demonstrating an early-season peak, because all that grows is taken away prematurely. This can be remedied by N-fertilizer, by delaying the next harvest or by both. The most obvious strategy is to employ the peak period to allow the grass to rest and to rebuild its root mass (*7.2.5*; *13.2*). Spelling, for the better part of the rainy season, followed by a N-topdressing may be the best joint-strategy yet designed to maximize dry-season yield.

12.7.2.3. N-fertilizer and herbage nutritive value

N-fertilizer has a substantial and linear effect on the % crude protein, over a range of harvest intervals (Fig. 12.4). At Kawanda (U), Stephens (1967a) concluded that some 250 kg N annually was needed to raise the % crude protein from 7 to 10%. Wouters (1985) showed at Naivasha that N-fertilizer improved the nutritive value of

Table 12.9. The effect of 75 kg/ha N on quality parameters in 8-weeks old regrowth of Elephantgrass (after Wouters, 1985).

	Nitrogen (kg/ha)	
	0	75
Yield of dry matter (kg/ha)	4250	5600
% Dry matter	18.5	18.3
% Crude protein	7.5	10.6
% Ash	20.7	18.1
Digestibility (% DOM)	68.8	71.0
% Leaf at 4250 kg yield level	59.0	70.7

Elephantgrass also in terms of digestibility and % leaf (Table 12.9).

Whilst having considerable and positive effects on DM-yield and % crude protein, the "diluting" effect consequent upon increased DM-yield is borne out in lower levels of % ash, including P and Ca. Elephantgrass is deficient in P and Ca (*12.7.3*; *12.7.5*). Also DM % is negatively affected by N-application. Table 12.9 shows further that, at a given yield level, the leaf:stem ratio is increased considerably by N. Leaf has a higher % crude protein and % DM than stem (Fig. 11.1; Tables 13.2 & 13.3).

At a given yield level, the % crude protein is much higher in herbage at the beginning of the rains than in the middle of the season (Fig. 6.5), which is due to N-mineralization following the dry season. Although this "Birch effect" (*5.3.1*) is most marked at the start of the main rains, smaller flushes occur after dry periods throughout the rainy season. High nitrate contents of grasses are rare but have been recorded, notably in Elephantgrass, at Entebbe (Harker & Kamau, 1961).

12.7.2.4. N-fertilizer requirements under zero-grazing

Table 12.5 has demonstrated how large the amounts of uptake by Elephantgrass are of N, P and K. Some 150 kg N are removed in zero-grazed Elephantgrass and even when the manure is returned the effect is not greater than of 20 kg N applied as a fertilizer (*12.7.1*). Recycling faeces and urine has a negligible N-effect, in contrast with that of P and K (see below).

Under zero-grazing more N-fertilizer is needed than under direct grazing. As a rule of thumb the following is postulated, again for Zone III. When 50 kg of N is applied after every cut a yield of 4–5 ton/ha is achieved when harvested at a height of 125 cm, within 10 weeks when growing conditions are favourable. This herbage has 50% leaf, 9% crude protein and a D-*vitro* of 55. When Elephantgrass is fed at a residue of 15% and with a supplement of a molasses-urea-mineral mixture, the milk yield per animal should be 10 kg/day. A gift of 50 kg N/ha/cut should be standard practice in zero-grazing.

12.7.2.5. N-fertilizer requirements under direct grazing

Much lower levels of fertilizer are needed when Elephantgrass is grazed directly. More than 75% of P and K is returned by the grazing animal in urine and faeces. Losses of N will occur, but P and K are held strongly in the soil complex and are recycled more efficiently the more intensive the grazing is. The evidence available suggests that only half the amount of N is required by grazed grass to give the same animal output as under a cutting regime (Fig. 7.1; Table 13.6; *13.3*).

12.7.3. P-fertilizer

At the DM-yield level of 12.5 ton/ha/year some 30 kg phosphorus (P), *i.e.* 70 kg phosphate (P_2O_5) is removed (Table 12.5). The P-responses observed in practices involving Elephantgrass or similar fodder grasses are often at variance with those observed in natural grassland (*4.2*; *4.3*) or in sown grasses in mixed-farming rotations (*7.1.8*; *7.2.4.3*). Although no response may be apparent, to apply 40–50 kg/ha/year P_2O_5 has been standard recommendation. Now that P-prices on the world market have developed unfavourably (Fig. 7.3) it is necessary to study P-requirements more carefully, also on a longer term.

It is usually recommended to apply phosphate in the planting hole, but the beneficial effect is less consistent than when applied in the seedbed of sown grasses. Since P is concentrated at the actual site of planting, small quantities of 10–20 kg P_2O_5 are adequate. Dicotyle weeds are greatly stimulated, in and around the planting hole, which is best overcome, apart from hand-weeding, by an application of 2,4-D or Atrazine (*12.2.2*).

As regards the effect of topdressing of Elephantgrass variable responses have been recorded in East Africa. Considerable work was done on the P-requirements for mulching purposes on vlei or other difficult soils (Child *et al.*, 1955; Gosnell, 1963; Gosnell & Weiss, 1965; Robinson & Brooke, 1959; Jones & Robinson, 1965; Boswinkle, 1961). The responses observed fluctuated wildly but yields were generally very low, indicating that the soils involved were unusually poor. The high response to P obtained by Child *et al.* (*op. cit.*), whether N was applied or not, appears to be exceptional. These results helped create the impression that P-topdressing of grasses was essential in spite of the earlier more specific grassland work mentioned above. P-fixation was traditionally believed to be characteristic of tropical soils. However, although acid, red, lateritic clay soils in particular fix this element, an initial remedial application is frequently adequate (de Geus, 1973); grass production can be maintained by relatively small dressings, at least under grazing. Only on-the-spot-fertilizer trials directed at specific conditions of soil and utilization can give the correct answer as to whether topdressing is likely to be successful.

At Kawanda, positive P-effects were observed but ancillary observations indicated that P limited Elephantgrass yields and following crop yields, if N-fertilizer was applied (Stephens, 1967a; Foster, 1969). In Zimbabwe, Addison (1956) also found a marked N × P interaction, under cutting regimes; with heavy N-dressings also heavy dressings of P were necessary if maximum yields were to be maintained. At Kitale, only very limited responses have been observed to P-topdressing of Elephantgrass and other grasses, whether grazed or cut and whether fertilized with N or not (*7.2.4.3*). Soils at research stations may not be the most appropriate to assess P-requirements but there is increasing evidence that tropical grasses in East Africa, unlike maize, wheat and legumes (*10*), are capable of extracting P from soils which to any soil analyst appear P-deficient. Tropical grasses can bring up to 30% of their P-uptake, about 15–30 kg/ha P_2O_5 annually, from below 30 cm depth (*5.3.2*; Nye & Foster, 1961). Mobilization of P and K from lower soil layers by deep roots and their accumulation in the topsoil after decomposition of plant matter was observed in Elephantgrass by Stephens (1967b) at Kawanda.

Sight of the long-term soil-P status should, however, not be lost (*7.2.4.3*). It is important to maintain an adequate P-level in the soil. The role of grasses and the grazing animal in the soil-P economy deserves more attention. P taken up by the grazing animal is largely returned through faeces and only little P is removed in animal produce. High grazing-intensity increases available soil-P by encouraging the accumulation of P in the surface soil and by promoting more frequent cycling of P than at a low intensity. Obviously, a quite different picture arises under zero-grazing.

12.7.3.1. The soil-P status under zero-grazing

Alarming losses of soil-P occur under zero-grazing. Wouters (1987a) monitored the fate of soil-P over 3 years of zero-grazed Elephantgrass in various districts of Kenya (Table 12.6). Both with manure and with P-fertilizer, P-losses from the soil due to zero-grazing were large and rapid. The decline in P was even more serious in the subsoil than in the topsoil. A loss of 10 ppm P over a depth of 30 cm of soil with a specific gravity of 1.3 implied the removal of 100 kg P_20_5. This, added to the gift of 130 kg of P_20_5 that was given over 3 years, meant that a gift of 77 kg P_20_5/ha/year was necessary to replenish the reserves in the topsoil. This corresponds with the amounts removed by cutting (Table 12.5) but only if the uptake from the subsoil is ignored. The stocking rates are mentioned in Table 12.6 to show that most zero-grazing units are heavily stocked and that grass is cut and removed at a young stage, rich in minerals.

The fields of Wouters survey had been zero-grazed for 3 years only and one is left to wonder what the longer-term effects of zero-grazing on the soil-P status would have been. It would seem that an annual application of 75 kg P_2O_5 to counter P-uptake (Table 12.5) is not adequate if also the uptake from the subsoil is to be replenished. As de Geus (1973) and Sanchez (1976) indicated, 100–150 kg/ha/year P_20_5 may be needed to match the removal of P in intensively zero-grazed grasses.

12.7.3.2. Animal requirements of phosphate

The mineral status of the animal relying on Elephantgrass should also be taken into account. Dougall (1960) investigating the P-content of the grass at a height of not more than 75–90 cm found only 0.18% P (0.42% P_20_5, see also Tables 7.4, 7.5 & 12.5). These levels decline with age and with N-application. According to Vicente-Chandler *et al.* (1974), a level in 60-days old grass of less than 0.17% P suggests a deficiency. Increasing the level of P-application increases the P-content of the herbage, often without raising the herbage yield. It may have a negative effect on the N-content of the herbage, but in general it is found that increasing the P-fertilizer level has no effect on the contents of other important minerals in the herbage. Long *et al.* (1970; 1972) investigating the mineral status of Uganda dairy farms based on grasses, including Elephantgrass, reported that P was often marginal or deficient. It should be added that leaves contain usually 50% more P than stems do. P-deficiency is unlikely to occur under grazing regimes but may become acute under zero-grazing. The rising P-cost (Fig. 7.3) is now focusing attention on direct P-supplementation of cattle (*10.2.1*).

The general conclusion on the role of P-topdressing in tropical grasses is that under grazing regimes P-fertilizer strategies should aim at maintaining the soil-P status at the level required to sollicit a high response in the cropping phase within the rotation. When grazed, grass rarely responds to P-topdressing. In zero-grazing systems, however, applications of 100–150 kg/ha/year P_2O_5 are necessary not only to prevent P-deficiency and to sustain grass productivity but also to maintain soil fertility and prevent soil depletion.

12.7.4. K-fertilizer

Assuming again a DM-yield of 12.5 ton/ha/year of Elephantgrass and a K-concentration of 3%, some 375 kg K (450 kg K_20) are removed annually (Table 12.5).

Elephantgrass is a luxury consumer of K, like Setaria (*6.7.3*). The requirement in edible dry matter for an adult dairy animal is probably not higher than 0.5%, but 3–4% is commonly found in Elephantgrass. Some 1–1.5% K is considered to be the critical level, above which the yield response to further fertilization of that nutrient is doubtful (Vicente-Chandler, *op. cit.*). Foster (1969) found a range of 3.07–6.70% K at Kawanda, which was likely to produce a cation imbalance. Mulching coffee with grass with such K-values induces Mg-deficiency, which affects the quality of the berry (*11.3.3*). Little is known of the effects of unbalanced K/Mg ratios on dairy cattle under tropical conditions, but in temperate countries hypomagnesaemia is a notorious nutritional disorder.

In most East African soils K is available in adequate amounts, except in some areas near Lake Victoria and the coast (*5.2.2.1*), in continuously cropped soils and when grass is cut and removed (Stephens, 1969; 1970). In the Kitale experiment (*12.7.2.1*) to assess the Elephantgrass requirements of muriate of potash at N-rates up to 550 kg N/ha/year, a K-response was only observed at the highest N-levels, from the 2nd year onwards. By the 4th year, a K-response was evident at the lower N-levels also. However, the response was small and in no comparison with the effect of N-fertilizer (Fig. 12.3).

In the Kawanda study on N, P, K and Mg responses of cut and grazed grass, Foster (*op. cit.*) concluded that exchangeable K in the soil was markedly increased under grazing (Table 13.5). However, if the grass was cut and only N and P were applied, K was likely to become limiting. In an ancillary study of field crops that followed the various Elephantgrass treatments, Foster (1971) concluded that soil-K supplies may be depleted to such an extent that crop yields are severely depressed (*5.5*; Table 5.7).

The situation in the survey (Table 12.6) by Wouters (1987a) was different for K from that for P in the sense that in spite of massive removal of K from the soil, reserves is the soil were generally large enough not to show a deficiency, within the 3 years that Elephantgrass had been zero-grazed. It was assumed that the topsoil K-levels had been replenished from further below in the rooting zone as had been observed by Stephens (1967b) at Kawanda (*5.3.2*; *12.7.3*).

It is evident that the quantities of K removed in zero-grazed Elephantgrass are even larger than those of P. The K-removal is usually increased by N-application. It is also obvious that with continued removal of N-fertilized grass, quantities of over 500 kg/ha K_2O are drawn up from the soil annually. Even though most soils are well supplied with K, such nutrient drains will sooner or later result in a shortage, as shown by Foster (1971). Under such circumstances it will be inevitable to replenish the K-reserves of the soil, especially in the event of subsequent cropping. Under direct grazing this question would not arise since most K consumed by animals is returned to the soil via excreta, especially urine.

12.7.5. Other minerals

Apart from K, Elephantgrass has a high % iron (Fe) and sulphur (S) which together with high % silica explains the high ash content of this grass (when young > 20%) and, therefore, the relatively low OM-digestibility. Vicente-Chandler (*op. cit.*) recorded a much higher silica content (3.5%) in Elephantgrass than in Guineagrass or Stargrass (see also Dougall, 1963b).

The high % K of Elephantgrass is associated with a low % Ca, of about 0.30, which is only half of what is found in other grasses (Table 12.5). Mg-content is also inversely related to K-uptake (*12.7.4*). Like Setaria and allied genera, Elephantgrass has a low cell-sap pH (Dougall & Birch, 1966). It accumulates oxalate up to levels which may prove fatal for livestock (Ndyanabo, 1974). Organic acids, *e.g.* oxalic acid, accumulate as a result of excess cation concentration in the leaves. This would link the excess contents of K, nitrate (Harker & Kamau, 1961) and oxalate with the deficiency of Ca and Mg encountered in Elephantgrass. High contents of K and oxalate have also been found in Setaria and allied genera and hypocalcaemia and hypomagnesaemia may occur (*6.7.3*; *12.7.4*). Supplementation with bonemeal or mineral licks seems desirable. A molasses-urea-mineral mixture as developed for commercial feedlots (Mulder *et al.*, 1973) is recommended.

Low levels of cobalt (Co) were found to occur in Elephantgrass (Ssekalo, 1972). Co is not essential for plant growth but in livestock this deficiency has been associated with enzootic marasmus or "Nakuruitis" because the main incidence of the disease was in areas of volcanic soil near Nakuru where copper deficiency occurs also (Howard, 1963). These deficiencies have become rare where mineral supplementation is practised.

Sufficient Ca as a nutrient is supplied in the form of single superphosphate (*7.1.8*). Liming is rarely beneficial except to reduce Al-toxicity on continuously cultivated soils with very low levels of pH and exchangeable-Ca (Foster, 1970). On most soils, however, liming may induce Zn-, Mn- and Fe-deficiencies and reduce OM and availability of soil-P. It is usually difficult to isolate the effects of pH from its confounding with Ca-supply, irrespective of the tolerance of tropical grasses to low pH.

Direct grazing has generally a beneficial effect on pH (Tables 5.3 & 13.5). Heavy applications, however, of residually acid fertilizer result in a decline of exchangeable bases and of pH. Especially sulphate of ammonia (SA) has been found to lower pH (Dougall, 1954c). Acidity may affect subsequent crops rather than the grass itself but sandy soils along the East African coast are particularly vulnerable to SA, which was found to have a negative effect on Elephantgrass (personal observation, 1975). At Tanga a single SA-application at 100 kg/ha N led to a drop in pH from 5.6 to 5.2 (Anderson, 1968b). In Zimbabwe Elephantgrass benefited by liming after 2 years of applying SA (Addison, 1958a). Dougall (1954a) indicated that 100 kg of SA increased the loss of lime by 80 kg of CaO. Increases in soil acidity are easy to prevent by liming but are difficult to correct once acidity develops deep in the profile. In addition to SA, N-fertilizers available in East Africa are urea, ammonium sulphate nitrate (ASN) and calcium ammonium nitrate (CAN); on soils tending towards acidity, lime should be applied at 1.2, 0.9, 0.6 and nil kg, respectively, per kg of fertilizer. Acidification is not confined to N-fertilizer. Soil-pH was lowered also by double superphosphate at Ruiru near Nairobi (Robinson, 1956). Soil acidification under P-, S- and Mo-fertilized legume-based pastures has been observed in Australia (*10.2.1*).

CAN is the N-fertilizer commonly recommended. Urea is less reliable. High losses due to ammonia volatilization are found when applied after heavy rain and before dry weather with an occasional shower. It should be applied when sufficient rain is expected to wash the urea into the soil. The advantage of urea are, however, its low cost and its high content of N which alleviates transport costs.

12.7.6. Zero-grazing and subsequent cropping

As mentioned above, when Elephantgrass is cut and removed, soil-P and K supplies may be depleted to an extent that crop yields are severely depressed. The effects may even be worse than after continuous cropping (Table 5.8). Elephantgrass gives better subsequent crop yields when grazed than when cut (Table 5.7). Foster (1971) concluded that if N only is applied, soil-P may become deficient and if NP-fertilizer is applied and the grass is cut, subsequent crops are likely to suffer from K-deficiency.

The negative effects on subsequent cropping are, however, not limited to mineral depletion. Rainfall acceptance and moisture status are reduced on bare soil; repeated interrow cultivation reduces the OM-status and destroys soil structure (*5.2*; *5.5*).

These combined effects may explain why crop yields were lower after Elephantgrass, when cut and removed, than after continuous cropping (Table 5.8). In the zero-grazing unit at Naivasha lower yields were recorded when maize followed Elephantgrass (Snijders *et al.*, 1992a).

12.8. Association with legumes

Demand for legumes has been very low compared with grasses (*10.10*). Tropical legumes do not form stable mixtures with sown grasses under current, however lenient, grazing regimes (*10.4*). The growth habit of the sown grasses, perhaps, as well as the need for frequent and low grazing to keep the grass itself in a nutritious state (rational grazing), make the sown-grass regime generally incompatible with tropical legumes. Especially in semiarid areas and under the high grazing pressures on small farms, tropical legumes have little chance of survival. Another deterrent to the use of legumes is their high requirement of P-fertilizer, which may wipe out their N-saving effect (*10.2.1*).

Elephantgrass seems a more suitable partner (Fig. 10.6). It produces an open clump, with bare ground, which should allow better for a legume associate. On the other hand also Elephantgrass benefits by infrequent defoliation (*12.5*). Consequently, the fodder-grass regime fits most tropical trailing legumes, *e.g.* Desmodium and Centro, the 2 most reliable legumes found thus far (*10.8*).

Ideally, legumes enrich the associated grass with nitrogen and stimulate grass growth. When this cannot be achieved, legumes should still be of value to halt the decline in nutritive value of herbage, when past the young stage (*10.2.2*). This is relevant for fodder grasses and tropical legumes because both are sensitive to harvesting in the young stage. Even when legumes do not achieve these goals adequately, their presence is still desirable to maintain some soil cover between the clumps and also as a criterion to guide the farmer not to harvest his Elephantgrass too frequently. Decline of the legume should be a signal for the farmer that both the legume and grass have been defoliated too often.

12.8.1. A review of research work in East Africa

The mixed cultivation of Elephantgrass and legumes was pioneered at Kitale by R. Strange in the 1950s. The legumes tried with some success were *Dolichos sericeus* (K5086 ex Kilima Kiu), *Dolichos lablab*, *Dolichos biflorus*, *Trifolium rueppelianum*, *Glycine javanica* and *Melilotus alba*, but considerably more success was achieved with *Desmodium uncinatum* (Strange, 1955c). Strange suggested to sow the legume first and to plant root splits of the fodder grasses later in the same row.

At Kawanda, Tiley (1969) found Desmodium, Centro and Puero to associate well with Elephantgrass. The following seed rates were recommended (kg/ha): Desmodium 1–1.5; Centro 4; Puero 3. Tiley found Desmodium the best. Centro and Puero proved less palatable. Tiley apparently had difficulty with keeping the legume component down to reasonable proportions and this may have been the reason why he recommended to sow the legume 2–3 weeks after planting Elephantgrass.

Consequently, both at Kitale and Kawanda best results were obtained with Desmodium (Fig. 12.6). *Desmodium intortum* is at times considered more vigorous, but seed is much more difficult to produce and is often not available (*10.8.2*).

When Elephantgrass is planted with Desmodium, the 1st-year yields of the grass can be reduced to 50% and the mixture yields are lower also, even when weeding is practised (Table 12.10; Suttie, Kitale Ann Rep, 1966). Less vigorous fodder grasses such as Guineagrass, Guatemalagrass and Giant Setaria are especially weak in competition with Desmodium (Table 14.3). Desmodium itself is not much affected by differences in companion grasses.

After the 2nd year, grass yield declines progressively and continues to yield less in mixture than in monoculture. In fact, the legume depresses and replaces part of the grass. Grass yield may be reduced by 25–50%, compared with monoculture, if Desmodium is allowed to dominate (Snijders *et al.*, 1992a). However, the substitution effect of the legume is such that the combined yields are higher than those of the grass itself in monoculture (Table 12.10; Suttie, Kitale Ann Rep, 1966; Thairu, Kitale Ann Rep, 1968; 1970).

Table 12.10. Total dry matter yield (ton/ha) of French Cameroons Elephantgrass without or with *Desmodium uncinatum* (Kitale Ann Rep 1966; 1968; 1970).

	Total dry matter yield (ton/ha)		
	Year 1	Year 3	Year 5
Grass (without Desmodium)	5.8	12.1	5.3
Grass plus Desmodium	4.2	11.3	11.2

As the age of the field increases, yields of Elephant + legume are generally 10–50% higher than the grass in monoculture, under infrequent cutting. The yield increase of the mixture consists usually exclusively of Desmodium (see also Fig. 12.5); grass yield stays the same or is reduced, especially in the dry season.

A typical grass-Desmodium response as compared with N-fertilized grass under cutting is illustrated by Fig. 12.5, due to Thairu (1972). Desmodium represented 33% of the mixture.

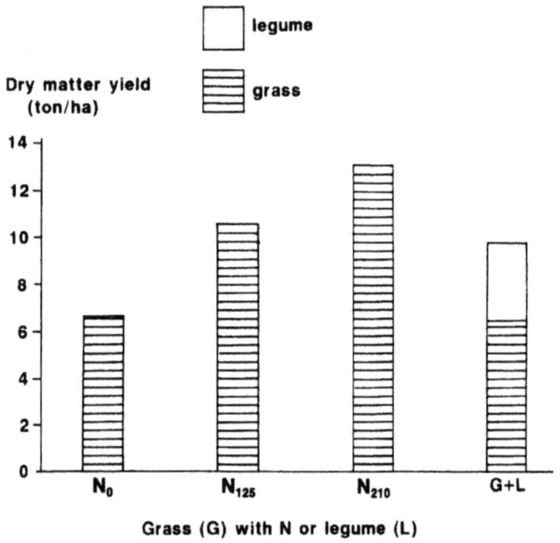

Fig. 12.5. Dry matter yields of grass (G) and legume (L) in relation to nitrogen-fertilized grass (after Thairu, 1972).

Desmodium hardly affected the yield of the associated grass but increased the yield of the mixture almost to the level of grass fertilized with 125 kg/ha N. Experience has shown that this is more or less the maximum that can be achieved with a legume in East Africa; see also Fig. 6.5, which related to the same experiment, and Fig. 10.6.

That the contribution of Desmodium to the combined grass-legume yield may be lower than the equivalent of 100 kg N-fertilizer given to grass is evident from Fig. 12.6 which Mbugua (1976b) computed on the basis of data of P.K. Kusewa of Embu Station.

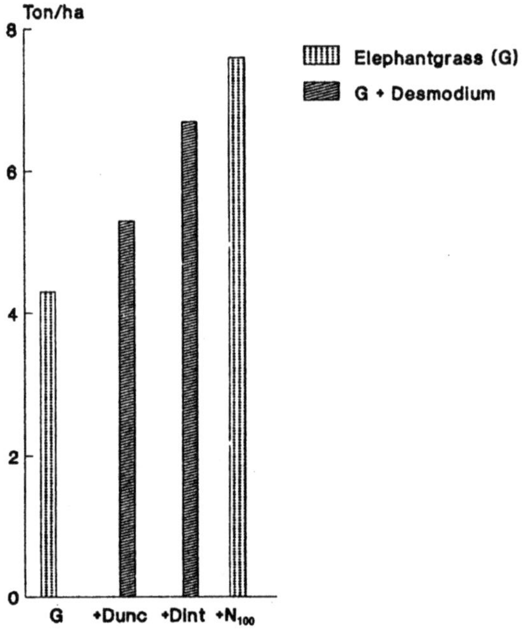

Fig. 12.6. Dry matter yields of Elephantgrass in association with *Desmodium uncinatum*, *D. intortum* or nitrogen fertilizer (after Mbugua, 1976b).

In stands which are subject to regular, but moderately infrequent cutting as commonly applied in agronomy experiments, Desmodium may overrun Elephantgrass, especially if Desmodium is sown first. Bauer (Kitale Ann Rep, 1975) planted French Cameroons Elephant- and Guatemalagrass at various spacings within 100-cm wide rows of legumes, sown 10 weeks earlier. At the end of the trial the grasses were still severely weakened by dominance of the legumes and suffered more from drought stress and even from wilting in the dry season than when grown in monoculture. Such mixtures may, therefore, present little of an advantage during the dry season and none at all in the 1st year.

Severe moisture stress in Elephantgrass is commonly observed in Elephant-legumes mixtures with a high proportion of legume at the East African coast. Soil moisture records showed that crops were under moisture stress when associated with kudzu (Puero) cover crops (Pereira & Hosegood, 1962; see also *5.3.3*).

12.8.2. The legume-grass balance

Fig. 12.5 suggests that to benefit by the legume without a substantial reduction of the grass it is necessary to aim at a 65/35 grass-legume balance, in the years after establishment. Desmodium suffers more from repeatedly frequent cutting than Elephantgrass (Fig. 6.1). On most farms practising zero-grazing, cuts are taken too frequently. This is the reason why the results in practice have been disappointing (Wouters, 1987a). This is in sharp contrast with what is found in agronomy experiments, cut less frequently.

Desmodium maintains itself well when the mixture is cut at the height that is recommended for Elephantgrass, 125–150 cm (*12.5*). Desmodium tends to overtop the grass, however, if the latter is harvested repeatedly at 150–200 cm. Beyond this height Desmodium is set back again, because it rarely climbs higher than 200 cm in tall grass. Besides, when harvest intervals are very long Elephantgrass seems to benefit more from N fixed by the legume (Table 12.11). Consequently, if the proportion of Desmodium is to be restored, harvest intervals should be such that the foliage height is between 150 and 200 cm.

Alternatives to reduce the legume component are, apart from frequent cutting: to cut the legume only; to apply N-fertilizer to the individual grass plants only; to withhold P-application; to graze the mixture. It is generally easier to increase the proportion of grass than that of the legume.

Much depends on the grass-legume balance achieved at the start. Desmodium develops better in the grass rows since it benefits by the shading and the lifting afforded by the grass which gives the twining vines more room for vertical expansion. In such "hedgerows" Desmodium is better protected from damage by treading. Interrow cultivation is possible with the cultivator straddling the grass rows. The surest method to start with a good Elephant-Desmodium balance is to sow the legume first and to plant grass root splits in the rows as soon as the seedlings emerge. A wide rowwidth of 200 cm should be adopted to maximize productivity in the dry season (*12.5*).

Neither Elephantgrass nor Desmodium, when grown by themselves, are easy to maintain free from weeds, grass weeds in particular. The mixture is liable to invasion by African Couchgrass and by sedges. Fortunately, Desmodium can tolerate high doses of Dalapon when her shoots are still in a very young stage, so that spot-application against grass weeds is possible (*12.2.2*).

An inconvenience when handling the mixture manually is the fact that Desmodium with its tough, twining stems is difficult to cut. The mixture is also hard to pass through a hand chopper, so that wastage during feeding is difficult to avoid.

12.8.3. Animal performance and nutritive value

Reference has already been made to the highly similar responses in terms of liveweight gain both per ha and per animal by including *Desmodium uncinatum* in a sward of either Elephantgrass or Rhodesgrass in the rainy season (Fig. 10.6; *10.6*). The improvement over unfertilized grass was, however, substantial in Elephantgrass but nil or even negative in Rhodesgrass. Thus, the liveweight gains achieved with Elephantgrass plus Desmodium were the same as those with Rhodesgrass alone. Rhodesgrass, however, was grazed 1 week out of 4, while Elephantgrass was grazed 1 week out of 8. Obviously, the effect of Desmodium was to compensate for the lower nutritive value of Elephantgrass at 8 weeks while it did not contribute to the value of 4-weeks old Rhodesgrass.

Similarly, in the zero-grazing unit at Naivasha a change of fodder supply from young Elephantgrass to mixtures with Desmodium had only a small or no effect on the milk production (Snijders, 1992a; see also *10.2.2*).

An increase in the % crude protein of fodder grasses when grown in association with Desmodium was observed by Suttie & Moore (1966), Dougall (Kitale Ann Rep, 1966) and Birch & Dougall (1967). In less vigorous fodders such as

Guatemalagrass and Guineagrass these effects were more marked as they naturally allow in more legume. Dougall (Kitale Ann Rep, 1966) and Birch & Dougall (1967) attributed the higher N-transfer from the legume to the grass, which they observed in the early compared with the later parts of the season (Table 12.11), to differences in nitrification.

good support for a shade-tolerant and trailing type of legume like Desmodium as it lifts the vines off the ground to give more room for vertical expansion. In return, Desmodium keeps up the nutritive value of swards when grasses are at an advanced age. This effect is small, and in fact redundant, when swards are utilized at a much younger stage.

Table 12.11. Average % nitrogen in various fodder grasses grown with (+) or without (−) *Desmodium uncinatum* (after Dougall, Kitale Annual Report, 1966).

Grass species	Legume	March–April	May–June	July–September
Elephantgrass	−	1.53	2.31	2.35
Elephantgrass	+	2.24	2.32	2.44
Guatemalagrass	−	1.36	2.41	2.04
Guatemalagrass	+	2.13	2.54	2.56
Giant Guinea	−	1.69	1.99	2.14
Giant Guinea	+	2.83	2.34	2.54
Overall	−	1.52	2.24	2.18
Overall	+	2.40	2.40	2.51

The low % N in the grass in March-April is, however, unusual and open to suspicion. An alternative explanation is that the 1st cut concerned grass of advanced maturity in monoculture, having grown undisturbed for many months. Conversely, in mixture with vigorous Desmodium, the yield of Elephantgrass is often reduced by 25–50%. Grass itself is then in a younger physiological stage. This may explain why Elephantgrass in such mixtures shows an increase in % N from only 0.1 unit at the more frequent harvest regimes, later in the season, to 1 unit in the mature stands after the dry season. The comparison between protein-rich and protein-poor grass runs then, in fact, between low and high yield, or between physiologically young and mature grass, respectively, although both are at the same calendar age of regrowth. In the 2nd part of the season, however, there is clearer evidence of N-transfer from Desmodium to grass (see also Fig. 6.5), irrespective of the physiological age of the grass.

12.8.4. General considerations on Desmodium mixtures

Elephantgrass and Desmodium combine well to show the mutual effect of grasses and legumes in a manageable stable association. Both are favoured by long harvest intervals. Elephantgrass provides a

Defoliation of fodder grasses and legumes should be infrequent but swift. When cut too frequently, as is common in zero-grazing systems, Desmodium suffers even more than Elephantgrass. Resting favours the legume and restores the balance. The main benefit of the legume is achieved in the later part of the rainy season (Table 12.11; Fig. 6.5). However, the more severe the legume competition in the dry season the more Elephantgrass is reduced. As with Elephantgrass on its own (*12.5*), the surest regime would be to rest the mixture during the rainy season and to exploit it as much as possible in the dry season (*13.2*).

It is postulated that if standing herbage in the dry season is to be maximized, the rainy season is the period to be employed for the full recovery of grass and legume, in anticipation of the next dry season.

Grass yield is generally not increased by Desmodium (Fig. 12.5) and may even be reduced when Desmodium comprises 30–40% of the mixture, especially in the 1st year. As the grass is then younger physiologically than when in monoculture, better nutritive values of the grass show up in compensation and higher % N values are observed than in monoculture (Table 12.11). Higher % N observed in the grass should, therefore, not simply be attributed to N-transfer from the legume.

By all accounts, Desmodium contributes annually to the yield of the combined grass-legume mixture the equivalent of 100 kg of N-fertilizer, provided it represents 30–40% of that mixture. To that end extra P-fertilizer is required, and possibly nutrients such as sulphur (Keya, 1974a) which are not essential to the grass component of the mixture. Consequently, part of the saving on N-fertilizer is offset by the legume's need of additional fertilizer (*10.2.1*).

Grass-legume mixtures may benefit by N-topdressing. At Makerere, MacDonald (1966) applied N when the legume became too dominant and P when the legume had to be encouraged. At Kitale, N resulted in a positive response of both Elephantgrass and of Desmodium (personal observation, 1969). Tiley (1970a) suggested that Desmodium could provide extra N to the grass, when is was first given N-fertilizer itself (Fig. 12.7).

Alternatives are to grow both the legume and grass in monoculture and to topdress Elephantgrass with N-fertilizer separately.

The advantage of grass-legume mixtures should also be seen from the soil conservation point of view as Desmodium enhances a good soil cover and suppresses weeds. Desmodium-Elephant mixtures are to be recommended as a matter of routine where moisture conditions permit. The chances of Desmodium survival are much better with Elephantgrass, either cut or grazed, than with sown grasses.

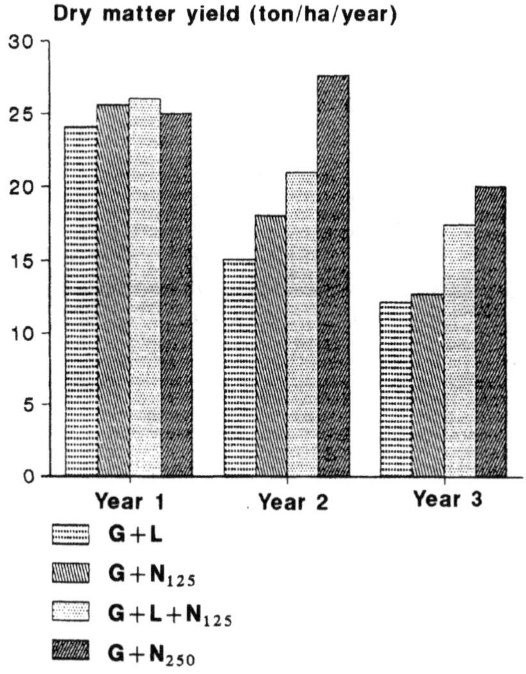

Fig. 12.7. Elephantgrass (G) and *Desmodium uncinatum* (L) with nitrogen fertilizer (after Tiley, 1970a).

12.9. Summary

Although some Elephantgrass varieties produce seed, establishment is poor. Vegetative propagation by root splits is the best method of establishment. Weed control, especially of African Couchgrass, is essential after every harvest. Planting at the recommended density of 100 × 50 cm involves 10 ton/ha planting material and 445 manhours. Plant density is less critical than has been commonly assumed, because the optimum densities of the wet and dry season are opposite. The lower the rainfall, the lower the plant density should be. In Eco-climatic Zone III, a density of 5000 plants (200 × 100 cm) is adequate. Stubble height is also less critical than often assumed. There is no harm in chopping Elephantgrass to ground level, provided that the grass is cut infrequently. Low cutting is less harmful than frequent cutting. The longer the harvest interval, the higher are the total annual yields, but the leaf yields stay the same and the leaf quality is reduced. When harvests are to be taken regularly in the rainy season, a height of 125–150 cm is a good compromise between quality and yield, also with the approaching dry season in mind. Longer harvest intervals in the wet season result also in higher yields in the dry season and this is an easy method to shift the availability of standing herbage from the wet to the dry season. N-fertilizer is essential if the dilemma between yield and nutritive value is to be broken. Herbage quality is at its best at the onset of the rains. The recycling of animal manure off and on Elephantgrass is useful to replenish the quantities of P and K removed by cutting but is not effective as an N-source since the N-losses in the processes between feeding and manure application are too large. The effect of 100 ton of animal manure is not greater than of 100 kg N as fertilizer. Elephantgrass crops of 12.5 ton/ha remove 150 kg N, 30 kg P and 375 kg K. Zero-grazing leads to a serious depletion of P and K in the top- and subsoil. When Elephantgrass is grazed, the effect of PK fertilizer is negligible and only half the rate of N is needed. N-fertilizer applied at 100 kg/ha/year N may double annual yields, while raising the % crude protein to 9–10. Smaller rates of N-fertilizer, applied strategically, are best given 2 months prior to the dry season. Legumes such as Desmodium can form stable mixtures with fodder grasses, since both groups of plants require long harvest intervals for their survival. The contribution of the legume is almost entirely due to its complementary yield and, especially, quality when the grass companion is relatively old. Legumes are useful also to protect bare soil. Extra P-fertilizer is, however, required for the legume to grow well. Also, when the legume attains the 30% level of the mixture required to produce a total yield equivalent to the level of grass + 100 N, grass yields are reduced, especially in the dry season. Dry season performance is enhanced by resting the mixture in the rainy season, followed by intensive exploitation in the dry season.

CHAPTER 13

Elephantgrass utilization

When a grass is cut and fed directly, different sets of rules apply than when it is grazed. Some aspects, discussed previously for sown and grazed grasses in general (*6*; *7*), are therefore dealt with here in some more detail.

13.1. Nutritive value

Nutritive value in herbage crops is a relative character (*6.3.4*). Factors of overriding importance are season, age, leaf:stem ratio and N-nutrition. Apart from a high silica content (*12.7.5*), there is little evidence to suggest that Elephantgrass is low in energy- or protein values compared with the other commonly used grasses. Extra care is needed in sample preparation for chemical analysis, in order to avoid inflated values. Dead plant material is all too easily lost from a sample of coarse grass, compared with finer grasses. A much longer drying time is also needed.

When cut at the same interval, Elephantgrass stays behind Rhodesgrass in digestibility in the early and late parts of the rainy season (Fig. 6.1). Only in the mid-season depression and during the dry season is Elephantgrass higher. This grass has the advantage of retaining a given level for a longer period of time. Whereas most other cultivated grasses were subject to a decline of 0.30 to 0.50 unit of dry matter digestibility (DMD), Elephantgrass only declined with 0.20 unit per day (Table 6.4; Reid *et al.*, 1973). Likewise, Odhiambo (1974) observed that the liveweight gain per animal remained unchanged for a range of growth stages up to 12 weeks and a range of heights up to 140 cm. Ware-Austin (1963) monitored the % crude protein of a Kitale Elephantgrass plot and recorded 16.7% as a year-round monthly average. Although these unusually high values are often quoted, no description of method of recording or of corresponding yields was given. The Elephantgrass survey by Wouters (1987a) indicated that 80% of the farms investigated did not reach the target of 10% crude protein, while more than 40% had less than 7.5%.

However, energy- or protein values are not the only determinants of nutritive value. Voluntary intake, although influenced by digestibility, is of paramount importance. The following factors would suggest that voluntary intake may be relatively low in Elephantgrass:

13.1.1. Dry matter (%)

The DM % of Elephantgrass is the lowest of all cultivated grasses in East Africa (Table 6.3). Under normal harvest regimes, every 2–3 months, the % DM does not exceed 16.5%, in contrast with Rhodesgrass and Guineagrass which reach over 20%. Stems in particular are responsible for the high contents of moisture (Fig. 11.1). The main features of the data are summarized in Table 13.1. The youngest leaf has a % DM twice that of the corresponding stem.

When leaf and stem are chopped in stall-feeding operations, intake may be lowered because of the high moisture content (Dirven & Ehrencron, 1963). Pre-wilting might be considered but Elephantgrass stems are thick and hard to cure.

13.1.2. Sward heterogeneity

In addition to the variation in % DM, tropical grasses in general are notorious for the large variation in nutritive value within and between leaf and stem components of the sward (Table 6.5). Table 13.2 shows the variation that Tiley (1969)

Table 13.1. Dry matter (%) in Elephantgrass in relation to age of regrowth and to plant component (Boonman, 1979b).

Age of regrowth (weeks)	Dry matter (%)			
	Youngest fully expanded leaf (leaf blade)	Leaf blade (all)	Whole plant (all)	Stem (all)
3	15	12	11	7
6	18	16	13	9
12	27	25	19	14

Table 13.2. Crude protein (%) in relation to leaf position in unfertilized Elephantgrass (after Tiley, 1969).

Leaf position	Crude protein (%) (in the leaf blade)
2nd–4th leaf	9
4th–5th leaf	7
11th–14th leaf	0
Stem fraction	3–5

found in % crude protein between leaves of different ages.

Table 13.2 illustrates how much variation exists in a sward of Elephantgrass, even within the leaf fraction itself. The data are in agreement with those of the survey of Wouters (*op. cit.*).

It was shown in Fig. 11.1 that the % leaf is about 40 at the normal grazing or cutting stage, *i.e.* at 2–3 months regrowth and at a yield level of 4–5 ton/ha DM. Tiley (1965b) found 35–50% edible leaf at the 125–150 cm stage, but in a later study (Tiley, 1969) lower values were reported, at 125–200 cm height with a yield level of only 5 ton (Table 13.3).

edible stem lies 40–50 cm from the top of the rolled leaf sheaths or 20–30 cm below the growing point. He found that shoots tend to be eaten a little lower in the field as the stems are firmly anchored.

It is, therefore, imperative to think in terms of yield of edible leaf and to maximize leaf production (Fig. 13.1). An advantage of ungrazed stem residue is that it generates more sites for leafy regrowth from axial buds on the stem nodes (*13.2*).

An additional complication arising from the nature of the sward profile of Elephantgrass is the difficulty grazing animals have in prehension of the loosely packed foliage. Leaf-density is very low:

Height of Elephantgrass at 6000 kg/ha	150 cm
Leaf yield at 50% leaf	3000 kg
Leaf density (kg/ha/cm)	20 kg

It is, therefore, difficult for the animal to grab enough leaf with the tongue, while additional bites are needed to bring in the leaf, which is often longer than 100 cm. Fatigue limits the time that can be spent grazing to a maximum of

Table 13.3. Ranges within Elephantgrass at a height of 125–200 cm (after Tiley, 1969).

	Plant part (%)	Dry matter (%)		Crude protein (%)
		Range	Mean	
Leaf blade	25–40	19–26	24	6–9
Stems + sheaths	60–75	8–15	12	3–5
Whole plant		11–21	16	4–7

This heterogeneity causes a high degree of selectivity in animal intake and, consequently, of wastage of less acceptable leaf and stem of more advanced age (*13.3.4*). Tiley (1969) found with stall-fed animals that the upper limit of

12–13 hours/day. This complication lends a further argument to the thesis that the management of Elephantgrass should be geared to maximizing production of leaf with a dense spatial distribution, as epitomized in Fig. 13.1.

Said (1976) compared the intake by sheep of Elephantgrass with that of other grasses at the University of Nairobi (Table 13.4). Intake was lower than of Kikuyu- and Rhodesgrass, although digestibility data were similar.

Table 13.4. Dry matter intake and digestibility of 7–10 weeks old regrowth with wether sheep (after Said, 1976).

	Dry matter intake (g/kg $w^{0.75}$/day)	Organic matter digestibility
Stargrass	30.6	58.2
Kikuyugrass	56.0	67.0
French Cameroons Elephantgrass	40.5	66.6
Pokot Rhodesgrass	47.7	64.3
Rhodes + *Desmodium uncinatum*	53.7	68.8

Odhiambo (Kitale Ann Rep 1967) confirmed the general farm experience that "milk production fell at a much faster rate in the rainy season on Elephantgrass than on ley pasture" (*11.3.2*). Mbugua (1976b) compared liveweight gain of 7–9 months old steers on Elephantgrass and Rhodesgrass. His conclusion was "that although the crude protein content and the digestibility of Rhodes- and Elephantgrass are comparable, the performance on Rhodesgrass is higher"; see also Fig. 10.6.

13.2. Leaf stripping in the dry season

Long resting periods are needed to sustain yield in Elephantgrass but the yield increment is progressively composed of inedible stem (*12.5*). The purpose of management should be to produce a maximum of leaf, preferably in the dry season. In Elephantgrass it is possible to make stems useful not so much as edible material, but rather as a framework to extend the tillering zone vertically on which more leaf can be produced and harvested. To illustrate this point evidence is available from work done in Florida (Blaser *et al.*, 1955; Fig. 13.1).

Three management systems were compared for their effect on yield of edible leaf:

A: Whenever the herbage reached a height of 90 cm, grazing was simulated by leaf stripping. New leaves developed rapidly from the axial buds at the nodes along the defoliated stems. The herbage thus harvested consisted of 88% leaf.

B: Whenever the herbage reached a height of 90 cm, it was cut back to 50 cm. The proportion of leaf in the yield was 67%.

C: Whenever the herbage reached a height of 90 cm, it was cut to 10 cm stubble height. This resulted in only 50% leaf.

D: Cutting only twice annually. A very high total yield of herbage was obtained but the % leaf was only about 40, and the leaf yield was lower than with stripping [A].

The row width was 200 cm and N-fertilizer was applied. The leaf yields decreased in the following order: A > B > D > C.

System A warrants attention to make Elephantgrass produce as much leaf as possible. Leaves should be picked as if the plant were a tea bush: tillers arising from the nodes are picked from a semi-permanent stubble. This stubble is left to produce consecutive crops of leaves until it becomes overcrowded with stems no longer capable of producing new tillers; it is then cut back to ground level to start a completely new cycle. Thus, defoliated stems function as an extended base of the plant's reserves and as a framework to produce aerial tillering as the harvestable part. Regrowth not only depends on the amount of photosynthetic area left but also on the presence and growth of terminal, axillary and basal meristems. Only the best part of the plant, young leaf, is harvested; see also *15.4.2*. Moreover, the rhythm of growth cycles can be timed in such a way that a maximum of leaf is produced in the dry season (*13.2.1*).

The Florida experiment showed further that when conventional harvesting was resumed in the following year, residual yields were much higher after stripping [A]. The yields of treatments [A] and [B] in the residual harvest were even higher than the yield of [D] in the main trial year. The maintenance of a lot of the stem base, *i.e.* a high stubble, is clearly beneficial to residual regrowth.

In order to maximize dry season yields under Kitale conditions, Strange (1959a) suggested taking only one grazing or cut in the rainy season, in August or September, followed by N- topdressing. At Serere (U), Horrell (1958a) left Elephantgrass

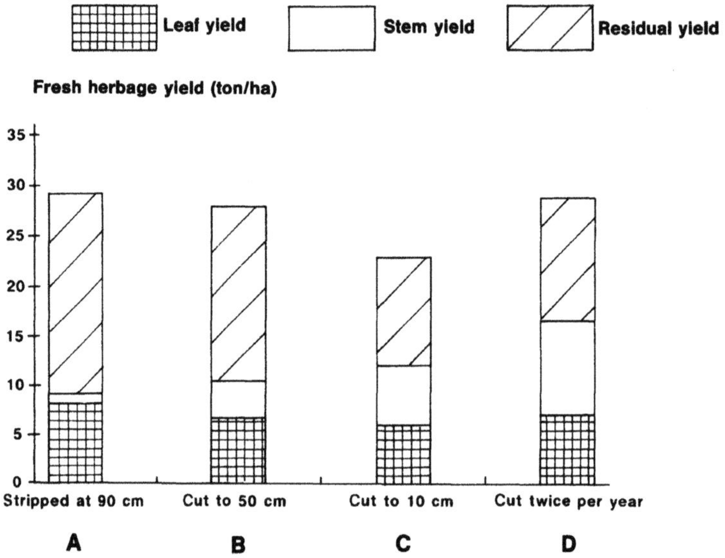

Fig. 13.1. Method of defoliation and yields in Elephantgrass (after Blaser *et al.*, 1955).

ungrazed until 1 month before the dry season and found it could then be grazed 2–3 times. This could be combined with the Florida method of leaf stripping, in the dry season. To this end the following system would seem appropriate for Kitale:

August–September:	silage cut followed by N-topdressing
December–March:	repeated stripping
April:	cut back to low level after last stripping
August–September:	silage cut etc.

13.2.1. Spelling during the rainy season

It has been emphasized (*12.5*) that the longer the preceding harvest interval, the higher the yield in the subsequent cut. By spelling (resting) a field, herbage yield can be shifted from one period to the next, from wet to dry season. Conversely, heavy utilization in the dry season requires resting over the larger part of the rainy season as well as N-fertilizer to rebuild root mass in favour of the next dry season and to suppress weeds.

The extra production capacity (Fig. 6.1) offered by growing conditions at the beginning of the rains, when the soil is still warm, is then converted by the plant in rebuilding its reserves and root system. The milder the exploitation has been during the dry season the larger the "extra" will be.

The best use in August would seem to be silage, as suggested by Maher (1936a). Elephantgrass does not, however, make high quality silage (Davies, 1963; Blaser *et al.*, 1955), due to losses associated with the low % DM (Table 6.3) and loose packing. An alternative use would be to cut it in the rainy season for mulch in order to preserve moisture for the dry season (Table 13.6; Tiley, 1965b; 1970a) and to prolong the longevity of the grass (Boshoff, 1966). Boshoff found at Makerere (U) that with mulching after 9-months of rainy season followed by strip grazing in the 3-months dry season, more than 5 beasts/ha could be grazed. The beneficial effect of mulching on rainfall acceptance has been amply demonstrated (*5.2.2.1*; *11.3.3*).

Farmers may find this hard to accept, but the value of Elephantgrass in the wet season should not be overestimated. Growth rates at Kitale declined to below 20 kg/ha/day in the middle of the wet season, far below that of Rhodesgrass or Setaria (Fig. 6.1). As already mentioned (*13.1.2*), livestock drops in production when turned into Elephantgrass from other good pasture during the wet season; see also Fig. 10.6. Consequently, in the wet season Elephantgrass is better spared. Essential is that, if dry season production is to be maximized, cuts are taken at long intervals in the wet season.

If Elephantgrass is to be used for feeding in the wet season, it should be allowed to grow to a height of 125–150 cm (*12.5*). In fact, this height seems to be the best compromise to achieve yields which are high and constant throughout the year, including the dry season, and still of a reasonable quality. With the aid of N-fertilizer it should in this way be possible to harvest grass with a crude protein of 10% and a digestibility of the organic matter (D-*vitro*) of 55.

13.3. Cutting versus grazing

There has been a persistent belief among agriculturalists, in the livestock branch in particular, that direct grazing is a less efficient technique of grass exploitation than when the grass is cut, carried and fed indoors. The available evidence is, however, clearly to the contrary. The impact should be viewed from four angles: [1] the soil; [2] herbage yields; [3] economics; [4] wastage.

13.3.1. Soil mineral composition after cutting or grazing

Grazing gives higher yields in subsequent crops than cutting and removal (*5.5*; *12.7.6*).

Data on the direct comparison of grazing versus cutting and their effects on soil conditions are available for Elephantgrass from Kawanda (Stephens, 1967a; Foster, 1969; 1971) and for Nandi Setaria from Kitale (Keya, 1974a). Although the Kitale values were low, the similarity of responses of the two locations makes it possible to present the data in one table (Table 13.5).

The general trend was that values in the soil of pH, P, K, and Ca were higher under grazing than under cutting. Especially pH and K were badly affected by cutting regimes (see also *12.7*). Foster (*op. cit.*) found that the direct return of dung and urine by the grazing animal itself significantly increased soil-N and exchangeable K.

These data confirm that soils are depleted of essential nutrients by zero-grazing (Table 12.6) and they explain why zero-grazing practices used to be found only on the most fertile soils. Reviewing work done in South Africa, Scott (1967) concluded that although zero-grazing had been attempted, it had not been a success because nutrient removal far outweighed replacement through fertilization.

13.3.2. Yield and quality under cutting versus grazing

Stephens (*op. cit.*; Fig. 7.1) examined the effects of fertilizer on grazed and cut Elephantgrass. Under grazing the additional increment of 125 kg N yielded 12 ton extra dry edible leaf but under cutting only 5 ton. The residual effect was much larger under grazing than cutting (*12.7.2.1*). N at the rate of 250 kg/ha/year increased DM-yield 3 to 4-fold and yield of crude protein 5-fold. N-fertilizer increased the % crude protein in the edible leaf from 7 to 10. The % crude protein in the edible leaf was higher under grazing (8.9%) than under cutting (8.6%). High N also had an increasing residual effect on grazed grass as time progressed. The effect of N-fertilizer on yield increased with each further grazing cycle after the 1st year. This was not so under cutting (Fig. 7.1).

Considerably more grass is produced under a grazing regime than when Elephantgrass is cut and removed as illustrated by Table 13.6 due to Tiley (1965b; 1970a). Tiley's experiment lasted 3 years. Grazing was carried out at the same frequency as cutting (150–180 cm).

No other experiments are known to have been done on this aspect with Elephantgrass, but Keya (1974a), working with Nandi Setaria at Kitale, observed up to 60% greater DM-yields in grazed than in cut plots (Table 13.7).

Ogborn (1965) at Namulonge (U), comparing zero-grazing, simulated grazing and direct grazing on Rhodesgrass, found that grazing yielded more herbage which contained also a higher % crude protein and phosphorus (Table 13.8).

It is of interest to note that Ogborn's study was the first to be published on zero-grazing in East Africa, but dealt in fact with Rhodesgrass. More recently, rapid recycling of nutrients in dung and urine under grazing has been shown to be responsible for better growth of grazed compared with ungrazed systems in the Serengeti (McNaughton, 1985).

Table 13.5. The effect of a 3-year period of grazing or cutting on soil data at Kawanda (Foster, 1969) and Kitale (Keya, 1974a).

	Kawanda (Elephant)		Kitale (Setaria)	
	Grazed	Cut	Grazed	Cut
pH	5.65	5.25	4.80	4.20
% C	1.66	1.65	—	—
% N	0.133	0.122	0.07	0.16
P ppm (Truog)	21	14	16	14
Exchangeable Ca (me%)	6.4	5.1	0.8*	0.4*
Exchangeable Mg (me%)	1.0	1.1	0.2*	trace*
Exchangeable K (me%)	0.59	0.27	0.38	0.24

Kawanda, kg/ha/year: 250 N; 100 P_2O_5; 125 K_2O; 325 MgO
Kitale, kg/ha/year: 500 N; 50 P_2O_5; 60 K_2O
*Unlikely low values.

Table 13.6. Annual yields (ton/ha) of Elephantgrass at Kawanda (after Tiley, 1965; 1970a).

Treatment	Dry matter yield (ton/ha)
All crops removed	16.2
Crops alternately removed/mulched	18.4
All crops grazed	19.1
Crops alternately grazed/mulched	16.3
All crops mulched	16.8

Table 13.7. Dry matter yields of Nandi Setaria under grazing or cutting regimes (after Keya, 1974a).

Treatment	Year			
	1970	1971	1972	Mean
Cutting	4.6	4.8	5.1	4.8
Grazing	8.3	7.8	7.1	7.7

Table 13.8. Rhodesgrass under cutting or grazing (after Ogborn, 1965).

	Yield (kg/ha)	Crude protein (%)	P (%)
Cutting	5260	12.0	0.29
Grazing	6080	12.6	0.34

13.3.3. Economics of zero-grazing

A survey of ruminant production systems, including zero-grazing, in the high potential areas of Kenya was produced by Chudleigh (1974). Musangi (1969) commented that systems of zero-grazing can be operated only when 2 or 3 cows are involved because of the amount of labour involved in cutting and carting grass and in manure disposal; the system is then profitable when high-yielding cows are involved.

Stotz (1977) presented a model estimate of the activity net income per hour of work and per acre for milk production on either grazed natural pasture or zero-grazed Elephantgrass (Table 13.9). Many of Stotz' figures on the economic value of manure, the concentrate requirements and the cost of buildings were based on assumptions due to lack of direct data.

It had already been established at that time that milk yield per cow does not increase as the result of stall-feeding. Stotz, however, attributed a carrying capacity 3 times higher to cutting than to grazing and arrived in that way at an equally higher net income per acre. The estimated income per hour of work on the zero-grazing system was only one-quarter of that on grazed pasture. Stotz' zero-grazing referred to Elephantgrass, whereas natural Kikuyu-, Star- or African Couchgrass pasture stood model for the grazed pasture. This was done for the ostensible purpose of presenting the contrast of zero-grazed Elephantgrass against the traditional fallow pasture (*4.5*). The differences in income were, therefore, not due to either cutting or grazing, but to good and poor grass.

Since no data are available on economic comparisons of the 2 systems based on the same grass from East Africa, we have to resort to those of Vicente-Chandler *et al.* (1974), in the humid tropics of Puerto Rico. Table 13.10 illustrates the main findings of their experiment with Elephantgrass in which comparisons were made of costs (excluding buildings) and of liveweight gains between cutting

Table 13.9. Economic estimates of zero-grazed Elephantgrass (after Stotz, 1977).

	Grazing natural pasture	Elephant grass zero-grazing
Gross output		
Milk (2,500 kg/cow/year) @ 1 sh/kg	2,500	2,500
Calf (8-day old)	100	100
Culling	200	200
Manure (10 ton @ 60 Shs)	–	600
Gross return/cow/year	2,800	3,400 Shs
Production costs		
Concentrates and minerals	400	150
Veterinary costs and dipping	60	60
Building material	–	70
Miscellaneous	20	60
Cow replacement	200	200
Interest for cow (10%)	150	150
Risk of loss of cow (5%)	100	100
Sub-total	930	790 Shs
Forage production costs	–	300
Annual investment costs (interest, depreciation, repairs)	200	400
Activity inputs (excluding labour)	1,130	1,490 Shs
Activity income (excluding costs for land and labour)	1,670	1,910 Shs
Demand on labour	200 hrs	1,000 hrs
Demand on land/cow	1.5 acre	0.5 acre
Activity net income per hour of work	8.35	1.91 Shs
Activity net income per acre	1,113	3,820 Shs

Table 13.10. Liveweight gains and production costs with Elephantgrass, grazed directly or cut and fed, in Puerto Rico (after Vicente Chandler *et al.*, 1974).

Feeding regime	Fertilizer application (15:5:10)		Carrying capacity (SSU*/ha)	Production costs (US $/ha)	Liveweight gain per year (kg)	
	Ton/ha	Costs/ha (US $)			Per head	Per head
Cutting/feeding	5.0	400	5.5	1175	186	1022
Direct grazing	2.5	200	3.3	517	266	876

* SSU: standard stock unit (450 kg steer).

plus feeding and grazing-only systems. Milk yields of 7500 kg and liveweight gains of 1000 kg/ha were obtained on Elephantgrass rations. An important finding was that, to achieve the same yield, grass used for cutting needed twice as much fertilizer as when grazed, while the total production costs were also more than double. The carrying capacity was higher but was partly offset by a lower liveweight gain per head. The end result was a 16.5% higher output per ha in favour of the cutting plus feeding system, although at more than double the cost. The extra liveweight gain achieved would have had to fetch the unlikely price (in 1974) of US$ 4.50 per kg to put both systems economically on the same footing, building costs not considered.

In Thailand, in a comparison of cows grazed outdoors or fed indoors on grass cut and carried from a similar area, milk yields did not differ. Ambient temperatures were lower indoors but humidity was higher. Animal health and % milk fat were higher in the animals grazing outdoors. This was seen as a positive financial incentive in favour of direct grazing (Hongyantarachai *et al.*, 1989).

13.3.4. Wastage

It had been suggested that zero-grazing systems were an efficient system of growing and collecting fodder without losses; it would prevent the animals from wasting their own grazing by treading and fouling. However, cattle, whether grazing or stall-fed, display marked selective consumption (Bredon 1965; Bredon & Marshall, 1962; Bredon *et al.*, 1967). Experience has shown that animals reject a large amount of Elephantgrass, especially of the stem, even when chopped in small particles of 2.5 cm. Animal intake as well as animal production itself are often much lower than what farmers expect from the intensive methods employed. In stall-feeding, waste residues of 10–15% are to be taken for granted if a DM-intake of 10 kg/day and a milk yield of 6 kg/day are to be achieved (*12.7.2*).

For an assessment of the assumed losses in pasture due to fouling and treading no data are available. In East Africa in the rainy season, however, it is rare to find a dung pat of more than a few days old in pasture. Within that time, coprophagous insects have buried or moved it (*5.2.2.3*). On the other hand, the problems of handling manure in stall-feeding systems are well known (*12.7.1*). As far as the effects of treading are concerned, losses due to poaching in wet weather may be relevant in the 1st year when the swards have still few tillers and are still relatively open (*7.1.11*). Cattle tracks are muddy in the rainy season, but hard in the dry season. On denuded land, treading by cattle is extremely damaging to the soil (*4.3.2.2*; Thomas, 1960).

13.3.5. Grazing Elephantgrass

There is no evidence to suggest that animals perform better under stall-feeding conditions than under direct grazing (*13.3.3*). Harker & Bredon (1963) observed in Uganda that in-door fed bullocks gave very low liveweight gains on Elephantgrass. The same is found in the data of Mulder *et al.* (1973) comparing stall-fed Elephantgrass with maize stover (*13.4.2*). At Kitale, Odhiambo (1974) found average liveweight gains of 0.39 kg/day when feeding Elephantgrass *ad libitum* indoors. In grazing trials with this grass the average liveweight gain over all stocking rates was 0.52 kg/day (Fig. 10.6).

The practice of grazing Elephantgrass in Kenya has been mainly limited to the larger farms. It will stand grazing if properly managed (Strange, 1958b). Mbugua (1976b) suggested a minimum of 8 paddocks to facilitate a rotation of 8 days on and 56 days out.

For small farms in Uganda, by contrast, the management of Elephantgrass for grazing has been described by Tiley (1965b), Wendt *et al.* (1970) and Horrell & Tiley (1970). They recommend controlled strip grazing and a heavy grazing pressure without backgrazing in small paddocks so that defoliation is completed rapidly. Grazing at 1 beast/acre was found adequate at Namulonge (Harker & McKay, 1962). In other countries, grazing seems quite normal also, *e.g.* in Puerto Rico (Vicente-Chandler *et al.*, 1974) and in Hawaii (Takahashi *et al.*, 1966).

It has been argued that grazing is no alternative for the smallholder since he cannot afford the cost of fencing. It is doubtful, however, whether the type of cattle sheds that has been recommended for stall-feeding and manure-collecting are cheaper than fencing the very areas where the grass is cut. Tethering, commonly found in the Kitui area of eastern Kenya and parts of Ethiopia, is a practice adaptable to Elephantgrass. The possibilities of new techniques of temporary, electric fencing should also be explored.

There is no point in the farmer enslaving himself to cutting and carrying fodder in situations where the animals could be feeding on pasture; the motto should be "where possible let the animal do the work", leaving the farmer free to attend to other work (Bauer, 1975).

Stobbs (1969e) has shown convincingly how direct grazing improves subsequent crop yields on the same field, compared with grass that is not grazed. Night paddocking was highly effective in transferring fertility to arable cropping (*5.2.2.3*). He concluded that by rotating crops with grazed pastures more intensive systems of agriculture are possible.

If direct grazing is to be considered the question may rightly be asked whether Elephantgrass is the correct choice (Fig. 10.6). Equally valid is the question whether for zero-grazing preference should be given to Elephantgrass rather than to other grasses or fodders such as lucerne (*17.5*). In fact, most stall-feeding units depend only partly on Elephantgrass (*11.3.4*).

13.4. Alternative crops or grasses for zero-grazing

13.4.1. Green maize

A characteristic of stall-feeding which lends little support to the fodder-efficiency aspect of this practice is the fact that green maize comes close to Elephantgrass in importance of the crops specifically grown for stall-feeding (*11.3.4*). Green maize is harvested when it has passed already more than one-half of its growing period, yet has only achieved one-third of its potential silage yield (*19.2*). This premature harvesting of green maize is a wasteful practice from the fodder point of view. The farmer may have other reasons, of a quite different nature. Obviously, the arguments favouring stall-feeding are not necessarily those that can equally be defended for zero-grazing (*11.4*).

It should be noted that in contrast with many Mediterranean countries little use as a feed is made in East Africa of maize leaves plucked when reaching maturity. Similarly, cutting the stems above the cobs when the grains have reached physiological maturity has been shown to be beneficial to the maize crop because it speeds up the drying process and it reduces the risk of lodging.

13.4.2. Maize stover compared with Elephantgrass

Maize stover is a source of fodder available at the end of the maize growing season. Mulder *et al.* (1973) compared animal production on maize stover and Elephantgrass, supplemented with locally available concentrates such as maize germ, bran meal, wheat bran, rice bran and molasses. This was done in stall-feeding trials at Farmers' Training Centres in central Kenya. The investigations were initially designed to study the feasibility of producing beef from indigenous Zebu cattle on smallholder farms, using the knowledge gained from the feedlot experiments at the Beef Research Station at Lanet (*11.3.4*). All rations were mixed with a molasses-urea-mineral mixture developed at Lanet. Elephantgrass represented 20–70% and maize stover 30–40% of the total rations on a DM-basis. All rations were chopped and fed *ad libitum*. Results showed that liveweight gains were only half of those obtained with maize silage rations in the Lanet feedlots, the reason thought to be the low value of the roughages and the coarseness of the chop obtained with hand tools. The general conclusion was that with the cost of concentrate feeds and the relatively low return from meat in the rural areas the practice was not likely to be an attractive proposition for farmers. The technique was later extended to dairy cattle. The rations were so designed that differing proportions of the DM-intake would come from the roughage, *i.e.* Elephantgrass and maize stover. The daily consumption per animal was only 5.5 and 3.9 kg, respectively. The experiment was finally abandoned (Creek & Squire, 1976).

In a large-scale forage system exercise at Kitale (Sheldrick & Goldson, 1978) very poor liveweight gains were achieved with maize silage compared with routinely grazed Rhodesgrass (*19.2*).

13.4.3. Other grasses and fodders

Giant Setaria, Guatemala- and Giant Guineagrass are commonly considered as alternative fodder grasses, but their yields are generally very low (*14*). Grass Sorghum is a better alternative in drier areas and conditions (*15*). Oats (*16*) and Brassica's (*18*) are used as a fodder at the higher altitudes and lucerne (*17*) is an obvious choice wherever soils are suitable. Bye-products such as sweet potato vines (*18*) are commonly fed.

Sown grasses such as Setaria and Rhodesgrass lend themselves well to cutting and are used for that purpose (*11.1*; *11.3.4*). The first published experiment on zero-grazing was on Rhodesgrass in Uganda (Ogborn, 1965). Objections are commonly heard to the effect that short grasses are difficult to cut with the "panga" (machete) and that farmers are not used to cutting this type of grasses. This is surprising since people have been accustomed to cutting short grasses for thatch. Rhodesgrass and Setaria are best cut with the "ringa" (*8.4.1*; *12.6*). Incidentally, the introduction of this tool would be beneficial since the present use of the panga implies that cutting is done continuously at ground level (*12.6*).

A direct comparison of Elephantgrass and sown grasses under grazing and cutting is long overdue. Anindo & Potter (1986) described the effects of an interim-period of stall-feeding fertilized Elephantgrass on Friesian cows grazing Rhodesgrass (Fig. 13.2). Milk yields dropped consistently during the stall-feeding period of 4 weeks and did not recover subsequently to the yield levels of the pre-trial period.

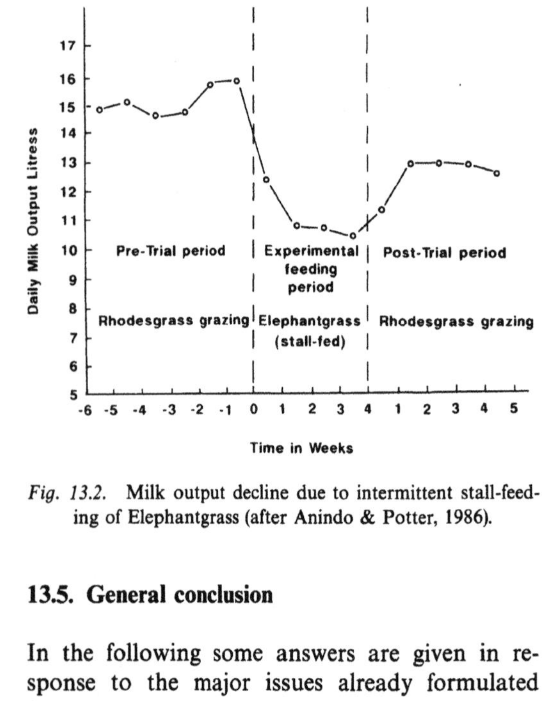

Fig. 13.2. Milk output decline due to intermittent stall-feeding of Elephantgrass (after Anindo & Potter, 1986).

13.5. General conclusion

In the following some answers are given in response to the major issues already formulated (*11.4*).

13.5.1. Elephantgrass

This grass seems to have been credited at times with properties it does not have, while not enough has been made of its main virtue: to produce a maximum of green leaf in the dry season in the subhumid Eco-climatic Zone III. In the wet season, growth rates (Fig. 6.1) and palatability are low. Elephantgrass is unique it that it can be exploited to the fullest in the dry season, if the wet season is made to function as an intermezzo to (re)build the reserves tapped in the dry season. Leaf stripping off an elevated stubble is the best strategy yet designed to utilize this grass in the dry season. If year-round exploitation is envisaged, N-fertilizer is necessary.

The drought tolerance Elephantgrass was claimed to have in the semiarid Zone IV has not been confirmed. Only when repeatedly rested over long periods will it be capable to survive, but yields are very low. Consequently, greenness in the dry season of subhumid zones should not be mistaken for possible adaptation to semiarid zones.

The yielding ability in subhumid zones is also less than what has been assumed on the basis of its giant stature. Elephantgrass is very low in % DM and in sward density. This explains why the annual DM-yields are not higher than those of Rhodesgrass. Only in the warm and humid Zone II, where sugarcane thrives also, can Elephantgrass achieve its high yield potential.

These varied responses explain why Elephantgrass is a powerful and at the same time such a controversial grass. It has virtues and vices, unknown in sown grasses. From the latter it differs mainly in that it is so costly in terms of the amount of labour required, *e.g.* for planting and weeding. In the warm and humid Zone II Elephantgrass yields well but, without a pronounced dry season, cannot display its comparative advantages. The major role of this grass lies in the dry season of Zone III. Here it can complement sown grasses when no other provisions are made for the dry season.

Elephantgrass is an extremely useful grass if and when its leaf production capacity can be maximized to suit the dry season and if the costs of operations can be reduced to a minimum. In most other circumstances it will prove to be out of place and to be too costly in terms of planting, weed control and eradication.

13.5.2. Stall-feeding

Stall-feeding is essentially a livestock system with which Elephantgrass has increasingly become associated. It was practised for a long time in some areas but it really began to take shape commercially near the cities in the 1960s to meet the increasing demand for fresh milk. Stall-feeding circumvents some of the problems typical of livestock husbandry in the tropics but the main advantage is that it can be founded on purchased feeds and on roughage brought in from outside. Although the commercial beef feedlots on large farms were soon abandoned as uneconomic, stall-feeding for dairy purposes continued as a commercial proposition on farm units with plenty of labour but with little (near-by) land. Typical of stall-feeding is the continuous need to supply labour, water and feeds. Where such conditions can be met and where fresh milk fetches a high price, stall-feeding is likely to remain an attractive proposition for small units which have not enough land of their own. Otherwise, it has proved to be quite unnecessary and even undesirable for tropical dairying (Humphreys, 1991).

13.5.3. Zero-grazing

Most of the Elephantgrass grown in Kenya is zero-grazed for the purpose of stall-feeding, but its share in the feed supply is smaller than what may appear from the roadside. Green maize is a good second, but arable bye-products and industrial wastes form frequently the bulk of the feeds used.

Stall-feeding of products otherwise wasted is a commendable practice but caution is needed with zero-grazing, especially of Elephantgrass. Fields subjected to zero-grazing look generally very poor, even in the middle of the wet season at a time when one would expect to see fields with grass 150 cm high. Unless N-fertilizer is added in substantial quantities yields drop very rapidly, which is the short-term effect of zero-grazing. On a longer term, the subsoil is seriously depleted of P, K and other elements, while the pH is decreased also. Since stall-feeding originated in regions with very fertile soils, the consequence of soil mining was overlooked in the early days of the zero-grazing campaigns. When problems began to be encountered in areas with more marginal soils, it was realized that soil fertility was seriously reduced. This can only partly be remedied by the return of animal manure to the same field because of the losses involved in the process, especially of N. Although legumes such as Desmodium are compatible with Elephantgrass and are helpful if supported by P- and possibly other fertilizers and correct husbandry strategies, very little use has been made of this resource.

Milk production per head is not higher in stall-fed animals than in those grazing the same grass in the open field. Attention has to be given to supplementation. The chemical composition as regards protein and energy values is as good in Elephantgrass as in any other grass in East Africa, if utilized young, but the mineral composition is unbalanced. It is high in K and silica, but low in P and Ca. A mixture of molasses-urea-minerals may usefully supplement the roughage, if the necessary precautions against possibly toxic side effects of urea and molasses are taken.

Where the farmer has the choice between direct or zero-grazing, there is little to commend the latter. Zero-grazing has little to justify it on agronomic grounds. Not surprisingly, it is often practised on land to which there is no title, *e.g.* roadsides.

Although many farmers practising zero-grazing have no option of direct grazing, Uganda experience has shown that under direct grazing more grass is produced, at a much lower level of fertilizer requirement, while a higher plane of plant nutrition and soil fertility, including pH, is maintained.

Unfortunately, in spite of all the work done to promote zero-grazing no research reports have become available on the comparative economic advantages, if any, over direct grazing.

It goes without saying that zero-grazing, as a farming system, is a fragile one, unless measures are taken to maintain or restore soil fertility. Sustainable agriculture does not seem possible with zero-grazing alone. The best solution lies perhaps in a combination with direct grazing and supplementary feeding, which was named "semi-zerograzing" by Stotz (1977).

The question remains why farmers practise zero-grazing with Elephantgrass and not with, for instance, lucerne which is well-known and which does well on good soils also. The answer probably is that lucerne dies rapidly under the severe frequencies of cutting normally employed in zero-grazing regimes. Elephantgrass suffers also and yields are reduced rapidly, but seems to survive with at least a few shoots reappearing.

Where attractive milk prices prevail, an excellent chance exists to introduce proper husbandry techniques. Elephantgrass is highly responsive to good husbandry.

13.5.4. *A husbandry package for Elephantgrass*

The main virtue of Elephantgrass is that it can be managed in such a way as to produce a maximum of leaf in the dry season of the zone where it is wanted most, in Zone III. Such management consists of:

1. A much restricted use (spelling) during the wet season.
2. A low plant density, *i.e.* wide rowwidth.
3. N- and other fertilizer as required.
4. Clean weeding.

Where Elephantgrass is needed throughout the year, the best combination seems to be:

1. Harvest at such intervals that the grass is 125–150 cm high.
2. N-fertilizer at 50 kg/ha/cut of N.
3. PK-fertilizer as required.
4. Plant density at 5000 plants/ha (200 × 100 cm).
5. Desmodium, to cover the bare soil.

If the aim is to produce a maximum of leaf in the dry season, leaf stripping off an elevated long-term stubble should be preferred (*13.2*; Fig. 13.1). When this is combined with spelling in the early part of the wet season and with N-fertilizer, a maximum amount of leaf is produced in the dry season.

Elephantgrass, when frequently cut, leaves the soil bare for most of the year. A legume cover is desirable. Both grass and Desmodium benefit by infrequent defoliation. Desmodium complements grass yield, although it replaces part of it.

N-fertilizer is a potent and flexible management tool. It is the only input with which the chronic shortages of fodder in zero-grazing can be combated instantly. It should be introduced as a strategic application of 50 kg/ha N, 2 months prior to the dry season. As experience builds up and as the farm system adapts itself to the new input, N-application can be extended to a point that N is applied at 50 kg/ha after every cut in the rainy season. With 100 kg N annually, yields of Elephantgrass can easily be doubled on most farms. P-, K- and other fertilizers are needed to offset soil depletion, although no direct yield response may be noticed.

A combination of sown grasses for grazing in the wet season with Elephantgrass for supplementation in the dry season seems the most productive and least expensive system yet proposed for dairy production in the medium and low altitude areas of East Africa, in particular in Zone III. Existing stall-feeding units will continue to have to rely largely on bye-products and fodders brought in or purchased from outside.

13.6. Summary

Since the dry season is the main bottleneck in the year-round feeding system and since Elephantgrass is of relatively limited value in the rainy season, management should be geared to a maximum utilization in the dry season. This requires resting in the rainy season so that the root mass can be rebuilt before the dry season. The grass is cut 2 months before the end of the rains and is fertilized with N. When in the dry season the foliage reaches a height of 100 cm, the leaves are stripped off the young stems which are left standing to produce further crops of leaves from the axial buds in the nodes. The process is repeated until the onset of the rainy season. The grass will then require a large part of the rainy season for its rehabilitation. Although many farmers practising zero-grazing of Elephantgrass have no option of direct grazing, Uganda experience has shown that under direct grazing more grass is produced, in fact at lower levels of fertilizer while a higher plane of plant nutrition and soil fertility, including pH, is maintained. Milk production per head is not higher in stall-fed animals than in those grazing the same grass in the open field. Little data are available on the comparative economic advantages of zero-grazing. Stall-feeding is an attractive proposition as a livestock system when relatively cheap feeds can be brought in from outside and when labour is cheap but zero-grazing Elephantgrass leads to a rapid depletion of soil fertility, unless restored by fertilizer.

CHAPTER 14

Alternative fodder grasses

Although Elephantgrass is by far the most productive and widely used fodder grass, other species are regularly mentioned in publications. No series of fodder demonstration plots seems complete without Giant Setaria (*Sspl*), Guineagrass (*Pmax*) and Guatemala grass (*Tand*). An overview of countrywide yields achieved in the Kenya District Trials (Table 6.2) shows the following (DM ton/ha/year):

French Cameroons Elephantgrass	13.34
Gold Coast Elephantgrass	13.52
Giant Setaria	9.69
Guatemalagrass	6.89

The expense of time and labour, of planting and maintaining these grasses, is even greater than with Elephantgrass because plants are smaller and high plant densities are required. Only root splits can be used in vegetative propagation. The relatively attractive method of propagation by cane cuttings (*12.1.1*) cannot be applied.

On the positive side, compared with Elephantgrass, mention can be made of the following:

1. The stem fraction is more acceptable to the animal.
2. The % DM, except of Giant Setaria, is higher (Table 6.3).
3. They are more pleasant to handle manually.
4. They are more suitable as bund grasses.
5. Giant Guinea and Giant Setaria may be grown from seed.
6. Giant Guinea performs better than Elephantgrass in semiarid areas.

Of these 3 species, only Giant Setaria and Guatemalagrass are grown on a commercial scale.

14.1. Giant Setaria

Giant Setaria (*S. sphacelata* var. *splendida*; syn. *S. splendida*) is grown in patches and as a bund grass on small farms on the lower slopes of Mount Kilimanjaro and Mount Meru in northern Tanzania and central Kenya. It is grown between coffee and bananas and is stall-fed (Naveh & Anderson, 1966) in preference to Elephantgrass (van Rensburg, 1969a); it is less competitive when planted in proximity to other crops and coffee. This grass is also of some importance in Rwanda, Burundi and Kivu (Vancoppenolle, 1988; Risopoulos, 1966). Its use on high mountain slopes would suggest that this grass is adapted to cool conditions and to the infertile and acid soils of the "Bracken belt" (*3.4.2.1*; *4.2.1.1*), but from the evidence of the Kenya District Trials (Table 6.2) it appears as if the *relative* performance of Giant Setaria is better under conditions of high temperature, rainfall and fertility. The only location where it was outstanding was Baraton which had, however, the lowest location average. Incidentally, Baraton was the Veterinary Centre where Nandi Setaria was first grown (*6.7*). It did very well at Lyamungu, on Mount Kilimanjaro (1200 m altitude; 1500 mm rainfall); see Table 14.1. Lyamungu is a very fertile site (Anderson & Naveh, 1965).

Table 14.1. The effect of harvest interval on Giant Setaria at Lyamungu (T) (after Lugenja *et al.*, 1977).

	Cutting interval (weeks)		
	3	6	9
Dry matter yield (ton/ha/year)	2.0	9.9	14.0
Crude protein (%)	15.0	10.2	9.0

Giant Setaria has more difficulty than Elephantgrass in surviving long dry seasons in Zone III (*3.4*). It usually regrows fast after the onset of the rains. Bumpus (Kitale Ann Rep 1958) found that the dry season yields of Giant Setaria were lower than of most of the other grasses under trial. This is an important point, since dry season performance should be a major criterion why to choose Giant Setaria in preference to, for instance, Nandi Setaria, that can be established easily from seed. However, where farmers have taken to this grass it is either because they find this grass more pleasant to handle or more acceptable as a bund grass compared with Elephantgrass (*11.3.4*).

Giant Setaria has a low % DM (Table 6.3), but is well liked by stock (Risopoulos, 1966). Oxalate contents can be high (*6.7*) and its digestibility is relatively low (Reid *et al.*, 1973). It is a tufted grass with broad, flat shoot bases and broad leaves; the majority of varieties have little stem. Planting is done with root splits which take relatively easy. Because of the smaller tuft size more plants need to be established than in the case of Elephantgrass (*12.4*). Giant Setaria may be grown from seed (J.B. Hacker, private communication). Viable seed is set, but seed production is not a commercial proposition. Infection of the spikelets by bunt, *Tilletia echinosperma*, is common (*8.7.2.5*). White mould (*Beniowskia sphaeroidea*) is widespread also, but is less damaging than on Elephantgrass (*11.8.1*).

Bogdan (1965b) noted the existence of several ecotypes, but in spite of the fact that this grass is used in actual practice in contrast with Giant Guinea no variety descriptions exist. Van Rensburg (1956) believed that material which he had collected from Kawanda (U) in 1946 had originated from Nyasaland (Malawi). Bogdan (1955) reported, in turn, that the Kitale material was derived from Tanganyika selections. The introduction K53768 was the best according to Strange (1958b). *Setaria splendida* is not found in Bogdan's (1958) "Revised list of Kenya grasses" and Rattray (1960) gives no mention of the species in his "Grass Cover of Africa". It seems likely that the natural distribution of this species was limited to southern Africa (Bogdan, 1977).

14.2. Giant Guinea

No use on farm-scale is made in East Africa of any of the Giant Guinea (*Pmax*) varieties. These robust types are quite unlike the sward-types such as Slender, Likoni and Makueni Guinea (*6.8*). Guineagrass is a variable species, especially in East Africa (Bogdan, 1977), and numerous ecotypes exist. Giant types of this species have some advantages over Elephantgrass:

1. Establishment from seed is a feasible proposition, although the % germination can be very low.
2. Giant Guineagrass performs better than Elephantgrass in Zone IV (Table 14.2).

The following varieties have been tested in Kenya (Strange, 1958a; Bogdan, 1965a, c; Sands *et al.*, 1970; Thomas, 1976a):

1. Mackinnon Road (K52129). Stems up to 400 cm high; leaves up to 90 cm, folded or rolled-in thus producing a "wilting" effect, hairless, slightly bluish in colour, with a thick and broad midrib. Large panicles. Collected by R. Strange from Mackinnon Road Station 90 km north west of Mombasa. It flowers and produces seed well at low altitudes, but not at Kitale. This variety performed well in the dry lower areas of Machakos and Kitui, but Sigor Hairless yielded more (Table 14.2). At Kawanda, Tiley (1970a) found this a coarse variety.
2. Sigor Hairless (K52531). Of smaller proportions than Mackinnon Road, hairless and bright green in colour. A small blade just below the panicle is characteristic. Collected by R. Strange near Sigor in north west Kenya. This variety sets seed well at Kitale. See Table 14.3.
3. Tengeru (K5599). Stems up to 200 cm, but leaves are long and numerous at the basis. This variety can be high yielding under Kitale conditions compared with Elephantgrass. Its drought tolerance is said to be poor. Originates from Tengeru near Arusha (T).
4. Ntchisi (K5918). Low growing, very broad tufts made up mainly of large basal leaves. Flowering stems are few and they only sligthly exceed the basal leaves. Sheaths are hairless; blades with numerous short hairs on both sides. A leafy variety which is suited to moderately dry areas. Originates from Lilongwe (M), where it does well (Thomas, 1976a; Dzowela, 1985).

For more morphological details see Bogdan (1965a). This list does not include the introductions tested later at the coast, *e.g.* cv M8 ex Kikambala (Goldson, 1976b).

Strange (1959a) studied the distribution of herbage yield of Sigor Guinea over the wet and

Table 14.2. Dry matter yields (ton/ha) at Katumani, 1958–1961 (after Sands *et al.*, 1970).

Variety	Year (with rainfall in mm)			
	1958–1959[a] (810)	1959–1960 (590)	1959–1961 (580)	Mean (660)
Elephantgrass local	9.1	—	1.2	3.43
Elephantgrass Ghana[b]	10.2	—	1.2	3.80
Elephantgrass Cameroons	9.6	—	0.9	3.50
Giant Guinea Sigor	9.7	3.7	1.6	5.00
Giant Guinea MacKinnon Road	7.9	2.9	1.0	3.93
Giant Guinea local	6.5	4.1	1.1	3.90
Rhodesgrass Chepararia	7.7	3.0	1.9	4.20

[a] 1st year yields not reported due to uneven establishment
[b] Gold Coast (*11.7.3*)

dry season at various rowwidths (Table 14.3). Interestingly, the wider rowwidths were more productive, not only in the dry seasons but also year-round, with the exception of the establishment year. It is important to add that this trial of Strange was established from seed. Some gapping was needed to fill in the empty spaces where germination had failed.

Table 14.3. The distribution of dry matter yield (kg/ha) of Sigor Guineagrass at different rowwidths at Kitale (after Strange, 1959a).

	Whole season (April–April)				Dry season (December–March)			
	Rainfall (mm)	Rowwidth (cm)			Rainfall (mm)	Rowwidth (cm)		
		120	180	240		120	180	240
1956–1957	1060	6600	6240	5930	130	960	930	1080
1957–1958 (+25 N)	1250	7530	8200	8370	360	2900	3050	3170
1958–1959 (+60 N)	1100	5000	5940	6170	130	2410	2620	2490
Mean	1140	6380	6790	6820	210	2090	2200	2250

Ntchisi Guinea was chosen by Suttie (Kitale Ann Rep 1966) for a study on the date of sowing Desmodium in relation to the type of fodder grass with which the legume was to be associated (Table 14.4). Ntchisi Guinea established very slowly in Desmodium sown 3 months earlier. Also in subsequent years legume competition remained strong (*12.8.1*). Desmodium was not capable of raising the yield of the Ntchisi Guinea + legume mixture to the level of that of Elephantgrass + legume.

Mention was made of the fact that little or no use is made of these varieties under farming conditions. It would seem, however, that varieties such as Sigor and Mackinnon Road are better suited to the semiarid areas than Elephantgrass. Sands *et al.* (1970), in the only variety trial published of this species, found Sigor Guinea to be more productive at Katumani near Machakos than any of the entries of Elephantgrass (Table 4.2). Also Thomas (1960) reported from Machakos that Giant Guinea varieties such as Mackinnon Road and Sigor had advantages over Elephantgrass. They are not hairy, are more drought tolerant and much less liable to termite attack when weakened by drought. Thomas also noted that Mackinnon Road can be established from seed, although Sigor produces more flowering heads and is more drought tolerant than Mackinnon Road.

Table 14.4. Dry matter yields (ton/ha) of Ntchisi Guineagrass, without or with *Desmodium uncinatum* in the establishment year; legume sown 3 months before the grass planting (after Suttie, Kitale Ann Rep 1966).

	Dry matter yield		
	Grass	Legume	Total
Grass (without Desmodium)	1.0	—	1.0
Grass plus Desmodium	0.1	1.2	1.3

It would appear that these varieties should be suitable for some of the areas in Zone IV. In addition to being easier and less costly in labour and time required, establishment from seed is more reliable in the event of a prolonged drought after establishment. It does appear, however, that high seed rates are necessary as the % germination of the seed is erratic and normally very low. Alternatively, seedlings can be raised to be transplanted with the first rains, which prolongs the growing season. If the grass is rested during the rainy season in order to maximize dry season production (*13.2.1*), seed collecting is an extra bonus.

14.3. Guatemalagrass

Guatemalagrass (*Tripsacum Andersonii*; syn. *T. Laxum*) is a relatively recent introduction (K58110). It is a native of Guatemala and neighbouring countries (Whyte *et al.*, 1959).

The maize-like appearance and the almost complete absence of true stems appear to account for the recurrent interest shown in this grass. Compère (1960b) suggested that Guatemalagrass would match the yield potential of Elephantgrass with the leafiness and nutritive value of maize. Guatemalagrass used to be grown on some government farms in Teso (U) and most of the larger farms in Kivu (Zaïre) have a small fodder bank of this grass in a nearby valley bottom (personal observation, 1992). This grass is also drawing attention in Burundi (Vancoppenolle, 1988). It was at one time suggested that it performs relatively better on waterlogged soil.

It would appear, however, that most farmers who grow this grass are not aware of the much higher potential offered by Elephantgrass. Guatemalagrass is generally low yielding and of low nutritive value (Whyte *et al.*, 1959). In spite of its leafiness it has at times been found to be totally refused by the grazing animal. The low yield potential of Guatemalagrass, as borne out in the Kenya District Trials (Table 6.2), applied to all locations except Baraton which had, however, the lowest location average. Only 2.4 ton was produced at the lowest-rainfall location, Naivasha, as against 16.7 ton of Boma Rhodes. It has no drought tolerance. Weed control is a problem because of low growth and thin stands. Vegetative propagation is rarely more than 75% successful. Establishment is slow.

Mwakha (1972), in an attempt to produce a maximum yield of leaf, harvested Elephantgrass (Table 12.3) and Guatemalagrass at 2, 4 and 8 months interval. At all intervals Guatemalagrass produced a lower total yield but the yield of leaf was higher at the 4 and 8 months interval than of Elephantgrass. However, the % crude protein of the leaf was lower in Guatemalagrass at all intervals. Thus, although the herbage consists predominantly of leaf, this does not necessarily imply a high nutritive value, as ageing of the leaf and also the level of insertion of the leaf have a bearing on the chemical composition (Table 13.2).

Whether this grass is acceptable to the grazing animal or not is a controversial issue. In a cafetaria-test of various fodders, Guatemala was eaten down to the root (personal observation, 1967). This was frequently observed and the conclusion was reached of a grass of high acceptability. In subsequent grazing experiments (Awili-Jandara, unpublished), however, the opposite was found. Even when animals were confined to the paddocks, grasses along the perimeter and the weed grasses were grazed but Guatemala was ignored. Similar reports were also received from Kisii.

Of the alternative "giant" fodder grasses it seems that Giant Guinea deserves more interest owing to its better performance in the semiarid areas and its ability to produce seed.

14.4. Summary

Giant Setaria has been traditionally planted at higher altitudes as a bund grass. Giant Guineagrass is little used, but it is a better choice in the dry areas than Elephantgrass. It can be raised from seed. Results with Guatemalagrass have been highly variable, although the grass continues to attract attention because of the absence of stems. In general, the yields of these fodder grasses are considerably lower than of Elephantgrass.

CHAPTER 15

Grass sorghum

This is an important group of fodder plants, widely used for short-term pastures, fodder-, break- or catch crops. Special interest is justified because it is the only species cultivated in the region that resists drought well in semiarid zones, while growing more actively in the dry season of the subhumid zones than perennial grasses. An estimated 10.000 ha is sown annually to Grass Sorghum in Kenya alone. The term "Grass Sorghum" is adopted here to encompass all Sorghum species in actual use, not only "Fodder Sorghum" (*S. bicolor*) but also Sudan- and Columbusgrass and various Sorghum hybrids.

15.1. Origin and classification

Sorghum has its greatest variability of both cultivated and wild types in north east Africa, especially Ethiopia. Wild Sorghum occurred as a weed in cereals and adaptation to become a crop plant probably began in the 4th or 3rd millennium B.C. (Doggett, 1970). In the later part of the 1800s Sorghum was pushed back as the staple food by maize to northern Uganda, the eastern and southern parts of the Lake Victoria basin and central Tanzania. Maize has the unique advantage of being much less vulnerable to diseases, pests and birds, while its storage qualities are also better. These characters are of such overriding significance that even in the low rainfall areas of Zone IV (*3.4.4*) recent efforts to reintroduce Sorghum as a grain crop have largely failed.

Although having lost terrain to maize as the peoples' main staple food, Sorghum has gained a great reputation over the past half century as a source of grazing or fodder. In this capacity the advantages of Sorghum (prolific tillering, drought resistance) are disentangled from the disadvantages since it is grazed or cut before the major diseases, pests and seed-eating birds take their toll. Thus, Grass Sorghum combines advantages from 2 graminaceous groups, annual cereals and perennial grasses. The seeding qualities and the seedling vigour of the former are coupled with the regrowth- and grazing potential of the latter. Grass Sorghum is capable of producing a lot of herbage in a short time and it can do so under genuinely dry conditions.

15.2. Sorghum types

In accordance with Bogdan (1977) we will follow the concepts of taxonomy described by Doggett (*op. cit.*), in which the subgenus Eu-Sorghum is subdivided as follows:

- *Arundinacea*: annuals or tufted perennials without rhizomes, 2n = 20, wild or cultivated.
- *Halepensia*: perennial wild grasses with more or less elongated rhizomes, 2n = 40.

Sorghum seedlings can be recognized by the leaf sheaths which encircle the culm, have overlapping margins and are not hairy.

15.2.1. Fodder Sorghum

Sorghum bicolor (syn. *S. vulgare*); group *Arundinacea*. Cultivated for grain or for fodder. Sorgo (a sweet type) is not cultivated in East Africa but is believed to be a parent of Sudangrass and has been used in Sorgo × Sudangrass crosses (Table 15.1).

The traditionally best known Fodder Sorghum in East Africa is Velvet Sorghum (Bogdan, 1965c), of unknown origin but brought into cultivation by farmers in the Nakuru area. It is tolerant to leaf blight (*15.6.1*) and is best suited to infrequent cutting. Under that regime it outyielded all USA hybrids in the amount of N recovered (Table 15.1). Only 150 kg/ha N was applied as fertilizer, so that the better entries, with the exclusion of Piper Sudan, displayed a net uptake on average of more than 275 kg/ha N, under infrequent defoliation. The high year-round yield and % crude protein should be noted.

Table 15.1. Dry matter yield (May 1968–May 1969), % crude protein and N-yield of Sorghum entries at Kitale (after Boonman & Chemorei, Kitale Ann Rep 1968).

Entry and origin* of seed source		Type	Cutting regime			
			Frequent		Infrequent	
			Dm (ton)	N (kg)	DM (ton)	N (kg)
Haygrazer	(USA)	Sorghum × Sudan	18.6	487	22.4	454
Sudax SX 16	(USA)	Sorgo × Sudan	16.0	427	18.6	381
Columbusgrass	(K)	S. almum	15.5	398	21.0	487
Excel Chow	(USA)	Sorghum × Sudan	14.9	399	20.3	426
Velvet Sorghum	(K)	S. bicolor	14.7	394	22.1	515
Sweet Sioux	(USA)	Sorgo × Sudan	14.4	370	23.3	513
Sudangrass	(K)	S. sudanense	11.2	271	12.7	274
Piper Sudan	(USA)	S. sudanense	9.3	223	8.6	185
Crude protein (%)				15.6		13.1

* USA hybrids or varieties, or local (K) seed source.

High-altitude entries from Kigezi (U) and Ethiopia, like E6518 and E1291, have figured as possible substitutes for maize in beef feedlot silage rations of the Lanet Station (van Arkel, 1977). Some of these entries (*e.g.* E6518, a fodder type) were highly impressive and yielded more DM than maize at Lanet but their nutritive value was markedly lower. This yielding superiority of Sorghum could not been confirmed for locations such as Kitale, where maize yielded much higher (Table 15.2). Lanet has only 850 mm rainfall and lower Lanet is semiarid (Zone IV).

High-altitude entries (*e.g.* E1291 and E5770) were used at Serere by B.N. Majisu to produce Sorghum × Sudangrass hybrids local (Table 15.3).

15.2.2. Sudangrass

Sorghum × Drummondii (syn. *S. sudanense*); group *Arundinacea.* Bogdan (1977) quotes reports that Sudangrass is a non-shattering segregate from hybrids involving S. virgatum and a Sorgo type. Sudangrass was sent from Sudan to USA in 1909 and from there to South Africa (Melle, 1918). It was well established in Kenya by the early 1930s (Maher, 1936b).

Sudangrass seeds have no disarticulation scar at the spikelet bottom, as found in Columbusgrass (*15.2.3*), but the piece of broken raceme attached to the spikelet is usually longer. The seed is also of a light colour. Sudangrass is more frost-, salt- and blight susceptible than Columbusgrass, which also applies to the Piper variety brought in from USA in the mid 1960s.

Table 15.2. A comparison of fodder crops at Kitale in 1975, including high-altitude (H.A.) Sorghum (after D. Enserink, unpublished; van Arkel, 1977).

	Yield of dry matter (ton/ha)
Maize (hybrid H 613)	22
Sunflower (Kenya White)	15
H.A. Sorghum (fodder type e.g. E6518)	11–13
H.A. Sorghum (grain type e.g. E1291)	7–8
Pearl millet (S10, S14)	6

Sudangrass is often lower yielding than Columbusgrass (Tables 15.1 & 15.5), but seems better adapted to the low, warm and dry areas of Zones IV and V (*6.5*). Sudangrass rivalled the best hybrids in persistence. Some 11 cuts were taken in all over 18 months at Katumani (Table

15.5). Also Bogdan (1977) noted that it is more adapted to a climate of hot and dry seasons. Another reason for Sudangrass' success is the lower incidence of leaf blight in dry areas, so that it can hold its own in comparison with Velvet Sorghum and Columbusgrass, which are more resistant to the disease in more humid zones. Also Sands *et al.* (1970) reported from Katumani that Sudangrass was more reliable than other Sorghum types. It is suitable as a grazing type (Table 15.1). Sudangrass is believed to be lower in prussic acid than most Sorghum × Sudangrass hybrids (*15.5.1*).

Sudangrass was used as the pollinator in the local hybrid crosses made at Serere (Table 15.3).

15.2.3. Columbusgrass

Sorghum × *almum*; group *Halepensia*. Columbusgrass was found in Argentinia and described by L.R. Parodi in 1943 (Roseveare, 1948). It is considered to be a hybrid between Johnsongrass (*Sorghum halepense*) and either a cultivated fodder Sorghum or a grain Sorghum. It became an important fodder crop in South Africa. Strange (1958a) reported its introduction at Kitale.

Columbusgrass had been tested earlier, at Kabete in 1947, but was eradicated (Edwards, 1950b) on suspicion of a weed potential similar to that of Johnsongrass. For the same reason it was temporarily banned in parts of USA and Australia; it may be found volunteering in arable crops. Columbusgrass is, however, now the most commonly used Grass Sorghum. It is valued for its seed yield, ease of establishment, drought- and salt tolerance, reasonably high yields of good herbage quality and persistence. If well managed, Columbusgrass can remain productive for over 2 years. It yielded less than some of the Sorghum × Sudangrass hybrids (Tables 15.1 & 15.5) but had the same resistance to leaf blight. Seed has been considerably cheaper and easier to obtain than seed of imported hybrids. In recent years it has regained most of the terrain it lost in the late 1960s to the USA-hybrids (Boonman, 1969). It is the best choice for the medium altitudes, both for the drier areas and for the dry season in the wetter zones. Around Lake Naivasha and Nakuru Columbusgrass is sometimes grown in combination with lucerne (*17*).

Velvet Sorghum, Columbusgrass and Sudangrass are supposedly represented by only one local variety each. Local varieties may have developed some separate identity under seed multiplication, but subsequent physical mixing of seed lots may have taken place at the premises of the seed companies.

15.2.4. Other natural hybrids

Sorgrass, supposedly a hybrid between Sudan- and Johnsongrass, but without rhizomes (Doggett, *op. cit.*), and Perennial Kavirondo Sorghum were widely acclaimed in East Africa but met finally with little practical success (Bogdan, 1965c). The story of Perennial Kavirondo Sorghum is worth reporting as an example how misleading over-optimistic first impressions and hybrid names of allegedly new plant types can be (see also *11.7.3*).

15.2.4.1. Perennial Kavirondo Sorghum

Perennial Kavirondo Sorghum, probably a hybrid between *S. bicolor* and *S. verticilliflorum*, was collected by Edwards (1941b) near Maseno in western Kenya (Kavirondo) in 1935 as seed from a single plant. It was similar to Sudangrass but plants, 5-years old, were still found to be vigorous. It was named Perennial Kavirondo Sorghum and recommended for the drier fringe of the mixed-farming areas for silage and aftermath grazing. In seed multiplication the hybrid segregated in distinct types on which new selections (*e.g.* R.2S42) were based (Edwards, 1948).

Not only the hybrid status but also its alleged perenniality and drought resistance caused considerable excitement. Its fame reached overseas journals like "Country Life" and "World Crops" with articles by Haarer (1948; 1951) after whom, incidentally, *Pennisetum haareri* had been named. Haarer termed Perennial Kavirondo Sorghum the "salvation of the desiccated soils of the groundnut areas in Tanganyika". Henderson & Preston (1959) devoted their discussion of Sorghum for silage entirely to Perennial Kavirondo Sorghum. However, Perennial Kavirondo Sorghum quickly lost to Columbusgrass, which appeared about that time, but which had earlier been banned (see above). Bogdan (1965c) reported that Perennial Kavirondo Sorghum was not as good as other

Sorghum types in either yield, nutritive value or ease of establishment.

15.2.5. Artificial hybrids

Sorghum bicolor × *S. sudanense.* Following the discovery of cytoplasmatic male sterility in Sorghum in 1952 in USA many artificially made hybrids became available which had a great impact on the grain and cattle industry in that country. The interest in Kenya started around 1965 when cooperatives began importing various hybrids for grazing purposes especially in the Nakuru area. Farmers, *e.g.* C.D. Hill of Glany-oro near Nakuru, were once again the pioneers of a new development (*2.2.2*; *6.1.2*). The same hybrids were tested also in Tanzania and Uganda.

Different types of hybrids exist, depending on whether the parents used in the crosses are Sorghum, Sorgo or Sudangrass. The hybrids imported into Kenya were mostly Sorghum × Sudangrass and Sorgo × Sudangrass but some Sudan- × Sudangrass hybrids (*e.g.* Trudan) have also been tried. Seed yields are higher and hybrids have more seedling vigour than Sudangrass, but the Sudangrass parentage transmits the ability to recover faster after cutting or grazing and also to have a larger proportion of leaf. Of equal importance is that hybrids are more resistant to leaf blight.

Table 15.3 demonstrates the potential of local hybrids, based on crosses made by B.N. Majisu at Serere between locally-proven high-altitude grain Sorghum and Sudangrass. Insufficient seed was available to conduct regional testing, but trials showed the superiority of the crosses involving the grain type E5770 × Sudangrass, compared with the best USA hybrid "Grazer A" (see also Table 15.5). Further hybrids were then made with E1291, which had been found to be the best grain Sorghum for Kitale conditions (Table 15.2). This parent was used because male sterility functions better at high altitudes. Unfortunately, no commercial hybrid seed production was undertaken, as the potential market was considered too small.

Broad adaptation is typical of hybrids, in graminaceous and other crops. This may be accentuated when high-altitude Sorghum is crossed with Sudangrass, the latter being more adapted to low altitudes.

Table 15.3. The performance at Kitale of Sorghum × Sudangrass hybrids developed by B.N. Majisu at Serere (after Boonman, 1979b).

Sorghum parent	Yield of dry matter (kg/ha)			
	1st cut Nov 1968	2nd cut Jan 1969	3rd cut March 1969	Total
E5770	3470	3660	1880	9010
E5771	2330	2200	1260	5790
E5751	1140	1580	870	3590
Grazer A (control)	2910	2670	1770	7350

15.3. Adaptation and use

The major feature of Sorghum is its ability to withstand dry conditions, which it does better than any other cultivated annual or perennial grass. Glover (1948) working at Amani (T) found that a Sorghum plant transpired 0.34 kg/day water between the 11th and 16th week, flowering taking place about the 14th week after planting. Maize transpired almost twice as much water at the equivalent stage of development. Later, Glover (1959) would stress the slower wilting of Sorghum and its power of recovery after wilting. He found that maize leaves subjected to wilting for a week or more suffered permanent damage to the stomata. Although the leaves would recover turgidity and regain their usual appearance, the normal diurnal pattern of stomatal changes was not resumed. By contrast, Sorghum proved well able to recover from prolonged wilting. After severe drought with wilting for 2 weeks leaves quickly recovered after watering and within 5 days the normal diurnal rhythm of change in stomatal aperture was resumed. CO_2-intake showed that stomata had been scarcely affected.

No conclusive evidence has been established to support the contention that the root distribution system of Sorghum is more efficient (see also *5.3.4*) but Doggett (*op. cit.*) referred to heavy silica deposits in the endodermis of the Sorghum root, forming a complete silica cylinder in the mature root. This would act as a mechanical reinforcement against collapse under drought stress. Maize roots showed very light silica deposits.

As Doggett pointed out it is usual to think of Sorghum as a drought resistant crop and, therefore, unable to cope with wet conditions. In fact, many xerophytic characters are of value also to plants growing in wet conditions under which root functioning may be temporarily impaired. Wild Sorghum is often found near swamps and in heavy clays of valley bottoms and on Vertisols. Doggett reported that at some of the very high rainfall sites where Sorghum outyielded maize, unfavourable wet soil conditions gave Sorghum the advantage. Sorghum is popular on heavy, poorly drained soils, *e.g.* the Kano plains near Kisumu.

The drought resistance of Grass Sorghum can be put to use in East Africa's agriculture in the following ways:

1. In Zones IV and V as a main-season crop, in which Sorghum is given full-crop treatment: early sowing, wide rowwidth, low seed rate, interrow cultivation, and hay or silage making rather than grazing.
2. In the mixed-farming areas of Zones II and III to fill the fallow gap between 2 main seasons' arable crops or between pasture and following crop. If crop land or pasture is ploughed well before the end of the rains, it is better and more economical to sow Sorghum, or oats in the areas above 2250 m (*16.1*), rather than to leave the soil bare. In addition to the obvious benefits of dry-season fodder, a temporary soil cover with Sorghum or oats implies less preparatory land cultivation. Some farmers sow Sorghum straight into ploughed pasture. As Sorghum develops the break down of the grass sod is enhanced, weeds are suppressed and the OM-content and nutrient status of the soil are maintained if not improved, especially if the Sorghum is subsequently grazed (*5.3.1*; *5.5*).

Sorghum × Sudangrass hybrids, grown and grazed as main-season crops have done well at the coast (Wilson, Kilifi Plantations Rep. 1970). Reports from the lower regions of Uganda are generally less optimistic. At Serere, both Columbusgrass and Sudangrass suffered from leaf diseases (Horrell, 1958a). On small farms in central Kenya, Sorghum × Sudangrass hybrids have been grown for fodder (Chudleigh, 1974), chiefly as a main-season crop. In the truly semiarid areas the best strategy, as already mentioned, is to grow Sudangrass or a hybrid and exploit the rainy season to the fullest by early sowing etc. In bimodal low-rainfall areas it is probably better to sow at the start of the short rains. As source of fodder there is probably no species of comparable abilities, but moisture depletion of the soil should always be weighed against the fodder benefit.

In the mixed-farming areas of Zones II and III, it is obviously better to have the fallow period covered with a Grass Sorghum than with weeds and this is the reason why the system is not necessarily less effective in terms of moisture conservation.

Double cropping, *i.e.* Grass Sorghum following an early crop of wheat, maize or a broken pasture, is also desirable from the view point of the growth cycle of the Sorghum itself. Sorghum can then be sown to make its peak potential coincide with the dry season. A reliable rule of thumb is to sow between 2 and 4 months before it is most needed. At Kitale (Fig. 3.3) Sorghum is planted from August till October. The seedling is strong enough to make its roots follow the receding soil water profile. If more paddocks are due for ploughing, sowing dates can be staggered so that some Sorghum is sown in August, some in October.

In the following, attention will focus on Grass Sorghum as a short-term crop in the mixed-farming systems: either in the semiarid areas as a main-season crop or as a late-season crop in the subhumid zones.

15.4. Husbandry

15.4.1. Establishment

Sorghum × Sudangrass hybrids have a 1000-seed weight of 25 g. Sudangrass seeds are only half as heavy and the seedlings are less vigorous also. Seed rates (5–25 kg/ha) depend, therefore, rather on the rowwidth employed, which again is a function of the rainfall expected. Depending on the rowwidth maize drills, wheat drills or fertilizer spreaders are used.

Nicholson (1971), working from Eldoret, laid down rowwidth trials at Eldoret, Njoro and Kitale in May 1970 and measured the fresh herbage yields in the 1st year (Table 15.4).

Narrow rowwidth of 20 cm nearly doubled the yield of fresh herbage. Nicholson concluded that

Table 15.4. The effect of rowwidth on fresh herbage yields (ton/ha) of Grazer A (USA hybrid) sown in May 1970 (Nicholson, 1971).

Row width (cm)	Location (and number of cuts)			
	Kitale (2)	Eldoret (2)	Njoro (3)	Mean
20	30.1	35.6	42.7	36.1
50	26.4	26.9	35.2	29.5
100	18.5	16.8	23.4	19.6

this was largely due to a better suppression of weeds. At Kitale, weeds formed 85% of the herbage at 100 cm rowwidth at the 1st cut, compared with 46% at 20 cm. Nicholson recommended 20 cm, the width of wheat drills. By contrast, if the aim is to produce a maximum of forage in the dry season, a rowwidth of 50 cm is more desirable. In areas of very low rainfall even wider rowwidths may be advantageous. Seed rates must be reduced and rowwidth increased as the season advances and rainfall declines, in order to reduce competition for moisture. To this end widening the rowwidth is more effective than lowering the seed rate, because compensatory tillering evens out the number of tillers within the row. Wide rows facilitate interrow cultivation also.

Atrazine or Propazine at 2–3 kg/ha *a.i.* are useful against annual weeds when applied pre-emergence. Atrazine can also be applied post-emergence at 2–3 kg in 50–100 l of water on moist ground. At Lanet, van Arkel (1979) observed better yields of herbage and grain at 4 kg. Application rates should be lowered on sandy soils, because of residual effects. 2,4-D amine, at 0.25–0.50 kg/ha, can be applied when Sorghum is 10–20 cm tall. Herbicides are not without risks in Sorghum and crop damage may be incurred.

15.4.2. Utilization and choice of type of Sorghum

The choice of species, variety or even the type of hybrid depends on the purpose for which the Sorghum is used, on how long the stand is supposed to remain productive and on the susceptibility to leaf blight. In the first varietal comparison made at Kitale in 1968 the observation was made that the ranking order depended on the frequency of cutting (Table 15.1). Although hybrids were generally superior at both frequencies, Velvet Sorghum was among the best at the infrequent cutting regime. DM-yields were higher when cut infrequently (at 150–175 cm) than at 50–100 cm, except for Piper Sudan, which was most of all entries infested with leaf blight at a greater height of harvesting. The hybrid Sweet Sioux was the highest yielder but its % crude protein was slightly lower than of Velvet Sorghum probably due to leaf blight also. It was striking to note that Haygrazer, Sudax SX-16 and Piper Sudan, unlike the other entries, produced a lower N-yield at 150–175 cm.

The conclusion was that there was little advantage in having hybrids for systems of infrequent cutting, e.g. silage making, a conclusion confirmed by the work done at Lanet (Miles *et al.*, 1972; van Arkel, 1977). Local varieties such as Velvet Sorghum, E6518 and E1291 are then superior to hybrids. Only Haygrazer appeared useful as a dual-purpose hybrid.

Table 15.1 illustrates that the strength of hybrids lies in their capacity to yield better under a multi-harvest system. Numerous additional hybrids were screened at Kitale for late-season yields under a system of 2–3 cuts over a 4–6 month period (Boonman *et al.*, Kitale Ann Rep 1969). The best of these entries, e.g. Grazer A, were tested further in November 1969 at Embu, Katumani and Eldoret. The results of these trials under standard husbandry (50 cm rows; 15 kg seed; 50 kg P_20_5; cutting at 50 cm) are shown in Table 15.5 for the top hybrids and local varieties only.

The superiority of the hybrid Grazer A was confirmed. Sudax SX-16, Haygrazer, Excel Chowmaker and Sweet Sioux performed also well again. Other screening trials were reported by Kusewa (Embu Ann Rep 1968) and Awili-Jandara (Kitale Ann Rep 1971).

It has already been noted that the main purpose of Grass Sorghum in the mixed-farming areas is to provide a productive fallow with a view of maximum dry-season fodder production. At Katumani the trial was continued for another year and 7 additional cuts were harvested (J.B.S. Makatiani, unpublished). After every 2nd cut 50 kg/ha N was applied; only 510 mm rainfall was recorded in this year. The DM-yields are shown in the last column of Table 15.5; the hybrids Sweet Sioux, Haygrazer and Chowmaker-C and Sudangrass proved the most persistent. It is of interest to note that Sudangrass was among the top yielders. This confirms its

Table 15.5. Dry-season fresh yields (ton/ha) of Sorghum entries in 1969/1970; residual yields at Katumani in 1970/1971 (Kusewa, Makatiani & Nicholson, unpublished).

Entry	Dry-season fresh yields				
	Location				
	Embu 2 cuts	Katumani 4 cuts	Eldoret 2 cuts	Mean	Katumani (residual yield)
Grazer A	17.8	20.0	26.1	21.3	3.4
Sudax SX 16	14.5	18.4	31.0	21.3	3.5
Haygrazer	12.3	17.5	21.3	17.0	4.3
Chowmaker C	12.0	19.8	18.1	16.6	4.2
Sweet Sioux	11.9	19.2	17.8	16.3	4.6
Gahi millet	14.7	10.6	15.0	13.4	0.0
Trudan 2	10.6	13.3	10.8	11.6	3.0
Velvet Sorghum	6.3	11.0	12.8	10.0	1.7
Columbusgrass	3.6	14.6	18.0	12.1	2.5
Sudangrass	4.5	13.6	4.4	7.5	4.0

yielding ability in the warm areas of low rainfall where it remains relatively free from leaf blight (*15.2.2*). Significant also was that these trials confirmed the value of Sweet Sioux, Haygrazer and Chowmaker-C which had become popular with Kenya farmers. These hybrids combine high dry-season production with a fair degree of persistence.

15.4.3. Fertilizers

Nicholson (1971) assumed that the fertilizer requirements would be the same as for maize and wheat. His trials with May-sown Sorghum at Eldoret, Kitale and Njoro demonstrated only a small effect of N-fertilizer. Sorghum has a general reputation of not responding markedly to fertilizer applied at or soon after sowing, which agrees with what is found in tropical grasses (*7.1*). However, approximately 30–50 kg/ha P and 150–200 kg K can be removed with a high yielding Sorghum crop when cut (see also *12.7*). Sorghum has, therefore, a reputation of depressing yields of following crops through not only moisture- but also nutrient depletion (Doggett, *op. cit.*).

To maintain soil fertility, P and eventually also K need to be replenished, except under grazing. It seems advisable not to attempt to save on P-fertilizer, when Sorghum is grown as a break- or catch crop before maize. Some 40 kg P_2O_5 applied in the seedbed should be adequate.

Sorghum seems capable of extracting soil-N reserves. Over 500 kg N was taken up within 12 months (Table 15.1), although only 150 kg N had been applied as a split-topdressing. Under grazing N-depletion is less likely to occur (*12.7*).

15.5. Nutritive value

Sorghum is usually rated as herbage of moderate to high quality and is relished by the grazing animal. Combinations of high yield (> 15 ton DM) and high % crude protein (15%) are possible (Table 15.1) which makes Sorghum so exceptional.

Sorghum is a fast growing plant and the farmer may easily find himself confronted with a crop already too advanced for grazing. It is essential to schedule the peak production of a given planting to match the time when the fodder is needed most. An early hay-cut must be anticipated in the event of surplus.

15.5.1. Toxicity

Asphyxiation, trembling, staggering and in many cases death are the signs of toxicity due to nitrite and prussic acid. These 2 causes are best dealt with together as no certainty exists as to which is responsible for the deaths which are from time to time recorded from animals grazing Sorghum. The animal's blood becomes a dark chocolatc brown colour with nitrite poisoning, and bright cherry in colour with prussic acid. Dark blue colouring on the linings of mouth and eyes is also typical of the latter. Deaths which occur to animals grazing Sorghum are usually explained as prussic acid poisoning. However, diagnosing the genuine cause of death is a tedious task which is rarely undertaken. Some deaths may even be common cases of bloat.

Nitrate can accumulate in young Sorghum. Nitrate is converted to nitrite which is very poisonous. However, nitrate generally accumulates less in the leaves so that under grazing the risk is not so great, as long as the contents are < 0.5% nitrate. Hay contains as much nitrate after curing as it does at cutting.

Prussic acid (hydrogen cyanic acid, HCN) is released when the plant wilts or is frozen, damaged or eaten. The cyanogenetic glucoside dhurrin occurs in all Sorghum potentially. Chopping Sorghum can increase HCN for a few days after harvest but fully cured hays or silages contain only low levels, contrary to the nitrate levels referred to above. Forage with less than 200 ppm HCN is considered safe to feed while those with more than 600 ppm are potentially very toxic. Much depends on what component of the plant is eaten and what proportion of the plant is represented by this component. Decreases in HCN of the total plant as maturity progresses are due mostly to the proportional increase in weight of low-HCN containing plant parts. HCN is always highest in the immature plant parts. New shoots are potentially dangerous until they reach a height of 40–50 cm. The leaf blade not including the midrib is high in HCN, followed by stem, leaf sheath and midrib. The HCN-content of the leaf remains high throughout, but that of the other plant parts declines as maturity progresses. New leaves contain more HCN than older leaves.

Braithwaite (1952) analyzed the R.2S42 strain of Perennial Kavirondo Sorghum (*15.2.4.1*). His results are illustrated in Fig. 15.1. HCN-content decreased with age but increased as the season progressed. Braithwaite concluded that HCN-content was related to the dryness of conditions. Any Sorghum (with more than 200 ppm HCN) is potentially dangerous, but Sorghum with this quantity of HCN is not consumed over a very short period and HCN is not immediately liberated from the glucoside; on the other hand, HCN is eliminated fairly rapidly from the body. Braithwaite

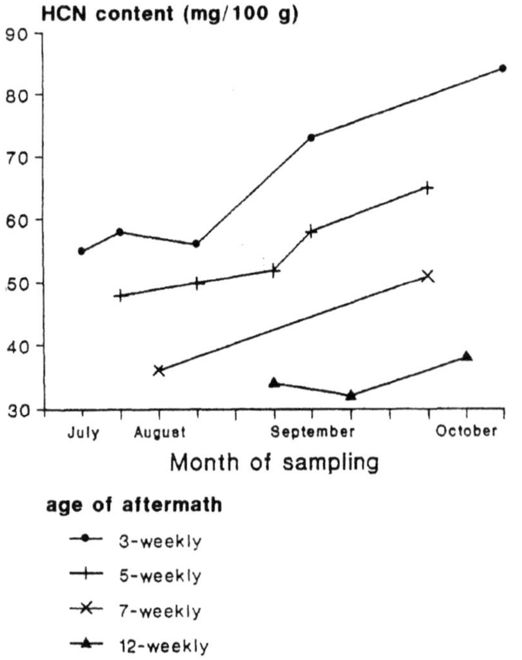

Fig. 15.1. Cyanide content in the aftermath growth of Perennial Kavirondo Sorghum at Kabete (after Braithwaite, 1952).

concluded that a HCN-content below 600 ppm should be safe. It is important to ensure that the grazing animal is not given access to young leafy regrowth or to wilted material, either standing or when fed as green chop or as hay. Ensiled material should be left for 2–3 weeks after filling because HCN, which develops through heating, needs time to escape. Silage is safer than hay. As Wheeler & Mulcahy (1989) concluded, only Sorghum considered safe for grazing should be used for hay.

Stock should be removed immediately from a suspect field. If immediate action is taken, cattle can be cured with an oral dosis of "hypo" (photographic sodium thiosulphate) at 60 g in 500 cc water, repeated at half-hourly intervals if necessary. Sheep should receive one-quarter of the above rate. In severe cases the solution can be injected intra-venously or into the paunch. Nitrate poisoning can be treated in cattle with 2 g methylene blue dissolved in water and injected into the jugular vein. Sheep should receive one-eighth of the dose.

In a recent review (Wheeler & Mulcahy, *op. cit.*) the following factors affecting HCN are listed: age;

genotype; temperature; high N-availability; P-deficiency; wilting. Age and genotype are the most important. The authors quote reports that low HCN, found in Piper Sudangrass, is associated with susceptibility to leaf blight and also with lack of vigour; Piper Sudan is one of the entries most affected by leaf blight (*15.4.2*). The authors state that the numbers of deaths from poisoning are not large and that the negative consequencies appear to be the fear, rather than actual poisoning. As a consequence farmers often allow access to the crop only when it is too mature.

HCN is by no means confined to Sorghum. It is known to reach high levels in Stargrass (*4.2.2.1*) and legumes such as *Lotononis* spp. and especially white clover.

15.5.1.1. Sulphur (S)

HCN is detoxified in the rumen and liver by the formation of thiocyanate from reactions with sulphide or cystine that require about 1.2 g of S per g HCN. As S is usually low in Sorghum (0.10–0.15%), a large portion of ingested S is made unavailable. White clover has a much higher % S. S may be incorporated in a salt lick at the ratio of 1:12. However, responses are not always apparent (Said *et al.*, 1983) and uncertainty remains whether the rate of detoxification is restricted by the supply of a S-donor (Wheeler & Mulcahy, *op. cit.*). For suspected S-deficiencies in soils and various crops, see *10.2.1* & *17.3*.

15.6. Diseases and pests

Many of the Sorghum diseases appear at advanced stages of grain maturity and are little cause of concern in Grass Sorghum, when these are grazed or otherwise utilized at early stages of growth. Obviously, diseases of the panicle should be of even less concern under such regimes.

15.6.1. Leaf blight

An important disease is leaf blight (*Helminthosporium turcicum*) (Nattrass, 1961). This fungus causes a seed rot and a seedling blight but is known in East Africa mainly in the leaves of older plants. Small reddish purple or tan spots enlarge to form elliptical patches, up to 12 mm wide and 3–8 cm long. The ends of the lesions are narrowed or ragged. The margin is fairly definite, while the central portion of the spot is typically dirty-straw coloured. The patches may coalesce and cover large areas of the leaf. In serious attacks as found at Kitale in mature Sudangrass and especially in Piper Sudan, leaves brown off prematurely. Warm humid conditions are conducive to a rapid development of the disease. Very little is found in the dry season, neither in the semiarid areas, which may explain why Sudangrass does so well there (*15.2.2*).

Leaf blight is found in the humid tropics and subtropics, also in the USA. Special attention was given to blight resistance in the USA hybrid breeding programmes, as a result of which resistant hybrids have become available. The hybrids Grazer A, Haygrazer, Chowmaker and Trudan were found to remain healthy in plantings in which Sudangrass and Piper Sudan succumbed fully to the disease. Also Columbusgrass and Velvet Sorghum as well as the hybrids developed by Majisu (Table 15.3) remained free from the disease in the Kitale trials. The physios occurring on maize are not pathogenic on Sorghum, and vice versa (Doggett, *op. cit.*).

Leaf blight can be avoided primarily by choosing entries which are resistant but also by not allowing susceptible varieties or hybrids to grow tall. It has been observed that entries such as Sweet Sioux, Piper Sudan and Sudangrass are more infested with leaf blight as plants mature. No leaf blight was observed in Sudangrass at Kitale until 3 months after sowing when foliage was already more than 90 cm high. Early grazing is important to evade the disease. This should not be difficult as Sudangrass is considered to be low in HCN. However, blight reduces the yield of seed, which farmers often want to harvest when Sudangrass is not needed for grazing.

Leaf blight is perhaps primarily responsible for the lack of persistence found in Sudangrass, in contrast with Columbusgrass, in the wetter areas. At Katumani (Zone IV), however, Sudangrass was healthy and persistent (Table 15.5).

The reader is referred to Doggett (*op. cit.*) for a discussion of other diseases affecting Sorghum, but some fungi affecting the spikelets will be discussed below (*15.7*).

15.6.2. Pests

Shootfly, aphid and stem-borer are important pests affecting Sorghum in East Africa (Doggett, *op. cit.*), but the damage in Grass Sorghum destined for grazing is often negligible. Shootfly (*Atherigona varia soccata*) is commonly found in seedlings, but some Sorghum types have a better recovery (Doggett, *op. cit.*). Shootfly is also found in Elephant- (*11.8.3*), Rhodes- and Wimbigrass (*6.6.2*) but the species or physios may be different. Larvae sever the growing point of the seedling, causing the inner leaf whorl to rot at its base, resulting in the typical "dead-heart" appearance, the newest leaf drying and turning brown. The plant may recover by the development of new tillers. Doggett noted that shootfly damage can be evaded by early sowing at the beginning of the rains. However, end-of-season plantings of Grass Sorghum do not necessarily carry excessive shootfly.

15.6.3. Witchweed (Striga)

Striga hermonthica is serious on Sorghum, but less so if at all above 1500–2000 m. *Striga* spp. are found on other graminaceous species also (*6.7.2*). It is a prolific seeder and seeds can remain dormant in the soil for many years. Doggett (1970) quoted reports from the Sudan, whereby trap-cropping for 5 weeks with Sudangrass greatly reduced the incidence of witchweed in successive Sorghum crops. Crop rotation is the best answer to the serious crop losses otherwise incurred. No witchweed was observed in any of the trials at any of the locations mentioned above, either because Sorghum is little grown or because the altitude was too high.

15.7. Seed production

Commercial production of seed of the hybrids developed by Majisu (Table 15.3) has not been undertaken. Also the import of USA hybrid seed has gradually diminished. In the late 1960s annual imports stood at some 40 ton in Kenya (R. Combes, private communication). Imported seed is more expensive and does not keep well in storage, when treated with biocides before shipment.

Seed of Sudangrass, Columbusgrass or Velvet Sorghum is often home-grown and saved, when a field is not required for grazing. Grass Sorghum produces seed in quantity and quality, yields of 1000 kg/ha being common, except in Sudangrass when attacked by leaf blight. Under humid conditions honeydew disease (*Sphacelia sorghi*) develops in the panicles, especially in male-sterile lines. A honeydew-like secretion is most conspicuous as sticky droplets with a typical pink colour (see also *8.7.2.5*). This is the imperfect stage of ergot, found under dry conditions. Under wet conditions *Sphacelia* itself is invaded by a parasitic fungus, *Cerebellum sorghi-vulgaris*, producing a black surface.

15.8. Silage

Local varieties of Sorghum, especially those of higher altitudes, received attention as silage crops for beef feedlots at Lanet (van Arkel, 1977). High yields were obtained but the grain:stover ratio and quality were lower than in maize; Sorghum silage was poorer than that of maize but better than of Elephantgrass. The best silage varieties are generally those that are recommended grain varieties in adjacent zones having a longer growing season (*19.2*). Maize is preferred for silage, but it is obvious that when Grass Sorghum is grown as a break crop, silage making is an alternative to grazing. Sorghum silage needs supplementation with protein, minerals (Ca and P) and vitamin A.

15.9. Pearl millet

No special mention has been made of the pearl millets (*Pennisetum glaucum*), which were included in some of the trials reported (Tables 15.2 & 15.5). Millets were exceptionally productive in the 1st and 2nd cut, but were very shortlived. In cafetaria grazing tests these entries were eaten down to the roots before Sorghum was even touched. At Serere, Horrell (1963) found that pearl millet was able to give grazing very quickly after sowing. At Mpwapwa, French (1946a) recorded 6% SiO and 2.5% K_2O in millet, which indicates that values are high both in millet and in Elephantgrass (*12.7.5*). Millets have been used in crossing programs with Elephantgrass (*11.7*).

15.10. Conclusion

On the variety issue the following can be recommended:

1. In semiarid areas below 1500 m as main-season crop: Sudangrass or Sorghum-Sudangrass hybrids.
2. In subhumid zones at altitudes of 1500–2000 m as break crop: Columbusgrass or Sorghum-Sudangrass hybrids.
3. In subhumid zones for silage: Velvet Sorghum or E6518.

15.11. Summary

The use of Grass Sorghum was pioneered by farmers. Sudangrass and Velvet Sorghum as well as the later introduced Columbusgrass and Sorghum × Sudangrass hybrids, of foreign and local origin, are suitable as short-term fodder crops to bridge the dry season between 2 main-season crops or as a break crop following pasture. As main-season fodder, in semiarid lowland areas, Sudangrass or Sorghum × Sudangrass hybrids may be employed. Grazing should be late enough to avoid possible cyanide (prussic acid) toxicity but early enough to evade leaf blight, to which Sudangrass is particularly susceptible in humid conditions. In semiarid lowland areas Sudangrass can remain productive for 18 months. Velvet Sorghum is as good as any hybrid when cut at a height of 150–175 cm, but hybrids are more suitable under multi-harvest (grazing) systems. Columbusgrass is the best all-round choice for medium altitudes.

CHAPTER 16

Fodder oats

Oats (*Avena sativa*) were among the first seed imports brought in by farmers (*2.2*). Its value was appreciated from the start of agricultural settlement of the Molo highlands, from 1905 onwards. Oats were part of the first recorded herbage variety trial carried out by Johnston (Kabete Ann Rep 1910).

Oats perform the same functions as Grass Sorghum (*15.3*) but then at higher altitudes, as either main-season or late-season crops, to provide complementary grazing and fodder in the dry season. Oats have become a popular fodder crop on farms throughout the medium and high altitudes of Kenya, yet it is worth noting that apart from Dougall (1954d) no research report on oats as a fodder crop can be found in the *East African Agricultural and Forestry Journal* since its inception in 1935. Apparently, oats were easy to grow and presented little challenge to be improved upon.

The survey of Chudleigh (1974) indicated that the most common fodder on small farms in Nyandarua district in central Kenya was oats. Sorghum × Sudangrass hybrids were also grown on small farms in the same region (*15.3*) which indicates that annual cereal fodder crops play an important part on small farms. Oats are also grown for fodder in Ethiopia (Gebrehiwot & Tadesse, 1984).

16.1. Versatility of oats for fodder or grain

Oats owe their reputation to their versatility as they grow well from 1750 to 3000 m, for grain, hay, silage or direct grazing. At lower altitudes, oats are less suitable as tillering is limited and no dense canopy is formed. Farmers are then better off with Grass Sorghum as fodder-, break- or catch crops (*15.3*) Thus, the concept of Grass Sorghum within the farm rotation applies also to oats, although more specifically for the higher altitudes. However, as with Grass Sorghum in the drier areas, it is common to find oats grown intensively as main-season crop for dairy animals towards the upper limits of high altitudes. In these areas grassland production is severely hampered (*3.4.2.1*; *4.2.1*). Kikuyugrass is unproductive, yet it behaves as a weed on cultivated land. Oats are then the best alternative, since they help to suppress Kikuyugrass.

Oats are grown for grain or for grazing, but also for both in sequence. The oats crop is valued for its flexibility and has importance as a strategic crop. A field destined for grain production can be diverted to grazing if rains are inadequate. On the other hand, it is not uncommon to find farmers graze oats (or even wheat) to reduce straw length and increase the time before maturity. This facilitates seed harvesting, especially towards the upper limit of the high altitudes (3000 m), where small cereals tend to run to excessive herbage and lodging. After each grazing, piping (heading) takes place at a lower herbage height than previously until, eventually, heading occurs at straw lengths already of 15 cm, however small the head.

Different management sets are possible (Henderson & Preston, 1959):

1. 2–4 grazings.
2. 2 grazings followed by a hay crop.
3. 1–2 grazings, followed by a hay crop and a grazing.
4. 1–2 grazings, followed by a grain crop.

These treatments apply when oats are sown as a main season crop, like wheat, *i.e.* at least 4 months before the anticipated end of the rainy season. If 4–6 weeks of rain is all that follows sowing, oats will not grow beyond the piping stage and remain in a green, dormant stage for another 4–6 weeks before wilting.

Oats are relatively at their best in cool and moist climates. Soil requirements are low, and oats will tolerate a very wide pH range and soils of poor drainage, fertility and texture.

16.2. Varieties

A high growth rate in the early part of the vegetative phase is characteristic of oats. Grazing is possible within 6–8 weeks, by which time secondary roots have anchored the plant adequately.

Three varietal characteristics largely determine the suitability of an oats variety for grazing:

1. Tillering capacity.
2. Stage of head initiation at time of grazing. If the heads are removed by stock, regrowth is retarded because further growth depends on the stimulation and subsequent growth of dormant lateral tiller buds. Thus, the requirement is for a variety which has a long vegetative phase before apex initiation and a high growth rate of vegetative tissue in this phase.
3. Tolerance to rust, crown rust (*Puccinia coronata*) in particular. Rust attacks have been at times so severe that interest is being shown in Triticale.

The latest KFA-catalogue described the following oats varieties:

Suregrain	Outyields all other varieties. Very resistant to rust. Has a bold grain and high bushel weight. Has performed very well for grain or fodder at the lower altitudes.
Lampton	Primarily a feed oat. High yielder but very susceptible to rust. Very suitable as fodder oat, giving a heavy yield of green material.
Grey Algerian	A popular porridge oat in Kenya. Not a heavy yielder and susceptible to rust.
Jasiri	Has a better rust resistance than Suregrain. Relatively low bushel weight.

This description illustrates that farmers have been mainly interested in dual-purpose varieties. On the Kitulo Plateau (2750 m) in southern Tanzania, Suregrain was quicker maturing, grew up to 100 cm height and was the best grain variety. Grey Algerian produced thicker culms, broader leaves and reached 140 cm (Prins, 1969). Suregrain confirmed its superior rust resistance in trials at Uyole (T) in the early 1980s, carried out together with tests on Triticale. Even at stages beyond 5 months after sowing, Suregrain displayed superior digestibility and voluntary intake as compared with Elephantgrass (Myoya & Mukurasi, 1988). Lampton and Jasiri were among the best entries in a large-scale screening of oats varieties in the central highlands of Ethiopia (Ibrahim & Orodho, 1983).

Oats can be distinguished from other small cereals because the ligule is well developed and toothed and there are no auricles.

16.3. Establishment and husbandry

Oats are normally sown with a wheat drill in 20–25 cm rows at rates varying from 60–80 kg, the high rates being recommended for poorer soils and for fodder production if the latter is the primary aim. Seed dressing is recommended against seed-borne and soil-borne diseases and pests such as the dusty brown beetle (*Gonocephalum simplex*), of which the adults eat the leaves and cut the stems but the larvae, known as false wireworms, feed on seeds and roots (Wheatley, 1965). Rolling after sowing helps to control cutworm (*Agrotis* sp.).

Oats are usually grown with very little fertilizer, but both P- and a starter dosis of N-fertilizer are beneficial, if not much time is left before the onset of the dry season. In particular, acid soils at high altitudes benefit by a P-dressing in the seedbed. Purple vetch at 15 kg/ha, peas or sweet lupins at 50 kg/ha, fodder rape at 4 kg/ha and serradella at 10 kg/ha are suitable companions with oats sown at 40–50 kg/ha (*18*; Henderson & Preston, *op. cit.*; *Kenya Dairy Farmer*, July 1965).

Grazing is the cheapest method of utilization. Grazing will be the rule, while hay and silage or grain production are the logical alternatives if grazing is not required. Henderson & Preston (*op. cit.*) have thought out systems of strip-grazing with the aid of electric fences. Animals should not be grazed until the dew had dried off; introduction should be gradual to prevent bloat. Young oats are low in fibre content (Table 16.1) andsome roughage in the form of hay may prove useful.

Grazing should be swift and clean. Long duration of grazing has a more negative effect on regrowth than high grazing pressure. N-fertilizer is effective in stimulating vegetative growth. Oats lose powers of regrowth once piping sets in (*16.1*).

16.4. Nutritive value

Data on the nutritive value of green oats grown on the Kinangop are shown in Table 16.1 (Dougall, 1954d).

Green fodder oats are extremely rich in crude protein and its digestibility is very high. The crop is also rich in ether-extract, contains high levels of Ca and P and is not fibrous. As the crop matures there is a progressive decline in protein, ether-extract and minerals.

Table 16.1. The composition of oats with advancing maturity at South Kinangop in Kenya (after Dougall, 1954d).

	Date			
	12-12-53 (30 cm)	7-01-54 (50 cm)	29-01-54 (early flowering)	12-02-54 (full flowering)
Ash	10.5	11.5	8.5	5.4
Crude protein	35.3	20.2	9.1	8.1
Crude protein (% digestible)	–	84.9	74.6	46.7
Ether-extract	5.7	3.7	3.2	2.4
Crude fibre	16.9	23.3	23.3	28.1
Carbohydrate	31.6	42.0	54.8	56.0
CaO	0.61	0.48	0.42	0.40
P_2O_5	0.68	0.56	0.33	0.35

Oats hay is highly valued. High digestibility is attained in the early-piping stage, which is the best stage for making hay. Silage is made when the oats have passed the piping stage. Then the % DM and carbohydrate content are higher so that oats can be ensiled without chopping or without the addition of molasses (Henderson & Preston, *op. cit.*).

Grain yields of 1500–2000 kg/ha are common. Oats seeds volunteer easily in subsequent arable crops. Oats have been suggested as a possible grazing companion-crop for undersown grasses (Kitale Ann Rep 1955; Goldson, 1967). Considerably higher yields of total herbage were harvested in the 1st year. The practice does not seem to have taken on, as there is a considerable risk of oats smothering the young grass seedling.

16.5. Summary

Oats are a useful and versatile source of grazing and fodder at high altitudes of 2000–3000 m, where Grass Sorghum thrives less well. The same management principles apply. Rust resistance is a principal criterion of variety assessment.

CHAPTER 17

Lucerne

Lucerne (*Medicago sativa*) is the only herbage legume in East Africa that has been grown consistently on farm scale for many decades. Lucerne's perenniality; tolerance of moisture stress; ability to draw moisture from deep in the profile; adaptation to climate; soil improving properties; ability to produce a highly nutritious fodder over long periods of the year as well as its persistence have rightly earned lucerne the title of "queen of the fodders" throughout the temperate and subtropical climates of the globe. Lucerne is world's oldest cultivated fodder.

Medicago is represented by about 20 relatively common species. The name "medics" or "burclovers" is used for annual species, native to the Mediterranean region, which have naturalized in large parts of southern Australia and which regenerate in crop rotations. Despite considerable effort it has not been possible to reproduce the Australian success of such adventitiously introduced medics within the systems of wheat rotations in Africa.

The *Medicago sativa* complex comprises 2 interfertile species: *M. sativa* (purple-flowered) of south west Asian origin and *M. falcata* (yellow-flowered), a cold-hardy native of Siberia. Many of the present varieties developed from these 2 species. Lucerne was introduced into Greece as early as 490 B.C. by the invading Medes and Persians, who used lucerne as a fodder for their horses. From Greece it spread over North Africa and Europe and with the early Spanish conquerors also to Latin America, from where Hairy Peruvian would later originate. In the late 1700s lucerne was also brought to south east Australia where Hunter River, another variety successfully deployed in East Africa, developed in the flats of the river of that name. Both varieties illustrate the rapidity with which introductions became adapted in the New World, without direct man-made selection or breeding, to yield varieties that excelled also when brought back to the Old World.

Samples of lucerne were brought to Kenya in 1904/5 (Huxley, 1935) and to Uganda in 1906 (Thomas, 1940b; Tiley 1970a). Johnston (Kabete Ann Rep 1910) found lucerne performing well. The superiority of Hunter River and Hairy Peruvian was already established back in 1929 by Lyne Watt (1929) at the Agricultural Laboratories of Nairobi. Ball (1935) showed that lucerne was a well established fodder crop in Kenya at a time when pasture research had barely started. Ball's notes were largely drawn from farmers' experience and it is worthwhile noting his observations as regards the need for lime, phosphate and inoculation as well as for the optimum rowwidth and seed rate. Obviously, much was and can still be learnt from experience in other countries.

In Kenya, lucerne is mostly grown around Lake Naivasha (1900 m) either irrigated or on the alluvial soils with a high water table. Much of it is grown for sale as hay or for dehydration and processing to produce lucerne pellets. This high quality product is used in mixed feeds, in which it serves as protein supplement containing vitamins A, E, K and xanthophylls, if properly stored. Lucerne has also been found to grow well in the areas of Mount Kenya, Kilimanjaro and Elgon, as well as in Kivu.

Around Lake Naivasha lucerne is also grown in combination with Columbusgrass (*15.2.3*). The same mixture was tested under direct grazing at Njoro, west of Nakuru, by Thomas (1970; 1972) under dryland conditions with a great deal of success, the main draw-back being the drying-out of the soil prior to cropping. The combined grazing of lucerne and pasture grasses such as Nandi Setaria, as suggested by Strange (1958a), has been practised on the lower slopes of Mount Kenya. At Makerere (U) lucerne grew well with Setaria, Rhodes- and Guineagrass and contributed more to some of the mixtures than *Desmodium intortum* (Olsen & Tiharuhondi, 1972). In fact, Tiley (1966) found that the yield of Rhodesgrass was increased by the presence of lucerne (Fig. 17.1), which is rarely observed among her tropical counterparts (*10.3*; Fig. 12.6).

Lucerne has disadvantages which militate against widespread use. Depredation of game, antilopes in particular, can cause havoc. Reports of cattle deaths due to bloat are sufficient to prevent many farmers even from trying. On the other hand, on many mixed-farms with a fairly wide range of fodder alternatives, farmers are usually not too keen on crops requiring a lot of attention all year round, which lucerne does. In pure stand lucerne needs repeated cultivation whereas haymaking requires proper timing and care to avoid the shedding of valuable leaf. Also, on larger farms it may seem irrational to spend a lot of time and expense on a small acreage of lucerne while the rest of the grazing resources receives so little.

Lucerne might fit in well with smallscale dairy farms used to zero-grazing, provided it is not cut too frequently (*13.4.3*). In this latter respect it does not differ much from most tropical legumes. Lucerne is worthy of labour-intensive feeding systems, as it combines yield with a high nutritive value. A reappraisal is, therefore, justified.

17.1. Varieties

After Lyne Watt's (1929) confirmation that Hunter River and Hairy Peruvian were the best introductions, Ball (1935) also mentioned Provence.

In spite of many subsequent screening trials in the region, in different zones and over a wide range of altitudes (1000–2750 m), the choice of recommended varieties is still very much the same. A large screening exercise was undertaken at Kitale (Zone III) in the mid 1950s. A lucerne collection of no less than 122 entries was obtained from Australia, the Americas, India, Iraq, Turkey, North Africa etc. A short-list of 16 was tested further for resistance to leaf-spot disease (*17.7*), drought tolerance, productivity and seed production (Kitale Ann Rep 1956). Hairy Peruvian, Saladina and other introductions from South America compared well with Hunter River and Provence and work was continued with Hairy Peruvian (Kitale Ann Rep 1956; 1957). In later Kitale reports mention is often made of "lucerne", without specifying variety names, apparently because of the small differences found (Table 17.1). Strange (1958a) and Bogdan (1965c) recommended Hairy Peruvian, Hunter River, Booborowie (Australian selection out of Hunter River) and Saladina, to which Strange (1958a) added Du Puits as of mixed *M. sativa* and *M. falcata* parentage, which had shown promise in upper Molo at 2750 m (Birch, 1959a). For Machakos (K; Zone IV) Sands *et al.* (1970) suggested Hunter River, Saladina and Booborowie at favoured sites. In a recent variety trial with some new accessions at Naivasha (Zone IV), the superiority of Hunter River was confirmed (Snijders, Naivasha Report 1989). At Kawanda (U; Zone II), J.R. Simpson achieved higher yields with some varieties than with Hairy Peruvian and Provence but Hairy Peruvian remained the recommended lucerne variety since it was commercially available (Tiley, 1970a). At Tengeru near Arusha (T; Zone III), at an equally low altitude, Anderson & Naveh (1965) worked mainly with Hunter River, but Hairy Peruvian and Saladina were equally outstanding (Naveh & Anderson, 1967). More recently, Hairy Peruvian was reported to do well at Uyole (T; 1800 m) by Myoya & Mukurasi (1988).

The most recent seed catalogues of Kenya still list Hunter River, Hairy Peruvian and Provence, of which the first two also happen to have the cheapest seed. A more recent variety being marketed in Kenya is Caliente. Hunter River and, to a lesser extent, Hairy Peruvian have been important in Australia also and a full description of these and other varieties mentioned above was given in the Register of Australian Herbage Plant Cultivars (Barnard, 1972). In brief, the habit of Hairy Peruvian is erect and compared with Hunter River the stems are coarser and less leafy, but leaflets are larger and thicker while the seeds are slightly larger. Provence, although lower yielding at times (Table 17.1), is more suitable for drying owing to its thin stems. The creeping habit of *M. falcata* material (Cancreep) aroused some interest in the 1960s in Australia and elsewhere, but none of the bred varieties was successful; productivity was too

Table 17.1. Dry matter yields (ton/ha) of lucerne varieties at Kitale (after Poultney, Kitale Ann Rep 1963).

Variety	Year			
	1961 (5 cuts)	1962 (6 cuts)	1963 (4 cuts)	Mean
Saladina (selected)	6.7	9.3	3.5	6.5
Hunter River	5.3	9.5	4.3	6.4
Booborowie	5.0	8.7	4.3	6.0
Saladina (commercial)	6.3	8.0	3.3	5.9
Hairy Peruvian	5.0	7.3	4.7	5.7
Provence	3.0	7.7	3.5	4.7

low while persistence was frequently inferior to Hunter River (Clements, 1989). Often no creeping is observed even at low plant densities (Barnard, 1972). It may be concluded that it is truly remarkable that despite outbreaks of diseases and pests (*17.7*) Hunter River has been able to hold out against newly bred varieties since it was recommended (Lyne Watt, 1929), 60 years ago.

Attempts to produce seed of Hairy Peruvian and Saladina were made at Kitale around 1960, but seed yields were even lower than those of white clover and further efforts were abandoned. A high degree of precision is required to promote pollination, which can best be obtained under irrigation. To set seed all flowers must be tripped by insects. Lucerne is not an attractive pollen source for honey-bees, so that little seed is set unless colonies are placed and moved throughout the field and competing sources of pollen are avoided. Attempts to improve seed setting with pollinating agents were made at Uyole (1978). Imported lucerne seed is only half as expensive as locally produced clover seed, so that reliance is entirely on import. Annual imports in Kenya average about 5 ton.

17.2. Establishment

The sowing rates practised are variable. Ball (1935) and Henderson & Preston (1959) recommended 20 kg/ha in rows of 15–30 cm under irrigation, the objective of high rates being weed control and hay quality (finer stems). Such high rates at current seed prices put the seed costs 2 to 3 times above those of Rhodesgrass pasture. For dryland, Ball (1935) recommended 8–10 kg in 50–60 cm rows. At Uyole (1977) no difference was found between seed rates of 4, 8, 18, 32 kg/ha in any of the 4 cuts that were harvested. Anderson & Naveh (1965) successfully grew Hunter River at Tengeru in 60-cm rows at 5 kg/ha.

Strange (1959b) ran plant density trials with dryland lucerne at Kitale for 4 years. In the establishment year the yields were significantly higher at 30 cm rowwidth. In the 3 subsequent years no significant differences were found between 30, 60, 90 and 120 cm, the general trend being in favour of wider rows (Table 17.2).

Table 17.2. The effect of rowwidth in post-establishment years on the yield of dry matter (kg/ha) of lucerne at Kitale; no gypsum applied (after Strange, 1959b).

Year	Rowwidth (cm)			
	30	60	90	120
1955	6570	7030	8250	8230
1956	4370	4830	4770	4830
1957	1990	2500	2160	2130
Mean	4310	4790	5060	5060

The variety of lucerne was not specified. A clean-weeding regime was practised and the conclusion was reached that wider rows were more convenient for interrow cultivation and also gave a more vigorous and healthy stand (less leaf-spot and leaf-rust). Strange also observed a density × gypsum interaction (Table 17.3) and commented that nodulation had been ineffective. Lucerne is usually regarded as being highly specific in its Rhizobium requirements and lime-pelleting is advisable in acid soils (Bumpus, 1957; Table 10.1). No inoculation is needed in land which has carried lucerne before.

Tables 17.1 & 17.2 show that yields drop markedly in the 3rd year.

17.3. Soils and fertilizers

Lucerne is usually associated with a preference for deep alluvial soils or loams, its taproots being able to reach great depth. The soil should be well supplied with lime, neither too acid nor too alkaline. Black-clay soils (Vertisols) are problematic. Anderson & Naveh (1965) grew lucerne successfully on soils in northern Tanzania, the majority of which were rich in P, K and bases. Good results are often achieved in dry climates with irrigated fertile soils. This may account for the popularity of lucerne as a fodder crop on the shores of Lake Naivasha, where most of the lucerne of Kenya is grown. At Kitale, Strange (1958b) and Poultney (1963b) illustrated that dryland lucerne can be relatively successful on moderately acid soils also, if proper precautions are taken as regards P, Ca and S (sulphur). The fertilizer used at that time was triple superphosphate, containing 18% CaO and no S (Foster, 1972). In the late 1950s S-deficiency was widely suspected (Strange, 1959b). Responses to S were observed in legumes at Serere (*10.2.1*). Lucerne- and other legume trials at Kitale were, therefore, modified to include gypsum ($CaSO_4$) which was also applied in the plant density trial already reported (Table 17.2). The effect of gypsum itself is presented in Table 17.3. Gypsum was applied at about 200 kg/ha in July 1956 and in March 1957. The data of Table 17.3 cover the period July 1956 to August 1957.

Table 17.3. The interaction between gypsum and rowwidth in their effect on lucerne yields (after Strange, 1959b).

Treatment	Rowwidth (cm)				
	30	60	90	120	Mean
Control	3480	4330	4080	4100	4000
Gypsum applied	5800	5680	5350	4870	5430

An increase of nearly 70% due to gypsum was observed at 30 cm rowwidth, the effect becoming progressively less in the wider rowwidths. Strange also observed that the effect of gypsum was most marked after the initial dressing, the response tending to decrease, allegedly due to other unidentified soil nutrient shortages. Agricultural lime had been applied at sowing at the rate of 6 ton/ha, while triple superphosphate and muriate of potash had been applied both in the seedbed and as annual topdressings at the rate of about 200 kg/ha. However, on the basis of other trials Strange (1959b) finally concluded that the responses to gypsum by pure stands of lucerne had been rather variable on the Kitale soils. The cooperative experiments (Kitale Ann Rep 1957) showed that gypsum responses had been found on legumes in a belt stretching from central Cherangani through Hoey's (Moi's) Bridge to the Lugari area and in one instance on Mount Elgon near the Suam River. No responses had been obtained on the soils of Endebess, Soy, Eldoret or Ainabkoi.

Further nutrition experiments were carried out by Poultney (1960; 1963b), which illustrates the amount of attention paid to mineral nutrition and legumes at the time. More detailed data are given in the Kitale Ann Rep 1961. The response to Ca and S and to the combination of Ca and S in the form of gypsum was not consistent, but when Ca and S were given in the form of $CaCO_3$ and Na_2SO_4, respectively, the response was significant, on average some 22% above the P-treatment. At Kawanda, J.R. Simpson found some indication of a better response to inoculation when either S or S+P were applied, but differences were small; single superphosphate at 75 kg P_2O_5 had been applied at planting (Tiley, 1970a). Poultney (1963b) emphasized that for Kitale soils which often have no more than 10–15 ppm P, rates exceeding 60 kg P_2O_5 were needed to obtain a sustained response beyond the 1st and 2nd cut in the 1st year and also to make lucerne responsive to other major elements. Yellowish-green leaves are often indicative of S-deficiency, not unlike N-deficiency although the latter shows up first in the lower, older leaves and not in the younger leaves as is the case with S-deficiency. Allan (1976) reported S-deficiency in maize in the Kitale area. He put the S-requirements of a good maize crop at 20–40 kg/ha which were more than covered by the recommended fertilizer rates for maize of 200 kg single superphosphate (21% P_2O_5; 21–31% CaO and 13% S) and 300 kg ammonium sulphate nitrate (26% N; 12% S).

Single superphosphate, manufactured from rock-phosphate near Tororo (U), became widely available in the early 1960s to become the standard phosphate fertilizer on account of its valuable proportions of Ca and S. It would seem that both the Ca- and S-requirements of the plant can be met at the same time as the P-requirements. On a very P-deficient soil (4 ppm P-Truog) at Makerere (U), Olsen & Moe (1971) recorded a response up to 450 kg P_2O_5 applied as single superphosphate. Even liming at 2.5 ton/ha of lime (80% $CaCO_3$) increased productivity and persistence; lime contains 1% P (*10.2.1*). In a critical review of S-responses found in Uganda, Foster (1972) concluded, however, that even in groundnuts with their high demand for Ca and S crops are unlikely to suffer from deficiency although soils of low OM may require extra Ca and S if yield potential is high.

It can be concluded that, in lucerne, P is more critical than either Ca or S and that applications of single superphosphate, at 200–300 kg/ha/year, should be adequate to meet the requirements of all three elements. On the alluvial soils of Lake Naivasha very little P is applied. Some starter N can increase initial production (D.M. Rocco, private communication).

17.4. Weed control

The soils around Lake Naivasha, in particular, carry a heavy seed burden of dicotyle weeds, also after the ploughing of natural Stargrass pastures (*3.4.2.2*; *4.2.2.2*). Rocco (1958) gave a description of the principal weeds. The herbicides which were then thought of as effective were "activated" dinoseb and 2,4-DB, applied to offset the high costs of manual weed control: 150–250 mandays/ha after sowing and another 100 mandays for maintenance. 2,4-DB can still be recommended as it controls many annual broad-leaved weeds, while doing little or no damage to the lucerne if the precautions listed in Table 17.4 are followed. 2,4-DB is effective against *Amaranthus* spp. (Rocco, 1958). However, *Malva* spp. and *Galinsoga parviflora* are hardly checked, so that other herbicides are also needed. A mixture of 1.2 kg *a.i.* of 2,4-DB and DNBP plus 8 kg of sulphate of ammonia, (as "activator" to reduce alkalinity of the lake water used for spraying) is still the standard application (D.M. Rocco, private communication).

In recent years new chemicals have entered the scene such as Bentazon, which gives some control of sedges (Veenstra & Boonman, 1974), but no reports on their efficiency on lucerne in Kenya are available so that it may prove useful to reproduce the control guide (Table 17.4) of Queensland, where the same varieties were recommended. Benfluralin, Chlorthal and Atrazin were effective. EPTC was recommended for the control of annual grasses. It was also said to delay emergence of sedges, but some damage to the lucerne seemed inevitable. Henderlong (1973) noted a beneficial effect with EPTC on lucerne in Uganda. A relatively new product applied with success in more temperate climates is Carbetamide (Legurame).

No herbicide can make up for bad management. From the 3rd year onwards, however, encroachment of perennial weed grasses such as African Couch-, Star- and Kikuyugrass is bound to occur. Rocco (1960) observed that 60 mandays/ha/year were needed to keep lucerne free from encroaching Star- and Kikuyugrass. Dalapon was tried in 6 bi-monthly applications of 5 kg/ha in 180 l water on grass patches in a 4th-year stand with reasonable success but the number of sprayings needed to be brought down (Rocco, 1960). The Queensland control guide of Table 17.4 shows that this is possible. Annual applications of 8 kg/ha used to be recommended for the Lake Naivasha area (D.M. Rocco, private communication). A relatively new product applied with success against grass weeds in Europe is Sethoxydim (Fervinal) at 3 kg/ha a.i. in 5–10 l mineral oil. There can be no doubt that the developments in chemical weed control of the past decades will open new avenues to the continued success of this highly valued legume in East Africa.

17.5. Husbandry and utilization

The highest yields are obtained with irrigation and some 10 cuts can be achieved totalling some 15 ton/ha/year DM. In dryland lucerne under Kitale conditions not more than 4–6 cuts each of 1 ton/ha DM are achieved. Yields drop markedly in the 3rd year (Tables 17.1 & 17.2). In the Mount Meru and Mount Kilimanjaro area similar yields were obtained (Anderson & Naveh, 1965). Olsen & Moe (1971) reported yields of 10 ton/year at Makerere, although the soils were much poorer than those of northern Tanzania.

Timing of defoliation is critical as it is in most legumes (*10.4*). There is general agreement that cutting should not be carried out before early flowering. Repeated cutting at early stages inevitably leads to serious grass weed encroachment and prematurely kills the stand. If longevity of the stand is no prime consideration, cutting may well take place at an earlier and more protein-rich

Table 17.4. Chemical weed control in lucerne (after *Queensland Agricultural Journal*, July–August, 1976).

Herbicide	Trade name	Product/ha	Time of application	Weeds controlled	Remarks
Pre-emergence					
Benfluralin	Balan	5.6–8.4 l	Pre-planting incorporated	Most annual grasses and some broad leaved weeds	May be applied from a few weeks to immediately before planting. Refer to manufacturer's label for methods
Chlorthal-dimethyl	Dacthal W75	11 kg	Post-planting pre-emergence	Most annual grasses and broad-leaved weeds	Apply soon after planting before weeds germinate. Follow with an irrigation of 10–25 mm. May also be applied to emerged lucerne before weed emergence
Post-emergence					
Atrazine	Several products	0.7 kg or 1 l	Post-emergence. Do not spray lucerne stands under one year old		Some residual control. Slight leaf yellowing of the lucerne may result
2,4 DB	Several 40% products	3–4 l		Many annual broad-leaved weeds	Seedling lucerne should be sprayed in the 1–8 trifoliate leaf stage. Established lucerne to be grazed or mown first before spraying when 10–12 cm regrowth is present
2,2-DPA (Dalapon)	Several products mostly 74%	5.5 kg	Established lucerne	Grasses	Add wetting agent. Repeat application as required but do not exceed 2 applications in one season. Mow or graze before spraying. Some damage of lucerne may occur
		1.0 kg	Seedling lucerne	Seedling grasses	Add wetting agent. Do not spray earlier than 2 weeks after lucerne emergence

stage, with the additional advantage of less leaf shedding due to leaf-spot disease. When grazed in combination with grasses lucerne needs a strict regime of deferred grazing. Continuous grazing weakens the stand rapidly. By contrast, height of cutting appears inconsequential for lucerne persistence.

A mixture of grass or Sorghum with lucerne markedly reduces the risk of bloat (Thomas, 1970). A common practice has been to accompany livestock whenever a bloat situation is likely to occur and administer a dose of turpentine orally (see also *4.2.1*; *16.3*). Tropical legumes such as Desmodium generally cause less bloat, which has been attributed to tannins, compounds which are, however, not conducive to high digestibility (*10.2.2*).

At Kitale, Strange (1958a) was an advocate of growing lucerne with grasses in alternate rows, 60 cm apart. Large increases in herbage yield were achieved by growing it in mixture with Nandi Setaria (Table 17.5). It was not clear from the data whether the increase was due to extra grass or lucerne or both (see also Fig. 17.1). Strange observed that these mixtures could last 4 years. At Uyole (1977) a mixture of 3 kg Pokot Rhodes and 4 kg Hairy Peruvian lucerne was more productive than either component in monoculture. Olsen & Tiharuhondi (1972) found at Makerere that lucerne maintained a higher legume content in mixtures with grasses than *Desmodium intortum* did, even at a seed rate of 1 kg/ha. Henderlong (1973), also at Makerere, observed that the regrowth of

Table 17.5. The combined fresh yield (ton/ha) of lucerne and Nandi Setaria grown in alternate rows, 60 cm apart (after Strange, Kitale Ann Rep 1955).

	Year		
	1954	1955	Mean
Lucerne	16.5	24.5	20.5
Lucerne with Nandi Setaria	33.7	37.5	35.6

lucerne was more rapid than of any of the tropical legumes. Within 14 days after defoliation, lucerne had accumulated 20 cm of growth, whereas Desmodium, Centro, Stylo and Siratro were only just beginning.

Of particular interest is the observation by Tiley (1966; 1970a) at Kawanda, in experiments initiated by J.R. Simpson in 1957 and conducted under grazing, that Provence lucerne in the presence of Rhodesgrass increased the yield not only of the grass-legume mixture, but also of Rhodesgrass itself by nearly 50% (Fig. 17.1) in the last 2 years. By contrast, grass yield was reduced, or at best maintained, in the presence of tropical legumes (Tiley, 1966), which is commonly observed in East Africa (*10.3*; Fig. 12.6). It would appear from these observations that lucerne is a legume of great interest, and should be part of any program intent on legumes.

Lucerne may not be suitable for late-season pastures. It will not hold its leaf when required to stand for any length of time after the end of the rains (Strange, 1958a).

For stall-feeding on small farms lucerne hay is purchased but surprisingly little is grown by smallholders. This is probably due to the fact that home-grown plots are under enormous pressure of utilization. Elephantgrass suffers from frequent defoliation but seems to survive with a few shoots, whereas lucerne dies rapidly when cut repeatedly before the flowering stage (*13.5.3*).

17.6. Nutritive value

Dougall (Kitale Ann Rep 1961) sampled the variety Provence whenever the early flowering stage was reached, over 8 harvests of an irrigated stand at an altitude of 2500 m. Some 35 kg P_2O_5 was applied on two occasions and K was also given. Table 17.6 shows that the leaf:stem ratio was 54:46 on a DM-basis and that the crude protein was much higher and the crude fibre much lower in the leaf than in the stem. P was at very much the same level in leaf and stem, but Ca was appreciably less in the latter. Dougall's report also referred to ash, N, S and carbohydrate as affected by triple superphosphate, potassium sulphate and gypsum but the effects of the different treatments were small and were dominated by age effects (Table 17.6). The content of silica-free ash, N, P and S decreased as the plants grew to maturity and the % carbohydrate increased.

This report confirms the exceedingly high quality of lucerne not only in terms of a high content of N, P, Ca and S but also in respect of the low level of crude fibre. Generally, voluntary intake of lucerne is high, although *in-vitro* digestibility data may seem relatively low. Reference has already been made to the high vitamin content.

17.7. Diseases and pests

Although management itself is most critical, poor growth of lucerne has traditionally been attributed to disease. Diseases do play a role, especially in the lower and humid areas below 1500 m. Of the fungi, leaf-spot (*Pseudopeziza medicaginis*) and rust (*Uromyces striatus*) are the most common, but anthracnose (*Colletotrichum trifolii*), downy mildew (*Peronospora trifoliorum*), burn (*Pseudoplea trifolii*) and root rot (*Rosellima necatrix*) are also of importance (Nattrass, 1961); for diseases in Australia, the cradle of Hunter River, see Irwin (1989).

The spotted alfalfa aphid (*Therioaphis maculata*) and the blue aphid (*Acyrthosiphon kondai*) became serious pests in the USA around 1975 and spread over Australia in a few months in 1977. In California, resistant material was rapidly identified and released (Lehman *et al.*, 1985) which, together with biological control, greatly helped to reduce the impact of the pests. As a result, Hunter River which is aphid-susceptible but persistent has regained much of its former popularity (Clements, 1989).

17.8. Summary

Lucerne is a valuable high-quality fodder plant which warrants continuous attention, since there

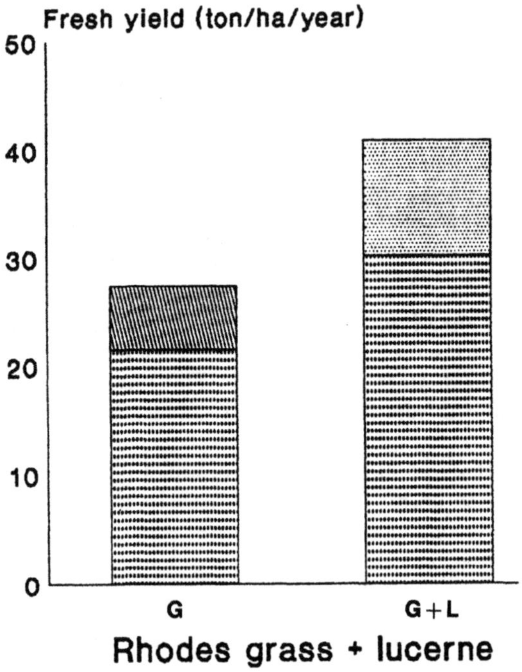

Fig. 17.1. The yield of Rhodesgrass (G) in the absence or presence of lucerne (L) (after Tiley, 1966). Lucerne; Volunteering legumes; Rhodes.

Table 17.6. Composition of Provence lucerne leaf and stem in relation to age at sampling (after Dougall, Kitale Ann Rep 1961).

	Plant component		Age at sampling (weeks) of whole plant								
	Leaf 53.8%	Stem 46.2%	1	2	3	4	5	6	7	8	9
Silica-free ash	10.67	11.05	11.13	11.07	10.73	10.61	10.31	9.52	9.96	10.15	9.43
N	5.59	3.53	5.00	4.73	4.26	4.00	3.64	3.76	3.93	3.41	3.17
P	0.38	0.36	0.46	0.40	0.35	0.31	0.30	0.28	0.29	0.27	0.24
Ca	1.06	0.62	1.11	1.02	1.03	1.04	1.08	1.24	1.10	0.99	0.93
S			0.30	0.28	0.29	0.26	0.25	0.25	0.24	0.23	0.21
Ether extract	3.95	1.86									
Crude fibre	13.06	31.40									
N-free extract	37.39	33.60									
Total carbohydrate	50.45	65.00	52.44	54.73	58.46	60.62	63.36	63.35	61.82	64.71	67.29

are very few other legumes of similar performance. Hunter River, Hairy Peruvian and Provence have dominated the variety scene for the larger part of this century and are not likely to be dislodged easily by other varieties. Chemical weed control has done away with many of the weed problems. Also, with judicious application of single superphosphate, containing Ca and S, many of the soil and nutrition problems that have been suspected can find a solution. Much work has been done around Naivasha in Kenya but also in Uganda and Tanzania. Lucerne is still the best allround legume available for intensive livestock production in East Africa.

CHAPTER 18

Other (non-graminaceous) fodder crops

The highlands of East Africa can aptly be described as the testing ground of fodder crops from all over the world, from cold temperate climates (*e.g.* Brassica) to the hot humid tropics, as represented by Leucaena. With fair confidence it can be said that no fodder crop, having performed well in some corner of the world, has not been brought to East Africa at some stage. This was especially true in the 1910s and 1920s when farmers and officials alike endeavoured to sample the world's crops, in their search for what might suit local requirements. Introductions were already made before 1920 of Atriplex saltbush (Johnston, Kabete Ann Rep 1910) and of Leucaena in 1915 in Uganda (*2.1.1*; Thomas, 1940c; Tiley 1970a). Many proved outright failures and were rejected with often no records left behind to indicate what the main drawbacks were. In many cases, however, the reasons of failure were pests and diseases.

Some crops were brought in by settlers and officials because of their familiarity with these crops overseas. The chance of success depends, however, on the merits relative to the local alternatives available. Beets may be the best choice in northern climates but may be less preferable in East Africa than local root crops. Fodder crops should also prove their value as supplement to or replacement of sown grasses. The interest in fodder crops preceded the development of cultivated pastures (*2.2.2*).

It is significant that most of the fodder crops listed in Table 18.1 were or are used in the cool highlands above 2250 m. These crops were readily available through seed imports from temperate countries and many of them performed equally well as in the home countries (kale, fodder rape). Varieties of Ryegrass and Cocksfoot did not become widely accepted until the 1950s (*6.15*) but temperate fodder crops had been used from the early days.

The primary requirements of a good fodder are high yield and resistance to pests and diseases. The second requirement is that it should be able to provide nutritious fodder at specified times of the year, directly or after storage. A third consideration is ease of accession to the harvestable part and ease of management.

Table 18.1. Fodder crops listed in the seed catalogue of the Kenya Farmers' Association in 1967 versus 1972.

1967	1972		
+	+	Kale	*Brassica oleracea*
+	+	Rape	*Brassica napus*
+	+	Mangolds	*Beta vulgaris*
+	+	Fodder beets	*Beta vulgaris*
+	+	Lupins	*Lupinus* spp.
+	+	Vetch	*Vicia benghalensis*
+	+	Chicory	*Cichorium intybus*
+	+	Field peas	*Pisum arvense*
+	–	Turnips	*Brassica rapa*
+	–	Sainfoin	*Onobrychis viciifolia*
+	–	Cowpeas	*Vigna unguiculata*
+	–	Cattle melons	*Cucurbita* spp.

A comparison of consecutive seed catalogues shows how preferences for certain fodder crops may change (Table 18.1). Turnips, sainfoin, cowpeas and melons had disappeared from the catalogue by 1972.

Turnips (*B. rapa*) and especially swedes (the rutabaga type of *B. napus*) are susceptible to the turnipfly, which is probably the same as the cabbage sawfly (*Athalia* spp.) with oily black or green leaf-eating larvae up to 2,5 cm long with black head (Wheatley, 1965). At Uyole (1977) considerable damage occurred due to the cabbage sawfly in the rainy season. It was recommended to sow under irrigation before the rains and to transplant later to maximize yield, but the results were not encouraging. Matters were further complicated because of the short period of possible utilization. The plants flower and go hollow too quickly. This would confirm Henderson & Preston's (1959) statement that turnips and swedes had been tried without much success, but promising results have recently been reported from Rwanda by Potts & Biranguza (1989). Turnips require less fertile soils and are more suitable for short growing seasons than kale (*18.2*).

Giant Sainfoin was suggested for use as hay, silage or as an alternative to cereals within the arable break, whereas Common Sainfoin was to be tried in grazing mixtures (Strange, 1958a). It was said to be highly palatable and easily eaten out if not allowed sufficient time of regrowth between two grazings.

Cowpea is a pulse crop grown for human consumption. Horrell (1963) reported that several types had been introduced from Lesotho, which produced seed crops in 70–80 days from sowing at Serere, but none were sufficiently productive in foliage to be considered good fodder crops. In subsistence farming, yields are as low as 150 kg/ha grain. Bruchids are a major pest.

Cattle melons used to enjoy some popularity in the low and dry areas to supply a succulent (over 90% moisture) fodder which could at times act as a source of drinking water. The melons have to be cut open to enable cows to get at the inside. The seed was sown at 90 × 60 cm. Each plant would normally give 3–5 melons about 25 cm in diameter (Henderson & Preston, *op. cit.*).

Some measure of the extent to which some of these crops are grown in Kenya was given in the survey of Chudleigh (1974). Below follows a description of the most important fodders, which have been in actual cultivation. Comparative nutritive values have been given by Dougall (1960).

18.1. Fodder rape

Fodder rape (*B. napus*) has been widely grown at high altitudes for grazing with sheep and pigs and is valued for its fast establishment and compatibility in mixtures with oats (*16.3*) or Italian Ryegrass. At one time it was also valued for feeding ostriches raised in captivity for their feathers (*2.2.2*). In crop-grass rotations it is successfully employed as a break crop to suppress Kikuyugrass before the field is resown to crops or to pastures (*4.2.1.2*). Fodder rape grows at temperatures which are too low even for Ryegrass. The seed rate is 3–5 kg/ha sown through the wheat drill or broadcast. Cotyls appear above the ground within 2 days. Drought tolerance is poor. Insect damage is common but less than in turnips (see above). Rape regrows quite well after grazing in the younger stages. Care should be taken when grazing cattle on rape as it can cause bloat. Rape has a very low % DM which often does not exceed 10%. The "Essex" varieties used to be well known. In oil-seed rape there has been a rapid change-over in recent years to varieties containing low levels of erucic acid and glucosinolates ("double-zero") in the seed. This has little relevance in terms of fodder quality so that conventional varieties are still as good as any. Rape seed stays viable in the soil for many years and is known to present a weed problem in arable crops.

18.2. Kale

Kale (*B. oleracea*) is commonly grown above 2000 m altitude. Thousand-headed kale leaves used to be plucked for poultry or rabbit farming, but marrow-stem kale was more productive and used to be grazed direct. Rowwidth of direct sown kale should be 50 cm at a seed rate of 7 kg/ha, but transplanting from nurseries was practised to economize on the expensive seed and to produce a more even population. Inter-cropping with maize was also common. Kale is a highly palatable fodder. Animals will graze the leaves and eat the softer top of the stems: the stumps can be left to give regrowth. Kale is a useful fodder for smallholdings, by plucking and feeding the leaf. Drawbacks are the price of seed, high soil-fertility requirements, cattle bloat, tainting of milk, diseases and pests, in particular aphids, cutworm, cabbage sawfly and root-knot nematodes. Kale (Sukuma ya wiki) is also used for human consumption.

18.3. Fodder beets and mangolds

Fodder beets (*Beta vulgaris*) have been grown successfully at higher altitudes, but also in the

Kitale region (Maher, 1936b). Mangolds used to be more popular but their dry matter and sugar contents are lower. Bolting (stem elongation) is rarely seen except at the highest altitudes. Transplanting from nursery beds gives the highest yields because mangolds have the ability to grow as long as the rainy season lasts; transplanting adds some 2 months to the growing period. "Multi-germ" seed was sown with a seed drill at 75–100 cm at 10 kg/ha, after which the seedlings were singled to 15 cm within the rows; "monogerm" varieties are now available which reduces the seed rate and the need for singling. Mechanical and hand weeding is essential, but effective herbicides are now available. Henderson & Preston (*op. cit.*) reported that the usual practice in Kenya was to allow animals to feed direct. Mangolds can remain in the ground until required, as bolting occurs rarely. Also, after a month of drought mangolds start growing again. Mangolds rise up in the ground as the result of swelling; cattle knock them over and feed on them. Mangolds are recommended for pigs and also for contention feeding during milking. Fields close to the milk shed with a high fertility would satisfy the high nutrient requirements of this crop. The protein content is, however, low. Otofte and Pajbjerg varieties were well known. Results of variety trials conducted in southern Tanzania were reported from Uyole (1977). The annual imports of beet seed in Kenya stand at 2–3 ton/year.

18.4. Sweet potato

Sweet potatoes (*Ipomoea batatas*) are found from sealevel to 2000 m altitude. The duration of this perennial depends on variety, climate and local custom, cultivation being permissible on sites liable to erosion, since the sweet potato gives a rapid and dense ground cover. The response to fertilizer has been variable. Plant densities varying from 12.000 to 50.000/ha gave similar yields (Acland, 1971). Rather than tubers, vine tops are commonly used as propagation material; the cut end is stuck in the ground so that one-half or two-third is buried. Labour requirements at planting are high but weeding demands are low in spite of the low-growing habit of the crop. The vines are high in % crude protein and are an excellent contention feed at milking time. Fed in large quantities, however, they cause scouring.

The main limitations are virus diseases and the sweet potato weevil (*Cylas formicarius* and *C. puncticollis*) which tunnels inside vines and tubers, also during storage. Dipping the material in an insecticide reduces the level of infestation. The sweet potato viruses cause stunting and are the most relevant threat to sweet potatoes. Varieties differ widely in their virus susceptibility so that resistance breeding has become important. Breeding is based on variation within seed progenies (MacDonald, 1965), a procedure also used in Elephantgrass (*9.7*). Table 18.2 gives the results of variety trials carried out at Embu (Embu Ann Rep 1968; 1969).

Table 18.2. Fresh yields of sweet potato varieties at Embu in ton/ha (Embu Ann Rep 1968; 1969).

Variety	1968		1969	
	Tubers	Vines	Tubers	Vines
Namunjuna	8	77	17	50
Gikanda	7	73	34	43
Bisale	14	68	35	23
Musinya	9	64	27	41
Muibai	13	44	33	21

Namunjuna was brought in from Uganda on account of its virus resistance. It remained virus-free throughout the experiment and yielded more vine than any other entry but its tuber production was exceedingly low. The opposite applied to Bisale. Apparently, high tuber and high vine yields are mutually exclusive in sweet potatoes. Gikanda and Musinya offer a compromise but it seems more appropriate to choose the variety which is best for the purpose for which it is wanted most. A very high % leaf-K was observed by Karachi (1982) which was feared to lead to soil depletion (see also *5.2.2.1*, *12.7.4* & *13.3.1*).

18.5. Leguminous fodder crops

For livestock purposes, very few legumes apart from lucerne are grown specifically but the research interest is always there. In this section some species are discussed. This list does not include the perennial pasture legumes discussed earlier (*10*; *12.8*).

18.5.1. Lablab or Njahi beans

Njahi beans (*Lablab purpureus*) are a traditionally important pulse crop in central and eastern Kenya and the highland areas of Tanzania, now losing ground to Phaseolus beans. It is a hardy crop in the 1250–2250 m altitude range (Acland, 1971). Njahi beans were used in some Kenya areas, *e.g.* Kitale for fodder purposes, grown together with maize for silage. It can also be grazed and 2–3 grazings are possible. Stems cure slowly and leaf shedding is common, so that haymaking is problematic. A white-flowered variety of ssp. *purpureus* was grown on some scale in the Rongai area of Kenya and the herbage was mixed with Sudangrass and ensiled (Bogdan, 1977). This variety was introduced into Australia in 1952 as CPI 16883 and is registered there as Rongai (Barnard, 1972). Known as Lubia, it has been a major fodder crop grown in rotation with cotton and Sorghum in the Sudan Gezira (Ishag, 1965). At Mount Makulu in Zambia, van Rensburg (1967) harvested 5400 kg/ha DM with 850 kg/ha crude protein. It nodulates well (de Souza, 1969). At Lanet (K) high levels of nutritive value were recorded by Karachi (1988). It is more cold tolerant than velvet beans or cowpeas. Njahi beans can replace lucerne where the latter does not do well. It is extraordinary that with so much focus on fodder legumes, so little attention has been spent on finding more room for this well adapted, nutritious and high yielding local pulse crop. For more details see Bogdan (1977) and Skerman (1977).

18.5.2. Velvet beans

Velvet beans (*Mucuna pruriens*) are in many respects comparable with Njahi beans but are more adapted to hotter areas (*2.1.1*; Horrell, 1958b; Henderson & Preston, *op. cit.*), where it can assume weed-like proportions (Ivens, 1971). For an account of the nutritive value, see French (1938b; 1940). The grain contains anti-nutritive compounds which are toxic to mono-gastric animals.

18.5.3. Lupins

Known as a cultivated plant under the following species:

L. albus	White lupin	(cultivated)
L. angustifolius	Narrow-leaved lupin	(cultivated)
L. luteus	Yellow lupin	(cultivated)

Native species are *L. princei* in Ethiopia, Kenya and Tanzania; *L. somaliensis* in the highlands of Somalia. Naturalized species are *L. pubescens* in Kigezi (Uganda) and Zaire; *L. ehrenbergii* in Mbeya district (T) and Malawi (Bauer, 1976). These are vigorous annual legumes which are easy to grow and are adapted to acid soils (*10.2*). Lupins have been the subject of study since the early work by Lyne Watt (1929). Bitter varieties, which have toxic levels of alkaloids, were tried as green manure crops. At Serere (U), Horrell (1958b) reported that lupins wilted badly even during short dry spells, were susceptible to nematodes and termites, and appeared to be of no use at lower altitudes. In the 1950s low-alkaloid varieties (sweet lupins) became available. The latter have drawn attention again in recent years for grain production as possible substitutes for soyabean and cotton seed cake. A comprehensive review of the early work done with lupins in both Kenya and southern Tanzania was prepared by Bauer (*op. cit.*), who proposed their use after breaking a pasture or as a nurse crop as had been tried by Poultney (1963b). Considerable work has since been conducted at Kitale on variety testing, inoculation- and fertilizer needs and animal performance. The Ultra variety gave the highest yield, while there was no need for inoculation or for P-fertilizer (Muyekho, 1988). Unfortunately, high proportions of lupin meal in rations led to lower intake and lower milk yields (Mukisira, 1988). Recent experience at the Kitale Station has shown that wilting due to fungi can be severe.

18.5.4. Vetches

Purple vetch (*Vicia benghalensis*) is a vigorous annual legume with very low requirements of soil fertility and fertilizer. Vetch was tried as a green-manure crop (*5.2.1*; Ball, 1937). In the problem areas of the Kinangop it was of some value in wheat rotations (J. Nightingale, private communication). It was also used in mixture with oats (*16.3*) but palatibility is low, at least in the wet season. Vetches are regarded by some as weeds. Purple vetch seed persists in the soil for many years and volunteers wherever it has been sown before. In one field at Kitale vetches have been known to re-establish themselves from seed with every ploughing, ever since its planting in the early 1960s, even after a break under pasture of 5 years. Herbicides are effective but, once buried, seeds survive in the soil for many years. *Vicia villosa* was reported to do well above 1850 m in Ethiopia (Ibrahim & Orodho, 1983).

18.5.5. Hubam clover

Hubam is one of the annual types of sweet clovers (*Melilotus alba*) and attained some success in the medium altitude range in East Africa, where it grows with vigour and ease and played some role as a potential crop for green manure (Ball, *op. cit.*). At Tengeru (T) it was outstanding in early development and yield (Naveh & Anderson, 1967). The typical odour of the sweet clovers comes from the coumarin which makes it not only unpalatable but also toxic during heating and spoilage in hay and which causes serious bleeding from slight wounds or internal haemorrhages. This is called sweet-clover bleeding disease. It is no wonder that this has held farmers back from growing what would otherwise have been a useful legume (Strange, 1958a; Horrell, 1958b; 1963).

18.5.6. Leucaena leucocephala and other leguminous shrubs

Leucaena is a shrub high in protein, palatability and, once the upper canopy gets beyond the reach of livestock, in persistence. This may explain the renewed interest of scientific circles in shrub legumes since it was recognized that herbaceous legumes lacked persistence under grazing (*10.10*).

It should be noted, however, that although Leucaena had been well known in the region since its introduction in 1915 (Thomas, 1940c), *e.g.* at Machakos (Maher, 1939), very little use has been made of this plant, in spite of the traditional interest in shrubs for browsing and lopping (see below). In some parts of Tanzania, near Amani and Morogoro, it was considered a nuisance (Parry, 1954) and in Zanzibar a breeding site for the tse-tsefly (Nat. Acad. Sci. Washington, 1977). Also in Uganda Leucaena tended to get out of hand (Horrell, 1963). In old coffee plantations in Kivu, in which it functioned as a shade tree (van den Abeele & Vandenput, 1951) but which have since been abandoned to grazing, Leucaena is still widespread but is not afforded special attention by graziers or foresters (personal observation, 1992).

It has been customary to blame the high mimosine contents for the lack of popularity of Leucaena. This uncommon amino acid comprises 3–5% of the protein. In livestock, mimosine causes hair to fall out and general ill-health when eaten as the sole feed. This was first observed in material introduced from Java in 1941 into Kivu (Damseaux, 1956; Compère, 1959). Since then reports have appeared of reduction in calving and of increased abortion, possibly due to mimosine metabolites. The toxic effect may persist after Leucaena feeding ceases. Attempts at breeding for low mimosine resulted in no success (Clements, 1989). It would be prudent, however, not to overemphasize the toxicity aspects, because fodder legumes should not be grazed or fed in large quantities, but rather as a supplement.

Its possible role in alley-farming has been tested at the Kenya coast (Jama *et al.*, 1986). Combinations of Leucaena and Elephantgrass have shown promise but, where crops are already marginal, moisture and nutrient stresses are accentuated by the presence of Leucaena (see also *12.8*). Maize growth is usually depressed in the first 5 rows neighbouring a Leucaena hedge, whether pruned or not. At Kiboko (K; Zone V) Leucaena had to be grown under irrigation (Child *et al.*, 1982).

Leucaena's effect on soil structure was studied by Pereira *et al.* (1954) in comparison with other legumes, grasses and crops (Table 5.4). Although its vigorous root system improved soil aeration, its lack of soil cover made it hard to resist the impact of heavy rainfall, so that there was no improvement in the rainfall infiltration rate in the soil.

Leucaena drew the attention of the European feedmilling industry in the 1970s. Leucaena meal contains carotene and xanthophyll, which could replace synthetic products in the event of them being prohibited in poultry rations to improve yolk colour in eggs. A pilot project was set up in southern

Malawi (Savory *et al.*, 1980). Very high plant populations, up to 200.000 plants/ha were used. Harvesting was done by leaf stripping. When the branches were cut and left to dry on the stumps leaf losses were considerable. Sun-drying reduced the % carotene. The yield achieved was far below target and the project was abandoned. All dietary levels of Leucaena induced a severe depression of growth in young chicks (d'Mello & Thomas, 1978).

Leucaena imparts a flavour to the milk of cows, which is an important aspect when its use is to be promoted among smallscale dairy farmers.

In spite of these disadvantages, the major drawback to a more widespread planting of Leucaena seems to have been the extremely slow rate of establishment and subsequent development, especially at altitudes above 1500 m and on less fertile, acid soils (*10.2*). Particular attention has to be given to seed scarification, nutrition and inoculation (Table 10.1), and to possible damage by feral animals. Another drawback is insect damage, of termites and more recently in many countries of psyllids (*Heteropsylla cubana*), which seriously affect Leucaena. Slow establishment should not be a problem where the shrub is grown on terrace banks, along boundary perimeters or as a hedge or shade tree on the homestead, since it can be transplanted from seedlings. For these purposes Leucaena should definitely have a place in the humid lowlands (van Rensburg, 1965; 1967). However, although Leucaena is widely publicized as a multi-purpose tree and a favourite research subject, it does not seem to have much appeal as a fodder crop to the farmer abroad either. Leucaena seed production in Australia, where the plant has drawn more scientific interest than anywhere else, stands at only 5 ton/year (Walker & Weston, 1990).

A species that has shown for decades such a high potential for protein production but is not being utilized to any great extent for animal production, must have obvious limitations that prevent this potential being fulfilled. Unless the advantages claimed for Leucaena are documented in terms of increased animal production and economic gain there will be little incentive for a wider utilization of the plant for livestock.

Leucaena may be unique in its combination of properties but alternative species have also been tested for a long time: Prosopis (Maher, 1939), Gliricidia and especially Sesbania. This matter is presently being pursued again in various parts of Africa (Onim & Dzowela, 1988), However, the poor rate of adoption has already become of major concern (Frances & Atta-Krah, 1989). *Sesbania sesban* is a useful natural browse in Kenya (Dougall & Bogdan, 1958a) but when established artificially it was found to give much lower yields of edible DM than Leucaena (Wandera *et al.*, 1991). *S. grandiflora* was even slower growing and the least productive. It should be noted that while farmers may show little interest in shrubs for fodder purposes, they may plant others as living hedges. Farmers in Kivu commonly establish shrubs such as *Erythrina abyssinica* to separate their paddocks.

It must be recalled that shrubs and trees and their use as browse or fodder had been an early subject of study in East Africa by many investigators (Maher, 1939; Thomas, 1940c; French, 1949; van Rensburg, 1947b; 1955a; Whyte, 1947; Whyte *et al.*, 1953; Dougall & Bogdan, 1958a; Payne & MacFarlane, 1963; Wilson & Bredon, 1963; Dougall *et al.*, 1964; Glover *et al.*, 1966; Lawton, 1967; McKay & Frandsen, 1969). The subject has been given renewed attention (le Houérou, 1980; IBPGR, 1984; Otsyina & McKell, 1985; Young, 1989). As French (1949) remarked: "For grassland investigators it is a humbling fact that probably more of the world's animal population feeds on shrubs and trees than on true grass or grass-legume pastures". The importance at that time attached to forestry in farming was borne out by the appropriate title given to the *East African Agricultural and Forestry Journal.*

Where they occur naturally, valuable trees and shrubs are treasured for browsing and lopping, but actual plantings are still largely experimental. Apart from a very slow start, many species are uncapable of withstanding regular cutting. Woody legumes can be high in N but low in digestibility and are often unpalatable. The capacity to supply protein would appear to be discounted due to low digestibility (Thomas & Schulze-Kraft, 1990). Based on work in the Sudan (Harrison & Jackson, 1958) it was concluded that although leguminous species may be numerous there is little special about these plants that makes them better browse or fodder than non-leguminous species (M.H. Harrison, private communication, 1969).

18.5.7. Additional legumes in East African literature

Serradella (*Ornithopus sativus*) was reported on by Dougall (1954d), Bogdan (1955) and Strange (1958a); Sulla (*Hedysarum coronarium*) by Horrell (1963) and Strange (1958a, b). The same publications of Horrell, Bogdan, Strange and van Rensburg describe a range of other legumes (*10.8.12*).

18.6. Miscellaneous fodder crops

Herbs are acclaimed at times on account of their mineral content or other virtues.

18.6.1. Chichory

Chichory (*Cichorium intybus*) is a herb which made quite an impression in the late 1950s (Strange, 1961b) on account of its high nutritive value and drought resistance and was regarded as the most important herb and superior to Ribgrass or Plantain (*Plantago lanceolata*) or Burnet (*Poterium sanguisorba*) (*2.2.2*; Anon., 1959). It was at times included in grass seed mixtures.

18.6.2. Russian Comfrey

Russian Comfrey (*Symphytum peregrinum*) is planted from tubers and can be harvested for both leaves and tubers. Van Rensburg (1956) and Strange (1959c) confirmed that Russian Comfrey was high in % ash (20% or more) and very low (10–13%) in crude fibre (Table 18.3) and, therefore, suitable for feeding pigs and poultry. Russian Comfrey did not respond to an application of gypsum, in contrast with lucerne (Table 17.3).

Table 18.3. Analysis of lucerne and Russian Comfrey herbage (after Strange, 1959c).

	August 1955		August 1956	
	Lucerne	Comfrey	Lucerne	Comfrey
Ash	10.4	19.7	9.6	22.1
Crude protein	30.5	23.5	20.2	17.3
Ether extract	3.4	4.1	2.6	2.5
Crude fibre	27.8	9.9	19.9	12.6
Carbohydrate	28.0	42.9	47.7	45.5
Phosphorus	0.29	0.34	0.33	0.47
Calcium	1.43	1.35	0.96	1.56
Silica	–	–	0.67	5.78

Strange further observed that Russian Comfrey was very susceptible to nematodes and was also weakened by the plant parasite *Alectra asperrima*. Owing to the hairiness of the leaves, soil contamination was considerably greater than on lucerne, as illustrated in Table 18.4.

Even without soil, Russian Comfrey had a silica content of nearly 6% which largely accounted for the high mineral content (Table 18.3) for which Russian Comfrey had been favoured. Comfrey further had a low yield and % DM. Work on Russian Comfrey was resumed at Uyole (T) in 1978, and at Kitale (Wanjala, 1981) together with Canna.

Table 18.4. Soil contamination on lucerne and Russian Comfrey (after Strange, 1959c).

Species	% Soil in herbage	% Silica (without soil)
Lucerne	1.72	0.67
Russian Comfrey	13.51	5.78

18.6.3. Canna

Canna (*Canna edulis*) can be grown over a wider altitude range than kale or beets, but the fertility requirements are similar. Planting is by tubers. Once established, Canna is almost impossible to eradicate. Yields can be very high (Pereira *et al.*, 1967). Canna tubers were at one time popular as a feed for pigs, and leaves were cut for cattle to provide some succulent feed during the dry season. Canna leaves were preferred by cattle above leaves of kale, Russian Comfrey and of sweet potatoes at Kitale (Wanjala, 1981). The nutritive value of Canna was described by French (1938a; 1940). A report on its use in Burundi was given by Scaillet (1965).

18.6.4. Amaranthus spp.

Interest was shown in the early 1970s at Makerere (U) in *Amaranthus* spp. (Dodo) as a "specialty" for dairy cows and young calves (Henderlong, 1973; Mugerwa & Bwabye, 1974). The yields and % crude protein were well above of what is normally encountered but curing the material before feeding was problematic. Amaranthus is also reported to have high levels of oxalic acid (*6.7.3*; *12.7.4*).

18.7. Summary

Fodder legumes which have been grown in East Africa include: Lablab beans; velvet beans; lupins; vetch; Hubam clover and Leucaena. Fodder crops which have played a role in farming in East Africa and which are described in some detail are: sweet potato; fodder rape; kale; mangolds; chichory; Russian comfrey; Canna and Amaranthus.

CHAPTER 19

Hay and silage

19.1. Hay and silage of grasses

Throughout this book emphasis has been laid on mowing, ancillary to grazing, in grassland husbandry to sustain productivity and to match supply and demand more evenly over time:

- Mowing can be designed in frequency and intensity so that periods of peak growth are shifted (*7.2.3*; *12.5*).
- Regrowth is accelerated after a long period of undisturbed growth (*13.2*).
- Mowing can be used as a tool to put the brake on herbage moving too fast to maturity (*15.3*; *16.1*).
- Mowing prevents losses due to decay and insect damage (*4.3.4*).
- Mowing helps in controlling bush and weeds (*4.3.4*).
- Mowing primary growth in the 1st year aids in establishment of the sward (*7.1*; *7.2.3*).
- Many species, *e.g.* Common Guineagrass, fodder-type grasses and legumes lend themselves better to (infrequent) cutting than to grazing.
- Mowing and grazing are mutually beneficial (Table 7.9).

Although more hay and silage are made than is commonly estimated, much more should be done to encourage their use (Bijman & Davis, 1973). A number of arguments can be adduced which have held the farmer back. Hay and silage presuppose a good deal of planning. Initiatives have to be taken months ahead of when they might bear fruit. Another objection usually raised against conservation is that the farmer doesn't have the machinery: mower, baler and trailer. It is unfortunate that hay and silage have become associated with mechanized largescale farming and with government stations. However, the hand-cutting of grass for stall-feeding has been widely promoted. Proponents of stall-feeding should naturally be major promoters of herbage conservation also. Neither can it be argued that a tradition of cutting and curing grass does not exist; the cutting of grass for thatching is too common a skill. On the other hand, encouragement from research circles has usually been small or absent. Feeding conserved grass was considered ineffective and uneconomic or it was proposed to grow fodder crops. Also, mowing surplus growth was considered less conducive to legume persistence than leaving the sward undisturbed (*10.4*).

The main impediment seems to be the farmer's optimism to the effect that "the next dry season will not be so dry". And indeed, in many of the subhumid regions one does run into the occasional dry season with enough rain to keep the pastures green. In fact, in semiarid areas such as Machakos and Baringo droughts are more consistent and this has forced farmers to adopt haymaking. Measures against losses due to termite damage are taken also (Pereira, 1959). Termites are equally a problem in "standing hay" and in Elephantgrass crops weakened by drought (*4.3.4*; *11.9*).

A practical objection is that "summer's bounty" is rarely there to take, as most of the pasture is chronically overgrazed. Under good husbandry pastures should give a surplus within 2–3 months after the start of the rains (Fig. 6.1), well in excess of the requirements of the grazing animals. It is this flush which signifies the pasture's recovery from the dry season. Growth is more vigorous and more nutritious (Fig. 6.5) than later in the year, even at the same temperature and rainfall. By contrast, when pasture has been overgrazed, the early-season flush will not be much in evidence.

Some species are more amenable to hay or silage on account of their inherently high % DM and thin stems, *e.g.* Rhodesgrass and Solai Coloured Guinea (Table 6.3). Stargrass has the highest % DM but stems are tough. At the other extreme, Elephantgrass not only has a very low % DM, especially in the stems (Fig. 11.1) but the stems are too thick to cure readily.

19.1.1. Haymaking

Haymaking is easier and cheaper, whether in hand labour or mechanized, but more weather dependent than silage. The best opportunity is at the beginning of the dry season in the year of sowing, of newly established pastures (*7.1.11*; *7.2.3*). First-year swards are invariably more vigorous, while they can be induced to be in the right stage for haymaking by the beginning of the dry season when sown 3–4 months before the expected end of the rains. In older pasture, heavy hay crops of relatively young grass cannot be expected unless N-fertilizer is applied. N should be applied to young grass. Late-applied N increases % crude protein of hay (Owen, 1964) but is not as effective (*7.2.4*).

In bi-modal rainfall areas the time of surplus grass coincides with the end of the rains. In uni-modal rainfall areas, except in the most humid climates, rainfall is rarely continuous. For instance at Kitale it is quite feasible to make effective use of the frequent 3–5 day dry-weather periods which occur in June. In fact, these seem more reliable than those encountered in summer in temperate countries where haymaking is so well entrenched. Rapid curing is important because hay which is spoilt by rain retards the regrowth of the sward over which it lies.

At Kitale it has been found that 2 dry, sunny days are adequate to make Rhodesgrass hay at a yield level of 4 ton/ha DM. On the 1st day the grass is mown in the late morning when the dew has dried off. The hay is turned over in the early afternoon. By evening time the hay is collected into windrows or large heaps so as to protect it from heavy dew and a possible shower of rain. On the morning of the 2nd day the windrows are spread out again followed by a turning-over in the early afternoon. By evening time the hay can be gathered and stacked. Operations should be planned so that any given area can be cut, handled and stacked in 2 days. Light hay crops can often be cut and carted on the same day, in sunny and windy weather, whereas a heavy crop may require a total of 3 days. The principle of haymaking is to cure the grass as rapidly as possible in order to stop respiration. This requires regular turning over, to begin within 2 hours of mowing. Hay crushing did not result in shorter curing because it was found to take up more moisture at night.

Mechanized haymaking techniques in Kenya have been described by Henderson & Preston (1959). However, the same principles apply in handmade hay. The use of tripods has been suggested, but these have not been adopted and are not really necessary. Windrows or heaps opened up when conditions are dry are also effective. Haymaking by hand can be speeded up with the use of scythes, although it has proved difficult to introduce alien tools. The reaping hook ("ringa") is effective but slow.

The nutritive value of hay as determined by laboratory methods is usually on the low side, as hay is often made at a mature stage. It is baffling, however, to see how eagerly low-quality hay is eaten also when there is other feed on offer. For instance, hay made of herbage residue after seed harvesting is well eaten, in fact better than when the animals are turned into the same field for grazing. An extreme example of unexpected behaviour of cattle is their appetite for old grass-thatch roofs, when within reach. At the beginning of the rains when the pasture grass is young and lush, supplementation with hay is most helpful in reducing scouring and "green flush losses" (*4.3.4*).

19.1.2. Grass silage

More than one-third of Henderson & Preston's (*op. cit.*) "Fodder Farming in Kenya" is devoted to silage making. The picture in this booklet illustrating a span of oxen on a grass silage clamp shows what can be achieved with low technology; this in contrast with the heavily mechanized maize ensilage of the beef feedlot systems promoted in the early 1970s (*19.2*). When hay or silage is identified with heavy expenditure, its feasibility will not improve in the eyes of the smallholder.

Pit silos, lined with bricks and covered with thatched roof have been used since the 1940s, *e.g.* at Mukumu Mission Farm (*4.2.2.2*). No molasses or any other additive is applied, neither is the grass chopped. Instead, compaction is ensured by having deep pits with heavy rocks piled on top. The silage preserves well for a number of years and is well eaten in the dry season.

In a Kitale experiment, Rhodesgrass silage was fed in the dry season *ad libitum* to 7-months old Friesian steers with or without concentrate supplement (Thomas, 1977a). When the steers were subsequently turned into pasture in the wet season those which had been fed on grass silage alone during the dry season gained more than those which had received concentrate supplements, probably because of compensatory growth (French & Ledger, 1957). The liveweight gain of the steers given grass silage was above maintenance (0.10 kg/head/day) in the dry season. The Rhodesgrass used was cut at a relatively mature stage of growth (8 weeks), wilted for 6 hours and chopped. The silage was well preserved as indicated by its pH of 4.5. Livestock production on Kitale grass silage has been found to be similar to that achieved with temperate grass in Europe (Thomas, unpublished). It was concluded that wilting was essential for adequate preservation in the absence of an additive (see also Snijders & Wouters, 1990).

The general experience at Kitale suggests that, apart from the pit silo, excavated slope trenches are to be preferred rather than clamps. The latter gave spoilage penetrating more than 50 cm into the sides unless coated with polythene sheeting. Compaction is also easier in pits and trenches.

Good compaction is important in tropical grasses as their DM-content is low and their packing is loose. This applies in particular to coarse grasses such as Elephantgrass. In Zambia, Davies (1963) found that with a consolidation pressure of just over 30 g/cm^2 it took the silage made from this grass 10 days to settle to three-quarters of the original height. The coarser and older the material, the more important it will be to chop the material finely and to consolidate it as heavily as possible. Wilting is advantageous the younger the grass material is. When the material to be ensiled is too dry, however, temperatures rise too high. Such material is springy and does not pack well.

Addition of molasses is rarely necessary and can be detrimental unless applied in quantities of up to 80 kg/ton (Catchpoole & Henzell, 1971). These quantities are needed to produce lactic-acid silage. The acetic-acid silage is more typical for tropical herbage and can be stable although the pH may be as high as 5. Recently developed commercial enzyme cultures appear to hold much promise.

An important saving by conserving surplus grass growth is that no extra land needs to be set aside for the growing of fodder crops. For pasture, instead of being subjected to grazing only, a rest followed by a hay or silage cut at a more advanced stage of growth is beneficial. Hay or silage making is, therefore, not only a cheap source of fodder for the dry season, but also a useful tool in pasture husbandry (see above). Numerous articles of French (1935; 1938c; 1939; 1950; 1956) refer to the value of grass hay and silage. French (1950) strongly argued against "standing hay", stating that making hay or silage at the appropriate stage ensured a better nutritive value.

19.2. Maize silage

A wide range of crops can be ensiled, but the major ones are maize and Sorghum (Thomas, 1961; Sheldrick, 1975; van Arkel, 1982; Dixon, 1982). In general the following applies:

1. Late-maturing maize and Sorghum varieties are capable of accumulating growth over 6–7 months at an almost linear rate. Very high yields can be harvested in one single operation.
2. Maize silage is easy to make. If protein, minerals and some vitamins are added maize silage is a most valuable and reliable source of fodder.
3. Land has to be set aside specifically to grow these crops.
4. Maize and Sorghum are East Africa's staple food crops. A traditional aversion prevails against "burying food grain".

The economics of maize or Sorghum silage must be viewed against: [1] animal output; [2] returns from maize or Sorghum as grain crops; [3] alternatives such as hay and silage made of surplus grass; [4] other fodder crops.

Miles *et al.* (1972) put the standard costs for Kenya feedlot operations of maize silage of 28% DM at Kshs 48/- per ton which was about the official price farmers received for 100 kg of maize grain in 1972. Miles concluded, therefore, that feeding maize grain was as advantageous. If the benefit of stover utilization after the maize grain harvest is included, the balance tips even further in favour of maize grain.

In a large-scale forage system exercise at Kitale very poor liveweight gains were achieved with maize silage compared with routinely grazed Rhodesgrass (Sheldrick & Goldson, 1978). Contrary to what is commonly believed, the nutritive value of maize is not necessarily higher than of tropical grasses. Following the work of Deinum (1966) it is now well understood that the chemical composition of a herbage is influenced by prevailing temperature and sunlight intensity. High growth temperatures consistently reduce digestibility of herbage, be it tropical or temperate, grass or legume, leaf or stem (Dirven & Deinum, 1977). Maize grown under high temperature regimes near the equator has a lower nutritive value than when grown under temperate conditions.

A second important factor reducing the energy value of maize in East Africa is the low crop index of the available varieties and hybrids. In temperate countries the crop index is often above 55% but in East African material the grain does not comprise more than 40% of the total DM-yield. This was confirmed by Cooper & Law (1977) and is illustrated in Fig. 19.1. The same study also refuted earlier opinion that maize for silage can safely be sown late, unlike maize sown for grain. Not only is the yield reduced by late sowing, but also the crop index is then not more than 25%. This late-planting effect is mainly due to the decrease in soil temperature as the rainy season proceeds (Fig. 3.4).

Fig. 19.2 refers to the work done at Kitale by Sheldrick (1975) on the optimum cutting date for silage maize made from hybrids of the "600" series. The peak yields of DM coincided with those of maximum yield of digestible matter and occurred some 6.5 months after sowing when the % DM was 30–35. The % crude protein was typically low, 6–7%, at that stage but the % DODM (digestible organic matter in the dry matter) was not much below 65. Sheldrick concluded that DM-yields of 16 ton/ha should be feasible in commercial practice. The commercial yields achieved in the feedlots at Lanet were 10 ton DM on average.

Maize silage becomes of interest economically only when inputs such as fertilized pastures – complemented with grass hay, grass silage, Sorghum or oats as break crops–have been exploited to their fullest. For instance, on the Mukumu Mission Farm (*4.2.2.2*) 10% of the acreage is set aside for maize silage.

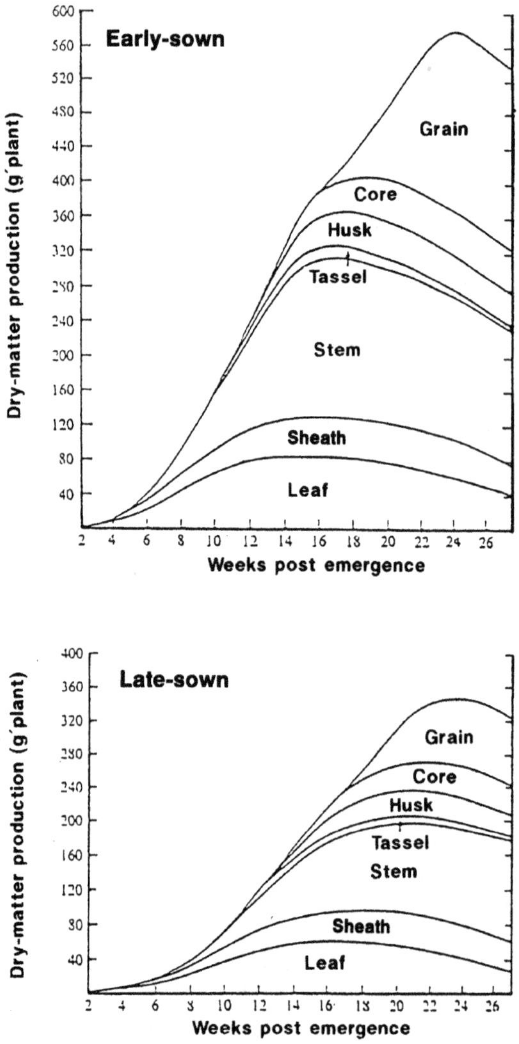

Fig. 19.1. The accumulation and partition of dry matter in maize at Kitale (Cooper & Law, 1977).

Fig. 19.2 shows also why the practice of feeding green maize at or just after tasseling (*13.4.1*) is inefficient. Green maize is cut when it has completed two-third of the period required to achieve maximum yield, yet it then yields only one-half of what would have been possible when harvested mature.

It has been suggested that varieties and hybrids should be bred specifically for the purpose of

Fig. 19.2. Maize crop profile at Kitale (after Sheldrick, 1975).

silage, *i.e.* for the whole maize plant, in terms of yield and nutritive value. However, experience in regions where maize is grown both for grain and for silage suggests that the commercially most successful silage varieties or hybrids are those primarily developed and selected for grain. The development of silage-specific inbred lines and the breeding for components of nutritive value will continue to meet with little commercial success, unless such lines or hybrids are competitive in terms of grain potential at the same time. There is an additional aspect of overriding importance why breeding for silage in or for a specific region is unlikely to be competitive. Silage maize can do with a shorter growing season than grain maize, so that later-maturity grain material from elsewhere can be adopted. The best silage varieties and hybrids are usually those which are the best for grain in adjacent zones with more "heat units", *i.e.* a more favourable zone with a longer growing period. Thus, the best silage hybrids at Kitale (Zone III) are the grain hybrids recommended for Kakamega (Zone II).

19.3. Summary

Hay and silage made off surplus grass growth in the wet season is the cheapest source of fodder for the dry season and is conducive to good grassland husbandry; no special fields, as with fodder crops, need to be set aside. Silage of maize is simple to make but is expensive.

19.4. General conclusion on fodders

Fodder crops are in essence complementary and are grown for special purposes, *e.g.* stall-feeding or for the dry season. The most commonly cultivated grass on small mixed-farms engaged in stall-feeding in East Africa is Elephantgrass, although the bulk of what is fed consists of other feeds. Elephantgrass yields well in warm and humid climates, but is disappointing in subhumid zones and does not survive in semiarid areas. When zero-grazed, huge quantities of potassium, phosphate and other minerals are removed, which should be returned with the manure if soil fertility is to be maintained. N-deficiency becomes rapidly acute and can only be remedied by N-fertilizer because animal manure is not effective or not available in sufficient quantity.

Grass Sorghum and oats function as main season fodder or grazing crops, where normal sown pasture is less feasible, but also as fodder catch crops in more amenable zones, following broken pasture, to provide fodder in the dry season. Lucerne is an ideal fodder legume there where it is well adapted. Lablab beans could take over this role elsewhere. Apart from Brassica, there are very few other fodder crops that can play a role that cannot be achieved more simply by other means. The cheapest source of fodder for the dry season is provided by hay and silage made off peak growth of pasture in the wet season.

The general farm type, whether large or small, in East Africa is that of mixed-farming. Cattle and crops are the two pillars of the farm economy and are mutually beneficial. Under such a system a farmer has many alternatives to see his animals, his dairy in particular, through the dry season. One of those are the bye-products of cropping: reject grain, straw and even the grass weeds of the arable phase.

Management should be such as to make the herbage supply match demand at the most critical times of the year. Maximum benefit must be sought from sown and grazed grasses, since it is by this method that soil fertility is sustained. The following alternatives have been discussed:

1. First year's sown pastures produce vigorously deep into the dry season and are ideal for an early crop of hay (7).
2. Grass Sorghum or oats, sown as break crops into mid-season ploughed pasture, to be utilized in the dry season (*15*; *16*).
3. Grass hay or silage taken off peak growth (7).
4. N-fertilizer to extend late-season grazing (7).
5. Elephantgrass managed specifically to maximize leaf yield in the dry season (*13*).

The ideal farm will seek a optimum mix of these alternatives within the objective of integrating pasture and crop farming. The pivot is the grazed pasture itself. Grass is mown and conserved when producing in excess and N-fertilized when in short supply. An important N-source, however, is soil-N and this is available by alternating crops and pastures in short rotation.

Laus Deo!

Glossary (Main Species)

For the authorities of botanical names of other grasses referred to in the text, see the Flora of Tropical East Africa by E.S. Clayton and co-workers (1970; 1974; 1982). In order to avoid ambiguity all common names begin with capital letters: Rhodes-, but also Elephantgrass.

Bothriochloa insculpta (A. Rich.) A. Camus	
Brachiaria decumbens Stapf (syn. *B. ruziziensis*)	Signal
Cenchrus ciliaris L.	Buffel
Chloris gayana Kunth	Rhodes
Cymbopogon nardus (L) Rendle	Citronella
Cynodon aethiopicus Clayton & Harlan	Giant Star
Cynodon nlemfuensis Vanderyst	Common Star
Cynodon plectostachyus (K. Schum.) Pilg.	Naivasha Star
Dactyloctenium geminatum Hack.	Crowfoot
Digitaria abyssinica (A. Rich.) Stapf (syn. *D. scalarum; D. vestita*)	African Couch; Chiendent
Digitaria macroblephara (Hack.) Stapf	
Digitaria milanjiana (Rendle) Stapf (syn. *Digitaria mombasana*)	
Digitaria velutina (Forssk.) P. Beauv.	
Eleusine coracana L.	Finger millet; Wimbi
Eleusine indica (L.) Gaertn.	Wimbi
Eleusine jaegeri Pilg.	Manyatta
Entolasia imbricata Stapf	Bungoma
Eragrostis racemosa (Thunb.) Steud. (syn. *E. boehmii*)	
Eragrostis superba Peyr.	Masai Love
Eragrostis tenuifolia (A. Rich.) Steud.	
Exotheca abyssinica (A. Rich.) Anderss.	
Heteropogon contortus (L.) Roem. & Schult.	Spear
Hyparrhenia cymbaria (L.) Stapf	
Hyparrhenia filipendula (Hochst.) Stapf	
Hyparrhenia rufa (Nees) Stapf	
Hyperthelia dissoluta (Steud.) W.D. Clayton	
Imperata cylindrica (L.) Raeuschel	Imperata
Loudetia kagerensis (K. Schum.) Hutch.	Russet
Loudetia simplex (Nees) C.E. Hubbard	Russet
Melinis minutiflora P. Beauv.	Molasses
Panicum coloratum L.	Coloured Guinea
Panicum maximum Jacq.	Common Guinea
Paspalum scrobiculatum L. (syn. *P. commersonii*)	Paspalum
Pennisetum glaucum (L) R.Br.	Pearl millet
Pennisetum clandestinum Chiov.	Kikuyu
Pennisetum purpureum Schumach	Elephant; Napier
Pennisetum mezianum Leeke	
Pennisetum sphacelatum (Nees) Th. Dur & Schinz (syn. *P. schimperi*)	Wire
Pennisetum stramineum Peter	
Pennisetum unisetum (Nees) Benth. (syn. *Beckeropsis uniseta*)	
Rhynchelytrum repens (Willd.) C.E. Hubbard	
Setaria sphacelata (Schumach.) Moss var. *sericea* (Stapf) W.D. Clayton	Setaria

Setaria sphacelata var. *splendida* (syn. *S. splendida*)	Giant Setaria
Sorghum × *almum* Parodi	Columbusgrass
Sorghum bicolor (L.) Moench (syn. *vulgare*)	Sorghum
Sorghum × *drummondii* (Steud.) illsp. & Chase (syn. *S. sudanense*)	Sudangrass
Sporobolus pyramidalis P. Beauv.	
Themeda triandra Forssk	Themeda; Red Oat

Exotic grasses

Avena sativa L.	Oats
Lolium perenne L.	Perennial Rye
Dactylis glomerata L.	Cocksfoot
Festuca arundinacea Schreb.	Tall Fescue
Festuca rubra L.	Red Fescue
Paspalum dilatatum Poir.	Dallis
Paspalum notatum Flügge	Bahia
Tripsacum andersonii *J.R. Gray* (syn. *T. laxum*)	Guatemala

Legumes

Centrosema pubescens Benth.	Centro
Desmodium intortum (Mill.) Urb.	Greenleaf
Desmodium uncinatum (Jacq.) DC.	Silverleaf
Lablab purpureus (L.) Sweet	Lablab
Leucaena leucocephala (Lam.) De Wit	Leucaena
Lotononis bainesii Baker	Lotononis
Macroptilium atropurpureum (DC.) Urb.	Siratro
Medicago sativa L.	Lucerne
Neonotonia wightii (Wight & Arnott) Lackay (syn. *Glycine wightii*)	Glycine
Pueraria phaseoloides (Roxb.) Benth.	Puero; Kudzu
Stylosanthes fruticosa (Retz.) Alston	
Stylosanthes guianensis (Aubl.) Sw. (syn. *S. gracilis*)	Stylo
Stylosanthes humilis H.B.K.	Townsville Stylo
Trifolium repens L.	White clover
Trifolium rueppellianum Fresen	Rueppels clover
Trifolium semipilosum Fresen	Kenya white clover

Abbreviations and Units

a.i.	Active ingredient
Ann Rep	Annual Report
CAB	Commonwealth Agricultural Bureaus, Farnham Royal
CIAT	Centro Internacional de Agricultura Tropical, Colombia
CSIRO	Commonwealth Scientific and Industrial Research Organization of Australia
D-value	% *In-vitro* digestible organic matter
DM	Dry matter
EARCC	East African Range Classification Committee
FAO	Food and Agriculture Organization, Rome
ICRAF	International Centre for Research in Agroforestry
ILCA	International Livestock Centre of Africa
INEAC	Institut National d'Etude Agronomique du Congo
IHE	Initial Head Emergence; 5–10 heads emerged per m^2
(K)	Kenya
KARI	Kenya Agricultural Research Institute
KDA	Kenya Department of Agriculture
LWG	Liveweight gain
L.U.	Livestock Unit (450 kg liveweight)
N,P,K	Nitrogen, Phosphorus and Potassium
N°; S°	Northern, Southern latitude
OHD	Optimum Harvest Date
OM	Organic Matter
PGS	Pure Germinating Seed
(T)	Tanzania
(U)	Uganda
UDA	Uganda Department of Agriculture
UNEP	United Nations Environment Programme

References

Abate, A., H. Kayonge-Male & C.N. Karue, 1981. Dry matter, protein, energy and fibre intake by dairy heifers grazing a Rhodes (*Chloris gayana*) pasture. *Anim. Feed Sci. Techn.* 6: 15–26.

Abate, A., H. Kayonge-Male, A.N. Abate & J.D. Wachira, 1984. Chemical composition, digestibility and intake of Kenyan feedstuffs by ruminants – a review. *Nutr. Abstr. Revs* 54: 1–13.

Abeele, M. van den & R. Vandenput, 1951. De Voornaamste Cultures van Belgisch-Congo (Les Principales Cultures du Congo Belge). Brussel.

Acland, J.D., 1971. East African Crops. Longman.

Addison, K.B., 1956. The effect of various cultural and manurial treatments on Napier fodder. I. Effect of nitrogen and phosphate on the production of Napier. *Rhod. agric. J.* 53: 491–506.

Addison, K.B., 1958a. The effect of various cultural and manurial treatments on Napier fodder. III. The effect of nitrogen and lime on the production of Napier fodder. *Rhod. agric. J.* 55: 33–39.

Addison, K.B., 1958b. The effect of various cultural and manurial treatments on Napier fodder. IV. The effect of deep cultivations on the production of Napier fodder. *Rhod. agric. J.* 55: 187–191.

Addy, B.L. & D. Thomas, 1977. Intensive fattening of beef cattle on Rhodes grass pastures on the Lilongwe plain, Malawi. *Trop. Anim. Hlth Prod.* 9: 99–106.

Agricultural Research Conference, Nairobi, 1947. *East Afr. agric. J.* 13: 65–87.

Agyemang, K., L.P. Nkhonjera, M.H. Butterworth & J. McIntire, 1988. Productivity and profitability of cattle stall-fed for beef on smallholder farms in Malawi. *ILCA Bull.* 32.

Ahn, P.M., 1974. Some observations on basic and applied research in shifting cultivation. In: Shifting cultivation and soil conservation in Africa. *FAO Soils Bull.* 24.

Allan, A.Y., 1972. The influence of agronomic factors on maize yield in Western Kenya, with special reference to time of planting. Thesis Makerere.

Allan, A.Y., 1976. The effects of sulphur on maize yields in western Kenya. *East Afr. agric. For. J.* 41: 313–322.

Anderson, G.D., 1965. Estimates of the potential productivity of some pastures in the Mbulu Highlands of Tanganyika by application of fertilizers. *East Afr. agric. For. J.* 30: 206–218.

Anderson, G.D., 1968a. Promising pasture plants for northern Tanzania. VI. Practical suggestions for pasture improvement. *East Afr. agric. For. J.* 34: 106–116.

Anderson, G.D., 1968b. Effects of fertilizers on botanical composition and productivity of pasture on the sandy soils of the Tanganyika coast. *East Afr. agric. For. J.* 34: 207–216.

Anderson, G.D., B.G. Houston & P.J. Northwood, 1966. Effects of soil, cultivation history and weather on response of wheat to fertilizers in Northern Tanzania. *Exp. Agric.* 2: 183–200.

Anderson, G.D. & Z. Naveh, 1965. The influence of climate and soil on the productivity of certain promising grasses and legumes in northern Tanzania. *Afr. Soils* 10: 241–255.

Anderson, G.D. & Z. Naveh, 1968. Promising pasture plants for northern Tanzania. V. Overall comparisons of promising plants. *East Afr. agric. For. J.* 34: 84–105.

Anindo, D.O. & H.L. Potter, 1986. Milk production from Napiergrass (*Pennisetum purpureum*) in a zero-grazing feeding system. *East Afr. agric. For. J.* 52: 106–111.

Anonymous, 1959. Notes on Kenya agriculture. VI. Grass leys and grassland plants. *East Afr. agric. J.* 24: 223–236.

Anslow, R.C., 1958. Controlling ant removal of Setaria sphacelata. *Emp. J. exp. Agric.* 26: 55–57.

Arkel, H. van, 1977. New forage crop introductions for the semi-arid highland areas of Kenya as a means to increase beef production. *Neth. J. agric. Sci.* 25: 135–150.

Arkel, H. van, 1979. Weed control in cold tolerant sorghum. *Trop. Agric. (Trin.)* 56: 105–113.

Arkel, H. van, 1982. The identification and introduction of a new crop in the highlands of Kenya. Thesis Wageningen.

Atlas of Kenya, 1959; 1970. Survey of Kenya, Nairobi.

Atlas of Uganda, 1962. Department of Lands and Surveys, Entebbe.

Baker, H.G., 1978. Invasion and replacement in Californian and neotropical grassland. In: J.R. Wilson (Ed.), Plant Relations in Pastures. CSIRO.

Ball, R.S., 1935. The cultivation of lucerne. *East Afr. agric. J.* 1: 29–36.

Ball, R.S., 1936. Mixed farming in East Africa. I. Grassland and arable dairying in certain parts of Kenya. Timboroa, Molo, Upper Gilgil, Thomson Falls, Njoro, Rongai, Subukia. *East Afr. agric. J.* 1: 399–411.

Ball, R.S., 1937. Experimental work with crops and pastures in the Rift Valley and Molo areas of Kenya 1934–36. *East Afr. agric. J.* 2: 254–269.

Banage, W.B. & S.A. Visser, 1967. Soil moisture and temperature levels and fluctuations in one year in a Uganda soil catena. *East Afr. agric. For. J.* 32: 450–455.

Barker, J.R. & M.M. Kidar, 1989. Herbage yields of *Cenchrus ciliaris* L. increase with clipping. *Trop. Grasslds* 23: 80–83.

Barnard, C., 1972. Register of Australian Herbage Plant Cultivars. CSIRO, Canberra.

Bartholomew, W.V., J. Meyer & H. Laudelout, 1953. Mineral nutrient immobilization under forest and grass fallow in the Yangambi (Belgian Congo) region. *Sér. techn. INEAC* 57: 1–27.

Bauer, P.J., 1975. Fodder crops for the small-scale farmer. Kitale report.

Bauer, P.J., 1976. Lupins in East Africa, a review. Kitale report (PRP/12).

Beckley, V.A., 1940. Compost-making in Kenya. *Emp. J. exp. Agric.* 8: 311–318.

Behaeghe, T., 1960. Conservation et utilisation des semences de graminées. *Bull. agric. Congo* 51: 1223–1240.

Behaeghe, T. & R. Blouard, 1962. Amélioration des semences et sélection des plantes prairiales au Congo, au Rwanda et au Burundi. *Bull. Inf. INEAC* 11: 307–338.

Bellis, E., 1958. Notes on Kenya agriculture. II. Tabulated Recommendations. *East Afr. agric. J.* 23: 234–235.

Bennison, R.H. & D.D. Evans, 1968. Some effects of crop rotation on the productivity of crops on a red earth in a semi-arid tropical climate. *J. agric. Sci.* 71: 365–380.

Bijman, P & J.B. Davis, 1973. Feeding of dairy cattle in the dry weather. *Kenya Farmer*, January: 21, 28.

Bille, J.C., J.P. Lebrun & R. Rivière, 1969. Pâturages et Cultures Fourragères en Afrique Intertropicale. IEMVT, Maisons-Alfort.

Birch, H.F., 1958. The effect of soil drying on humus decomposition and nitrogen availability. *Plant & Soil 10*: 9–31.

Birch, H.F., 1960. Soil drying and soil fertility. *Trop. Agric. (Trin.)* 37: 3–10.

Birch, H.F. & H.W. Dougall, 1967. Effect of a legume on soil nitrogen mineralisation and percentage nitrogen in grasses. *Plant & Soil 27*: 292–296.

Birch, H.F. & M.T. Friend, 1956. The organic matter and nitrogen status of East African soils. *J. Soil Sci.* 7: 156–167.

Birch, W.R., 1958a. The effects of weeds and their control by MCPA spraying on the establishment of perennial ryegrass and cocksfoot in the Kenya highlands. *J. Brit. Grassld Soc.* 13: 126–136.

Birch, W.R., 1958b. Observations on the flowering of exotic pasture grasses in Kenya. *East Afr. agric. J.* 24: 142–147.

Birch, W.R., 1959a. High altitude leys. *Kenya Farmer*, July.

Birch, W.R., 1959b. High altitude ley agronomy in Kenya. Part I. Observation on the climate and soil. *East Afr. agric. J.* 25: 35–41.

Birch, W.R., 1959c. High altitude ley agronomy in Kenya. Part II. The effects of phosphates and nitrogen. *East Afr. agric. J.* 25: 113–120.

Birch, W.R., 1960. High altitude ley agronomy in Kenya. Part III. The effects of lime, molybdenum and some observations on the use of legumes. *East Afr. agric. J.* 25: 156–164.

Birch, W.R., 1964. The germination purity of some Kenya grasses. The results of ten years of seed testing in Kenya. *East Afr. agric. For. J.* 30: 105–116.

Blaser, R.E., C.E. Richey, W.G. Kirk & P.T. Dix Arnold, 1955. Experiments with Napier grass. *Florida Agric. Exp. Sta. Bull.* 568.

Blouard, R. & T. Behaeghe, 1961. Etablissement et exploitation des pâturages en région forestière équatoriale. *Bull. Inf. INEAC* 10: 109–124.

Blouard, R. & L. Thuriaux, 1962. *Stylosanthes gracilis*, son comportement et son utilisation au Congo. *Bull. Inf. INEAC 11: 339–355.*

Bock, K.R., E.J. Guthrie & H.Y. Kulkarni, 1973. Notes on East African plant virus diseases. 3. Pea mosaic virus. *East Afr. agric. For J.* 39: 77–81.

Bogdan, A.V., 1949a. Observations on palatability of some leguminous plants of Kenya. *East Afr. agric. J.* 15: 38–40.

Bogdan, A.V., 1949b. Two species of Star grass in Kenya. *East Afr. agric. J.* 15: 102–103.

Bogdan, A.V., 1950. A list of weeds of the Kenya highlands. *East Afr. agric. J.* 15: 118–123.

Bogdan, A.V., 1954. Bush-clearing and grazing trial at Kisokon, Kenya. *East Afr. agric. J.* 19: 253–259.

Bogdan, A.V., 1955. Herbage plants at the Grassland Research Station, Kitale, Kenya. *East Afr. agric. J.* 20: 151–165.

Bogdan, A.V., 1956. Indigenous clovers of Kenya. *East Afr. agric. J.* 22: 40–45.

Bogdan, A.V., 1958a. A Revised List of Kenya Grasses with Keys for Identification. Govt Printer, Nairobi.

Bogdan, A.V., 1958b. Some edaphic vegetation types at Kiboko, Kenya. *J. Ecol.* 46: 115–126.

Bogdan, A.V., 1959a. The selection of tropical ley grasses in Kenya. General considerations and methods. *East Afr. agric. J.* 24: 206–217.

Bogdan, A.V., 1959b. Flowering habits of *Chloris gayana*. *Proc. Linn. Soc.* 170: 154–158.

Bogdan, A.V., 1960a. The breeding behaviour of molasses grass in Kenya. *East Afr. agric. For. J.* 26: 49–50.

Bogdan, A.V., 1960b. A molasses grass variety trial. *East Afr. agric. For. J.* 26: 132–133.

Bogdan, A.V., 1960c. Climate and grass breeding in Kenya. *Proc. 8th int. Grassld Congr., Reading*: 340–342.

Bogdan, A.V., 1961a. The depth of planting Rhodes grass seed. *East Afr. agric. For. J.* 26: 187.

Bogdan, A.V., 1961b. Breeding behaviour of *Cenchrus ciliaris* in Kenya. *East Afr. agric. For. J.* 26: 241.

Bogdan, A.V., 1961c. Intra-variety variation in Rhodes grass (*Chloris gayana* Kunth.) in Kenya. *J. Brit. Grassld Soc.* 16: 238–239.

Bogdan, A.V., 1961d. Hybridization in the "*Setaria sphacelata*" complex in Kenya. *Comptes Rendus 4e Réunion Plénière de l'AETFAT*: 311–317. Lisboa.

Bogdan, A.V., 1962. Grass pollination by bees in Kenya. *Proc. Linn. Soc.* 173: 57–61.

Bogdan, A.V., 1963a. A note on the breeding behaviour of some East African grasses. *J. Brit. Grassld Soc.* 18: 74.

Bogdan, A.V., 1963b. Three interesting introductions from the local grass flora at Kitale, Kenya. *J. Brit. Grassld Soc.* 18: 247–248.

Bogdan, A.V., 1963c. *Chloris gayana* without anthocyanin colouration. *Heredity* 18: 364–368.

Bogdan, A.V., 1963d. Notes on the damage by army worm (*Laphygma exempta*) to grass species at Kitale, Kenya. *East Afr. agric. For. J.* 29: 120–121.

Bogdan, A.V., 1963e. A note on the breeding behaviour of *Panicum maximum* in Kenya. *Trop. Agric. (Trin.)* 40: 313–314.

Bogdan, A.V., 1964a. A study of the depth of germination of tropical grasses: a new approach. *J. Brit. Grassld Soc.* 19: 251–254.
Bogdan, A.V., 1964b. Observations on Rhodes grass breeding. *East Afr. agric. For. J.* 30: 38–39.
Bogdan, A.V., 1965a. Cultivated varieties of tropical and subtropical herbage plants in Kenya. *East Afr. agric. For. J.* 30: 330–338.
Bogdan, A.V., 1965b. Vegetative propagation of fodder and pasture grasses. *Kenya Farmer*, April: 33–34.
Bogdan, A.V., 1965c. Pasture and fodder grasses and legumes for medium and low altitudes. *Kenya Farmer*, June: 30–35.
Bogdan, A.V., 1966a. Plant introduction, selection, breeding and multiplication. In: W. Davies & C.L. Skidmore (Eds), Tropical Pastures. Faber & Faber.
Bogdan, A.V., 1966b. *Glycine javanica* under experimental cultivation in Kenya. *Trop. Agric.* (*Trin.*) 43: 99–105.
Bogdan, A.V., 1966c. Seed morphology of some cultivated African grasses. *Proc. int. Seed Test. Ass.* 31: 789–799.
Bogdan, A.V., 1966d. Pollination and breeding behaviour in the *Trifolium rueppellianum* complex in Kenya. *New Phytol.* 65: 417–422.
Bogdan, A.V., 1966e. Weeds in herbage seeds in Kenya. *East Afr. agric. For. J.* 32: 63–66.
Bogdan, A.V., 1969. Rhodes grass. Review Article. *Herb. Abstr.* 39: 1–13.
Bogdan, A.V., 1971. Notes on bunt disease of setaria grass. *Kenya Farmer* 183: 81.
Bogdan, A.V., 1977. Tropical Pastures and Fodder Plants. Longman.
Bogdan, A.V. & E.M. Kidner, 1967. Grazing natural grassland in western Kenya. *East Afr. agric. J.* 33: 31–34.
Bogdan, A.V. & E. Mwakha, 1970. Observations on some grass/legume mixtures under grazing. *East Afr. agric. For. J.* 36: 35–38.
Bogdan, A.V. & D.J. Pratt, 1961. Common Acacias of Kenya. Govt Printer, Nairobi.
Bogdan, A.V. & D.J. Pratt, 1967. Reseeding Denuded Pastoral Land in Kenya. Govt Printer, Nairobi.
Bogdan, A.V. & A. Storrar, 1954. The control of Aristida and other annuals in Kenya Rift Valley pastures. *Emp. J. exp. Agric.* 22: 211–223.
Bohlen, E., 1973. Crop Pests in Tanzania and their Control. Paul Parey.
Boonman, J.G., 1969. Sorghum-sudan grass hybrids for dry season grazing. *Kenya Weekly News*, July.
Boonman, J.G., 1971a. Experimental studies on the seed production of tropical grasses in Kenya. I. General introduction and analysis of problems. *Neth. J. agric. Sci.* 19: 23–26.
Boonman, J.G., 1971b. Experimental studies on the seed production of tropical grasses in Kenya. II. Tillering and heading in seed crops of eight grasses. *Neth. J. agric. Sci.* 19: 237–249.
Boonman, J.G., 1972a. Experimental studies on the seed production of tropical grasses in Kenya. III. The effect of nitrogen and row width on seed crops of *Setaria sphacelata* cv Nandi II. *Neth. J. agric. Sci.* 20: 22–34.
Boonman, J.G., 1972b. Experimental studies on the seed production of tropical grasses in Kenya. IV. The effect of fertilizer and planting density on *Chloris gayana* cv Mbarara. *Neth. J. agric. Sci.* 20: 218–224.
Boonman, J.G., 1972c. Experimental studies on the seed production of tropical grasses in Kenya. V. The effect of time of nitrogen top dressing on seed crops of *Setaria sphacelata*. *Neth. J. agric. Sci.* 20: 225–231.
Boonman, J.G., 1973a. Experimental studies on the seed production of tropical grasses in Kenya. VI. The effect of harvest date on seed yield in varieties of *Setaria sphacelata*, *Chloris gayana* and *Panicum coloratum*. *Neth. J. agric. Sci.* 21: 3–11.
Boonman, J.G., 1973b. On the seed production of tropical grasses in Kenya. *Agric. Res. Rep.* 794. Wageningen.
Boonman, J.G., 1977. Report on a visit to Australia. Kitale report.
Boonman, J.G., 1978a. Rhodes grass breeding in Kenya. I. Intra-variety variation and character relationships. *Euphytica* 27: 127–136.
Boonman, J.G., 1978b. Rhodes grass breeding in Kenya. II. Clonal evaluation for seed yield within maturity classes of two varieties. *Euphytica* 27: 419–426.
Boonman, J.G., 1978c. Rhodes grass breeding in Kenya. III. Seed and herbage yield in selections of four maturity classes based upon intra-variety variation. *Euphytica* 27: 649–656.
Boonman, J.G., 1978d. Herbage quality in Rhodes grass (*Chloris gayana* Kunth). I. The effect of heading date on intra-variety variation in yield and in-vitro digestibility. *Neth. J. agric. Sci.* 26: 304–311.
Boonman, J.G., 1978e. Herbage quality in Rhodes grass (*Chloris gayana* Kunth). II. Intra-variety variation in yield and digestibility of plants of similar date. *Neth. J. agric. Sci.* 26: 337–343.
Boonman, J.G., 1979a. Tropical grass seed production in Africa with particular reference to Kenya. In: P.A. Sanchez & L.E. Tergas (Eds), Pasture Production in Acid Soils of the Tropics. Proc. CIAT Seminar, Cali, 1978. CIAT 03 EG-5.
Boonman, J.G., 1979b. Fodders of East Africa. Kitale report.
Boonman, J.G., 1980. Seed production of tropical grasses in Kenya. In: P. D. Hebblethwaite (Ed.), Seed Production. *Proc. 28th Easter School in Agricultural Science, Nottingham,* 1978. Butterworths.
Boonman, J.G., 1983. Forage production on small farms: lessons from East Africa. In: M.J. Chang (Ed.), Seminar Agricultural Research in Rwanda, Kigali. ISNAR report.
Boonman, J.G., 1984. Pastures. In: P.J. Terry (Ed.), Guide to Weed Control in East African Crops. Kenya Literature Bureau, Nairobi.
Boonman, J.G., D.J. Glas & A.J.P. van Wijk, 1979. Grass breeding. In: J. Sneep & A.J.T. Hendriksen (Eds), Plant Breeding Perspectives. PUDOC, Wageningen.
Boonman, J.G. & J. Goldson, 1980. Mission to Tanga. Report to the Tanzania Government.
Boonman, J.G. & A.J.P. van Wijk, 1973. Experimental studies on the seed production of tropical grasses in Kenya. VII. The breeding for improved seed and herbage productivity. *Neth. J. agric. Sci.* 21: 12–23.
Boonman, J.G. & A.J.P. van Wijk, 1989. Grass breeding beyond the year 2000. *Prophyta* International Issue (in English, French, Spanish and Dutch) 8: 29–46.

Booysen, P. de V., N.M. Tainton & J.D. Scott, 1963. Shoot-apex development in grasses and its importance in grassland management. Review Article. *Herb. Abstr.* 33: 209–213.

Borget, M., 1969. Résultats et tendances présentes des recherches fourragères à l'IRAT. *Agron. trop.* 24: 103–123.

Boshoff, W.H., 1966. Permanent Napier grass for the supply of mulch and dry season fodder in the long grass area of Uganda. In: R.S. Musangi (Ed.), Pasture Production and Utilization by Cattle in the Tropics with special Reference to Uganda. Seminar Makerere, 1966. *Techn. Bull.* 4: 19–20.

Boswinkle, E., 1961. Residual effects of phosphorus fertilizers in Kenya. *Emp. J. exp. Agric.* 29: 136–142.

Botha, J.P., 1945. Setaria grasses for Eastern Transvaal. *Fmg S. Afr.* 20: 273–280.

Boudet, G., 1970. Management of savannah woodland range in West Africa. *Proc. 11th int. Grassld Congr., Surfers Paradise*: 1–3.

Boudet, G., 1978. Manuel sur les Pâturages Tropicaux et les Cultures Fourragères. IEMVT, Maisons-Alfort.

Bouwer, W.A.C. & L.P.N. Strange, 1948. Temporary ley grass trials on the Uasin Gishu Plateau Kenya Colony. *East Afr. agric. J.* 13: 203–206.

Bowden, B.N., 1963. Studies on *Andropogon gayanus* Kunth. 1. The use of *Andropogon gayanus* in agriculture. *Emp. J. exp. Agric.* 31: 267–273.

Braithwaite, D.P., 1952. The hydrocyanic acid content of perennial Kavirondo sorghum. *East Afr. agric. J.* 18: 54–58.

Bradford, E.L., 1946. Bukura native agricultural school, Nyanza Province, Kenya. *East Afr. agric. J.* 11: 162–164.

Braun, H.M.H., 1980. Agro-climatic Zone Map of Kenya; scale 1:1.000.000. Kenya Soil Survey, Nairobi.

Braun, H.M.H., 1985. Use and improvement of poorly drained land in Uasin Gishu District, Kenya. ILRI Wageningen report.

Braun, H.M.H., 1986. Seasonal distribution of rainfall in Kenya. Kenya Soil Survey report M14.

Bredon, R.M., 1965. Factors affecting stall feeding of zebu cattle in Uganda. *Proc. 4th East Afr. Spec. Comm. Anim. Nutr. Physiol. Breed., EAAFRO.*

Bredon, R.M., K.W. Harker & B. Marshall, 1963. The nutritive value of grasses grown in Uganda when fed to zebu cattle. I. The relation between the percentage of crude protein and other nutrients. *J. agric. Sci.* 61: 101–104.

Bredon, R.M. & C.R. Horrell, 1961. Chemical composition and nutritive value of some common grasses in Uganda. I. General pattern of behaviour of grasses. *Trop. Agric. (Trin.)* 38: 297–304.

Bredon, R.M. & C.R. Horrell, 1962. Chemical composition and nutritive value of some common grasses in Uganda. II. The comparison of chemical composition and nutritive values of grasses throughout the year, with special reference to the later stages of growth. *Trop. Agric. (Trin.)* 39: 13–17.

Bredon, R.M. & C.R. Horrell, 1963. Management studies with *Panicum maximum* in Uganda. II. The effect of cutting interval and nitrogen fertilizer on chemical composition and nutritive value. *Emp. J. exp. Agric.* 31: 343–350.

Bredon, R.M. & B. Marshall, 1962. Selective consumption by stall-fed cattle and its influence on the results of a digestibility trial. *E. Afr. agric. For. J.* 27: 168–172.

Bredon, R.M. & J. Wilson, 1963. The chemical composition and nutritive value of grasses from semi-arid areas of Karamoja as related to ecology and types of soil. *East Afr. agric. For. J.* 29: 134–142.

Bredon, R.M., D.T. Torell & B. Marshall, 1967. Measurement of selective grazing of tropical pastures using oesophageal fistulated steers. *J. Range Management* 20: 317–320.

Brockington, N.R., 1960. Studies on the growth of a Hyparrhenia dominant grassland in Northern Rhodesia. I. Growth and reaction to cutting. *J. Brit. Grassld Soc.* 15: 323–338.

Brockington, N.R., 1961a. Studies on the growth of a Hyparrhenia dominant grassland in Northern Rhodesia. II. Fertilizer response. III. The effect of fire. *J. Brit. Grassld Soc.* 16: 54–64.

Brockington, N.R., 1961b. The growth, yield and protein content of twelve cultivated grasses in Northern Rhodesia. *Emp. J. exp. Agric.* 29: 197–211.

Brockington, N.R., 1962. Fertilizer trials on some cultivated grasses in Northern Rhodesia. *Emp. J. exp. Agric.* 30: 345–354.

Brockington, N.R., 1964. Fertilizer responses on giant Rhodes grass (*Chloris gayana* Kunth) in Northern Rhodesia. *Emp. J. exp. Agric.* 32: 76–83.

Brockington, N.R., T.H. Stobbs, P.W. Newhouse & G.A. Wadsworth, 1965. The effects of leys on soil fertility in the annual cropping areas of Uganda. *Afr. Soils* 10: 473–481.

Brockway, Lucille H., 1979. The Role of the British Royal Gardens. Academic Press Inc.

Brown, L.H., 1963. A National Cash Crops Policy for Kenya. Govt Printer, Nairobi.

Brzostowski, H.W., 1961. Establishment of *Cenchrus ciliaris* from caryopses. *East Afr. agric. For. J.* 26: 242–244.

Brzostowski, H.W., 1962. Influence of pH and superphosphate on establishment of *Cenchrus ciliaris* from seed. *Trop. Agric. (Trin.)* 39: 289–296.

Brzostowski, H.W. & M.A. Owen, 1964. Botanical changes in the sown pasture. *Trop. Agric. (Trin.)* 41: 231–242.

Bullock, J.A., 1965. The control of *Hylemya arambourgi* Séguy (*Dipt. Anthomyiidae*) on barley. *Bull. Entom. Res.* 55: 645–661.

Bumpus, E.D., 1957. Legume nodulation in Kenya. 1. Exploratory field experiments. *East Afr. agric. J.* 23: 91–99.

Bumpus, E.D. & R.G. Poultney, 1960. The problem of nitrogen fixation by legumes in Kenya in relation to soil-nutrient status. *Proc. 8th int. Grassld Congr., Reading*: 244–247.

Bunting, A.H. & J.D. Lea, 1956. Some observations on natural and established pastures and pasture plants in the Sudan rainlands. *Emp. J. exp. Agric.* 24: 185–191.

Burton, G.W., R.H. Hart & R.S. Lowrey, 1967. Improving forage quality in Bermudagrass by breeding. *Crop Sci.* 7: 329–332.

Cameron, D.G., R.M. Jones, G.P.M. Wilson, H.G. Bishop, B.G. Cook, G.R. Lee & K.F. Lowe, 1989. Legumes for heavy grazing in coastal subtropical Australia. *Trop. Grasslds* 23: 153–161.

Cassady, J.T., 1973. The effect of rainfall, soil moisture and harvesting intensity on grass production on two rangeland sites in Kenya. *East Afr. agric. For. J.* 39: 26–36.

Catford, J.R., 1952. Pasture management in Uganda in relation to Western Equatoria. *East Afr. agric. J.* 17: 183–187.

Catchpoole, V.R. & E.F. Henzell, 1971. Silage and silage-making from tropical herbage species. *Herb. Abstr.* 41: 213–221.

Chada, H.L. & E.A. Wood, 1960. Biology and Control of the Rhodesgrass Scale. *USDA Techn. Bull.* 1221.

Chheda, H.R., O. Babalola, M.A.M. Saleem & M.E. Aken'ova, 1983. Sown grassland as an alternative to shifting cultivation in lowland humid tropics. *Proc. 14th int. Grassld Congr., Lexington (Kentucky)*, 1981: 774–777.

Chema, S., 1984. Small-scale milk production in Kenya. In: J.A. Kategile (Ed.), The Potential for Small Scale Milk Production in Eastern and Southern Africa. IDRC Ottawa.

Child, R., N.A. Goodchild & J.R. Todd, 1955. A fertilizer experiment with Napier grass. *Emp. J. exp. Agric.* 23: 220–225.

Child, R.D., J.N. Njuki, R.M. Hansen & D.L. Whittington, 1982. Digestibility and protein content of *Leucaena leucocephala*. *East Afr. agric. For. J.* 48: 32–34.

Childs, A.H.B. & C.G. Groom, 1964. Balanced farming with cattle and coconuts. *East Afr. agric. For. J.* 29: 206–207.

Chippindall, Lucy K.A., 1955. A guide to the identification of grasses in South Africa. In: D. Meredith (Ed.), The Grasses and Pastures of South Africa. Central News Agency, Cape Town.

Chudleigh, P.D., 1974. A description of smallholder ruminant production systems in the high potential areas of Kenya. Kitale report (PRP/4).

Chudleigh, P.D., J.G. Boonman & P.J. Cooper, 1977. Environmental factors affecting herbage yield of Rhodes grass (*Chloris gayana*) at Kitale. *Trop. Agric.* (*Trin.*) 54: 193–204.

Clarke, R.T., 1962. The effect of some resting treatments on a tropical soil. *Emp. J. exp. Agric.* 30: 57–62.

Clatworthy, J.N., 1970. A comparison of legume and fertilizer nitrogen in Rhodesia. *Proc. 11th int. Grassld Congr., Surfers Paradise*: 408–411.

Clatworthy, J.N., 1984a. Pasture research in Zimbabwe: 1964–1984. In: J.A. Kategile (Ed.), Pasture Improvement Research in Eastern and Southern Africa. Proc. Workshop Harare, 1984. IDRC-237e, Ottawa.

Clatworthy, J.N., 1984b. Effect of reinforcement of native grazing with Silverleaf desmodium (*Desmodium uncinatum*) on dry season performance of beef steers in Zimbabwe. *Trop. Grasslds* 18: 198–205.

Clatworthy, J.N. & C. Madakadze, 1988. Some thoughts on the collection, introduction and evaluation of pasture legumes in Zimbabwe. In: B.H. Dzowela (Ed.), African Forage Plant Genetic Resources, Evaluation of Forage Germplasm and Extensive Livestock Production Systems. Proc. Workshop Arusha 1987, ILCA, Addis Ababa.

Clayton, W.D., 1970. *Gramineae* (Part 1). In: Flora of Tropical East Africa. Crown Agents.

Clayton, W.D. & J.R. Harlan, 1970. The genus *Cynodon* L.C. Rich. in tropical Africa. *Kew Bull.* 24: 185–189.

Clayton, W.D., S.M. Phillips & S.A. Renvoize, 1974. *Gramineae* (Part 2). In: Flora of Tropical East Africa. Crown Agents.

Clayton, W.D. & S.A. Renvoize, 1982. *Gramineae* (Part 3). In: Flora of Tropical East Africa. Balkema, Rotterdam.

Clement, C.R., 1961. Benefit of leys-structural improvement or nitrogen reserves. *J. Brit. Grassld Soc.* 16: 194–200.

Clements, R.J., 1989. Developing persistent pasture legume cultivars for Australia. In: G.C. Marten *et al.* (Eds), Persistence of Forage Legumes. American Society of Agronomy.

Colloquium on Grassland Research, 1949. *East Afr. agric. J.* 15: 200–226.

Combes, D. & J. Pernès, 1970. Variations dans les nombres chromosomiques du *Panicum maximum* Jacq. en relation avec le mode de reproduction. *Comptes Rendus Acad. Sci. Paris* 270: 782–785.

Combes, R., 1976. Grass Varieties. *Kenya Farmer*, July.

Combes, R. & W.H. Verburgt, 1971. The commercial growing, cleaning and marketing of tropical and subtropical pasture seeds. Seminar Forage Crops Research, University of Ibadan.

Compère, R., 1959. Etude toxicologique de *Leucaena glauca* chez les bovins. *Bull. agric. Congo* 50: 1311–1320.

Compère, R., 1960a. Mise en valeur des pâturages improductifs à "Kikuyu" (*Pennisetum clandestinum*). Essai en région du Mulume-Munene (Kivu). *Bull. agric. Congo* 51: 923–945.

Compère, R., 1960b. Introduction au Kivu de *Tripsacum laxum* Nash., plante fourragère de soudure pour les élévages laitiers. *Bull. agric. Congo* 51: 1085–1103.

Compère, R., 1961a. Le comportement des mélanges herbages *Trifolium repens* – graminées, dans les régions d'altitude du Kivu. *Sér. techn. INEAC* 65.

Compère, R., 1961b. Productivité et caractéristiques bromotalogiques de l'herbe de quelques types de pâturages étudiés à la station de Mulungu (Kivu). *Bull. agric. Congo* 52: 62–93.

Compère, R., 1961c. Cinq ans d'exploitation de la culture fourragère de *Desmodium intortum* (Mill.) Urb. à la station de Mulungu (Kivu). *Bull. agric. Congo* 52: 572–590.

Compère, R., 1961d. Influence d'une fumure minérale rationelle sur la composition chimique de l'herbe de certains pâturages artificiels. *Bull. agric. Congo* 52: 1271–1278.

Compère, R., 1968. Exploitation rationelle de la prairie permanente d'altitude à *Exotheca abyssinica* Anders. et *Eragrostis boehmii* Hack. au Rwanda et au Burundi. *Bull. Rech. agron. Gembloux* 3: 583–604.

Compère, R., 1978. Agrostologie des régions chaudes. II. Gembloux.

Cooke, G.W., 1967. The Control of Soil Fertility. English Language Book Society.

Cooper, P.J.M. & R. Law, 1977. Soil temperature and its association with maize yield variations in the highlands of Kenya. *J. agric. Sci.* 89: 355–363.

Crauford, R.Q., 1978. Pasture grasses and legumes report on plant breeding at Mount Makulu 1967–1978. *Zambia Res. Memorandum* 26.

Creek, M.J. & H.A. Squire, 1976. Kenya feedlots: from pilot project to commercial application. *Span* 19: 96–99.

Crowder, L.V., H. Chaverra & J. Lotero, 1970. Productive improved grasses in Columbia. *Proc. 11th int. Grassld Congr., Surfers Paradise*: 147–149.

Crowder, L.V. & H.R. Chheda, 1982. Tropical Grassland Husbandry. Longman.

Crowder, L.V., 1984. *Clitoria ternatea* (L.) Due as a forage and cover crop – a review. *Nigerian agric. J.* 11: 61–65.

Dagg, M., 1962. Physical properties of the surface soils, infiltration and availability of soil moisture. *East Afr. agric. For. J.* 27 (special issue): 68–70.

Dagg, M., 1965. Climatic classification in East Africa. *Proc. III spec. Meet. appl. Met. East Africa.* EACSO, Nairobi.

Dagg, M., 1969. Hydrological implications of grass root studies at a site in East Africa. *J. Hydrol.* 9: 438–444.

Damseaux, J., 1956. Etude de trois légumineuses fourragères introduites au Congo Belge en vue de l'alimentation de bétail. *Bull. agric. du Congo Belge* 47: 93–111.

Davidson, B.R., 1958. The climatic limitations of subterranean clovers in Kenya. *East Afr. agric. J.* 24: 121–124.

Davies, T., 1963. Fodder conservation in Northern Rhodesia. *J. agric. Sci.* 61: 308–328.

Deinum, B., 1966. Climate, nitrogen and grass. I. Research into the influence of light intensity, temperature, water supply and nitrogen on production and chemical composition of grass. *Meded. Landb. Hogesch.* Wageningen 66–11.

Deinum, B., 1984. Chemical composition and nutritive value of herbage in relation to climate. Meeting of the European Grassland Federation. As, Norway.

Denny, P. (Ed.), 1985. The Ecology and Management of African Wetland Vegetation. Kluwer Academic Publishers.

Deschuytener, G., 1967. L'importance et les possibilités de l'élevage au Rwanda. *Ann. Gembloux* 73: 61–76.

Dewez, J., 1959. Le semis de *Brachiaria ruziziensis* dans la plaine de la Ruzizi. *Bull. Inf. INEAC* 8: 303–307.

Dirven, J.G.P. & B. Deinum, 1977. The effect of temperature on the digestibility of grasses. An analysis. *Forage Research* 3: 1–17.

Dirven, J.G.P. & V.K.R. Ehrencron, 1963. Dry matter percentages of grasses in the humid tropics. *Surin. Landb.* 11: 88–93.

Dirven, J.G.P. & H.A. van Hoof, 1960. A destructive virus disease of Pangola grass. *Tijdschrift v. Plantenziekten* 66: 344–349.

Dirven, J.G.P., L.J.M. van Soest & K. Wind, 1979. The influence of photoperiod on head formation in some *Brachiaria* species and *Chloris gayana* cv Masaba. *Neth. J. agric. Sci.* 27: 48–59.

Dixon, H.N., 1982. The growing and harvesting of forage crops for silage in northern Tanzania. *East Afr. agric. For. J.* 45: 22–35.

d'Mello, J.P.F. & D. Thomas, 1978. The nutritive value of dried Leucaena leaf meal from Malawi: studies with young chicks. *Trop. Agric. (Trin.)* 55: 45–50.

Döbereiner, Johanna, J.M. Day & P.J. Dart, 1972. Nitrogenase activity and oxygen sensitivity of the *Paspalum notatum Azotobacter paspali* complex. *J. gen. Microbiol.* 71: 103–116.

Doggett, H., 1970. Sorghum. Longman.

Donald, C.M., 1975. Temperate pasture species. In: R. Milton Moore (Ed.), Australian Grasslands. Australian National University Press.

Dougall, H.W., 1953a. Fertilizer experiments on ley establishment, established leys and natural pastures in the Kenya Highlands. *East Afr. agric. J.* 19: 36–37.

Dougall, H.W., 1953b. Kikuyu grass. I. Seasonal variation in the composition of degenerate and productive pastures. *East Afr. agric. J.* 19: 74–80.

Dougall, H.W., 1954a. Kikuyu grass. II. An attempt to improve a worn-out pasture. *East Afr. agric. J.* 19: 212–218.

Dougall, H.W., 1954b. Fertilizer experiments on grassland in the Kenya highlands. *East Afr. agric. J.* 19: 171–178.

Dougall, H.W., 1954c. Effect of ammonium and nitrate nitrogen on a temporary molasses grass pasture. *East Afr. agric. J.* 19: 261–262.

Dougall, H.W., 1954d. The composition of green oats for forage and silage. *East Afr. agric. J.* 20: 118–119.

Dougall, H.W., 1954e. Some fertilizer aspects of grassland husbandry in the Kenya highlands. *J. Brit. Grassld Soc.* 9: 141–148.

Dougall, H.W., 1960. Average nutritive values of Kenya feeding stuff for ruminants. *East Afr. agric. For. J.* 26: 119–128.

Dougall, H.W., 1962a. The chemical composition of some species and varieties of Trifolium. *East Afr. agric. For. J.* 27: 142–144.

Dougall, H.W., 1962b. Ash distribution in Kenya roughages. *East Afr. agric. For. J.* 28: 87.

Dougall, H.W., 1962c. Crude fibre – crude protein relationship in Kenya roughages. *East Afr. agric. For. J.* 28: 88.

Dougall, H.W., 1963a. Changes in the chemical composition of plants during growth in the field. *East Afr. agric. For. J.* 28: 182–189.

Dougall, H.W., 1963b. Silica distribution in Kenya roughages. *East Afr. agric. For. J.* 29: 58–59.

Dougall, H.W. & H.F. Birch, 1966. pH of grasses in relation to genera. *Nature* 210 (5038): 844.

Dougall, H.W. & H.F. Birch, 1967. Further experiments on the acid grass *Setaria sphacelata. Plant & Soil* 26: 85–98.

Dougall, H.W. & A.V. Bogdan, 1958a. Browse plants of Kenya with special reference to those occurring in south Baringo. *East Afr. agric. J.* 23: 236–245.

Dougall, H.W. & A.V. Bogdan, 1958b. The chemical composition of the grasses of Kenya. I. *East Afr. agric. J.* 24: 17–23.

Dougall, H.W. & A.V. Bogdan, 1960. The chemical composition of the grasses of Kenya. II. *East Afr. agric. J.* 25: 241–244.

Dougall, H.W. & A.V. Bogdan, 1965. The chemical composition of the grasses of Kenya. III. *East Afr. agric. For. J.* 30: 314–319.

Dougall, H.W. & A.V. Bogdan, 1966. The chemical composition of some leguminous plants grown in the herbage nursery at Kitale, Kenya. *East Afr. agric. For. J.* 32: 45–49.

Dougall, H.W., V.M. Drysdale & P.E. Glover, 1964. The chemical composition of Kenya browse and pasture herbage. *East Afr. Wildl. J.* 2: 86–121.

Dougall, H.W. & P.E. Glover, 1964. On the chemical composition of *Themeda triandra and Cynodon dactylon. East Afr. Wildl. J.* 2: 67–70.

Dradu, E.A.A. & G.N. Harrington, 1972. Seasonal crude protein content of samples obtained from a tropical range pasture using oesophageal fistulated steers. *Trop. Agric. (Trin.)* 49: 15–21.

Dradu, E.A.A. & J. Ogwang, 1979. Influence of insecticide spraying on seed yield of *Macroptilium atropurpureum* (Siratro) at Serere, eastern Uganda. *East Afr. agric. For. J.* 44: 237–240.

Drysdale, V.M., 1965. The removal of nutrients by the maize crop. *East Afr. agric. For. J.* 31: 189–190.

Duvigneaud, P., 1958. La végétation du Katanga et de ses sols métallifères. *Bull. Soc. roy. Bot. Belg.* 90: 127–286.

Dzowela, B.H., 1985. Highlights of pasture research in Malawi: 1975–84. In: J.A. Kategile (Ed.), Pasture Improvement Research in Eastern and Southern Africa. Proc. Workshop Harare 1984, IDRC-237e, Ottawa.

Dzowela, B.H., 1988. African Forage Plant Genetic Resources, Evaluation of Forage Germplasm and Extensive Livestock Production Systems. Proc. Workshop Arusha 1987, ILCA, Addis Ababa.

East African Farmer and Planter, 1958, New hybrid from South Africa shows promise. July 1958.

East African Meteorological Department, 1975. Climatological Statistics for East Africa. I. Kenya. II. Uganda. III. Tanzania.

Ebbels, D.L. & D.J. Allen, 1979. A supplementary and annotated list of plant diseases, pathogens and associated fungi in Tanzania. *Phytopathological Papers* No 22.

Eberhart, S.A. & W.A. Russell, 1966. Stability perimeters for comparing varieties. *Crop Sci.* 6: 36–40.

Edwards, D.C., 1932. "Hard" seeds in *Panicum coloratum* Stapf. *Nature* 132: 209.

Edwards, D.C., 1934. Rhodes Grass for Hay and Pasture. Govt Printer, Nairobi.

Edwards, D.C., 1935a. The grasslands of Kenya. I. Areas of high moisture and low temperature. *Emp. J. exp. Agric.* 3: 153–159.

Edwards, D.C., 1935b. A further note on Rhodes grass. *East Afr. agric. J. 1: 233*–235.

Edwards, D.C., 1936. Report on grassland improvement in Kenya, 1934. *East Afr. agric. J.* 1: 269–277.

Edwards, D.C., 1937. Three ecotypes of *Pennisetum clandestinum* Hochst. Kikuyu grass. *Emp. J. exp. Agric.* 5: 371–376.

Edwards, D.C., 1939. National reversion to grass. *East Afr. agric. J.* 4: 411–414.

Edwards, D.C., 1940a. The reaction of Kikuyu grass (*Pennisetum clandestinum*) herbage to management. *Emp. J. exp. Agric.* 8: 101–110.

Edwards, D.C., 1940b. A vegetation map of Kenya with particular reference to grassland types. *J. Ecol.* 28: 377–385.

Edwards, D.C., 1940c. Pasture and fodder grasses of Kenya. *East Afr. agric. J.*: 5: 248–254.

Edwards, D.C., 1941a. The progress of grassland research in East Africa. *East Afr. agric. J.* 6: 141-143.

Edwards, D.C., 1941b. A new fodder crop: perennial Kavirondo Sorghum. *East Afr. agric. J.* 6: 183–185.

Edwards, D.C., 1941c. The possibility of establishing grass under maize. *East Afr. agric. J.* 6: 233–234.

Edwards, D.C., 1942. Grass-burning. *Emp. J. exp. Agric.* 10: 219–231.

Edwards, D.C., 1943. The Nzoia type of Rhodes grass (*Chloris gayana*) for temporary leys. *East Afr. agric. J.* 9: 62–68.

Edwards, D.C., 1945. Horn of Africa (including Kenya) Vegetation. Govt Printer, Nairobi.

Edwards, D.C., 1948. Kavirondo Perennial Sorghum: an improved type R2S42. *East Afr. agric. J.* 13: 202.

Edwards, D.C., 1950a. Grassland research in Kenya. Colloquium on Grassland Research, 1949. *East Afr. agric. J.* 15: 208–211.

Edwards, D.C., 1950b. Pasture and fodder. *Ann. Rev.* 1950, Kenya Dep. Agric. Nairobi.

Edwards, D.C., 1956. The ecological regions of Kenya: their classification in relation to agricultural development. *Emp. J. exp. Agric.* 24: 89–108.

Edwards, D.C. & A.V. Bogdan, 1951. Important Grassland Plants of Kenya. Pitman.

Eggeling, W.J., 1947. An Annotated List of the Grasses of the Uganda Protectorate. Govt Printer, Entebbe.

Eichinger, A., 1914. Die Weide und Futterverhältnisse von West Usambara. *Der Pflanzer* 10: 1–97.

Etherington, J.W., 1943. Trials with *Bromus marginatus* on the Kinangop, Kenya Colony. *East Afr. agric. J.* 9: 33–34.

Etherington, J.W., 1945. More about grass leys on the Kinangop, Kenya. *East Afr. agric. J.* 11: 67–68.

Evans, A.C. & H.W. Mitchell, 1962. Soil fertility studies in Tanganyika. I. Improvement of crop and grass production on a leached sandy soil in Bukoba. *East Afr. agric. For. J.* 27: 189–196.

FAO, 1969. Report of the ad hoc technical conference on grassland management and fodder production south of the Sahara. Nairobi. Report PL; 1969/M/6: 1–32.

FAO-UNESCO, 1974. Soil Map of the World. UNESCO, Paris.

FAO, 1982. Production Yearbook. FAO, Rome.

Farnworth, J., 1976. Irrigated forage production under extreme arid zone conditions in Saudi Arabia. *Exp. Agric.* 12: 177–187.

Ferguson, H., 1948. Equatoria Province. In: J.D. Tothill (Ed.), Agriculture in the Sudan. London.

Ferguson, J.E., 1990. Overview of seed supply of forage species in tropical Latin America. *Int. Herb. Seed Prod. Res. Group Newsl.* 13: 9–13.

Ferguson, J.E. & L.V. Crowder, 1974. Cytology and breeding behaviour of *Brachiaria ruziziensis* Germain et Evrard. *Crop Sci.* 14: 893–895.

Fernandes, A.B., 1971. Interspecific hybridization of *Pennisetum purpureum* Schumach. with *Pennisetum typhoides* Stapf and Hubb. Seminar Forage Crops Research, University of Ibadan.

Fiennes, R.N. T.-W.-, 1940. Grasses as weeds of pasture land in northern Uganda. *East Afr. agric. J.* 5: 255–258.

Foster, H.L., 1969. The effects of different management and fertilizer treatments on the soil and leaf nutrient status of an elephant grass ley. *East Afr. agric. For. J.* 34: 468–475.

Foster, H.L., 1970. Liming continuously cultivated soils in Uganda. *East Afr. agric. For. J.* 36: 58–69.

Foster, H.L., 1971. Crop yields after different elephant grass ley treatments at Kawanda Research Station, Uganda. *East Afr. agric. For. J.* 37: 63–72.

Foster, H.L., 1972. Comparison of the yield response of groundnuts to single and triple superphosphate in Uganda. *East Afr. agric. For. J.* 38: 103–113.

Frame, J. & P. Newbould, 1986. Agronomy of white clover. *Adv. Agron.* 40: 1–88.

Frances, P.A. & A.N. Atta-Krah, 1989. Sociological and ecological factors in technology adoption of fodder trees in south east Nigeria. *Exp. Agric.* 25: 1–10.

Fredericksen, J.H. & J.A. Kategile, 1980. The effects of nitrogen fertilization and time of cutting in first growth in *Brachiaria brizantha* on yield, crude protein content and in vitro digestibility. *Trop. Anim. Prod.* 5: 136–143.

Fredeslund, A. & J. Cassady, 1969. Seeding grasses on denuded Kenya bushland. *Kenya Farmer*, November; 7, 30.

French, M.H., 1935. The place of silage in the dietary of livestock. *East Afr. agric. J.* 1: 241–243.

French, M.H., 1938a. The composition, digestibility and nutritive value of the tops and tubers of the edible canna. *Ann. Rep. Dep. Vet. Sci. Anim. Husb. Tanganyika* 1938: 37–43.
French, M.H., 1938b. The nutritive value of velvet beans. *Ann. Rep. Dep. Vet. Sci. Anim. Husb. Tanganyika* 1938: 43–45.
French, M.H., 1938c. The conservation of green fodders for the dry-season feeding of stock. *East Afr. agric. J.* 4: 206–210.
French, M.H., 1939. The comparative feeding values of grass when fed green, as hay, and as silage. *East Afr. agric. J.* 5: 261–264.
French, M.H., 1940. Some recent observations on the feeding values of some local foodstuffs. *East Afr. agric. J.* 6: 87–90.
French, M.H., 1941. The nutritive values of some Tanganyika grasses. *Emp. J. exp. Agric.* 9: 23–28.
French, M.H., 1943a. The feeding value of green lucerne hay. *East Afr. agric. J.* 8: 241–243.
French, M.H., 1943b. Feeding value of stubble from maize, millet and bulrush millet. *East Afr. agric. J.* 9: 88–89.
French, M.H., 1945. The composition of different types of star grass. *East Afr. agric. J.* 11: 100–103.
French, M.H., 1946a. Composition and feeding values of green maize, millet and bulrush millet cut for soiling purposes. *East Afr. agric. J.* 11: 159–161.
French, M.H., 1946b. Mineral deficiencies and excesses in pastures, a review. *East Afr. agric. J.* 11: 189–191.
French, M.H., 1949. Some use and misuse of shrubs and trees as fodder, a review. *East Afr. agric. J.* 14: 157–165.
French, M.H., 1950. The nutritive value of East African grass and fodder plants. Colloquium on Grassland Research, 1949. *East Afr. agric. J.* 15: 214–220.
French, M.H., 1956. The nutritive value of East African hay. *Emp. J. exp. Agric.* 24: 53–60.
French, M.H., 1957. Nutritional value of tropical grasses and fodders. Review Article. *Herb. Abstr.* 27: 1–9.
French, M.H. & H.P. Ledger, 1957. Liveweight changes of cattle in East Africa. *Emp. J. exp. Agric.* 25: 10–18.
French, M.H. & J.T. Purvis, 1935. Maize silage in East Africa: Its preparation, quality and feeding value. *East Afr. agric. J.* 1: 229–232.
Fritz, J., 1971. Recherches sur les productions fourragères effectuées par l'IRAT-Réunion 1963–1970. *Agron. Trop.* 26: 1248–1269.
Fröman, B., 1975. Pasture Management in Ethiopia with special reference to the conditions in the Chilalo Awraja. *Reports and Dissertations* 32, Uppsala.
Fröman, B. & S. Persson, 1974. Grasses of Ethiopia. CADU, Asella.
Froment, D., 1960. Aménagement et exploitation des pâturages à dominance de "Hyparrhenia" de la région de Nioka. *Bull. Inf. INEAC* 9: 49–70.
Gaboryaheze, A., 1984. Pasture research in Burundi. In: J.A. Kategile (Ed.), Pasture Improvement Research in Eastern and Southern Africa. Proc. Workshop Harare 1984, IDRC-237e, Ottawa.
Gastel, A.J.G. van, 1978. Napier grass breeding. Kitale report.
Gathecha, T.W., 1970. The maintenance and improvement of soil fertility under arable crops and grass leys in the lst and 2nd rotation cycles of a fertilizer trial at Embu. *East Afr. agric. For. J.* 35: 246–253.
Gebrehiwot, L. & A. Tadesse, 1984. Pasture research and development in Ethiopia. In: J.A. Kategile (Ed.), Pasture Improvement Research in Eastern and Southern Africa. Proc. Workshop Harare 1984, IDRC-237e, Ottawa.
Germain, R., 1954. Considérations agrostologiques relatives au Congo Belge et au Rwanda Burundi. *Bull. Inf. INEAC* 3: 347–366.
Germain, R. & A. Scaut, 1960. Herbage and nutritional aspects of animal husbandry in the equatorial forests of the Congo. *Proc. 8th int. Grassld Congr., Reading*: 371–374.
Gethin Jones, G.H., 1942. The effect of a leguminous cover crop on building up soil fertility. *East Afr. agric. J.* 8: 48–52.
Geus, J.G. de, 1973. Fertilizer Guide for the Tropics and Subtropics. Centre d'Etude de l'Azote, Zurich.
Gihad, E.A., 1976. Studies on the nutritional value of pasture species in Zambia. *East Afr. agric. For J.* 41: 335–339.
Gildenhuys, P.J., 1950. A new fodder grass; a promising cross between Babala and Napier fodder. *Fmg S. Afr.* 25: 151–153.
Gillet, J.B., 1952. The genus Trifolium in southern Arabia and in Africa south of the Sahara. *Kew Bull.* 3: 367–404.
Gillman, C., 1949. A vegetation-types map of Tanganyika Territory. *Geogr. Rev.* 39: 7–37.
Glover, J., 1948. Water demands by maize and sorghum. *East Afr. agric. J.* 13: 171.
Glover, J., 1959. The apparent behaviour of sorghum stomata during and after drought. *J. agric. Sci.* 53: 412–416.
Glover, J. & W.R. Birch, 1962. The effect of rainfall and age on the yield of some unfertilised fodder crops in Kenya. *J. agric. Sci.* 58: 53–57.
Glover, J. & H.W. Dougall, 1961. Milk production from pastures. *J. agric. Sci.* 56: 261–264.
Glover, J., D.W. Duthie & H.W. Dougall, 1960a. A table of estimates of nutritive values of ruminant feeds. *East Afr. agric. For. J.* 26: 117–118.
Glover, J., D.W. Duthie & H.W. Dougall, 1960b. The total digestible energy of ruminant feeds. *J. agric. Sci.* 55: 403–408.
Glover, J., D.W. Duthie & M.H. French, 1957. The apparent digestibility of crude protein by the ruminant. I. A synthesis of the results of digestibility trials with herbage and mixed feeds. *J. agric. Sci.* 48: 373–378.
Glover, J. & M.H. French, 1957. The apparent digestibility of crude protein by the ruminant. IV. The effect of crude fibre. *J. agric. Sci.* 49: 78–80.
Glover, P.E., J. Glover & M.D. Gwynne, 1962. Light rainfall and plant survival in E. Africa, II. Dry grassland vegetation. *J. Ecol.* 50: 199–206.
Glover, P.E., Joyce Stewart & M.D. Gwynne, 1966. Masai and Kipsigis notes on East African Plants. *East Afr. agric. For. J.* 32: 184–207.
Goldson, J.R., 1963. Observations on the grazing behaviour of grade dairy cattle in a tropical climate. *East Afr. agric. For. J.* 29: 72–77.
Goldson, J.R., 1967. Under sowing as a means of establishing the ley in western Kenya. *East Afr. agric. For. J.* 32: 274–281.
Goldson, J.R., 1968. Standard recommendations for the Coast Province of Kenya (Napier grass). Kikambala report.
Goldson, J.R., 1976a. The effect of the night boma on the pasture and farm fertility on the smallholder farm. Kitale report (PRP/16).

Goldson, J.R., 1976b. The present state of knowledge on pastures and fodders in Taita district. Kitale report (PRP/18).

Goldson, J.R., 1977. Napier grass in East Africa, a review. Kitale report (PRP/24).

Goldson, J.R., P. Bauer & A.G.O. Okech, 1977. A one year study of the dairy enterprise on six smallholdings in western Kenya. Kitale report (PRP/19).

Gosnell, J.M., 1963. Vleiland development in West Kenya. I. Pasture introduction. *East Afr. agric. For. J.* 29: 99–105.

Gosnell, J.M. & E.A. Weiss, 1965. Vleiland development in West Kenya. II. Pasture fertilizer and grazing trials. *East Afr. agric. For. J.* 30: 169–176.

Graham, M.D., 1945a. Some notes on soil fertility with particular reference to African farmers. *East Afr. agric. J.* 11: 3–11.

Graham, M.D., 1945b. Golf greens and lawns in East Africa. *East Afr. agric. J.* 11: 94–99.

Gramshaw, D., J.W. Read, W.J. Collins & E.D. Carter, 1989. Sown pastures and legume persistence: an Australian overview. In: G.C. Marten *et al.* (Eds), Persistence of Forage Legumes. American Society of Agronomy.

Greenway, P.J., 1933. The vegetation of Mpwapwa, Tanganyika Territory. *J. Ecol.* 21: 28–43.

Greenway, P.J., 1943. Second draft report on vegetation classification. East Afr. Pasture Res. Conf., Nairobi.

Griffith, G. ap., 1949. A note on the nitrate content of soil under Pennisetum purpureum. *East Afr. agric. J.* 14: 187–188.

Griffith, G. ap., 1951. Factors influencing nitrate accumulation in a Uganda soil. *Emp. J. exp. Agric.* 19: 1–12.

Griffiths, J.F. (Ed.), 1972. Climates of Africa. World Survey of Climatology, Volume 10. Elsevier.

Griffiths, R.G., 1990. Pasture seed production in eastern and southern Africa. *Int. Herb. Seed Prod. Res. Group Newsl.* 13: 14–16.

Grimes, R.C. & R.T. Clarke, 1962. Continuous arable cropping with the use of manure and fertilizers. *East Afr. agric. For. J.* 28: 74–80.

Grof, B., 1968. Viability of seed of *Brachiaria decumbens. Qld. J. agric. Anim. Sci.* 25: 149–152.

Grof, B. & W.A.T. Harding, 1970. Dry matter yields and animal production of Guinea grass (*Panicum maximum*) on the humid tropical coast of North Queensland. *Trop. Grasslds* 4: 85–95.

Gunn, H., 1930. Summary of Report on Pasture Investigations in Kenya, 1926–1929. Rowett Res. Inst.

Gusman, H.W., W.R. Cowley & J.H. Barton, 1954. Reaction of some grasses to artificial salinization. *Agron. J.* 46: 412–414.

Gwynne, M.D., 1966. Plant physiology and the future. In: W. Davies & C.L. Skidmore (Eds), Tropical Pastures. Faber & Faber.

Haarer, A.E., 1948. East African leys and pastures. *Country Life*, U.K. April.

Haarer, A.E., 1951. Grasses and grazing problems in East Africa. *World Crops* 3: 8.

Hacker, J.B., 1966. Cytological investigations in the *Setaria sphacelata* complex. *Aust. J. agric. Res.* 17: 297–301.

Hacker, J.B., 1985. Breeding tropical grasses for ease of vegetative propagation and for improved seed production. *Proc. 15th int. Grassld Congr., Kyoto*: 251–252.

Hacker, J.B. & R.J. Jones, 1969. The *Setaria sphacelata* complex, a review. *Trop. Grasslds* 3: 13–34.

Hacker, J.B. & R.J. Jones, 1971. The effect of nitrogen fertilizer and row spacing on seed production of *Setaria sphacelata. Trop. Grasslds* 5: 61–73.

Haile, A., 1983. Ecological factors influencing pasture and plant selection criteria in the Central Highlands of Ethiopia. *Eth. J. agric. Sci.* 5: 50–63.

Halcrow, M., 1937. Experimental work on pastures in the Trans Nzoia District, Kenya Colony, 1935–36. *East Afr. agric. J.* 3: 109–112.

Halcrow, M., 1938. Pasture problems in the Trans Nzoia. *East Afr. agric. J.* 3: 290–293.

Hall, M., 1966. Some economic aspects of dairy farming with reference to East Africa. In: R.S. Musangi (Ed.), Pasture Production and Utilization by Cattle in the Tropics with special Reference to Uganda. Seminar Makerere, 1966. *Techn. Bull.* 4: 40–57.

Hall, T.D. & B. Allen, 1938. Intensive grazing on Kenya veld. *East Afr. agric. J.* 4: 165–182.

Hanski, I. & Y. Cambefort, 1991. Dung Beetle Ecology. Princeton University Press.

Hanson, J. & J.R. Lazier, 1989. Forage germplasm at the International Livestock Centre for Africa (ILCA): an essential resource for evaluation and selection. *Proc. 16th int. Grassld Congr., Nice*: 265–266.

Harker, K.W., 1954. The establishment of *Chloris gayana* under a sorghum silage crop. *East Afr. agric. J.* 20: 54–56.

Harker, K.W., 1962a. Grassland development in the fertile crescent of Uganda. *J. Brit. Grassld Soc.* 17: 188–193.

Harker, K.W., 1962b. A fertilizer trial on *Paspalum notatum* pasture. I. The effects on yield. *East Afr. agric. For. J.* 27: 201–203.

Harker, K.W. & R.M. Bredon, 1963. The effect of elephant grass feeding and maize meal supplementation on indoor fed bullocks. I. Liveweight gains of cattle. *Trop. Agric.* (*Trin.*) 40: 307–312.

Harker, K.W. & A.K. Kamau, 1961. A preliminary survey of the nitrate content of various grasses. *East Afr. agric. For. J.* 27: 57–60.

Harker, K.W. & A.D. MacKay, 1962. A stocking rate trial on rough grazing in Buganda. *East Afr. agric. For. J.* 27: 220–222.

Harlan, J.R., 1970. *Cynodon* species and their value for grazing and hay. *Herb. Abstr.* 40: 233–238.

Harlan, J.R., J.M.J. de Wet & K.M. Rawal, 1970. Geographical distribution of the species of *Cynodon* L.C. Rich. (Gramineae). *East Afr. agric. For. J.* 36: 220–226.

Harrington, G.N., 1974. Fire effects on a Uganda savanna grassland. *Trop. Grasslds 8:* 87–101.

Harrington, G.N. & D.D. Thornton, 1969. A comparison of controlled grazing and manual hoeing as a means of reducing the incidence of *Cymbopogon afronardus* Stapf in Ankole pastures, Uganda. *East Afr. agric. For. J.* 35: 154–159.

Harrington, G.N. & D. Pratchett, 1974. Stocking rate trials in Ankole, Uganda. I. Weight gain of Ankole steers at intermediate and heavy stocking rates under different managements. *J. agric. Sci.* 82: 497–506.

Harris, W.V., 1949. Some aspects of the termite problem. *East Afr. agric. J.* 14: 151–155.

Harrison, M.N. & J.K. Jackson, 1958. Ecological Classification of the Vegetation of the Sudan. *Forests Dep. Bull.* 2, Khartoum.

Harrison, R.E., 1986. The role of improved pastures in commercial production in the tropics and sub-tropics. *Trop. Grasslds* 20: 3–17.

Harrop, J.F., 1970. Soils. In: J.D. Jameson (Ed.), Agriculture in Uganda. Oxford University Press.

Hatch, M.D., C.R. Slack & H.S. Johnson, 1967. Further studies on a new pathway of photosynthetic carbon dioxide fixation in sugar cane and its occurrence in other plant species. *Biochem. J.* 102: 417–422.

Heady, H.F., 1960. Range Management in East Africa. Govt Printer, Nairobi.

Heady, H.F., 1966. The influence of grazing on the composition of *Themeda triandra* grassland, East Africa. *J. Ecol.* 54: 705–727.

Healey, J.S., 1975. Kikuyugrass poisoning. Kikuyu – a research report. *Agric. Gaz. of N. S. Wales* 86 (5): 18.

Heath, M.E., R.F. Barnes & D.S. Metcalfe (Eds), 1985. Forages. Iowa State University Press.

Hedberg, O., 1971. The high mountain flora of the Galama Mountain in Arussi Province, Ethiopia. *Webbia* 26: 101–128.

Henderson, G.R. & P.T. Preston, 1959. Fodder Farming in Kenya. East African Literature Bureau, Nairobi.

Henderlong, P.R., 1973. Pasture research in Uganda, past, present and future. Seminar Uganda Soc. Agron., May.

Hendy, K., 1975. Review of natural pastures and their management problems on the north coast of Tanzania. *East Afr. agric. For. J.* 41: 52–57.

Henrici, Marguerite, 1931. Rhodes Grass. *Fmg S. Afr.* 6: 210.

Henzell, E.F., 1968. Sources of nitrogen for Queensland pastures. *Trop. Grasslds* 2: 1–17.

Hiddink, B. & A.P. Wouters, 1987. The nitrogen effect of different types of slurry on the yield and nutritive value of Napier grass. Naivasha report.

Hochman, Z. & K.R. Helyar, 1989. Climatic and edaphic constraints to the persistence of legumes in pastures. In: G.C. Marten *et al.* (Eds), Persistence of Forage Legumes. American Society of Agronomy.

Holme, R.V. & E.G.P. Sherwood, 1954. The fertilizer requirements of the Kenya Highlands. Colonial Res. Study 12. HMSO, London.

Hongyantarachai, S., G. Nithichai, N. Wongsuwan, S. Prasanpanich, S. Siwichai, S. Pratumsuwan, T. Tasapanon & B.R. Watkin, 1989. The effects of grazing versus indoor feeding during the day on milk production in Thailand. *Trop. Grasslds* 23: 8–14.

Hopkinson, D., 1969. Leguminous cover crops for maintaining soil fertility in sisal in Tanzania. *Exp. Agric.* 5: 283–294.

Hopkinson, D., 1970. The potential of some pasture plants in the sisal areas of the Tanga region of Tanzania. *East Afr. agric. For. J.* 35: 299–310.

Hopkinson, D. & J. Breitenstein, 1969. Tropical Kudzu as a cover crop in sisal. *World Crops* 21: 256–258.

Hopkinson, J.M., 1977. Siratro seed production. *Trop. Grasslds* 11: 33–39.

Hopkinson, J.M. & B.H. English, 1982. Spikelet population dynamics in seed crops of *Panicum maximum* Gatton. *Seed Sci. Techn.* 10: 379–403.

Hopkinson, J.M., English B.H. & R.L. Harty, 1988. Effect of different drying patterns on quality of some tropical pasture grasses. *Seed Sci. Techn.* 16: 361–369.

Horrell, C.R., 1958a. Herbage plants at Serere Experiment Station, Uganda, 1954–1957. I. Grasses. *East Afr. agric. J.* 24: 41–46.

Horrell, C.R., 1958b. Herbage plants at Serere Experiment Station, Uganda, 1954–1957. II. Legumes. *East Afr. agric. J.* 24: 133–138.

Horrell, C.R., 1963. Herbage plants at Serere, Uganda, 1957–1961. *East Afr. agric. For. J.* 28: 174–180.

Horrell, C.R., 1964. Effect of two legumes on the yield of unfertilised pastures at Serere. *East Afr. agric. For. J.* 30: 94–96.

Horrell, C.R. & R.M. Bredon, 1963. Management studies with *Panicum maximum* in Uganda. I. Effect of cutting interval and nitrogen fertilizer on yield. *Emp. J. exp. Agric.* 31: 334–342.

Horrell, C.R. & M.N. Court, 1965. Effect of the legume *Stylosanthes gracilis* on pasture yields at Serere, Uganda. *J. Brit. Grassld Soc.* 20: 72–76.

Horrell, C.R. & P.W. Newhouse, 1965. Yields of sown pastures in Uganda as influenced by legumes and fertilizers. *Proc. 9th int. Grassld Congr., Sao Paulo*: 1133–1136.

Horrell, C.R. & G.E.D. Tiley, 1970. Grassland. In: J.D. Jameson (Ed.), Agriculture in Uganda. Oxford University Press.

Hosking, H.R. & A.L. Stephens, 1941. Pastures and grasses on the Serere Experiment Station, Uganda. *East Afr. agric. J.* 6: 213–219.

Houérou, H.N. le, 1980. Browse in Africa. ILCA, Addis Ababa.

Houérou, H.N. le & G.F. Popov, 1981. An Eco-climatic Classification of Intertropical Africa. FAO, Rome.

Howard, D.A., 1963. Notes on animal diseases. XXIII. Mineral deficiency diseases. *East Afr. agric. For. J.* 28: 191–194.

Howard, D.A., M.L. Burdin & G.H. Lampkin, 1962. Variation in the mineral and crude-protein content of pastures at Muguga in the Kenya highlands. *J. agric. Sci.* 59: 251–256.

Hubbard, C.E., 1927. East African Pasture Plants. Crown Agents.

Huffnagel, H.P., 1961. Agriculture in Ethiopia. FAO, Rome.

Humphreys, L.R., 1967. Buffel grass (*Cenchrus ciliaris*) in Australia. *Trop. Grasslds* 1: 123–134.

Humphreys, L.R., 1980. Deficiencies of adaptation of pasture legumes. *Trop. Grasslds* 14: 153–158.

Humphreys, L.R., 1981. Environmental Adaptation of Tropical Pasture Plants. Macmillan, London.

Humphreys, L.R., 1991. Tropical Pasture Utilization. Cambridge University Press.

Humphreys, L.R. & F. Riveros, 1986. Tropical Pasture Seed Production. FAO, Rome.

Hutchinson, J., H.L. Manning & H.G. Farbrother, 1958. On the characterization of tropical rainstorms in relation to runoff and percolation. *Quart. J. Royal met. Soc.* 84: 250–258.

Hutchison, H.G., 1965. Ranching in Tanganyika. *East Afr. agric. For. J.* 30: 268–275.

Hutton, E.M., 1961. Intravariety variation in Rhodes grass (*Chloris gayana* Kunth). *J. Brit. Grassld Soc.* 16: 23–29.

Hutton, E.M., 1962. Siratro, a tropical pasture legume bred from *Phaseolus atropurpureus*. *Aust. J. exp. Agric. Anim. Husb.* 2: 117–125.

Hutton, E.M., 1970. Tropical Pastures. *Adv. Agron.* 22: 1–73.

Huxley, E., 1935. Lord Delamere and the Making of Kenya. Chatto & Windus.

Ibrahim, K.M. & A. Orodho, 1983. Test adaptation and evaluation trials of forage plants in major ecological regions of Ethiopia and Kenya. *Proc. 14th int. Grassld Congr., Lexington (Kentucky)*, 1981: 240–243.

Ikombo, B.M., 1984. Effects of farmyard manure and fertilizer on maize in semiarid areas of eastern Kenya. *East Afr. agric. For. J.* 44 (special issue): 266–274.

International Board for Plant Genetic Resources (IBPGR), 1984. Forage and Browse Plants for Arid and Semi-arid Africa. Kew.

International Seed Testing Association (ISTA), 1985. International Rules for Seed Testing. *Seed Sci. Techn.* 13 (2).

Irwin, J.A.G., 1989. Diseases of pasture legumes in Australia. In: G.C. Marten *et al.* (Eds.), Persistence of Forage Legumes. American Society of Agronomy.

Ishag, H.I., 1965. Fodder crops in the Sudan Gezira. *Afr. Soils* 10: 441–444.

Itulya, S., 1975. A summary of the experimental work carried out between 1954–1975 at Nyandarua Agricultural Research Station, Ol Joro Orok. Ol Joro Orok report.

Ivens, G.W., 1971. East African Weeds and Their Control. Oxford University Press.

Jackson, D.C., 1958. Some preliminary indications of soil moisture changes under arable crops. *Rhod. agric. J.* 55: 426–438.

Jackson, G., 1951. Pasture and grass testing investigations. *Nyasaland agric. quart. J.* 10: 110–116.

Jahnke, H.F., 1982. Livestock Production Systems and Livestock Development in Tropical Africa. Vauk, Kiel.

Jama, B., G. Amare, D. Ngugi & B. Macklin, 1986. Leucaena alley-cropping for Kenya Coast. In: R.T. Prinsley & M.J. Swift (Eds), Amelioration of Soil by Trees. ICRAF, Nairobi.

Jameson, J.D., 1970. Agriculture in Uganda. Oxford University Press.

Jameson, J.D. & R.K. Kerkham, 1960. The maintenance of soil fertility in Uganda. I. Soil fertility experiment at Serere. *Emp. J. exp. Agric.* 28: 179–192.

Jätzold, R. & H. Schmidt, 1982. Farm Management Handbook of Kenya. GTZ/Ministry of Agriculture, Nairobi.

Jewitt, T.N. & H.W.B. Barlow, 1949. Animal excreta in the Sudan Gezira. *Emp. J. exp. Agric.* 17: 1–17.

Joblin, A.D.H., 1960. The influence of night grazing on the growth rates of Zebu cattle in East Africa. *J. Brit. Grassld Soc.* 15: 212–215.

Joblin, A.D.H., 1962. The use of grazing-animal observations in the early stages of pasture evaluation in the tropics. *J. Brit. Grassld Soc.* 17: 171–177; 274–281.

Joblin, A.D.H., 1963. Strip versus paddock grazing under tropical conditions. *J. Brit. Grassld Soc.* 18: 69–73.

Jones, D.I.H., 1964. Mineral content of some cultivated grasses grown in Northern Rhodesia. *Rhod. J. agric. Res.* 2: 57–59.

Jones, P.A. & J.B.P. Robinson, 1965. Response of mature Napier grass to fertilizers and cattle manure in Kenya. *East Afr. agric. For. J.* 30: 276–285.

Jones, R.J., 1990. Phosphorus and beef production in northern Australia. 1. Phosphorus and pasture productivity – a review. *Trop. Grasslds* 24: 131–139.

Jones, R.J. & T.R. Evans, 1961. The establishment and production of a ley. *East Afr. agric. For. J.* 27: 85–90.

Jones, R.J. & R.M. Jones, 1971. Agronomic factors in pasture and forage crop production in tropical Australia. *Trop. Grasslds* 5: 229–244.

Jones, R.J. & R. Roe, 1976. Seed production, harvesting and storage. In: N.H. Shaw & W.W. Bryan (Eds), Tropical Pasture Research: Principles and Methods. CAB Bull. 51.

Jones, R.J., A.A. Seawright & D.A. Little, 1970. Oxalate poisoning in animals grazing the tropical grass *Setaria sphacelata. J. Aust. Inst. agric. Sci.* 36: 41–43.

Jones, R.M. & E.D. Carter, 1989. Demography of pasture legumes. In: G.C. Marten *et al.* (Eds), Persistence of Forage Legumes. American Society of Agronomy.

Jones, R.M., J.C. Tothill & R.J. Jones, 1984. Pastures and Pasture Management in the Tropics and Sub-tropics. Trop. Grassld Soc. Australia. Occasional Publ. No. 1.

Jordan, S.M., 1957. Reclamation and pasture management in the semi-arid areas of Kitui district, Kenya. *East Afr. agric. J.* 23: 84–88.

Juko, C.D. & R.M. Bredon, 1961. The chemical composition of leaves and whole plant as an indicator of the range of available nutrients for selective grazing by cattle. *Trop. Agric. (Trin.)* 38: 179–187.

Jurion, F. & J. Henry, 1967. De l'agriculture itinérante à l'agriculture intensifiée. *INEAC Hors Série*, Bruxelles.

Kahurananga, J. & A. Tsehay, 1991. Variation in flowering time, dry matter and seed yield among annual *Trifolium* species, Ethiopia. *Trop. Grasslds* 25: 20–25.

Karachi, M.K., 1982. The performance of sweet potato (*Ipomoea batatas* L.) Lam in western Kenya. *East Afr. agric. For. J.* 47: 55–67.

Karachi, M.K., 1983. Anthracnose attack, seed and dry matter production of *Stylosanthes* accessions at Alupe, Kenya. *East Afr. agric. For. J.* 48: 92–96.

Karachi, M.K., 1988. Yield and quality of Lablab purpureus ecotypes in high altitude semi-arid areas in Kenya. In: B.H. Dzowela (Ed.), African Forage Plant Genetic Resources, Evaluation of Forage Germplasm and Extensive Livestock Production Systems. Proc. Workshop Arusha 1987, ILCA, Addis Ababa.

Karanja, G.M., 1981. Bana grass. *Kenya Farmer*, October: 50–52.

Karanja, G.M., 1988. Napier grass for dry season use. *Proc. Anim. Prod. Soc. Kenya* 17: 1–6.

Karue, C.N., 1974. The nutritive value of herbage in semi-arid lands of East Africa. I Chemical composition. *East Afr. agric. For. J.* 40: 89–95.

Karue, C.N., 1975a. The nutritive value of herbage in semi-arid lands of East Africa. II. Seasonal influence on the nutritive value of *Themeda triandra. East Afr. agric. For. J.* 40: 372–387.

Karue, C.N., 1975b. Intake and digestibility of *Themeda triandra* hay by age-paired Boran and Hereford steers. *East Afr. agric. For. J.* 41: 35–41.

Kategile, J.A. (Ed.), 1984. Pasture Improvement Research in Eastern and Southern Africa. Proc. Workshop Harare 1984, IDRC-237e, Ottawa.

Kategile, J.A. & A.N. Said (Eds), 1981. Utilization of Low Quality Roughages in Africa. Proc. Workshop Arusha 1981. IDRC, Ottawa.

Kategile, J.A., A.N. Said & B.H. Dzowela (Eds), 1987. Animal Feed Resources for Small-Scale Livestock Producers. Proc. 2nd PANESA Workshop Nairobi, 1985. IDRC-MR165e, Ottawa.

Kayongo-Male, H., C.N. Karue & E.R. Mutiga, 1977. The effect of supplementation on the growth of dairy heifers grazed on medium quality pasture under East African conditions. *East Afr. agric. For. J.* 42: 432–440.

Keay, R.W.J., 1955. Montane vegetation and flora in the British Cameroons. *Proc. Linn. Soc. London* 165: 140–143.

Kennan, T.C.D., 1952. The improved West African Napier fodders. *Rhod. agric. J.* 49: 221–228.

Kennan, T.C.D., R.R. Staples & O. West, 1955. Veld management in Southern Rhodesia. *Rhod. agric. J.* 52: 4–21.

Kerkham, R.K., 1946. The carrying capacity of pastures at Serere Experiment Station. *East Afr. agric. J.* 11: 170–172.

Kerkham, R.K., 1947a. Grass fallow in Uganda. 1. Grazing during the fallow period. *East Afr. agric. J.* 13: 3–7.

Kerkham, R.K., 1947b. Effect of different systems of pasture management on yield of arable crops. Proc. agric. Res. Conf., Nairobi, 1947. *East Afr. agric. J.* 13: 78–79.

Kerkham, R.K., 1950. Recent pasture work in Uganda. Colloquium on Grassland Research, 1949. *East Afr. agric. J.* 15: 206–208.

Kerkham, R.K., H. Cronly & W.A. Allan, 1947. Live stock improvement in the Buganda province of Uganda. *East Afr. agric. J.* 13: 116–119.

Kerr, A.J., 1942. A new system of grass-fallow strip-cropping for the maintenance of soil fertility. *Emp. J. exp. Agric.* 10: 125–132.

Kevelenge, J.E.E., A.N. Said & B. Kiflewahid, 1983. The nutritive value of four arable farm by-products commonly fed to dairy cattle by small scale farmers in Kenya. *Trop. Anim. Prod.* 8: 162–186.

Keya, N.C.O., 1973. The effect of N.P. fertilizers on the productivity of Hyparrhenia grassland. *East Afr. agric. For. J.* 39: 195–200.

Keya, N.C.O., 1974. Grassland/legume pastures in Western Kenya. *East Afr. agric. For. J.* 39: 240–257.

Keya, N.C.O., 1975. Recommendations for growing pastures and fodder crops in Kenya. Kitale report.

Keya, N.C.O. & C.L.M. van Eijnatten, 1975. Studies on oversowing of natural grasslands in Kenya. *East Afr. agric. For. J.* 40: 261–263; 351–365.

Keya, N.C.O., F.J. Olsen & R. Holliday, 1971a. The role of superphosphate in the establishment of oversown tropical legumes in natural grasslands of western Kenya. *Trop. Grasslds* 5: 109–116.

Keya, N.C.O., F.J. Olsen & R. Holliday, 1971b. Oversowing improved pasture legumes in natural grasslands of the medium altitudes of Western Kenya. *East Afr. agric. For. J.* 37: 148–155.

Keya, N.C.O., F.J. Olsen & R. Holliday, 1972. Comparison of seed beds for over-sowing a *Chloris gayana* Kunth/*Desmodium uncinatum* (Jacq.) mixture in Hyparrhenia grassland. *East Afr. agric. For. J.* 37: 286–293.

Kidner, E.M., 1963. Milk yields in relation to pasture composition. *East Afr. agric. For. J.* 29: 125–130.

Kidner, E.M., 1966. Beef production from pasture. *East Afr. agric. For. J.* 31: 389–391; 32: 34–36 & 91–95.

Kiflewahid, B., G. Potts & R. Drysdale (Eds), 1982. By-product Utilization for Animal Production. Proc. Workshop Nairobi, 1982. IDRC, Ottawa.

Kolbe, L., 1961. Eradication of Kikuyugrass. Machinery tests in the Kiambu district. *Kenya Farmer*, May.

Kretschmer, A.E., 1989. Tropical forage legume development, diversity, and methodology for determining persistence. In: G.C. Marten *et al.* (Eds), Persistence of Forage Legumes. American Society of Agronomy.

Kulich, J., 1985. The potential for seed production of tropical pasture species in Zambia. *Proc. 15th int. Grassld Congr., Kyoto*: 262–263.

Kulich, J., 1988. The potential of pasture legumes and their role in improving farming practices under extensive livestock production systems in Zambia. In: B.H. Dzowela (Ed.), African Forage Plant Genetic Resources, Evaluation of Forage Germplasm and Extensive Livestock Production Systems. Proc. Workshop Arusha 1987, ILCA, Addis Ababa.

Kulich, J. & E.M. Kaluba, 1984. Pasture research and development in Zambia. In: J.A. Kategile (Ed.), Pasture Improvement Research in Eastern and Southern Africa. Proc. Workshop Harare 1984, IDRC-237e, Ottawa.

Kusewa, P.K., 1974. Review of pasture research work conducted from Embu Agricultural Research Station in Central Kenya. Embu report.

Lang Brown, J.R. & J.F. Harrop, 1962. The ecology and soils of the Kibale grasslands, Uganda. *East Afr. agric. For. J.* 27: 264–272.

Langdale-Brown, L., H.A. Osmaston & J.G. Wilson, 1964. The Vegetation of Uganda and its Bearing on Land-use. Govt Printer, Entebbe.

Langer, R.H.M. & G.J.A. Ryle, 1958. Vegetative proliferations in herbage grasses. *J. Brit. Grassld Soc.* 13: 29–33.

Langlands, B.W., 1967. Burning in Eastern Africa. *East Afr. Geogr. Rev.* 5: 21–37.

Lansbury, T.J., 1959. The composition and digestibility of some conserved fodder crops for dry season feeding in Ghana. *Trop. Agric. (Trin.)* 36: 305–308.

Larbi, A., P. Mislevy, M.B. Abdej & W.F. Brown, 1990. Seasonal herbage and animal production from three *Cynodon* species. *Trop. Grasslds* 24: 305–310.

Law, R. & P.J.M. Cooper, 1976. The effect and importance of soil temperature in determining the early growth rate and final grain yields of maize in western Kenya. *East Afr. agric. For. J.* 41: 189–200.

Lawton, R.M., 1967. The value of browse in the dry tropics. *East Afr. agric. For. J.* 33: 227–230.

Lazier, J.R., 1989. Forage legume productivity in a medium altitude acid soil area of Ethiopia. *Proc. 16th int. Grassld Congr., Nice*: 201–204.

Lebrun, J., 1956. La végétation et les territoires botaniques du Ruanda-Urundi. In: Le Ruanda-Urundi, Ses Resources Naturelles, Ses Populations: 22–48. Bruxelles.

Lebrun, J., 1960. Sur les horizons et étages de végétation de divers volcans du massif des Virunga (Kivu-Congo). *Bull. Jard. bot.*, Bruxelles. 30: 255–277.

Ledger, H.P., 1950. The effect of lime on a *Chloris gayana* ley. *East Afr. agric. J.* 16: 43–46.

Lehman, W.F., V.L. Marble & M.W. Nielson, 1985. Key factors in the rapid development and seed increase of alfalfa with insect resistance. *Proc. 15th int. Grassld Congr., Kyoto*: 232–233.

Le Mare, P.H., 1959. Soil fertility studies in three areas of Tanganyika. *Emp. J. exp. Agric.* 27: 197–221.

Le Mare, P.H., 1972. A long term experiment on soil fertility and cotton yields in Tanzania. *Exp. Agric.* 8: 299–310.

Le Pelley, R.H., 1959. Agricultural Insects of East Africa. East African Literature Bureau, Nairobi.

Lenné, Jillian M., 1986. Recent advances in the understanding of Anthracnose of Stylosanthes in tropical America. *Proc. 15th int. Grassld Congr., Kyoto*: 773–776.

Lenné, Jillian M. & R.M. Sonoda, 1985. Diseases of *Macroptilium atropurpureum* – a review. *Trop. Grasslds* 19: 28–34.

Lenné, Jillian M. & Julie M. Stanton, 1990. Diseases of Desmodium – a review. *Trop. Grasslds* 24: 1–14.

Léonard, A., 1962. Les savanes herbeuses du Kivu. *Sér. sci. INEAC* 95.

Lewis, A.H., J. Procter & A.E.M. Hood, 1960. A comparison of ley and arable farming systems. *J. agric. Sci.* 54: 310–317.

Lex-Blake, A.J. (Ed.), 1957. Gardening in East Africa. Longman.

Lind, E.M. & M.E.S. Morrison, 1974. East African Vegetation. Longman.

Loch, D.S., 1977. *Brachiaria decumbens* (Signal grass), a review with particular reference to Australia. *Trop. Grasslds* 11: 141–157.

Loch, D.S., 1980. Selection of environment and cropping system for tropical grass seed production. *Trop. Grasslds* 14: 159–168.

Lock, J.M., 1972. The effects of hippopotamus grazing on grasslands. *J. Ecol.* 60: 445–467.

Long, M.I.E., 1967. Investigations on the in vitro digestibility techniques used under East African conditions. *East Afr. agric. For. J.* 33: 166–169.

Long, M.I.E., D.D. Thorton & B. Marshall, 1969. Nutritive value of grasses in Ankole and the Queen Elisabeth National Park, Uganda. II. Crude protein, crude fibre and soil nitrogen. *Trop. Agric. (Trin.)* 46: 31–42.

Long, M.I.E., W.K. Ndyanabo, B. Marshall & D.D. Thorton, 1969. Nutritive value of grasses in Ankole and the Queen Elisabeth National Park, Uganda. II. Mineral content. *Trop. Agric. (Trin.)* 46: 201–209.

Long, M.I.E., D.D. Thorton, W.K. Ndyanabo, B. Marshall & H. Ssekaalo, 1970. The mineral status of dairy farms in the parts of Buganda and Busoga bordering Lake Victoria, Uganda. II. Nitrogen and mineral content of pastures. *Trop. Agric. (Trin.)* 47: 37–50.

Long, M.I.E., B. Marshall, W.K. Ndyanabo & D.D. Thorton, 1972. Mineral status of dairy farms in eastern Uganda. II. Nitrogen and mineral content of grasses and some mineral contents of bovine plasma. *Trop. Agric. (Trin.)* 49: 227–234.

Low, A.J., 1955. Improvements in the structural state of soils under leys. *J. Soil Sci.* 6: 179–199.

Low, A.J., F.J. Piper & P. Roberts, 1963. Soil changes in ley-arable experiments. *J. agric. Sci.* 60: 229–238.

Lugenja, M.M.S., E.W. Rowland & C. Senga, 1977. The effect of cutting interval on the yield and quality of *Setaria splendida* (Stapf) and *Medicago sativa* (L) under conditions prevailing in northern Tanzania. *East Afr. agric. For. J.* 43: 106–110.

Lyne Watt, W., 1929. Notes on trials of leguminous plants at the Scott Agricultural Laboratories. *Dep. Agric. Nairobi. Bull.* 1.

Lwoga, A.B., 1981. Nitrogen on East African pasture. *Proc. 3rd Conf. Ass. Adv. agric. Sci. Afr. (AAASA), Ibadan* 11: 133–162.

Lwoga, A.B., M.M.S. Lugenja & A.R. Kajuni, 1984. Pasture research in Tanzania. In: J.A. Kategile (Ed.), Pasture Improvement Research in Eastern and Southern Africa. Proc. Workshop Harare 1984, IDRC-237e, Ottawa.

McCown, R.L., G. Haaland & C. de Haan, 1979. The interaction between cultivation and livestock production in semi-arid Africa. In: A.E. Hall, G.H. Cannell & H.W. Lawton (Eds), Agriculture in Semi-Arid Environment. Springer.

McCown, R.L., B.A. Keating, M.E. Probert & R.K. Jones, 1992. Strategies for sustainable crop production in semi-arid Africa. *Outl. Agric.* 21: 21–31.

McCrae, A.W.R., 1975. Insect damage to exotic pasture in the southern highlands of Tanzania. *East Afr. agric. J.* 40: 335–336.

MacDonald, A.S., 1965. Variation in open pollinated sweet potato seedlings in Buganda. *East Afr. agric. For. J.* 31: 183–188.

MacDonald, A.S., 1966. Some aspects of ley establishment and management in the Kabanyolo Buganda environment. In: R.S. Musangi (Ed.), Pasture Production and Utilization by Cattle in the Tropics with special Reference to Uganda. Seminar Makerere, 1966. *Techn. Bull.* 4: 7–13.

McKay, A.D., 1970. Range management research at EAAFRO. *East Afr. agric. For. J.* 35: 346–349.

McKay, A.D., 1971. Seasonal and management effects on the composition and availability of herbage, steer diet and live-weight gains in a *Themeda triandra* grassland in Kenya. *J. agric. Sci.* 76: 1–26.

McKay, A.D. & P.E. Frandsen, 1969. Chemical and floristic components of the diet of zebu cattle (*Bos indicus*) in browse and grass range pastures in a semi-arid upland area of Kenya. I. Crude protein. *Trop. Agric. (Trin.)* 46: 276–292.

McKay, A.D., P.E. Frandsen, J.C. Odero, M. Sondergaard & S.P. Nganga, 1969. Chemical composition of herbage from Themeda grassland in Kenya before and after collection from oesophageal fistulated steers. *East Afr. agric. For. J.* 35: 190–196.

McNaughton, S.J., 1983. Serengeti grassland ecology: the role of composite environmental factors and contingency in community organization. *Ecological Monographs* 53: 291–320.

McNaughton, S.J., 1985. Ecology of a grazed ecosystem: the Serengeti. *Ecological Monographs* 55: 259–294.

McWalter, A.R. & R.H. Wimble, 1976. A long-term rotational and manurial trial in Uganda. *Exp. Agric.* 12: 305–317.

Maher, C., 1936a. Elephant grass (*Pennisetum purpureum*) as a cattle fodder in Kenya. *East Afr. agric. J.* 1: 340–342.

Maher, C., 1936b. Mixed farming in East Africa. II. Grassland and arable dairying in the Trans Nzoia District. *East Afr. agric. J.* 2: 12–27.

Maher, C., 1939. Hill culture. *East Afr. agric. J.* 5: 36–43.

Maher, C., 1950. Soil conservation in Kenya Colony. *Emp. J. exp. Agric.* 18: 137–149; 233–248.

Maitland, T.D., 1932. The grassland vegetation of the Cameroons Mountain. *Kew Bull.* 9: 417–425.

Maitland, T.D. & C.E. Hubbard, 1927. Uganda Grasses. *Kew Bull. Misc. Inf.* 272.

Mannetje, L. 't, 1966. A punched card key to species of *Trifolium* L. in African south of the Sahara excluding Ethiopia. *East Afr. agric. For. J.* 31: 261–270.

Mannetje, L. 't & R.M. Jones, 1990. Pasture and animal productivity of buffel grass with Siratro, lucerne or nitrogen fertilizer. *Trop. Grasslds* 24: 269–281.

Mannetje, L. 't & A.J. Pritchard, 1968. The effects of photoperiod and night temperature on flowering and growth in some African *Trifolium* species. *New Phytol.* 67: 257–263.

Mannetje, L. 't, R.J. Jones & T.H. Stobbs, 1976. Pasture evaluation by grazing experiments. In: N.H. Shaw & W.W. Bryan (Eds), Tropical Pasture Research: Principles and Methods. CAB Bull. 51.

Marandu, W.Y.F., A.P. Uriyo & H.O. Mongi, 1975. Residual effects of NPK fertilizer application on the yield of maize and pasture on a ferrisol at Morogoro. *East Afr. agric. For. J.* 40: 400–407.

Maranga, E.K., M.J. Tricla & F. Smeins, 1983. Water relations of *Panicum maximum* and *Digitaria macroblephara* on a semi-arid rangeland of Kenya. *East Afr. agric. For. J.* 48: 74–80.

Marshall, B. & R.M. Bredon, 1963. The chemical composition and nutritive value of elephant grass (*Pennisetum purpureum*). *Trop. Agric.* (*Trin.*) 40: 63–66.

Marshall, B. & R.M. Bredon, 1967. The nutritive value of *Themeda triandra*. *East Afr. agric. For. J.* 32: 375–379.

Marshall, B. & M.I.E. Long, 1971. Calcium intake and excretion of zebu cattle used for digestibility trials. *East Afr. agric. For. J.* 37: 46–48.

Marshall, B., R.M. Bredon & C.D. Juko, 1961. Nutrition of Zebu cattle. IV. The intake of dry matter by cattle with some notes on the digestibility of Rhodes grass and star grass hays. *J. agric. Sci.* 56: 191–196.

Marshall, B., D. Pratchett & W.K. Ndyanabo, 1976. The mineral status of beef and dairy farms in Acholi and Lango districts, Uganda. *East Afr. agric. For. J.* 41: 218–230.

Marten, G.C., A.G. Matches, R.F. Barnes, R.W. Brougham, R.J. Clements & G.W. Sheath (Eds), 1989. Persistence of Forage Legumes. American Society of Agronomy.

Martin, E.F. & W.J. Badcock, 1937. Preliminary studies on grasses and grazing made at Ngetta, Northern Province, Uganda. *East Afr. agric. J.* 3: 192–195.

Martin, W.S., 1944a. Grass covers in their relation to soil structure. *Emp. J. exp. Agric.* 12: 21–33.

Martin, W.S., 1944b. Soil structure. *East Afr. agric. J.* 9: 189–195.

Martin, W.S. & C.E.J. Biggs, 1937. Experiments on the maintenance of soil fertility in Uganda. *East Afr. agric. J.* 2: 371–378.

Masefield, G.B., 1946. Agricultural extension methods amongst African peasant farmers. *East Afr. agric. J.* 11: 214–221.

Mastenbroek, C., 1988. Plant breeders rights, an equitable legal system for new plant cultivars. *Exp. Agric.* 24: 15–30.

Mathieu, P., 1960. L'élevage en Urundi. *Bull. agric. Congo* 51: 885–922.

Matthews, G.A., 1984. Application equipment. In: P.J. Terry (Ed.), Guide to Weed Control in East African Crops. Kenya Literature Bureau, Nairobi.

Mays, D.A. (Ed.), 1974. Forage Fertilization. American Society of Agronomy.

Mbugua, P.N., 1976a. Review of role of Napier grass in ruminant feeding. Kitale report.

Mbugua, P.N., 1976b. Miscellaneous Napier grass reports. Kitale.

Mears, P.T., 1970. Kikuyu (*Pennisetum clandestinum*) as a pasture grass, a review. *Trop. Grasslds* 4: 139–152.

Mehlich, A., 1965. Mineral nutrient content of organic manures and mulching materials with particular reference to calcium, magnesium and potassium. *Kenya Coffee* 30: 152–161.

Melle, H.A., 1918. Agricultural Grasses and their Culture. *Pretoria Dep. Agric. Bull.* 5.

Menaut, J.C., R. Barbault, P. Lavelle & M. Lepaze, 1984. African savannas: biological systems of humification and mineralization. In: J.C. Tothill & J.J. Mott (Eds), Ecology and Management of the World's Savannas. Int. Savanna Symp. 1984. CAB.

Meyer, J.A., 1961. Cultures de sidération, cultures améliorantes et fertilité. *Bull. agric. du Congo* 52: 541–569.

Meyer, J.A., 1963. Ecologie et sociologie des microchampignons du sol de la Cuvette centrale congolaise. *Sér. sci. INEAC* 101.

Meyn, K., 1971. Beef Production in East Africa. Institut für Wirtschaftsforschung, München.

Middleton, C.H. & G.A. Barry, 1978. A study of oxalate content in five grasses in the wet tropics of Queensland. *Trop. Grasslds* 12: 28–35.

Miles, D.G., H.A. Squire, D. Destro & M.J. Creek, 1972. Studies in forage production. Lanet report.

Milford, R., 1960. Nutritional values for 17 subtropical grasses. *Aust. J. agric. Res.* 11: 138–148.

Millar, R.P., 1967. Oestrogenic activity in Central African highveld grasses. *Rhod. Zam. & Malawi J. agric. Res.* 5: 179–183.

Mills, W.R., 1953. A fertilizer trial with *Chloris gayana* and the yield of a subsequent cotton crop. *Proc. 2nd Conf. Interafr. Soils*: 785–794.

Minderhoud, J.W., P.F.J. van Burg, B. Deinum, J.G.P. Dirven & M.L. 't Hart, 1974. Effects of high levels of nitrogen and adequate utilization on grassland productivity and cattle performance with special reference to permanent pastures in the temperate regions. *Proc. 12th int. Grassld Congr., Moscow*, 1: 99–121.

Minson, D.J., T.H. Stobbs, M.P. Hegarty & M.J. Playne, 1976. Measuring the nutritive value of pasture plants. In: N.H. Shaw & W.W. Bryan (Eds), Tropical Pasture Research: Principles and Methods. CAB Bull. 51.

Moore, C.E.M., 1965. Production of pasture seeds from medium altitude. *Kenya Farmer*, April: 16–17.

Moreau, R.E., 1938. Climatic classification from the standpoint of East African biology. *J. Ecol.* 26: 467–496.

Morrison, J., 1966a. The nitrogen and phosphorus fertilization of perennial ryegrass grown at high altitudes in Kenya. *J. Brit. Grassld Soc.* 21: 218–223.

Morrison, J., 1966b. The influence of lime, lime pelleting and method of sowing on the nodulation and growth of subterranean clover. *J. Brit. Grassld Soc.* 21: 270–275.

Morrison, J., 1966c. The effects of nitrogen, phosphorus and cultivation on the productivity of Kikuyu grass at high altitudes in Kenya. *East Afr. agric. For. J.* 31: 291–297.

Morrison, J., 1966d. A note on the factors concerned in the establishment of grass/clover swards at high elevations in Kenya. *East Afr. agric. For. J.* 31: 445–448.

Morrison, J., 1966e. Productivity of grass and grass/legume swards in the Kenya highlands. *East Afr. agric. For. J.* 32: 19–24.

Morrison, J., 1966f. Effects of cover crop and sowing dates on grass/clover swards in the Kenya highlands. *East Afr. agric. For. J.* 32: 25–30.

Morrison, J., 1968. Report of a pasture tour of Central and Eastern Provinces. Kitale report.

Morrison, J., 1969. Grasses and legumes at high altitudes in tropical Africa. Review Article. *Herb. Abstr.* 39: 101–106.

Mugerwa, J.S. & Ruth Bwabye, 1974. Yield, composition and in vitro digestibility of *Amaranthus hybridus* subspecies incurvatus. *Trop. Grasslds* 8: 49–53.

Mugerwa, J.S. & B.H. Ogwang, 1976. Dry matter production and chemical composition of Elephantgrass hybrids. *East Afr. agric. For. J.* 42: 60–65.

Mugerwa, J.S., D.A. Christensen & S. Ochetim, 1973. Grazing behaviour of exotic dairy cattle in Uganda. *East Afr. agric. For. J.* 39: 1–11.

Mukisira, E.A., 1988. The effect of lupin/maize meal supplementation on the performance of lactating Friesian cows. In: B.H. Dzowela (Ed.), African Forage Plant Genetic Resources, Evaluation of Forage Germplasm and Extensive Livestock Production Systems. Proc. Workshop Arusha 1987, ILCA, Addis Ababa.

Mulder, J., C.M. Waweru & C.K. Kamau, 1973. Intensive cattle production on a smallholder basis. Lanet report.

Munyi, D.N. & C. Smit, 1973. Productivity of a Kisii grass ley at different ages. Naivasha report.

Musangi, R.S., 1965. Feed intake studies in ruminants. *Afr. Soils* 10: 313–336.

Musangi, R.S. (Ed.), 1966. Pasture Production and Utilization by Cattle in the Tropics with special Reference to Uganda. Seminar Makerere, 1966. *Techn. Bull.* 4.

Musangi, R.S., 1969. Dairy Husbandry in Eastern Africa. Longman.

Muyekho, F.N., 1988. A review on evaluation and agronomy of promising lupine varieties in Kenya. In: B.H. Dzowela (Ed.), African Forage Plant Genetic Resources, Evaluation of Forage Germplasm and Extensive Livestock Production Systems. Proc. Workshop Arusha 1987, ILCA, Addis Ababa.

Mwakha, E., 1969. Observations on the effect of temperature on the growth of *Trifolium semipilosum* (Pres.). *East Afr. agric. For. J.* 34: 289–292.

Mwakha, E., 1971. Germination depth of Bungoma grass *Entolasia imbricata* Stapf. *East Afr. agric. For. J.* 37: 26–28.

Mwakha, E., 1972. Effect of cutting frequency on productivity of Napier and Guatemala grasses in Western Kenya. *East Afr. agric. For. J.* 37: 206–210.

Myoya, T.J. & N.J. Mukurasi, 1988. A review of pasture and fodder crops germplasm evaluation at Uyole, Mbeya in Tanzania. In: B.H. Dzowela (Ed.), African Forage Plant Genetic Resources, Evaluation of Forage Germplasm and Extensive Livestock Production Systems. Proc. Workshop Arusha 1987, ILCA, Addis Ababa.

Nada, Y., 1985. Responses of tropical pasture grasses to photoperiod and temperature. *Proc. 15th int. Grassld Congr., Kyoto*: 1256–1258.

Nakajima, K., 1985. Isolation of sexual plants in Guineagrass and the potential of its hybridization breeding in Japan. *Proc. 15th int. Grassld Congr., Kyoto*: 258–259.

National Academy of Sciences, 1977. Leucaena, Promising Forage and Tree Crop for the Tropics. Washington D.C.

Nattrass, R.M., 1941. Notes on plant diseases. I. The white mould of Napier grass. *East Afr. agric. J.* 7: 56.

Nattrass, R.M., 1961. Host lists of Kenya fungi and bacteria. *CAB Mycol. Papers* 81.

Naveh, Z., 1966a. Selection of promising pasture plants for northern Tanzania. III. The early performance of drought-resistant *Glycine javanica* L. ecotypes. *East Afr. agric. For. J.* 32: 103–107.

Naveh, Z., 1966b. The determination of range condition and trend on East African range lands. *East Afr. agric. For. J.* 32: 159–161.

Naveh, Z., 1966c. Range research and development in the dry tropics with special reference to East Africa. Review Article. *Herb. Abstr.* 36: 77–85.

Naveh, Z. & G.D. Anderson, 1965. Some preliminary yields from promising grass/legume mixtures at Tengeru, northern Tanzania. *Afr. Soils* 10: 493–499.

Naveh, Z. & G.D. Anderson, 1966. The introduction and selection of promising pasture plants for the Arusha and Kilimanjaro Regions of northern Tanzania. *East Afr. agric. For. J.* 32: 41–44; 96–102.

Naveh, Z. & G.D. Anderson, 1967. Selection of promising pasture plants for northern Tanzania. IV. Legumes, grasses and grass/legume mixtures. *East Afr. agric. For. J.* 32: 283–304.

Ndawula-Senyimba, M.S., 1972. Some aspects of the ecology of *Themeda triandra*. *East Afr. agric. For. J.* 38: 83–93.

Ndyanabo, W.K., 1974. Oxalate content of some commonly grazed pasture forages of Lango and Acholi districts of Uganda. *East Afric. agric. For. J.* 39: 210–214.

Nicholson, D.D.A., 1971. Forage sorghum cultivation. Eldoret report.

Nightingale, J.M. & H.T. Lloyd, 1946. Conversion of veld grazing into ley pastures in the Kenya highlands. *East Afr. agric. J.* 12: 84–87.

Noirot, M., 1989. The morphological relation in the panicle of *Panicum maximum* Jacq. Consequences on the pollination and the fertility. *Proc. 16th int. Grassld Congr., Nice*: 479–480.

Noll, I.E. van der & B.H. Janssen, 1983. Cow dung slurry as an organic manure for fodder crops in Kilifi, Coast Province. Kilifi Pedology Project report.

Norris, D.O., 1967. The intelligent use of inoculants and lime pelleting for tropical legumes. *Trop. Grasslds* 1: 107–121.

Norris, D.O., 1975. Nodulation in pasture legumes. In: R. Milton Moore (Ed.), Australian Grasslands. Australian National University Press.

Norris, D.O. & L. 't Mannetje, 1964. The symbiotic specialization of African *Trifolium* spp. in relation to their taxonomy and their agronomic use. *East Afr. agric. For. J.* 29: 214–235.

Northwood, P.J. & J.C. Macartney, 1969. Pasture establishment and renovation by direct seeding. *East Afr. agric. For. J.* 35: 185–189.

Nye, G.W., 1937. Preliminary notes on the use of Elephant Grass as a fallow crop in Buganda. *East Afr. agric. J.* 3: 186–190.

Nye, G.W., 1940. Some results from Bukalasa experiment station. I. General. *East Afr. agric. J.* 5: 460–465.

Nye, P.H. & W.N.M. Foster, 1961. The relative uptake of phosphorus by crops and natural fallow from different part of the root zone. *J. agric. Sci.* 56: 299–306.

Nye, P.H. & D.J. Greenland, 1960. The Soil under Shifting Cultivation. Commonw. Bur. Soils, Harpenden Techn. Comm. 51.

Odhiambo, J.F., 1974. The nutritive value of various growth stages of *Pennisetum purpureum*. *East Afr. agric. For. J.* 39: 325–329.

Oenema, O., 1980. Combined application of organic manure and triple superphosphate to hybrid maize in South West Kenya. Kisii Pedology Project report.

Ogborn, J.E.A., 1965. Some effects of zerograzing on forage production. *Afr. Soils* 10: 337–341.

Ogwang, B.H. & J.S. Mugerwa, 1976. Yield response to nitrogen application and *in-vitro* digestibility of elephant grass and bulrush millet hybrids. *East Afr. agric. For. J.* 41: 231–242.

Okwaro, G.A., H. Kayongo-Male & C.N. Karue, 1978. The effect of variety and location on the nutritive value of nine sorghum varieties grown under Kenyan conditions. *East Afr. agric. For. J.* 43: 185–192.

Olsen, F.J., 1972. Effect of large applications of nitrogen fertilizer on the productivity and protein content of four tropical grasses in Uganda. *Trop. Agric.* (*Trin.*) 49: 251–260.

Olsen, F.J., 1973. Effects of cutting management of a *Desmodium intortum* (Mill.) Urb./ *Setaria sphacelata* (Schumach.) Stapf mixture. *Agron. J.* 65: 714–716.

Olsen, F.J., 1974a. Effects of fertilization and cutting management on yield and botanical composition of permanent pasture. *East Afr. agric. For. J.* 39: 391–396.

Olsen, F.J., 1974b. Effect of nitrogen fertilizer on yield and protein content of *Brachiaria mutica* (Forsk.) Stapf, *Cynodon dactylon* (L.) Pers. and *Setaria splendida* Stapf in Uganda. *Trop. Agric.* (*Trin.*) 51: 523–529.

Olsen, F.J. & P.G. Moe, 1971. The effect of phosphate and lime on the establishment, productivity, nodulation and persistence of *Desmodium intortum*, *Medicago sativa* and *Stylosanthes gracilis*. *East Afr. agric. For. J.* 37: 29–37.

Olsen, F.J. & E.R. Tiharuhondi, 1972. The productivity and botanical composition of some selected grass/legume pasture mixtures at different seeding rates. *East Afr. agric. For. J.* 38: 16–22.

Onim, J.F.M. & B.H. Dzowela, 1988. The distribution of Sesbania species in the Panesa region. In: B.H. Dzowela (Ed.), African Forage Plant Genetic Resources, Evaluation of Forage Germplasm and Extensive Livestock Production Systems. Proc. Workshop Arusha 1987, ILCA, Addis Ababa.

O'Rourke, J.T., P.J. Terry & G.W. Frame, 1976. Experimental results of *Eleusine jaegeri* Pilg. control of East African highlands. *East Afr. agric. For. J.* 41: 253–265.

Oram, R.N., 1990. Register of Australian Herbage Plant Cultivars. 3rd Edition. CSIRO, Melbourne.

Orr, J.B. & A. Holm, 1931. The mineral deficiencies of pastures in Kenya Colony. 6th Rep. Econ. Advnt Counc. Comm. 14–65.

Otieno, N.C. 1967. The anatomy of common pasture grasses in Kenya II: *Digitaria scalarum* Chiov. (African couch grass). *East Afr. agric. For. J.* 33: 23–30.

Otim, J.J., 1975. Comparative beef production from ley and improved natural pasture. *East Afr. agric. For. J.* 41: 45–51.

Otsyina, R.M. & C.M. McKell, 1985. Africa. Browse in the nutrition of livestock. *World Anim. Rev.* 53: 33–39.

Owen, M.A., 1964. A note on the effect of a late application of nitrogenous fertilizer on hay quality. *East Afr. agric. For. J.* 29: 322–325.

Owen, M.A. & H.W. Brzotowski, 1966. A grass cover for the upland soils of the Kongwa plain, Tanganyika. *Trop. Agric.* (*Trin.*) 43: 303–314.

Owen, M.A. & H.W. Brzotowski, 1967. Grass establishment under semi-arid conditions in Central Tanganyika. *Trop. Agric.* (*Trin.*) 44: 275–291.

Oyenuga, V.A., 1957. The composition and agricultural value of some grass species in Nigeria. *Emp. J. exp. Agric.* 25: 237–255.

Parijs, A. van, 1959. Maintien de la productivité des sols sous cultures continues en Ituri (Congo Belge). *Proc. 3rd Conf. Interafr. Soils*: 857–863.

Parry, M.S., 1954. Tree-planting in Tanganyika. IV. Species for coastal areas. *East Afr. agric. J.* 20: 49–53.

Parsons, D.J., 1958. Problems of increasing livestock productivity in the intensive farming areas of Uganda. *East Afr. agric. J.* 23: 167–171.

Parsons, J.J., 1970. The "Africanization" of the New World tropical grasslands. *Tübinger Geographische Studien* 34: 141–153.

Parsons, J.J., 1972. Spread of African pasture grasses to the American Tropics. *J. Range Management* 25: 12–17.

Paterson, D.D., 1938. Further experiments with cultivated tropical fodder crops. *Emp. J. exp. Agric.* 6: 323–340.

Pauw, E. de, 1984. Soils, Physiography and Agroecological Zones of Tanzania. Ministry of Agriculture, Dar es Salaam.

Payne, W.J.A. & J.S. MacFarlane, 1963. A brief study of cattle browsing behaviour in a semi-arid area of Tanganyika. *East Afr. agric. For J.* 29: 131–133.

Payne, W.J.A., 1971. Cattle Production in the Tropics. Longman.

Peat, J.E. & A.N. Prentice, 1949. The maintenance of soil productivity in Sukumuland and adjacent areas, Tanganyika. *East Afr. agric. J.* 15: 48–56.

Peers, A.W. & R.J. Jones, 1962. Establishment of grass leys by undergrowing to cereals. *East Afr. agric. For. J.* 27: 145–149.

Pereira, H.C., 1959. Lessons gained from grazing trials at Makavete, Kenya. *East Afr. agric. J.* 25: 59–62.

Pereira, H.C., 1962. Grazing control in semi-arid ranchland. *East Afr. agric. For. J.* 27 (special issue): 42–46.

Pereira, H.C. & V.R.S. Beckley, 1953. Grass establishment on eroded soil in a semi-arid African reserve. *Emp. J. exp. Agric.* 21: 1–15.

Pereira, H.C. & P.A. Jones, 1954. Field responses by Kenya coffee to fertilizers, manures and mulches. *Emp. J. exp. Agric.* 22: 23–36.

Pereira, H.C., E.M. Chenery & W.R. Mills, 1954. The transient effects of grasses in the structure of tropical soils. *Emp. J. exp. Agric.* 22: 148–160.

Pereira, H.C. & P.H. Hosegood, 1962. Soil moisture effects of Kudzu as a clove-orchard cover crop. *East Afr. agric. For. J.* 27: 225–229.

Pereira, H.C., P.H. Hosegood & M. Dagg, 1967. Effect of tied ridges and grass leys on a lateritic soil in Kenya. *Exp. Agric.* 3: 89–98.

Pereira, H.C., P.H. Hosegood & D.B. Thomas, 1961. The productivity of tropical semi-arid thorn-scrub country under intensive management. *Emp. J. exp. Agric.* 29: 269–286.

Pereira, H.C., R.A. Wood, & H. Brzotowski, 1958. Water conservation by fallowing in semi-arid tropical East Africa. *Emp. J. exp. Agric.* 26: 213–218.

Peterson, D.D. & B.S. McGinnes, 1979. Vegetation of south Masailand, Tanzania. *East Afr. agric. For. J.* 44: 252–271.

Phillips, J.F.V., 1929. Some important vegetation communities in the Central Province of Tanganyika Territory. *S. Afr. J. Sci.* 26: 332–372.

Phillips, J.F.V., 1931. A sketch of the floral regions of Tanganyika Territory. *Trans. Roy. Soc. S. Afr.* 19: 363–372.

Pichi-Sermolli, R.E.G., 1955. Tropical East Africa (Ethiopia, Somaliland, Kenya, Tanganyika) in Plant Ecology: reviews of research. *Unesco Arid Zone Res.* 6: 302–353.

Pole Evans, I.B., 1933. Agricultural possibilities of some of the African grasses. *Emp. J. exp. Agric.* 1: 341–347.

Pole Evans, I.B., 1939. Report on a Visit to Kenya. Govt Printer, Nairobi.

Pole Evans, I.B., 1948. Roadside Observations on the Vegetation of East and Central Africa on a Journey from Pretoria to Kenya. Govt Printer, Pretoria.

Pole Evans, I.B., 1950. Beef production possibilities in Southern Africa. *Emp. J. exp. Agric.* 18: 81–94.

Popay, A.I., 1974. Investigations into the behaviour of the seeds of some tropical weeds. *East Afr. Agric. For. J.* 40: 31–43; 408–415.

Popay, A.I. & G.W. Ivens, 1982. East Africa. In: W. Holzner & M. Numata (Eds), Biology and Ecology of Weeds. Dr W. Junk.

Potter, H.L., 1987. Inventory of feed resources for the smallholder farmer in Kenya. In: J.A. Kategile, A.N. Said & B.H. Dzowela (Eds), 1987. Animal Feed Resources for Small-Scale Livestock Producers. Proc. 2nd PANESA Workshop Nairobi, 1985. IDRC-MR165e, Ottawa.

Potter, H.L. & A.N. Said, 1986. Effects of defoliation on vegetation growth in a semi-arid area of Kenya. *East Afr. agric. For. J.* 52: 81–87.

Potts, M.J. & E. Biranguza, 1989. The evaluation of alternative fodder crops under African highland conditions. *Exp. Agric.* 25: 99–107.

Poultney, R.G., 1959. Preliminary investigations on the effect of fertilizers applied to natural grassland. *East Afr. agric. J.* 25: 47–49.

Poultney, R.G., 1960. The effect of potassium on the productivity of lucerne. *East Afr. agric. J.* 25: 266.

Poultney, R.G., 1963a. A comparison of direct seeding and undersowing on the establishment of grass, and the effect on the cover crop. *East Afr. agric. For. J.* 29: 26–30.

Poultney, R.G., 1963b. The effects of applied phosphate and gypsum on the productivity of lucerne and clover. *East Afr. agric. For. J.* 29: 106–109.

Pratt, D.J., 1963. Reseeding denuded land in Baringo district, Kenya. *East Afr. agric. For. J.* 29: 78–91.

Pratt, D.J., 1964. Reseeding denuded land in Baringo District, Kenya. II. Techniques for dry alluvial sites. *East Afr. agric. For. J.* 29: 243–260.

Pratt, D.J., 1968. Rangeland development in Kenya. *Ann. Arid Zone* 7: 177–208.

Pratt, D.J., 1969. Management of arid rangeland in Kenya. *J. Brit. Grassld Soc.* 24: 151–157.

Pratt, D.J. & J. Knight, 1964. Reseeding denuded land in Baringo District. Kenya. III. Techniques for capped red loam soils. *East Afr. agric. For. J.* 30: 117–125.

Pratt, D.J., P.J. Greenway & M.D. Gwynne, 1966. A classification of East African rangeland with an appendix on terminology. *J. appl. Ecol.* 3: 369–382.

Pratt, D.J. & M.D. Gwynne, 1977. Range Management and Ecology in East Africa. Hodder & Stoughton.

Preston, T.R. & R.A. Leng, 1987. Matching ruminant production systems with available resources in the tropics and subtropics. Penambul Books.

Prins, W.H., 1969. Final report on pasture agronomy. Kitulo, Tanzania.

Prins, W.H., 1975. Plant production and protection with special reference to pasture research and extension in Zambia. FAO report.

Quinlan, T.J., K.A. Shaw & W.H.R. Edgley, 1975. Kikuyugrass. *Advisory Leaflet* 1335, Dep. Prim. Ind., Brisbane.

Rattray, R.M., 1960. The Grass Cover of Africa. *FAO agric. Studies 49.*

Reed, J.D., B.S. Capper & P.J.H. Neate, 1988. Plant Breeding and the Nutritive Value of Crop Residues. ILCA, Addis Ababa.

Reid, R.J., A.J. Post, F.J. Olsen & J.S. Mugerwa, 1973. Studies on the nutritional quality of grasses and legumes in Uganda. I. Application of in-vitro digestibility techniques to species and stage of growth effects. *Trop. Agric.* (*Trin.*) 50: 1–15.

Rensburg, H.J. van, 1947a. The role of pasture development in soil conservation, Tanganyika territory. *East Afr. agric. J.* 13: 23–26.

Rensburg, H.J. van, 1947b. Notes on some browse plants. *East Afr. agric. J.* 13: 164–166.

Rensburg, H.J. van, 1948a. The cultivation of introduced pasture plants. *East Afr. agric. J.* 14: 3–9.

Rensburg, H.J. van, 1948b. Pasture management. *East Afr. agric. J.* 14: 26–28.

Rensburg, H.J. van, 1950. Tanganyika work on fodder plants and grazing management. Colloquium on Grassland Research, 1949. *East Afr. agric. J.* 15: 200–206.

Rensburg, H.J. van, 1952. Grass burning experiments on the Msima River stock farm, Southern Highlands, Tanganyika. *East Afr. agric. J.* 17: 119–129.

Rensburg, H.J. van, 1953. Encroachment and control of shrubs in relation to grassland development in Africa. *East Afr. agric. J.* 19: 61–65.

Rensburg, H.J. van, 1955a. Run-off and soil erosion tests, Mpwapwa, central Tanganyika. *East Afr. agric. J.* 20: 228–231.

Rensburg, H.J. van, 1955b. Land usage in semi-arid parts of Tanganyika. *East Afr. agric. J.* 20: 247–253.

Rensburg, H.J. van, 1956. Comparative values of fodder plants in Tanganyika. *East Afr. agric. J.* 22: 14–19.

Rensburg, H.J. van, 1958a. Gully utilization and erosion control. *East Afr. agric. J.* 23: 190–192.

Rensburg, H.J. van, 1958b. Grassland trials at Tarime, Lake province, Tanganyika. *East Afr. agric. J.* 24: 64–66.

Rensburg, H.J. van, 1960. Ecological aspects of the major grassland types in Tanganyika. *Proc. 8th int. Grassld Congr., Reading*: 367–393.

Rensburg, H.J. van, 1965. Development of planted fallows of short duration to replace long-term bush fallows. *Afr. Soils* 10: 381–384.

Rensburg, H.J. van, 1967. Pasture Legumes and Grasses in Zambia. Govt Printer, Lusaka.

Rensburg, H.J. van, 1968. Growth and seasonal composition of natural grassland in Zambia. *J. Brit. Grassld Soc.* 23: 51–52.

Rensburg, H.J. van, 1969a. Management and Utilization of Pastures of East Africa. FAO, Rome.

Rensburg, H.J. van, 1969b. Selection of Productive Strains of *Chloris gayana* in Zambia. Govt. Printer, Lusaka.

Rensburg, H.J. van, 1969c. Introduced Legumes in Natural Grassland and in Established Pasture in Zambia. Govt. Printer, Lusaka.

Rensburg, H.J. van, 1969d. Legume/grass pastures in Zambia. *Fmg Zambia* 4: 1–4.

Renson, C., 1928. Le zacate gordura (*Melinis minutiflora*). *Bull. agric. Congo Belge* 19: 451–460.

Richardson, E.F., 1965. Cover crop recommendations. *Kenya Sisal Board Bull.* 53.

Ridder, N. de & H. van Keulen, 1990. Some aspects of the role of organic matter in sustainable intensified arable farming systems in the West African semi-arid tropics (SAT). *Fert. Res.* 26: 299–310.

Risopoulos, S.A., 1966. Management and Use of Grasslands of the Congo. FAO, Rome.

Robertson, A.G. & J.P. Bernacca, 1958. Game elimination as a tsetse control measure in Uganda. *East Afr. agric. J.* 23: 254–261.

Robinson, J.B.D., 1956. The influence of fertilisers and manure on the pH reaction of a coffee soil. *East Afr. agric. J.* 22: 76–80.

Robinson, J.B.D. & E.M. Chenery, 1958. Magnesium deficiency in coffee with special reference to mulching. *Emp. J. exp. Agric.* 26: 259–273.

Robinson, J.B.D. & T.R. Brooke, 1959. Camber bed cultivation of ground water (vlei) soils. 1. Experimental crop yields. *East Afr. agric. J.* 24: 184–191.

Robyns, W., 1928. Plantes Congolaises pour engrais verts et pour couverture. *Bull. agric. Congo Belge* 19: 483–511.

Robyns, W., 1934. Flore Agrostologique du Congo Belge. Bruxelles.

Robyns, W., 1948. Les territoires phytogéographiques du Congo Belge et du Rwanda Burundi. Atlas Général du Congo. Bruxelles.

Rocco, D.M., 1958. Chemical weed control in lucerne by means of "activated" DNBP. *East Afr. agric. J.* 23: 262–264.

Rocco, D.M., 1960. Some trials on the use of Dalapon for the control of grasses in lucerne. Conf. Couch Contr. EAAFRO, Muguga.

Rodel, M.G.W., 1972. Effect of different grasses on the incidence of neonatal goitre and skeletal deformities in autumn born lambs. *Rhod. agric. J.* 69: 59–60.

Rogerson, A., 1955. Nutritive value of green *Indigofera subulata* herbage. *East Afr. agric. J.* 20: 240.

Rogerson, A., 1956. Feeding values of local barley, maize and oat straws. *East Afr. agric. J.* 21: 159–160.

Roseveare, G.M., 1948. The Grasslands of Latin America. *Bull. 36 Imperial Bur. Past. Field Crops.*

Rossiter, J., 1974. Familiarisation tour reports in eastern Kenya. Katumani reports.

Rovira, A.D., 1978. Microbiology of pasture soils and some effects of microorganisms on pasture plants. In: J.R. Wilson (Ed.), Plant Relations in Pastures. CSIRO.

Royle, R.B., 1961. Stall-feeding of native cattle in Cholo. *Nyasaland Farmer & Forester* 5.

Rijkenbusch, P.A.H., 1967. Notes on leguminous cover crops in sisal. Tanganyika Sisal Growers Association. *Sisal Res. Sta. Bull.* 44.

Rudert, C.P., 1975. Mineral supplementation of ruminants grazing star grass pastures. *Rhod. agric. J.* 72: 113–114.

Russell, E.W., 1973. Soil Conditions and Plant Growth. Longman.

Ruthenberg, H., 1976. Farming Systems of the Tropics. Clarendon Press.

Ryley, M.J., 1981. Ergot in *Panicum maximum* in Queensland. *Aust. Pl. Pathol. 10: 37.*

Sabiiti, E.N., 1984. Evaluation of *Indigofera hirsuta* L. in association with *Chloris gayana* Kunth. In: J.C. Tothill & J.J. Mott (Eds), Ecology and Management of the World's Savannas. Int. Savanna Symp. 1984. CAB

Said, A.N., 1971. *In-vivo* digestibility and nutritive value of Kikuyu grass (*Pennisetum clandestinum*) with a tentative assessment of its yield of nutrients. *East Afr. agric. For. J.*. 37: 15–21.

Said, A.N., 1976. Evaluation of feeds. Some results from the study of the nutritive value of Kenya grasses and fodder crops. In: C.T. Thomas (Ed.) Conference Kitale, 1976.

Said, A.N., 1984. Past and current trends of pasture research in Kenya. In: J.A. Kategile (Ed.), Pasture Improvement Research in Eastern and Southern Africa. Proc. Workshop Harare 1984, IDRC-237e, Ottawa.

Said, A.N., J.L. Wheeler & D.A. Hedges, 1983. Sulphur supplementation of Hereford × Boran steers grazing *Sorghum almum. Trop. Anim. Prod.* 8: 53–55.

Sanchez, P.A., 1976. Properties and Management of Soils in the Tropics. Wiley.

Sanchez, P.A. & L.E. Tergas (Eds), 1979. Pasture Production in Acid Soils of the Tropics. Proc. CIAT Workshop Cali, 1979. CIAT 03 EG-5.

Sands, E.B., D.B. Thomas, J. Knight & D.J. Pratt, 1970. Preliminary selection of pasture plants for the semi-arid areas of Kenya. *East Afr. agric. For. J.* 36: 49–57.

Santana, J.R. de, R.M.A. Moreno, N.G. Arruda, J.M. Pereira & E.M. Moreira, 1987. [Evaluation of Napier grass (*Pennisetum purpureum*) cultivars.] *Herb. Abstr.* 57: 553.

Savidan, Y.H., 1983. Genetics and utilization of apomixis for the improvement of Guineagrass (*Panicum maximum Jacq.*). *Proc. 14th int. Grassld Congr., Lexington (Kentucky)*, 1981: 182–184.

Savory, R. & D. Thomas, 1977. Pasture Handbook for Malawi. Lilongwe.

Savory, R., J.A. Breen & C.I.A. Beale, 1980. Leucaena as a forage crop on small farms in Malawi. In: H.N. le Houérou (Ed.), Browse in Africa. ILCA, Addis Ababa.

Scaife, M.A., 1971. The long-term effects of fertilizers, farmyard manure and leys at Mwanhala, western Tanzania. *East Afr. agric. For. J.* 37: 8–14.

Scaillet, M., 1960. Etude phytosociologique et bromatologique de quelques associations herbagères de la Luvironza. *Bull. Inst. agron. Gembl.* 3, *Hors Série*: 1412–1435.

Scaillet, M., 1965. Les cultures et jachères fourragères d'altitude au Burundi. *Afr. Soils* 10: 205–210.

Scientific Council for Africa South of the Sahara, 1956. Int. Conf. Phytogeography. Publ. No 22, Yangambi.

Schlippe, P. de, 1956. Shifting Cultivation in Africa. Routledge & Kegan.

Schmidt, P.E. & R. Watkins, 1968. Response of natural coastal pastures to fertilizers and some economic considerations in beef production. Tanga report.

Scott, J.D., 1967. Advances in pasture work in South Africa. Review Articles. *Herb. Abstr.* 37: 1–9; 159–167.

Scott, R.M., 1969. Soils. In: W.T.W. Morgan (Ed.), East Africa: Its People and Resources. Oxford University Press.

Seitzer, J.L., M.J. Butt & H.A. Mulamula, 1970. Fertilizer response to wheat in Kenya. *East Afr. agric. For. J.* 36: 131–138.

Semb, G. & J.B.D. Robinson, 1969. The natural nitrogen flush in different arable soils and climates in East Africa. *East Afr. agric. For. J.* 34: 350–370.

Shantz, H.L. & C.F. Marbut, 1923. The Vegetation and Soils of Africa. *Res. Ser. Am. geogr. Soc.* 13: 1–263. New York.

Shantz, H.L. & B.L. Turner, 1958. Photographic Documentation of Vegetational Changes in Africa over a Third of a Century. *Rep. Arizona agric. Exp. Sta.* 169: 1–158.

Sheldrick, R.D., 1975. Optimum cutting period for silage maize in Western Kenya. *East Afr. agric. For. J.* 40: 394–399.

Sheldrick, R.G. & J.R. Goldson, 1978. The results of a forage system exercise at Kitale. Kitale report (PRP/23).

Sheldrick, R.D. & D.M. Thairu, 1975. Seasonal patterns of grass growth at Kitale. Kitale report (PRP/3).

Silvey, M.W. & R.M. Jones, 1990. Permanent pastures on a brigalow soil: changes in pasture yield and composition during the first five years. *Trop. Grasslds* 24: 282–290.

Simpson, J.R., 1960. The mechanism of surface nitrate accumulation on a bare fallow soil in Uganda. *J. Soil Sci.* 11: 45–60.

Simpson, J.R., 1961. The effects of several agricultural treatments on the nitrogen status of a red earth in Uganda. *East Afr. agric. For. J.* 26: 158–163.

Simpson, J.R. & T.H. Stobbs, 1981. Nitrogen supply and animal production from pastures. In: F.H.W. Morley (Ed.), Grazing Animals. World Animal Science. Elsevier.

Skerman, P.J., 1977. Tropical Forage Legumes. FAO, Rome.

Skerman, P.J. & F. Riveros, 1989. Tropical Grasses. FAO, Rome.

Slagsvold, P., 1969. Some aspects of livestock feeding and fertility. *East Afr. agric. For. J.* 35: 174–180.

Smith, C.A., 1962. Tropical grass/legume pastures in Northern Rhodesia. *J. agric. Sci.* 59: 111–119.

Smith, C.A., 1963. Oversowing pasture legumes in the Hyparrhenia grasslands of Northern Rhodesia. *Nature* 200: 811–812.

Smith, C.A., 1964a. Studies on the Hyparrhenia veld of Northern Rhodesia. IV. The effects of nitrogen fertilizer and defoliation. *J. agric. Sci.* 62: 299–306.

Smith, C.A., 1964b. Studies on the Hyparrhenia veld of Northern Rhodesia. V. The yield of veld compared with Giant Rhodes grass (*Chloris gayana*). *J. agric. Sci.* 63: 75–79.

Smith, C.A., 1965. Studies on the Hyparrhenia veld of Northern Rhodesia. VI. The fertilizer value of animal excreta. *J. agric. Sci.* 64: 403–406.

Smith, C.J., 1966. Problems in establishing dry-land pastures in Rhodesia. *J. Brit. Grassld Soc.* 21: 238–244.

Smith, C.J., 1971. Seed dormancy in Sabi Panicum. *Proc. int. Seed Test. Ass.* 36: 81–97.

Snijders, P.J.M., A. Nahuis, F. Wekesa & A.P. Wouters, 1992a. Yield and quality of Napier grass and greenleaf Desmodium at two cutting regimes. Naivasha report.

Snijders, P.J.M., A.B. Orodho & A.P. Wouters, 1992b. Effect of manure application methods on yield and quality of Napier grass. Naivasha report.

Snijders, P.J.M. & A.P. Wouters, 1990. Silage quality and losses due to ensiling of Napiergrass, Columbus grass and maizestover under smallholder conditions. Naivasha report.

Snowden, J.D., 1933. A study in altitudinal zonation in South Kigezi and on Mounts Muhavura and Mgahinga, Uganda. *J. Ecol.* 21: 7–27.

Snowden, J.D., 1953. The Grass Communities and Mountain Vegetation of Uganda. Crown Agents, London.

Soest, L.J.M. van, 1976. Exploration and evaluation of grasses for seed production. TF-Ken 29 (Nor), FAO, Rome.

Sombroek, W.C., H.M.H. Braun & B.J.A. van der Pouw, 1982. Explanatory soil map and agro-climatic zone map of Kenya. Kenya Soil Survey report E1, Nairobi.

Soneji, S.V., R.S. Musangi & F.J. Olsen, 1971. Digestibility and feed intake investigations at different stages of growth of *Brachiaria ruziziensis, Chloris gayana* and *Setaria sphacelata* using Corriedale wether sheep. *East Afr. agric. For. J.* 37: 125–128; 267–271.

Souza, D.I.A. de, 1969. Legume nodulation and nitrogen fixation studies in Kenya. *East Afr. agric. For. J.* 34: 299–305.

Spurling, A.T. & Daphne Spurling, 1972. Stallfeeding experiments with beef cattle in southern Malawi. *East Afr. agric. For. J.* 38: 70–74.

Ssekaalo, H., 1972. Cobalt and molybdenum contents of pasture plants on dairy farms of the eastern region of Uganda. *Trop. Agric. (Trin.)* 49: 221–226.

Stapf, O. & C.E. Hubbard, 1934. Flora of Tropical Africa. IX. *Gramineae*. L. Reeve & Co.

Staples, R.R., 1931–1939. *Ann. Rep. Dep. Vet. Sci. Anim. Husb.*, Mpwapwa.

Staples, R.R., 1932. Pasture investigations. *Trop. Agric.* (*Trin.*) 9: 359–361.

Staples, R.R., 1945. Bush control and deferred grazing as measures to improve pastures. *East Afr. agric. J.* 10: 217–222.

Stephens, D., 1960. Three rotation experiments with grass fallows and fertilizers. *Emp. J. exp. Agric.* 28: 165–178.

Stephens, D., 1967a. Effects of fertilizers on grazed and cut elephant grass leys at Kawanda Research Station, Uganda. *East Afr. agric. For. J.* 32: 383–392.

Stephens, D., 1967b. Effects of grass fallow treatments in restoring fertility of Buganda clay loam in South Uganda. *J. agric. Sci.* 68: 391–403.

Stephens, D., 1969. The effects of fertilizers, manure and trace elements in continuous cropping rotations in southern and western Uganda. *East Afr. agric. For. J.* 34: 401–417.

Stephens, D., 1970. Soil fertility. In: J.D. Jameson (Ed.), Agriculture in Uganda. Oxford University Press.

Stewart, D.R.M., 1965. The epidermal characters of grasses, with special reference to East African plains species. *Botanische Jahrbücher* 84: 63–174.

Stobbs, T.H., 1962. The design, working and profitability of small stall-feeding units for cattle in Uganda. *East Afr. agric. For. J.* 27: 260–263.

Stobbs, T.H., 1965a. Animal production from pasture leys. *Afr. Soils* 10: 293–299.

Stobbs, T.H., 1965b. Beef production from Uganda pastures containing *Stylosanthes gracilis* and *Centrosema pubescens*. *Proc. 9th int. Grassld Congr., Sao Paulo*: 939–942.

Stobbs, T.H., 1969a. The effect of grazing management upon pasture productivity in Uganda. I. Stocking rate. *Trop. Agric.* (*Trin.*) 46: 187–194.

Stobbs, T.H., 1969b. The effect of grazing management upon pasture productivity in Uganda. II. Grazing frequency. *Trop. Agric.* (*Trin.*) 46: 195–200.

Stobbs, T.H., 1969c. The effect of grazing management upon pasture productivity in Uganda. III. Rotational and continuous grazing. *Trop. Agric.* (*Trin.*) 46: 293–301.

Stobbs, T.H., 1969d. The effect of grazing management upon pasture productivity in Uganda. IV. Selective grazing. *Trop. Agric.* (*Trin.*) 46: 303–309.

Stobbs, T.H., 1969e. The effect of grazing resting land upon subsequent arable crop yields. *East Afr. agric. For. J.* 35: 28–32.

Stobbs, T.H., 1969f. The influence of inorganic fertilizers upon the adaptation, persistency and production of grass and grass/legume swards in eastern Uganda. *East Afr. agric. For. J.* 35: 112–117.

Stobbs, T.H., 1969g. Animal production from Hyparrhenia grassland oversown with *Stylosanthes gracilis*. *East Afr. agric. For. J.* 35: 128–134.

Stobbs, T.H., 1969h. The value of *Centrosema pubescens* (Benth.) for increasing animal production and improving soil fertility in northern Uganda. *East Afr. agric. For. J.* 35: 197–202.

Stobbs, T.H., 1969i. The use of liveweight gain trials for pasture evaluation in the tropics. III. The measurement of large pasture differences. *J. Brit. Grassld Soc.* 24: 177–183.

Stobbs, T.H., 1970. The use of liveweight-gain trials for pasture evaluation in the tropics. *J. Brit. Grassld Soc.* 25: 73–78.

Stobbs, T.H., 1971. Quality of pasture and forage crops for dairy production in the tropical regions of Australia. *Trop. Grasslds* 5: 159–170.

Stobbs, T.H., 1975. Beef production from improved pastures in the tropics. *World Rev. Anim. Prod.* 11: 58–65.

Stobbs, T.H. & A.D.H. Joblin, 1966. The use of liveweight-gain trials for pasture evaluation in the tropics. *J. Brit. Grassld Soc.* 21: 49–55; 181–185.

Stotz, D., 1977. Zerograzing. Livestock Recording Centre, Naivasha.

Stotz, D., 1979. Smallholder Development in Past, Present and Future in Kenya. Thesis Hohenheim.

Strange, R., 1954a. The production and handling of herbage seeds. *Royal agric. Soc. Kenya* 3: 18–21.

Strange, R., 1954b. Ley establishment at medium altitudes. *Royal agric. Soc. Kenya* 4: 13–20.

Strange, R., 1955a. The management of permanent pastures in the high rainfall mixed farming areas. *Royal agric. Soc. Kenya* 7: 5–7.

Strange, R., 1955b. Forage legumes for medium altitudes in Kenya. *East Afr. agric. J.* 20: 221–224.

Strange, R., 1955c. Miscellaneous experiments. Kitale report.

Strange, R., 1958a. Pasture plants. *Kenya Farmer* (6 monthly installments), Jan-June 1958.

Strange, R., 1958b. Preliminary trials of grasses and legumes under grazing. *East Afr. agric. J.* 24: 92–102.

Strange, R., 1959a. Perennial late season pastures. Kitale report.

Strange, R., 1959b. Observations on the growth of non-irrigated lucerne. *East Afr. agric. J.* 24: 171–173.

Strange, R., 1959c. A comparison between Russian comfrey and lucerne. *East Afr. agric. J.* 24: 203–205.

Strange, R., 1960a. The effects of four management treatments on temporary leys. *East Afr. agric. J.* 25: 251–254.

Strange, R., 1960b. Observations on grasses and legumes under grazing. *East Afr. agric. For. J.* 26: 54–55.

Strange, R., 1961a. Effect of legumes and fertilizers on the yield of temporary leys. *East Afr. agric. For. J.* 26: 231–234.

Strange, R., 1961b. Chicory (*Cichorum intybus*): its use and limitations as a forage crop. *Farmer & Planter* 64: 13–14.

Strange, R., 1963. Grassland and forage farming in the high bracken, Kikuyu and star grass zones of Mount Kenya and Eastern Aberdares. Embu report 4/63.

Strange, R., 1980. An Introduction to African Pastureland Production. *Pasture and Fodder Crop Studies No.* 6, FAO, Rome.

Suttie, J.M., 1965a. The establishment of pastures at medium altitudes. *Kenya Farmer*: 16–17.

Suttie, J.M., 1965b. Some fodder crops of Kenya in relation to altitude. *Afr. Soils* 10: 217–224.

Suttie, J.M., 1967. Notes on Kenya Pastures. Kitale report.

Suttie, J.M., 1968. Pasture legume research in Kenya. *East Afr. agric. For. J.* 33: 281–285.

Suttie, J.M., 1970a. The effect of single superphosphate on the yield and chemical composition of a grazed/legume pasture at Kitale, Kenya. *East Afr. agric. For. J.* 35: 359–363.

Suttie, J.M., 1970b. A review of crop introduction in Kenya and a check list of crops. *East Afr. agric. For. J.* 35: 372–385.

Suttie, J.M. & C.E.M. Moore, 1966. *Desmodium uncinatum. Kenya Farmer*, March.
Swynnerton, R.J.M., 1949. Some problems of the Chagga on Kilimanjaro. *East Afr. agric. J.* 14: 117–132.
Taerum, R., 1970a. Comparative shoot and root growth studies on six grasses in Kenya. *East Afr. agric. For. J.* 36: 94–113.
Taerum, R., 1970b. A study of root and shoot growth on three grass species in Kenya. *East Afr. agric. For. J.* 36: 155–170.
Taerum, R., 1970c. A note of chemical content of some East African grasses. *East Afr. agric. For. J.* 36: 171–176.
Takahashi, M., J.C. Moowaw & J.C. Ripperton, 1966. Studies on Napier grass. III. Grazing management. *Hawaii Agric. Exp. Sta. Bull.* 128.
Takahashi, Y., S. Uozumu, S. Ono & Y. Yoden, 1985. Italian ryegrass response to allelopathic effects of stubble-residues from several tropical grasses. *Proc. 15th int. Grassld Congr., Kyoto*: 382–383.
Tainton, N.M., 1969. Environmental control of flowering in tropical-subtropical grasses. *Proc. Grassld Soc. S. Afr.* 4: 49–55.
Tempany, H. & D.H. Grist, 1958. An Introduction to Tropical Agriculture. London.
Terry, P.J., 1974. Field evaluation of glyphosate, asulam and dalapon on East African couch grass (*Digitaria scalarum*). *East Afr. agric. For. J.* 39: 386–390.
Terry, P.J. (Ed.), 1984. A Guide to Weed Control in East African Crops. Kenya Literature Bureau, Nairobi.
Thairu, D.M., 1970. The importance of pasture in Kenyan farming. *World Crops* 22: 318–319.
Thairu, D.M., 1972. The contribution of *Desmodium intortum* to the yield of *Setaria sphacelata. East Afr. agric. For. J.* 37: 215–219.
Thairu, D.M., 1973. The use of legumes in pasture production in Kenya. *Proc. Kenya Soc. Anim. Prod.* 6: 22–27.
Thomas, A.S., 1940a. Grasses and Grazing: Weeds. In: J.D. Tothill (Ed.), Agriculture in Uganda. Oxford University Press.
Thomas, A.S., 1940b. Grasses as indicator plants in Uganda. *East Afr. agric. J.* 6: 19–22; 77–80.
Thomas, A.S., 1940c. Cover crops and shade trees. In: J.D. Tothill (Ed.), Agriculture in Uganda. Oxford University Press.
Thomas, A.S., 1943. The vegetation of the Karamoja district, Uganda. *J. Ecol.* 31: 149.
Thomas, A.S., 1960a. The trampling animal. *J. Brit. Grassld Soc.* 15: 89–93.
Thomas, A.S., 1960b. Grass and fire. *Proc. 8th int. Grassld Congr., Reading*: 405–407.
Thomas, A.S., 1965. The importance of African aquatic grasses. *Proc. 9th int. Grassld Congr., Sao Paulo*: 1397–1399.
Thomas, A.S., 1966. Ecology and human influence. In: W. Davies & C.L. Skidmore (Eds), Tropical Pastures. Faber & Faber.
Thomas, C., 1977a. The effect of level of concentrate feeding on voluntary intake and performance by steers given silage of *Chloris gayana* during the dry season and their subsequent growth to slaughter. *East Afr. agric. For. J.* 42: 328–336.
Thomas, C., 1977b. The effect of supplementing hay with pyrethrum marc on voluntary intake and digestibility. *East Afr. agric. For. J.* 42: 337–341.
Thomas, C., P.K. Njoroge & J.S. Felon, 1980. Prediction of digestibility in three tropical grasses. *Trop. Agric.* (*Trin.*) 57: 75–81.
Thomas, D., 1973. Nitrogen from tropical pasture legumes on the African continent. *Herb. Abstr.* 43: 33–39.
Thomas, D., 1975. Establishing a mixed pasture under maize in Malawi. *Exp. Agric.* 11: 257–263; 273–276.
Thomas, D., 1976a. Evaluation of cultivars of Panicum on the Lilongwe Plain, Malawi. *Trop. Agric.* (*Trin.*) 53: 225–230.
Thomas, D., 1976b. Effects of close grazing on the productivity and persistence of tropical legumes with Rhodes grass in Malawi. *Trop. Agric.* (*Trin.*) 53: 321–327.
Thomas, D., 1976c. Effects of close grazing or cutting on the productivity of tropical legumes in pure stands in Malawi. *Trop. Agric.* (*Trin.*) 53: 329–333.
Thomas, D. & R. Schulze-Kraft, 1990. Evaluation of five shrubby legumes in comparison with *Centrosema acutifolium*, Carimagua, Colombia. *Trop. Grasslds* 24: 87–92.
Thomas, D.B., 1960. Perennial pasture plants for low rainfall areas of Ukambani. Katumani report.
Thomas, D.B., 1970. Lucerne good, but more research needed. *Kenya Farmer*, August.
Thomas, D.B., 1972. Observations on lucerne as a pasture legume on volcanic ash soil at Njoro, Kenya. *East Afr. agric. For. J.* 37: 336–340.
Thornton, D.D., 1970. A stocking rate trial on rough grazing in Buganda. III. Liveweight gains in the last two years. *East Afr. agric. For. J.* 35: 331–335.
Thornton, D.D. & G.N. Harrington, 1971. The effect of different stocking rates on the weight gain of Ankole steers on natural grassland in western Uganda. *J. agric. Sci.* 76: 97–106.
Thorp, T.K., 1979. A Zambian Handbook of Pasture and Fodder Crops. FAO, Rome.
Tieszen, L.L., M.M. Senyimba, S.K. Imbamba & J.H. Troughton, 1979. The distribution of C-3 and C-4 grasses and carbon isotype discrimination along an altitudinal and moisture gradient in Kenya. *Oecologia* 37: 337–350.
Tiharuhondi, E.R., F.J. Olsen & R.S. Musangi, 1973. Application of nitrogen and irrigation to pasture to enhance cattle production during the season in Uganda. *East Afr. agric. For. J.* 38: 383–393.
Tiley, G.E.D., 1959. Elephant grass in Uganda. Kawanda report.
Tiley, G.E.D., 1965a. The ecology of natural grasslands in Uganda and its bearing on planted pastures. *Afr. Soils* 10: 273–283.
Tiley, G.E.D., 1965b. The management and yields of elephant grass ley under cutting and grazing. *Afr. Soils* 10: 357–360.
Tiley, G.E.D., 1965c. The history of research into ley farming in western and central Uganda and its present aims. *Afr. Soils* 10: 401–404.
Tiley, G.E.D., 1965d. The effects of the grass ley on arable crops with special reference to the elephant grass areas of Uganda. *Afr. Soils* 10: 409–412.
Tiley, G.E.D., 1966. Some results of the study of legumes in planted pastures in central Uganda. In: R.S. Musangi (Ed.), Pasture Production and Utilization by Cattle in the Tropics with special Reference to Uganda. Seminar Makerere, 1966. *Techn. Bull.* 4: 14–18.
Tiley, G.E.D., 1969. Elephant grass. Kawanda Technical Communication 23.
Tiley, G.E.D., 1970a. Pasture research at Kawanda Research Station, Kampala. Kawanda report.

Tiley, G.E.D., 1970b. Weeds. In: J.D. Jameson (Ed.), Agriculture in Uganda. Oxford University Press.

Timberlake, J. & A.C. Dionisio, 1984. Review of the use of improved pasture species in Mozambique. In: J.A. Kategile (Ed.), Pasture Improvement Research in Eastern and Southern Africa. Proc. Workshop Harare, 1984. IDRC-237e, Ottawa.

Timson, S.D., 1943. Prussic acid content of Giant Star grasses (*Cynodon plectostachyus*) and of Kavirondo sorghum. *Rhod. agric. J.* 40: 371–373.

Todd, J.R., 1956. Investigations into the chemical composition and nutritive value of certain forage plants at medium altitudes in the tropics. *J. agric. Sci.* 47: 29–59.

Topps, J.H. & R.C. Elliott, 1965. Nitrogen metabolism of cattle and sheep in Southern Africa. *Outl. Agric.* 4: 310–365.

Tothill, J.D., 1940. Agriculture in Uganda. Oxford University Press.

Tothill, J.C. & J.J. Mott (Eds), 1984. Ecology and Management of the World's Savannas. Int. Savanna Symp. 1984. CAB.

Trapnell, C.G., 1953. The Soils, Vegetation and Agriculture of North Eastern Rhodesia. Rep. ecol. Surv. Govt Printer, Lusaka.

Trapnell, C.G., W.R. Birch, M.A. Brunt & D.J. Pratt, 1966. Kenya Vegetation: sheet l. Directorate of Overseas Surveys, Tolworth.

Trapnell, C.G., M.A. Brunt & W.R. Birch, 1970. Map of Vegetation – Land Use Survey of S.W. Kenya. Sheets 1–3. Survey of Kenya, Nairobi.

Trapnell, C.G. & J.M. Clothier, 1937. The Soils, Vegetation and Agricultural Systems of North Western Rhodesia. Govt Printer, Lusaka.

Trapnell, C.G., M.T. Friend & C.T. Chamberlain, 1976. The effects of fire and termites on a Zambian woodland soil. *J. Ecol.* 64: 577–588.

Trapnell, C.G. & J.F. Griffiths, 1960. The rainfall altitude relation and its ecological significance in Kenya. *East Afr. agric. J.* 25: 207–213.

Trapnell, C.G. & I. Langdale-Brown, 1969. Natural Vegetation. In: W.T.W. Morgan (Ed.), East Africa: Its People and Resources. Oxford University Press.

Tuley, P., 1968. *Stylosanthes gracilis.* Review Article. *Herb. Abstr.* 38: 87–94.

Urio, N.A., S.V. Sarwatt & E.J. Mtengeti, 1988. A review of the potential of *Brachiaria* species as forage crop for livestock in Tanzania. In: B.H. Dzowela (Ed.), African Forage Plant Genetic Resources, Evaluation of Forage Germplasm and Extensive Livestock Production Systems. Proc. Workshop Arusha 1987, ILCA, Addis Ababa.

USA Operations Mission of Ethiopia, 1954. The Agriculture of Ethiopia.

Uyole Agricultural Centre, 1977/1978. Uyole reports.

Vancoppenolle, R., 1988. Ecologie, phytotechnie, productivité et valeur alimentaire de quatre graminées fourragères au Burundi. *Ann. Gembloux* 94: 245–274.

Veenstra, T. & J.G. Boonman, 1974. Chemical weed control in tropical grasses and legumes. *Proc. 5th East Afr. Weed Control Conf.*: 139–151.

Verboom, W.C., 1965. Legumes in Zambia. *Trop. Agric.* (*Trin.*) 42: 229–242.

Verdcourt, B. & E.C. Trump, 1969. Common Poisonous Plants of East Africa. Collins.

Vesey-FitzGerald, L.D.E.F., 1963. Central African grasslands. *J. Ecol.* 51: 243–274.

Vesey-FitzGerald, L.D.E.F., 1965. The utilization of natural pastures in the Rukwa Valley, Tanganyika. *East Afr. Wildl. J.* 7: 131–145.

Vesey-FitzGerald, L.D.E.F., 1970. The origin and distribution of valley grasslands in East Africa. *J. Ecol.* 58: 51–76.

Vesey-FitzGerald, L.D.E.F., 1973. East African Grasslands. East African Publishing House, Nairobi.

Vicente-Chandler, J., 1974. Fertilization of humid tropical grasslands. In: D.A. Mays (Ed.), Forage Fertilization. American Society of Agronomy.

Vicente-Chandler, J., F. Abruna, R. Caro-Costas, J. Figarella, S. Silva & R.W. Pearson, 1974. Intensive Grassland Management in the Humid Tropics of Puerto Rico. *Puerto Rico University Bull.* 233.

Voorthuizen, E.G. van, 1971. A quality evaluation of four widely distributed native grasses in Tanzania. *East Afr. agric. For. J.* 36: 384–391.

Voorthuizen, E.G. van, 1972. The effects of cutting frequency and height on four naturally occurring grasses in Tanzania. *East Afr. agric. For. J.* 37: 258–264.

Walker, B., 1968. Grazing experiments at Ukiriguru, Tanzania. II. Comparisons of rotational and continuous grazing systems on natural pastures of hardpan soils using an "extra-period latin-square change-over design". *East Afr. agric. For. J.* 34: 235–244.

Walker, B., 1969a. Herbage plants for pasture improvement in western Tanzania. *East Afr. agric. For. J.* 35: 1.

Walker, B., 1969b. Effects of nitrogen fertilizer and forage legumes on a *Cenchrus ciliaris* pasture in western Tanzania. *East Afr. agric. For. J.* 35: 2–5.

Walker, B., 1969c. Effect of feeding hay on the early wet season weight loss of cattle in western Tanzania. *Exp. Agric.* 5: 53–57.

Walker, B. & G.D. Scott, 1968a. Grazing experiments at Ukiriguru, Tanzania. I. Comparisons of rotational and continuous grazing systems on pastures of hardpan soils. *East Afr. agric. For. J.* 34: 224–234.

Walker, B. & G.D. Scott, 1968b. Grazing experiments at Ukiriguru, Tanzania. III. A comparison of three stocking rates on the productivity and botanical compositions of natural pastures of hardpan soils. *East Afr. agric. For. J.* 34: 245–255.

Walker, B. & E.J. Weston, 1990. Pasture development in Queensland – a success story. *Trop. Grasslds* 24: 257–268.

Walker, B.H., 1984. Structure and function of savannas. In: J.C. Tothill & J.J. Mott (Eds), Ecology and Management of the World's Savannas. Int. Savanna Symp. CAB.

Wallis, J.A.N., 1962. Observations on the differential responses of 40 grass species and strains to low doses of Dalapon. *Pest Abstr.* 8: 200–202.

Wandera, F.P., B.H. Dzowela & M.K. Karachi, 1991. Production and nutritive value of browse species in semi-arid Kenya. *Trop. Grasslds* 25: 349–355.

Wanjala, B.W., 1981. Some four important feeding stuffs for ruminants for small-scale farmers. Kitale report.

Wapakala, W.W., 1976. Changes in some chemical properties of a red-clay soil: results of long-term fertilizer and rotational trials in Central Kenya. *East Afr. agric. For. J.* 42: 201–218.
Ware-Austin, W.D., 1948. Pasture management. *East Afr. agric. J.* 14: 26–28.
Ware-Austin, W.D., 1963. Napier grass for milk production in the Trans Nzoia. *East Afr. agric. For. J.* 28: 223–227.
Webster, C.C., 1954a. Land utilisation and soil fertility maintenance in Kenya. *Proc. 2nd Conf. Interafr. Soils, Léopoldville* 2: 1425.
Webster, C.C., 1954b. The ley and soil fertility in Britain and Kenya. *East Afr. agric. J.* 20: 71–74.
Weinmann, H., 1964. Response of grasses to fertilizers in Southern Rhodesia. A review. *Rhod. agric. J. Techn. Bull.* 2.
Wendt, W.B., 1970. Responses of pasture species in eastern Uganda to phosphorus, sulphur and potassium. *East Afr. agric. For. J.* 36: 211–219.
Wendt, W.B., 1971. Effects of inoculation and fertilisers on *Desmodium intortum* at Serere, Uganda. *East Afr. agric. For. J.* 36: 317–321.
Wendt, W.B., T.H. Stobbs, G.E.D. Tiley & G.G. Tucker, 1970. Pasture Handbook. Entebbe.
West, O., 1965. Fire in vegetation and its use in pasture management, with special reference to tropical and subtropical Africa. CAB.
Wheatley, P.E., 1965. Pest Handbook. Govt Printer, Nairobi.
Wheeler, J.L. & R.J. Jones, 1977. The potential for forage legumes in Kenya. *Trop. Grasslds* 11: 273–282.
Wheeler, J.L. & C. Mulcahy, 1989. Consequencies for animal production of cyanogenesis in Sorghum forage and hay. A review. *Trop. Grasslds* 23: 193–202.
Whiteman, P.C., 1980. Tropical Pasture Science. Oxford University press.
Whyte, R.O., 1939. Erosion and other surveys in East Africa. *Herb. Rev.* 7: 101–103.
Whyte, R.O., 1947. Africa. The Use and Misuse of Shrubs and Trees as Fodder. *Imperial agric. Bur. Joint Publication* 10: 94–109.
Whyte, R.O., 1958. Plant Exploration, Collection and Introduction. *FAO agric. Studies.* 41.
Whyte, R.O., 1962. The myth of tropical grasslands. *Trop. Agric.* (*Trin.*) 39: 1–11.
Whyte, R.O., 1968. Grasslands of the Monsoon. Faber & Faber.
Whyte, R.O., 1974. Tropical Grazing Lands. Communities and Constituent Species. Dr W. Junk.
Whyte, R.O., G. Nillson-Leissner & H.C. Trumble, 1953. Legumes in Agriculture. *FAO agric. Studies* 21.
Whyte, R.O., T.R.G. Moir & J.P. Cooper, 1959. Grasses in Agriculture. *FAO agric. Studies* 42.
Wigg, P.M., 1973. The role of sown pastures in semi-arid central Tanzania. *East Afr. agric. For. J.* 38: 375–382.
Wigg, P.M., M.A. Owen & N.J. Mukurasi, 1973. Influence of farmyard manure and nitrogen fertilizers on sown pastures and on seed yield and quality of *Cenchrus ciliaris* L., Kongwa, Tanzania. *East Afr. agric. For. J.* 38: 367–374.
Wijk, A.J.P. van, 1976. Herbage yield and quality relationships of three varieties of *Setaria sphacelata* (Schumach.) Stapf and Hubbard. *Neth. J. agric. Sci.* 24: 147–154.
Wijk, A.J.P. van, 1977. The breeding of Napier grass (*Pennisetum purpureum* Schumach.). Kitale report.
Wijk, A.J.P. van, 1980. Breeding for improved herbage and seed yield in *Setaria sphacelata* (Schumach.) Stapf and Hubbard ex Moss. *Agric. Res. Rep.* 900, Wageningen.
Wijk, A.J.P. van, J.G. Boonman & W. Rumball, 1993. Achievements and perspectives in the breeding of forage grasses and legumes. *Proc. 17th int. Grassld Congr., New Zealand.*
Wild, H., 1969. Geobotanical anomalies in Rhodesia. I. The vegetation of copper bearing soils. *Kirkia* 7: 1–71.
Wilkinson, H., 1940. Termites in East Africa. II. The biology and control of termites damaging grassland. *East Afr. agric. J.* 6: 67–72.
Williams, R.J., R.L. Burt & R.W. Strickland, 1976. Plant introduction. In: N.H. Shaw & W.W. Bryan (Eds), Tropical Pasture Research: Principles and Methods. CAB Bull. 51.
Wilson, A.D., 1970. Kilifi Plantations report.
Wilson, J.G., 1962. Opportunities for pasture improvement in the various ecological zones of Karamoja. *East Afr. agric. For. J.* 27 (special issue): 47–48.
Wilson, J.G. & R.M. Bredon, 1963. Nutritional value of some common cattle browse and fodder plants of Karamoja, Northern Provice, Uganda. *East Afr. agric. For. J.* 28: 204–208.
Wilson, J.R. & D.J. Minson, 1980. Prospects for improving the digestibility and intake of tropical grasses. *Trop. Grasslds* 14: 253–259.
Woolhouse, H.W. & S. Walker, 1981. The physiological basis of copper toxicity and copper tolerance in higher plants. In: J.F. Loneragan, J.F. Robson & R.D. Graham (Eds): Copper in Soils and Plants. Academic Press, Sydney.
Wouters, A.P., 1985. Uninterrupted growth of Napier grass (var. Banagrass) at two levels of nitrogen application. Naivasha report.
Wouters, A.P., 1986. The productivity and quality of Napier grass (var. Banagrass) during the dry season at three levels of N-application. Naivasha report.
Wouters, A.P., 1987a. Dry matter yield and quality of Napier grass on farm level, 1983–1986. Naivasha report.
Wouters, A.P., 1987b. Results of feed intake study on farms in Kiambu and Kericho. Naivasha report.
York, P.A., E. Nyamadzawo & E. Bemont, 1989. Forage and seed yield of *Chloris gayana* (Rhodes) varieties at 3 sites in Zimbabwe. *Proc. 16th int. Grassld Congr., Nice*: 367–368.
Young, A., 1989. Agroforestry for Soil Conservation. CAB.
Young, A. & P. Brown, 1962. The Physical Environment of Northern Nyasaland with Special Reference to Soils and Agriculture. Govt Printer, Zomba.
Young, C.C. & D.P. Bartholomew, 1981. Allelopathy in a grass-legume association. I. Effects of *Hemarthria altissima* (Poir.) Stapf. and Hubb. root residues on the growth of *Desmodium intortum* (Mill.) Urb. and *Hemarthria altissima* in a tropical soil. *Crop Sci.* 21: 771–774.
Zyl, L.G. van, 1970. Elephant grass in the south eastern Transvaal. *Fmg S. Afr.* 46 (July): 25, 27, 29, 31, 35.

Index

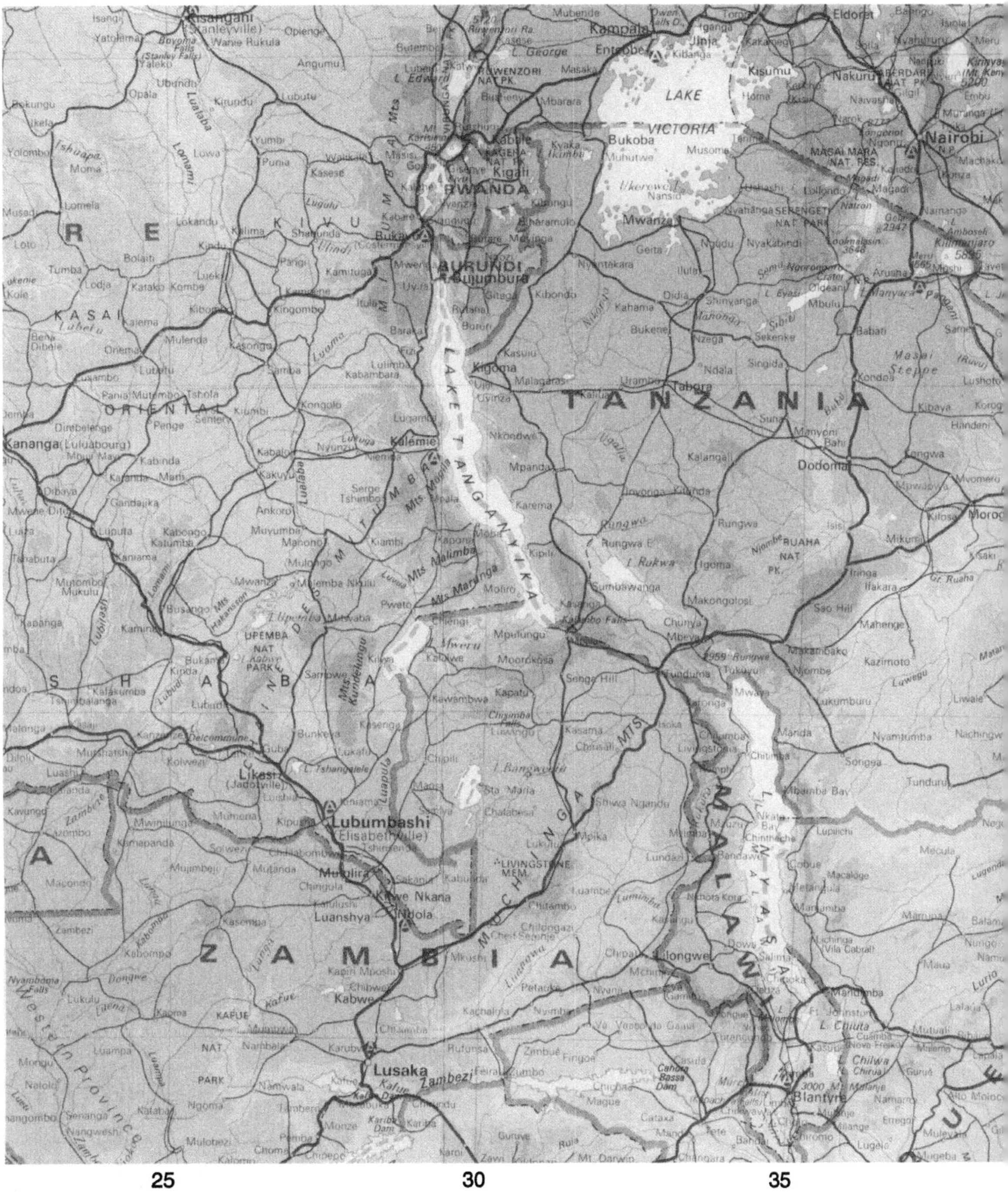

GANDAJIKA 07°S; 24°E 780 m - 31/18°C

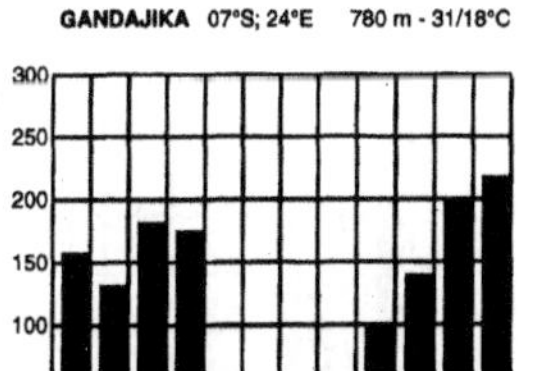

LUBUMBASHI 12°S; 27°E 1190 m - 28/14°C

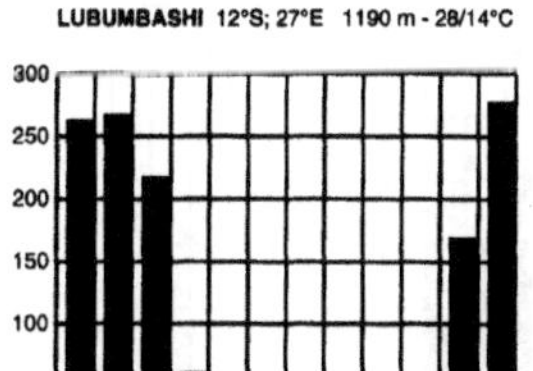

LUSAKA 15°S; 28°E 1270 m - 26/15°C

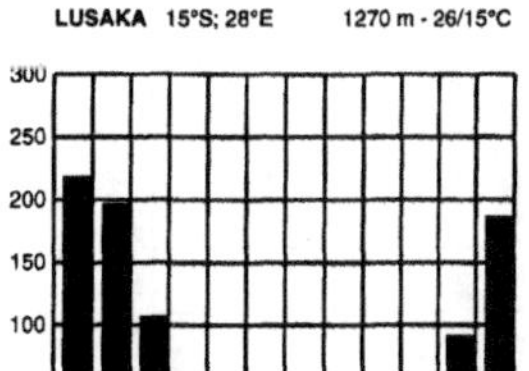

BUJUMBURA 03°S; 29°E 810 m - 29/19°C

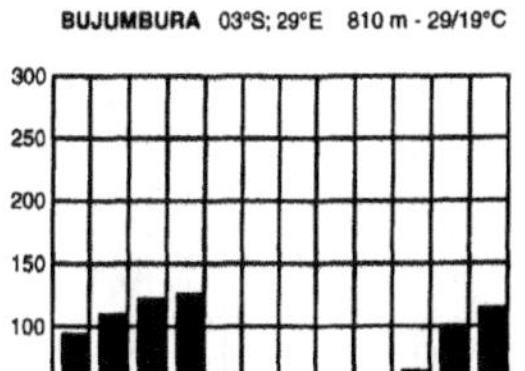

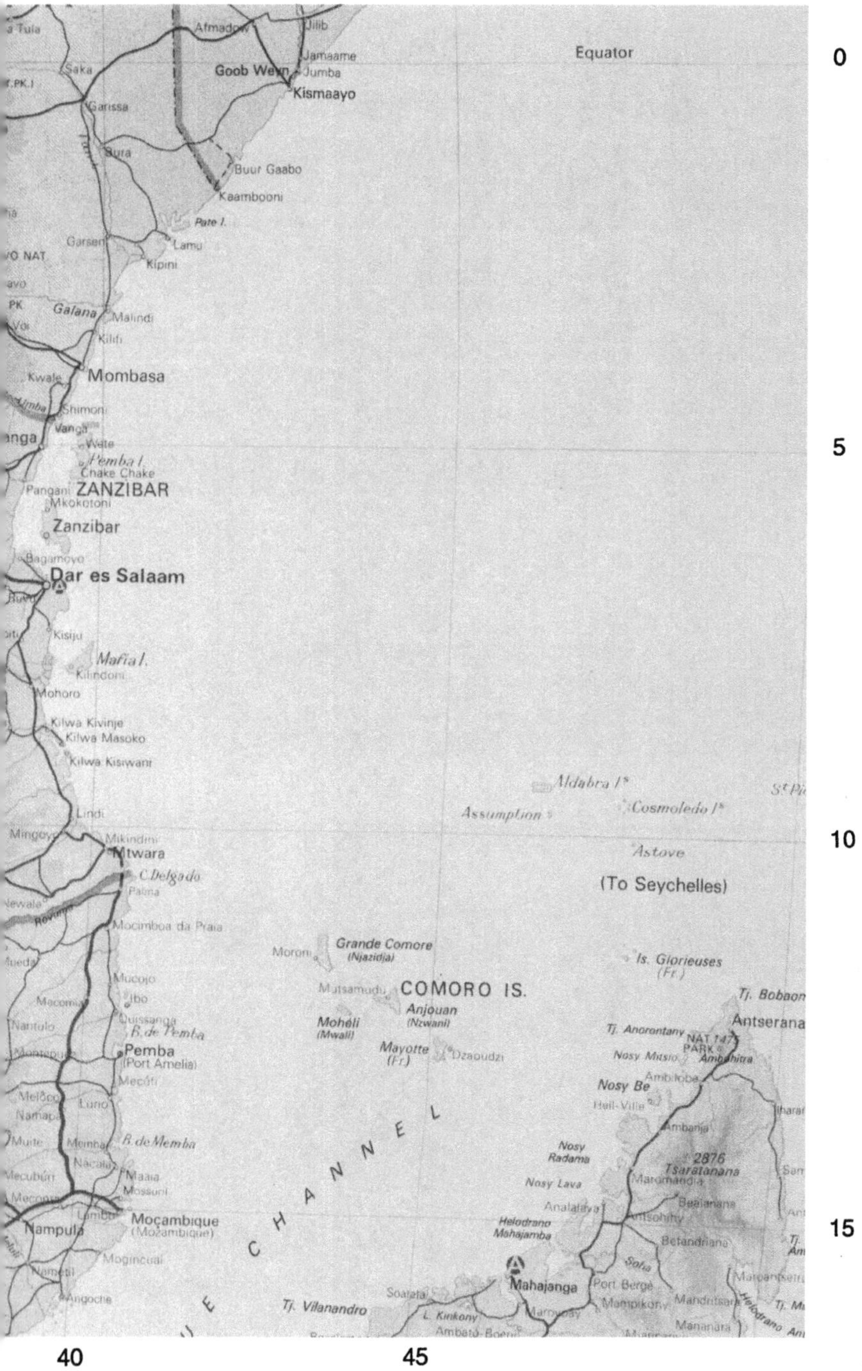

UKIRIGURU 3°S; 33°E 1200 m - 28/17°C
(MWANZA)

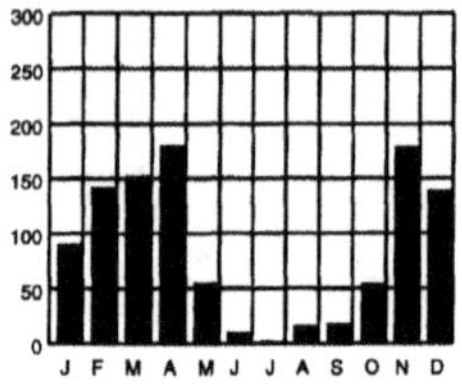

ARUSHA 3°S; 37°E 1390 m - 25/14°C

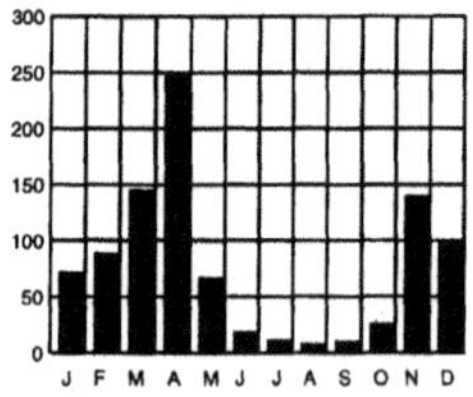

MOMBASA 4°S; 40°E 60 m - 30/22°C

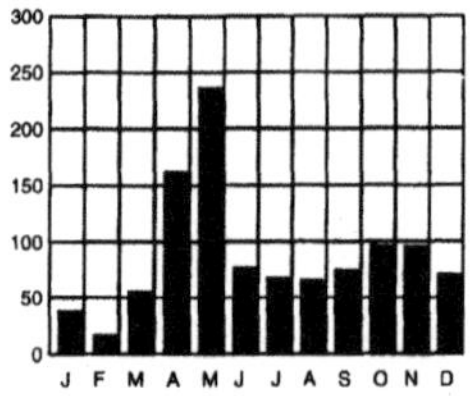

TABORA 5°S; 33°E 1270 m - 29/17°C

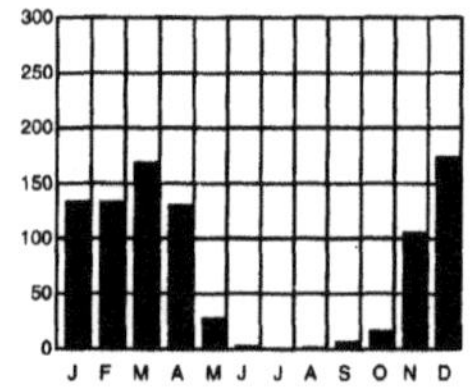

KONGWA 6°S; 36°E 1020 m - 30/15°C

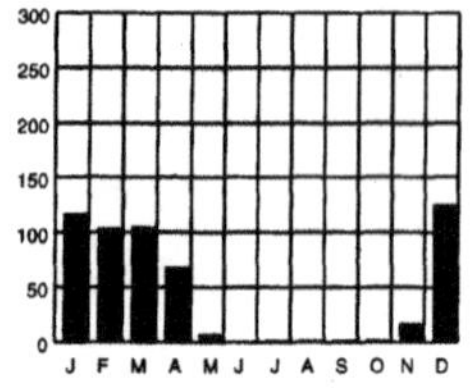

DAR ES SALAAM 7°S; 39°E 10 m - 30/22°C

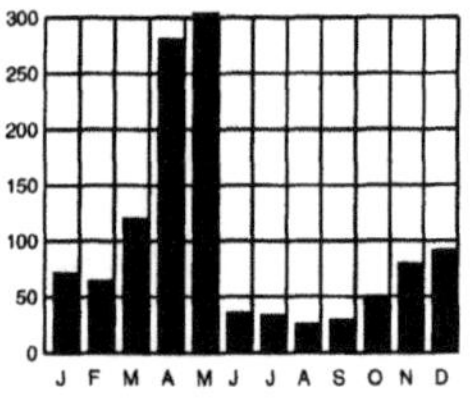

MTWARA 10°S; 40°E 110 m - 31/21°C

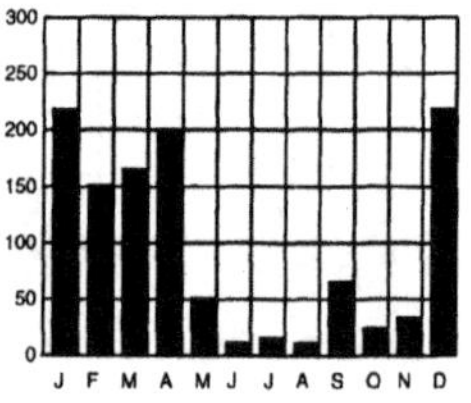

MBEYA 9°S; 33°E 1740 m - 23/12°C

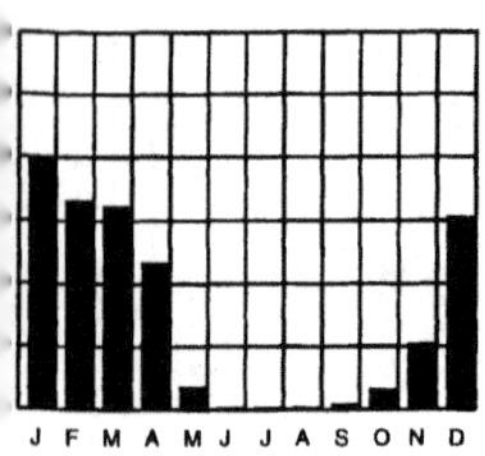

LILONGWE 14°S; 34°E 1130 m - 27/13°C

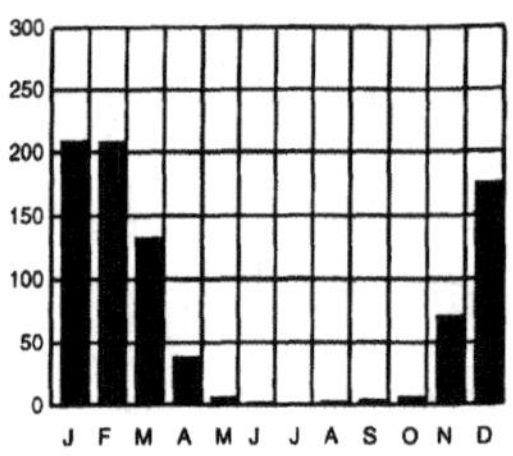

MOSSURIL 15°S; 41°E 10 m - 31/21°C

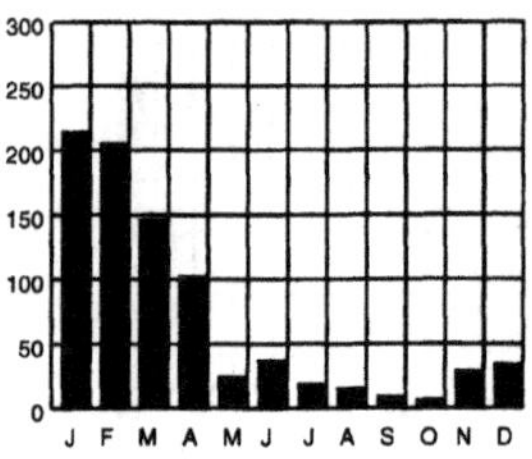

GPSR Compliance

The European Union's (EU) General Product Safety Regulation (GPSR) is a set of rules that requires consumer products to be safe and our obligations to ensure this.

If you have any concerns about our products, you can contact us on ProductSafety@springernature.com

In case Publisher is established outside the EU, the EU authorized representative is:

Springer Nature Customer Service Center GmbH
Europaplatz 3
69115 Heidelberg, Germany

Batch number: 10607920

Printed by Printforce, the Netherlands